National Engineering
Volume 1

National Engineering Mathematics

Volume 1

J.C. Yates
Senior Mathematics Lecturer
Wigan and Leigh College

First published 1993 by
MACMILLAN PRESS LTD
Houndmills, Basingstoke, Hampshire RG21 6XS
and London
Companies and representatives
throughout the world

ISBN 0-333-54854-X

A catalogue record for this book is available from the British Library.

10 9 8 7 6 5 4 3 2
04 03 02 01 00 99 98 97 96

Printed in Hong Kong

Contents

Acknowledgements

For the help I have received whilst writing this volume I am truly grateful. Without Alan Smalley and Alan Skinner it may never have been written. I offer my sincere thanks to Alan Smalley for his work with Chapter 17 and his review of Chapter 13. Also I wish to thank Alan Skinner for his review comments and his encouragement. Without wishing to detract from these contributions my deepest thanks are reserved for my wife, Veronica. Her encouragement and patience have been so remarkable I could have wished for no more.

The author and publishers are grateful to the following organisations and individuals for permission to reproduce illustrative material: Blue Circle Cement, British Aerospace Corporate Jets Ltd, Duraflex Systems Ltd, Farnell Instruments Ltd, J. H. Fenner & Co. Ltd, Ford Motor Co., Foundation and Exploration Services Ltd, Northern Electric, H. W. Peel & Co. Ltd, Val Randall, Saab Motor Co., Texas Instruments, Wickes.

To Veronica

Author's Note

This is the first of 2 volumes covering the mathematics required by an ONC/D student in engineering and science. It aims to cover the objectives of standard BTEC and college devised syllabi with an eye on practical applications of the subject. This means some readers may neither wish nor need to study all the chapters.

The continuity will not be lost if Chapter 8 or the latter parts of Chapter 9 are omitted at a first reading. The overall order of study is suggested by the numbered chapters. However, the order becomes less important towards the end of the book.

Each chapter is introduced with an Assignment. This is a specimen example of how the mathematics within the chapter can be applied. The Assignment provides a common thread for the chapter, linking together the theory and techniques as they develop. At appropriate stages each Assignment is re-visited for further attention.

An electronic calculator is discussed in Chapter 1 and used throughout the text.

Greek letters used

Mathematics needs more letters than those provided by the alphabet. That is why the following Greek letters have been used:

α alpha
β beta
γ gamma
δ delta
θ theta
μ mu
σ sigma
π pi
ϕ phi
ω omega
Δ delta – this is capital delta
Σ sigma – this is capital sigma
Ω omega – this is capital omega

1 The Electronic Calculator

The objectives of this chapter are to:

1. Use a calculator to carry out the basic arithmetic calculations of addition, subtraction, multiplication and division.
2. Use a calculator to find either the square root or the square of a number and understand that these operations are opposites of each other.
3. Understand the accuracy of decimal places and significant figures.
4. Write numbers in standard form.
5. Recognise a recurring decimal.
6. Appreciate the size of a number related to its quoted accuracy.
7. Make a rough check for a calculation before using the calculator.
8. Use a calculator's power and root keys.
9. Use a calculator's reciprocal key.
10. Draw up a table of values.
11. Work out calculations and formulae in a logical sequence.

Introduction

There are many different calculators currently available at relatively low cost. Each year, with the introduction of new function buttons, they appear to be capable of performing more and more complicated calculations. Which one should you choose? Because there are so many features for use in different applications there is no right or wrong answer to this question. Perhaps you should include some of the following points in your decision to buy:

1) Is there a recommended calculator for your course that will be suitable throughout all the subject areas?
2) Do you want power from a battery or a solar cell or a combination of these?
3) Is it a scientific calculator?
4) Price.

The majority of the calculators available today are produced by a small number of manufacturers. They have a history of reliability. A more expensive one need not be a better one for you. The **keys** (or buttons) on a calculator are either numbers or functions. Simply, if the key is not one of 0, 1, 2, 3, 4, 5, 6, 7, 8 or 9 then it is a function key. A **function** (or **operation**) carries out some task on the numbers in a calculation. More function keys and a different display are the most likely reasons for the extra cost. In fact a calculator that is too complicated may take longer to learn to use and cause you frustration. Do not attempt to learn about all the functions at once. Throughout this book with the aid of the manual that comes with your calculator you should learn about the majority of them.

ASSIGNMENT

During this chapter we will consider a vehicle travelling along a road with a particular speed, then the brakes are applied. We will use the calculator to work out the distances travelled from the braking point at various times. In fact our first attempt at the method will be with **whole numbers (integers)**. Once we have learned more about the importance of decimal places we will repeat the calculation.

Calculator layout

There are many different layouts for the calculator keyboard. The photograph shows part of a typical one. We will include other keys later in the book. The positions of the ON and OFF keys change from one calculator to another. Some will switch off automatically after lying unused for a number of minutes to save the battery.

The main part of the picture involves the figure keys of

0| 1| 2| 3| 4| 5| 6| 7| 8| 9|

together with the arithmetic keys of

×| ÷| +| −| =| .|

There are two other keys that, at first glance, look to be very similar. These are the AC| and C| keys. Pressing the AC| key will clear everything except the contents of the memory. The use of the C| key is less dramatic. It will only delete the effect of the last key pressed, so preserving the earlier part of the calculation.

Examples 1.1

i) 5×3 is evaluated by pressing the keys

5| ×| 3| =|

so that the final display is 15.

ii) 12÷3 uses the keys

12| ÷| 3| =|

to give a final display of 4.

iii) 7+9 uses the keys

7| +| 9| =|

to give a final display of 16.

iv) 11−4 uses the keys

11| −| 4| =|

to give a final display of 7.

v) Suppose that when attempting 11−4 you realise at a late stage that the calculation should have been 11−5. If you had pressed

11| −| 4|

and realised your mis-read now press C|. This will delete the 4| you entered and display 0. You may complete the calculation with 5| =| to display the correct answer of 6. The full order is

11| −| 4| C| 5| =|

+/– is the key that changes the **sign before a value.**

Example 1.2

To enter –5 you press 5 +/– which will change the initial display from 5 to the required –5.

Some calculators may not have function keys for brackets. You should work out that part of the calculation in the brackets before any other.

Examples 1.3

i) To work out (13+9)4 the order of key operations is

13 + 9 = × 4 =

After the first = key has been pressed the screen will display 22, finally displaying the correct answer of 88 once the second = key has been pressed.

ii) To work out 3(13–9) the order of key operations is

13 – 9 = × 3 =

Obeying the rule of working out the bracket first, the screen will display 4 after the first = key has been pressed and then 12 after the second = key has been pressed.

iii) In the evaluation of $\frac{96}{2 \times 6}$ even though no brackets are obvious it is understood that they exist around (2×6) in the denominator (bottom line of the fraction). It is this part of the calculation that is to be attempted first. The order of key operations is

2 × 6 =

The screen display of 12 is noted and the screen cleared using the function key AC. The calculation continues

96 ÷ 12 =

Now the final answer of 8 is displayed on the screen.

iv) We can use the answer from part iii) above in the evaluation of $\left(\frac{96}{2 \times 6} - 5\right)7$.
Already we have a solution of 8 from $\frac{96}{2 \times 6}$ so that the problem is reduced to (8–5)7. The order of key operations is

8 – 5 = × 7 =

3 will be displayed after the first = and the final answer of 21 after the second =.

EXERCISE 1.1

Use your calculator to work out the following:

1 15×6

2 -15×6

3 $213 \div 3$

4 $213 \div (-3)$

5 $(-213) \div (-3)$

6 $(10 + 5)6$

7 $6(10 + 5)$

8 $(10 + 5 - 4)2$

9 $(10 - 5 + 4)2$

10 $\dfrac{213 \times 6}{3}$

11 $\dfrac{75}{3 \times 5}$

12 $4 + \left(\dfrac{75}{3 \times 5}\right)$

13 $(-10 + 5 - 3)/2$

14 $\dfrac{10 + 5}{3}$

15 $\dfrac{120}{10 - 5}$

16 $\dfrac{120}{10 - 5} + 7$

17 $\dfrac{120}{10 - 5} - 7$

18 $\left(\dfrac{120}{10 - 5} - 7\right)3$

19 $160 \div \left(\dfrac{75}{3 \times 5}\right)$

20 $\dfrac{160 \times 3 \times 5}{75}$

---

x^2 is used to square whatever number is displayed.

Examples 1.4

i) To work out 6^2 the order of key operations is

6 x^2

displaying the answer 36.

ii) To work out $(-5)^2$ the order is

5 +/− x^2

displaying the answer 25.

iii) This example places the minus sign in a slightly different position to the previous one by leaving out the brackets. -5^2 means that only the 5 is squared and then the minus sign is put in front of the squared value. The full order of keys, to get a correct display of −25, is

5 x^2 +/−

Whenever a positive number or a negative number is squared the answer will always be a positive value. This means that the opposite operation of taking the square root of a positive value should give two answers: a positive number and a negative number. However the calculator is only able to display one of these numbers, usually the positive one. So, for example, the square roots of 49 are 7 and −7 though the display will only show 7.

Examples 1.5

i) To work out $\left(\frac{21-5}{8}\right)^2$ the order of key operations is

21 − 5 = ÷ 8 = x^2

After the first = the display will be 16 and after the second = the display will change to 2. Once all the operations are complete the final answer is 4.

ii) To work out $\left(\frac{21-5}{8}\right)^2 + 19$ the order of key operations uses those from part i) to obtain 4 and then simply adds 19 to give an answer of 23.

iii) To work out $\left(\frac{21-5}{8}\right)^2 + 19^2$ again the order of key operations uses the result from part i) which should be noted. Press the AC button to clear the calculator. Now we are in a position to work out 19^2 by pressing

19 x^2

The screen will display 361 and then we may add our earlier result of 4 to give a final answer of 365.

iv) To work out $19\left(\frac{21-5}{8}\right)^2$ it must be remembered that only the bracket is squared. Then that value is multiplied by 19. The order of key operations is

21 − 5 = ÷ 8 =

x^2 × 19 =

This means that the display from part i) of 4 is multiplied by 19 to give a final answer of 76.

The key to calculate the square root is [√]

Example 1.6

To work out the square root of 49 the order of key operations is

[49] [√]

displaying 7 and implying that −7 is also another answer.

Because √ is the opposite operation to square you can check on your calculator that both $7^2=49$
and $(-7)^2=49$.

The square root of a negative value does not exist in real terms and any attempt at this operation will display -E- on the calculator screen.

What appears under the √ should be treated as though it is in a bracket and so should be worked out first.

Examples 1.7

i) $\sqrt{67-3}$ is worked out using the key operations

[67] [−] [3] [=] [√]

to give the answer 8.

ii) $\sqrt{\dfrac{67-3}{4}}$ is worked out using the key operations

[67] [−] [3] [=] [÷] [4] [=] [√]

to give an answer of 4.

iii) $\dfrac{\sqrt{67-3}}{4}$ has the √ symbol in a slightly different position which needs a different order of key operations

[67] [−] [3] [=] [√] [÷] [4] [=]

to give an answer of 2.

[x^y] is the key that raises a number, x, to a power, y.
The order of pressing keys is important as the following examples show clearly.

Examples 1.8

i) 5^4 is worked out using the order of key operations

[5] [x^y] [4] [=]

to display the answer 625.

ii) 4^5 uses the key operations

4 x^y 5 =

to display the answer 1024.

iii) $(-5)^4$ involves a combination of several key operations that we have mentioned so far in this chapter

5 +/− x^y 4 =

to display the answer 625.

iv) -5^4 requires a slightly different order of key operations because the minus sign applies only after 5 has been raised to the power 4. The order is

5 x^y 4 = +/−

which displays the answer −625.

EXERCISE 1.2

Use your calculator to work out the following:

1 9^2

2 $9^2 \times 3$

3 3×9^2

4 $3 \times (-9)^2$

5 $3^2 \times 9^2$

6 $(3 \times 9)^2$

7 $\left(\frac{9+3}{2}\right)^2$

8 $7 + \left(\frac{9+3}{2}\right)^2$

9 $7^2 + \left(\frac{19+5}{2}\right)^2$

10 $4\left(\frac{19-3}{8}\right)^2$

11 $\sqrt{9}$

12 $\sqrt{81}$

13 $\sqrt{9^2}$

14 $\sqrt{3^2 \times 9^2}$

15 $\frac{\sqrt{19-3}}{4}$

16 $\sqrt{\frac{19-3}{4}}$

17 $14 + \sqrt{\frac{19-3}{4}}$

18 $14 - \frac{\sqrt{19-3}}{4}$

19 $2^2 - \sqrt{3^2 \times 9^2}$

20 $4 + 2^2 - \frac{\sqrt{19-3}}{4}$

21 6^4

22 $(-6)^4$

23 -6^4

24 $2 + 6^4$

25 $2 - 6^4$

26 2×3^3

27 $\frac{3 \times 4^5}{8}$

28 $21-(2\times 3^3)$

29 $21+(3\times 2^3)^3$

30 $\dfrac{21+(3\times 2^3)}{3}$

So far we have looked at only some of the calculator keys to attempt our first Assignment, though we shall return to the same basic theme later in this chapter.

ASSIGNMENT

Now we are in a position to attempt the assignment using only **whole numbers** (i.e. **integers**). Suppose that the vehicle is travelling at a speed of 30 ms^{-1} when the brakes are applied to produce a deceleration of 4 ms^{-2}. It is thought that once these brakes are applied the distance the vehicle travels is given by the formula

$$s=30t-2t^2.$$

Deceleration and retardation being negative acceleration mean a slowing down.

Unfortunately there has been an error of judgment of 1 m in the measuring so that the formula requires correcting to

$$s=1+30t-2t^2$$

where s is the distance travelled
and t is the time from when the brakes are applied.

To many people this formula means very little. A diagram or graph of the distance plotted against time would be more appealing. Before attempting this it is useful to have a **table of values** so that later on in various chapters we may consider the graph. Suppose that the measurements are made at the moment the brakes are applied (i.e. $t=0$) and then after 1, 2, 4 and 6 seconds. We will go through the calculations stage by stage, ending up with the table at these times, but leaving gaps so that you can work out the expected distance values at 3 and 5 seconds.

Firstly we need a line of values of t, i.e.

t	0	1	2	3	4	5	6

Now the expected value of s is made up of three different terms on the right-hand side. We will consider them in turn for each of these time values.
The first term on that side is 1. It is a constant value and so is *unaffected* by each value of t, i.e.

1	1	1	1		1		1

The next term is $30t$. Each value of t is considered in turn and multiplied by 30. For example when t is 4 the key operations are

30 × 4 =

to display an answer of 120. This line of the table is

$30t$	0	30	60		120		180

The final term on the right-hand side is $-2t^2$ and again each different value of t is used in the calculations. It is only the t that is squared. Once this operation has been performed, multiplication by -2 is necessary. For example when t is 6 the order of key operations is

6 x^2 × 2 = +/−

to display an answer of -72. The line of the table is

$-2t^2$	0	−2	−8		−32		−72

After the first line of t values we have a series of rows from each of the three terms of $s=1+30t-2t^2$. Having carefully drawn the table we need to look closely at the columns formed once these rows have been put together. Because we have already taken care of the minus sign all that remains is to add each column of values to get the expected value for s. The last of these columns is given as 1, 180 and -72.

These will add to 109 and it is this value that appears in the last column of the last row.

The complete table is given below:

t	0	1	2	3	4	5	6
1	1	1	1		1		1
$30t$	0	30	60		120		180
$-2t^2$	0	−2	−8		−32		−72
s	1	29	53		89		109

Row 1 shows the original values of t.
Rows 2–4 show all the workings for the various terms.
Row 5 is the row for the expected values of s, being the addition of row values 2, 3 and 4.

From this table we can check our practical distance measurements to see if they tally with our calculator values. The closer the tally the closer our formula is at predicting what will happen later to the vehicle, and eventually when it will stop.

Accuracy

A calculator will **not answer all your problems**. It is an aid to make the numerical calculations less tedious and speed up your work. Before you attempt to complete a calculation you need to understand the size and meaning of the figures you are working with. Often a value is quoted correct to a number of **decimal places** which means that it has been rounded or approximated to this figure. For example you

might read 53.46 correct to 2 decimal places,

perhaps shortened to 53.46 (2 dp),

because there are 2 **figures** (i.e. **digits**) to the right of the decimal point. This means that if we want accuracy to this number of decimal places then 53.46 is the closest figure. It is probable that we will have no idea of the true original figure.

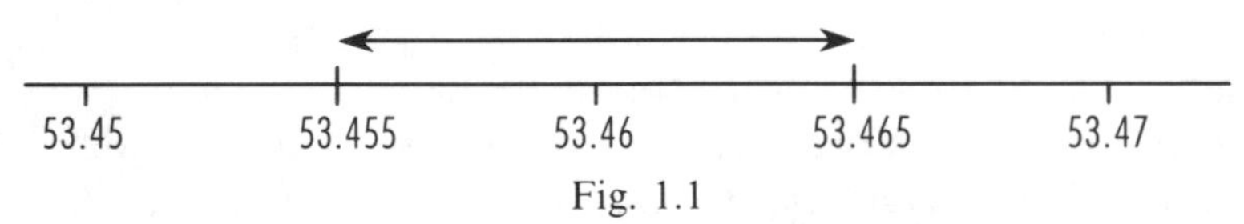

Fig. 1.1

Fig. 1.1 shows a magnified view of a number line. It is a simple scale measurement showing that any value between 53.455 and 53.465 is closer to 53.46 (2 dp) than any other value correct to 2 decimal places.

Examples 1.9

i) 53.458 could be rounded to 53.46 (2 dp).

ii) 53.454 could be rounded to 53.45 (2 dp).

If you have 53.455 you may round it up to 53.46 (2 dp) or down to 53.45 (2 dp). You could make a similar decision to round up 53.465 to either 53.47 (2 dp) or down to 53.46 (2 dp). Because numbers like 53.455 and 53.465 are on the borderline you may round either up or down. Whichever rounding decision you choose, be consistent in your calculations. For borderline decisions do *not* round up in one part of your calculation and then round down in a later part.

An alternative rounding to decimal places is a rounding to significant figures. For example 53.46 (2 dp) is the same as

53.46 correct to 4 significant figures,

perhaps shortened to 53.46 (4 sig fig),

or 53.46 (4 sf),

because there are 4 figures in significant positions within the overall number. You approximate significant figures by rounding up or down according to the same rules used for decimal places.

Examples 1.10

i) 1.07641 could be rounded to 1.0764 (5 sf).

ii) 1.07648 could be rounded to 1.0765 (5 sf).

Now consider rounding 1.07641 to only 3 significant figures. Firstly from the left we must consider the first 3 figures, i.e. 1.07, and ask ourselves whether 1.07641 is closer to 1.07 or 1.08. The borderline position is half-way between these values at 1.075. Fig. 1.2 shows that 1.07641 is slightly

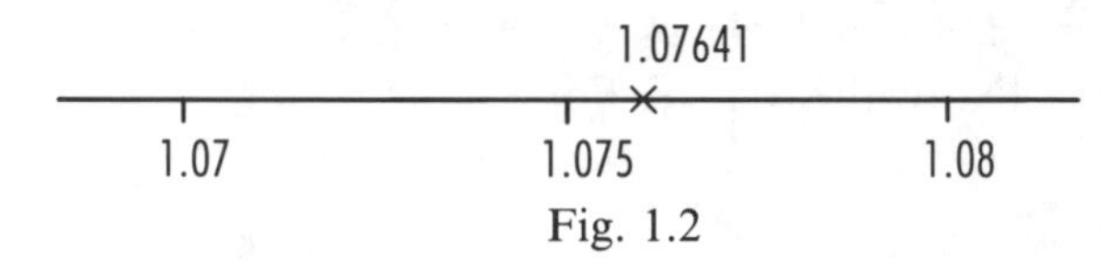

Fig. 1.2

bigger than the borderline value and therefore closer to 1.08 (3 sf) than any other number correct to 3 significant figures. Then 1.07641 may be rounded to 1.08 (3 sf).

If we are presented with a number correct to some significant figures it is unlikely that we would know of its true original value. Even then its 'true' value is only as accurate as the measuring device used. Suppose we are presented with a measurement of 76.2 mm. All we can say is that this is correct to 3 significant figures or 1 decimal place and that the true value lies somewhere between 76.15 and 76.25 mm.

Not all figures are significant. The general rule is that leading and trailing zeroes are not significant.

Examples 1.11

i) 0.00169 is correct to 3 significant figures because the zeroes at the beginning (leading) are not significant.

ii) 0.0016900 is still correct to only 3 significant figures because all the zeroes at the end (trailing) are not significant either.

iii) 0.0010900 is correct again to 3 significant figures. The zero sandwiched between the 1 and 9 is significant. It neither leads nor trails.

Example 1.12

A local engineering company made profits of £527 000 in the last financial year.

This profit is correct to 3 significant figures (the zeroes are trailing and so not significant) and has been rounded. It is very unlikely that the company can be exact when reporting its profit, and has said this is correct to the nearest thousand pounds.

Standard form

Numbers in standard form are all displayed in the same format: a value multiplied by 10 to a positive or negative whole number power. That value starts at 1 and almost, but not quite, reaches 10. We can use the numbers of Examples 1.11 and 1.12.

Examples 1.13

i) $0.00169 = 1.69 \times 10^{-3}$ in standard form.
The decimal point of 0.00169 has leap-frogged 3 places (this is the figure in the power of 10) to the right (this is the negative in the power of 10) from the original number to create the standard form.

> The calculator display will miss out the 10. If you enter 0.00169 and press [=] the display will show 1.69^{-03}. It is taken for granted that you understand this to mean 1.69×10^{-3} in standard form.

ii) $0.0010900 = 1.09 \times 10^{-3}$ in standard form.

iii) $0.00023 = 2.3 \times 10^{-4}$ in standard form, having leap-frogged 4 places to the right.

iv) $527\,000 = 5.27 \times 10^{5}$ in standard form, having leap-frogged 5 places to the left from the original number to create the standard form.

v) $527 = 5.27 \times 10^{2}$ in standard form, having leap-frogged 2 places to the left.

vi) 5.27 would usually remain unchanged without any power of 10.

EXERCISE 1.3

Round the following values to the stated number of decimal places or significant figures.

1	0.992	(2 dp)
2	0.38847	(4 dp)
3	17.38847	(2 dp)
4	5.69	(1 dp)
5	21.399	(2 dp)
6	4.32	(1 sf)
7	14.32	(1 sf)
8	217 000	(2 sf)
9	0.0037602	(3 sf)
10	3165	(2 sf)
11	62.04	(1 dp)
12	6.204	(1 dp)

13 20.3 (1 sf)
14 620.401 (4 sf)
15 620.461 (1 dp)
16 99.8 (1 sf)
17 5489.97 (3 sf)
18 1 400 000 (1 sf)
19 4239.06 (2 sf)
20 0.070629 (3 dp)
21–40 Using the original questions 1–20 above re-write each one in standard form.

ASSIGNMENT

Let us return to our original assignment. This time, with improved accuracy, we have managed to measure the speed at 30.75 ms^{-1} just before the brakes are applied to give the deceleration at 4.40 ms^{-2}. The 1 m error measurement has been re-checked and found to be 0.86 m.

It is thought that once these brakes are applied the distance the vehicle travels is given by the formula

$$s = 0.86 + 30.75t - 2.2t^2$$

where s is the distance travelled
and t is the time from when the brakes are applied.

Just like last time we will construct a table of values in stages based on times from when the brakes are applied (i.e. $t = 0$) and then after 1, 2, 4 and 6 seconds. Again some gaps will be left in the table to give you some practice with your calculator. The calculation stages are as before, but this time using slightly different numbers in the formula for the expected value of s.

Firstly we need a line of values of t, i.e.

t	0	1	2	3	4	5	6

Again s is made up of three different terms on the right-hand side which we will consider in turn for each of these time values. The first term on that side is 0.86. It is a constant value and so is unaffected by each value of t, i.e.

0.86	0.86	0.86	0.86		0.86		0.86

The next term is $30.75t$. Each value of t is considered in turn and multiplied by 30.75. For example when t is 4 the key operations are

30.75 × 4 =

to display an answer of 123. This line of the table is

$30.75t$	0	30.75	61.50		123.00		184.50

The final term on the right-hand side is $-2.2t^2$ and again each different value of t is used in the calculations. It is only the t that is squared. Once this operation has been carried out, multiplication by −2.2 is necessary. For example when t is 6 the order of key operations is

6 x^2 × 2.2 = +/−

to display an answer of −79.2. The line of the table is

$-2.2t^2$	0	−2.20	−8.80		−35.20		−79.20

After the first line of t values we have a series of rows from each of the three terms of $s = 0.86 + 30.75t - 2.2t^2$. Carefully drawn we need to look closely at the columns formed once these rows have been put together. Because we have already taken care of the minus sign all that remains is to add each column of values to get the expected value for s. The last of these columns is given as

0.86
184.50
and −79.20.

These will add to 106.16 and it is this value that appears in the last column of the last row.

The complete table is given below:

t	0	1	2	3	4	5	6
0.86	0.86	0.86	0.86		0.86		0.86
$30.75t$	0	30.75	61.50		123.00		184.50
$-2.2t^2$	0	−2.20	−8.80		−35.20		−79.20
s	0.86	29.41	53.56		88.66		106.16

The original figures of 30.75, 4.40 and 0.86 are correct to 2 decimal places. There is an argument for approximating all these calculated values of s to only 1 decimal place. This is left as a simple exercise for you.

Row 1 shows the original values of t.
Rows 2–4 show all the workings for the various terms.
Row 5 is the row for s, being the addition of row values 2, 3 and 4.

From this table we can check our practical distance measurements to see if they tally with our calculator values. The closer the tally the closer our formula is at predicting what will happen later to the vehicle, and eventually when it will stop.

Truncation

Some calculators do *not* round numbers properly, but cut off according to how many figures (**digits**) can be displayed. When the rounding is a rounding down there is no problem as the calculator is deciding to round correctly. Unfortunately it is the **rounding up** that needs care. Suppose we have a calculator with an 8 digit display. If you try 14 divided by 3 it may display 4.6666666, showing the 6 repeated many times (called **recurring**). A bigger capacity display may just show more 6s. In fact 4.6666666... is closer to 4.6666667 than the figure shown on the display and it is this latter display that is shown on the better calculators. You need less decimal places than those actually displayed on the calculator. The last figure in the display should not be a problem.

To save space we have a notation to represent a number that is recurring. We place a dot or a line over the figures that recur; e.g

114.6666666... is shown by $114.\dot{6}$ or $114.\bar{6}$,
and 21.09090909... is shown by $21.\dot{0}\dot{9}$ or $21.\overline{09}$.

The true value of a number

If you have a £10 note you understand that it is worth £10. If you buy something costing less than this note, say something worth £9.60, then you will get some change. Also if your purchase is more than £10 you will need to offer more money. This accuracy is not true for everything. We can return to our engineering company profit of £527 000 which has been quoted correct to 3 significant figures. This means that the true (probably unknown) profit lies somewhere between £526 500 and £527 500 i.e. £527 000±£500. There is a variation hidden in that 4th significant figure we did not quote originally.

The same idea works for small values. 0.00169 is correct to 5 decimal places, but the exact value might be somewhere between 0.001685 and 0.001695. This means that 0.00169 is correct only to 0.000005, i.e. 0.00169±0.000005.

The true value of a number may be important when attempting simple arithmetic calculations.

Think about 2 values, 69 and 33.
Now 69 may vary between 68.5 and 69.5
and 33 may vary between 32.5 and 33.5.
Then 69 + 33 might lie between some limits as well.
Addition of the large extreme values gives 69.5 + 33.5 = 103,
of the quoted values gives 69 + 33 = 102
and of the small extreme values gives 68.5 + 32.5 = 101.
Which answer is correct?

If we use the quoted values, which we believe to be true, then our answer is 102. It is possible that it may be slightly in error, perhaps as high as 103 or as low as 101 in the extreme cases. Overall we can quote our answer to be 102 ± 1.

Similarly we may subtract our values.

The biggest difference is $69.5 - 32.5 = 37$,
the difference using the quoted values is $69 - 33 = 36$
and the smallest difference is $68.5 - 33.5 = 35$.

Again we can think about our answers and overall quote our answer to be 36 ± 1.

When attempting either addition or subtraction the general rule is that the maximum error is the sum of the original errors.

In these cases that error is $\pm(0.5 + 0.5) = \pm 1$. In both our answers, 102 and 36, we were not confident of the last digits of 2 and 6 because this is where any change may occur.

What happens in multiplication?

Multiplying the large values gives $69.5 \times 33.5 = 2328.25$,
the quoted values gives $69 \times 33 = 2277$
and the small values gives $68.5 \times 32.5 = 2226.25$.

Again we must ask which of these answers is correct. If we know with absolute certainty that the original values of 69 and 33 are true then the correct answer is 2277. If we are unsure about our values then in reality we may say only that the answer is 2000 correct to 1 significant figure. This is because all 3 answers will round to the same answer of 2000 correct to that 1 significant figure.

What happens in division?

Dividing extreme values gives $\frac{69.5}{32.5} = 2.138461$ (as displayed),

the quoted values gives $\frac{69}{33} = 2.090909$ (as displayed)

and other extreme values gives $\frac{68.5}{33.5} = 2.044776$ (as displayed).

Our original values were quoted to 2 significant figures, though our answers still differ, being 2.1, 2.1 and 2.0 respectively to this degree of accuracy. Only to 1 significant figure are all the answers the same.

For multiplication and division the general rule is that answers are correct to one signifcant figure less than the original values.

It is interesting to notice the arithmetic operations and the use of these extreme values. When adding and multiplying the pairing is both large extreme values together and both small extreme values together. When subtracting and dividing the choice is one extreme large and one extreme small value together. This is necessary to discover the maximum errors possible. Other combinations do not give us these wide variations.

The 2 general rules are important where a complete problem has a variety of part calculations before the final answer. Do not round your part answers too soon or you may find that the final answer is wildly inaccurate.

Rough checks

It is important to **have some idea of the size of your final answer** before attempting to use your calculator. This allows you to check roughly that you have pressed the correct keys.

We shall return to our values of 69 and 33. Correct to the nearest 10 we may think of them as 70 and 30. When multiplied together, $70 \times 30 = 2100$ which is roughly in the region of $69 \times 33 = 2277$.

During division it is easier to think of 70 in place of 69 and 35 in place of 33 because 70 is exactly divisible by 35,

i.e. $\frac{70}{35} = 2$, which again is close to our answer of $\frac{69}{33} = 2.090909$ as displayed.

Examples 1.14

i) 1.694×0.42 may be thought of as approximately 1.7×0.4.
The approximate answer is 0.68 compared with the true displayed answer of 0.71148. Now 1.694 is quoted correct to 4 significant figures but 0.42 is quoted to only 2 significant figures. We pay attention to the lesser number of significant figures because this is where there is less accuracy. Then our answer must be to 1 significant figure less than this, i.e. $2 - 1 = 1$ significant figure,
i.e. $1.694 \times 0.42 = 0.7$ (1 significant figure).

ii) $(2.143)^2$ may be thought of as only 2^2. This gives an approximate answer of 4 compared with 4.59 (3 significant figures).

iii) $(2.515)^2$ is closer to 3^2 than to 2^2. It would be safe to suggest that the answer lies somewhere between 9 and 4. In fact even though the original 2.515 lies closer to 3 than to 2 this value squared lies closer to 4 than to 9. The reason for this is the effect of squaring a value.
$(2.515)^2 = 6.33$ (3 significant figures).

EXERCISE 1.4

For each question you should

i) estimate (roughly work out) the size of your answer with a rough check on paper, not using your calculator;

ii) work out the calculation based on those figures quoted being true, using your calculator.

1 3.12×7.90

2 4.35×1.6

3 $59.72 \div 4.11$

4 $3.672 \div 9.051$

5 $4.35^2 \times 1.6$

6 $(4.35 \times 1.6)^2$

7 $\dfrac{22.2 \times 3.9}{1.3}$

8 $\dfrac{14.72 \times 4.06}{1.28}$

9 $1.15 + \left(\dfrac{22.2 \times 3.9}{1.3}\right)$

10 $73.69 - (4.35 \times 1.6^2)^2$

11 $\dfrac{26.14 \times 958}{105.34 \times 9.26}$

12 $\dfrac{(153.4)^2 \times 7.85}{(295.7)^2 \times 6.91}$

13 $\dfrac{22.2^2 \times 3.9^2}{1.3^2}$

14 $\left(\dfrac{22.2 \times 3.9}{1.3}\right)^2$

15 $\dfrac{76.31 \times \sqrt{3.87}}{1.95 \div 15.03}$

16 $\dfrac{\sqrt{84.35} \times 12.62}{3.11 \times 25.09}$

17 $\dfrac{\sqrt{84.35 + 12.62}}{1.95 \div 15.03}$

18 $\sqrt{\dfrac{(84.35 \times 12.62)}{(3.11 \times 25.09)}}$

19 $\dfrac{(9.45)^2}{40.2} + \dfrac{(2.55)^2}{8.5}$

20 $\dfrac{12.35 \times 120.5 \times \sqrt{30.05}}{17.6 \times \sqrt{25.67} \times 3.45}$

Some other key operations

The majority of scientific calculators have many more function keys than we have discussed so far in this chapter. Because the models of calculators change quite often we cannot usefully consider all possibilities. In this section we will learn how to use some of them, but will leave more until later chapters.

$\sqrt[3]{\ }$ is the key that finds the cube root of a number.

Examples 1.15

i) $\sqrt[3]{64}$ is worked out using

64 $\sqrt[3]{\ }$

to display the answer 4.

ii) $\sqrt[3]{-64}$ is worked out using

64 +/-

in order to display −64 and then

$\sqrt[3]{\ }$

to display the answer of -4.

$x^{1/y}$ is the key that finds the yth root of a number x.
We may attempt the previous examples again, using this key, because the cube root has a power $1/3$.

Examples 1.16

i) $\sqrt[3]{64}$ may be thought of as $64^{1/3}$. It is understood that all of the number $(64)^{1/3}$ is raised to the power $1/3$.
The order of key operations is

64 $x^{1/y}$ 3 =

to give the correct display of 4.
After pressing $x^{1/y}$ it is the 3 that is the value used to replace y in the power.

ii) $\sqrt[3]{-64}$ is $(-64)^{1/3}$, emphasising the position of the minus (**negative**) sign.
The order of key operations is

64 +/– $x^{1/y}$ 3 =

to give a display of -4.

When y is an **odd value** (i.e. 1, 3, 5, . . .) there is no problem finding the root of a positive or negative number. If the original number is positive then its root is positive as well. Similarly if the original number is negative then its root is negative.

A little more care is needed when y is an **even number** (i.e. 2, 4, 6, . . .). If the original number is positive there will be 2 real roots, one being positive and the other negative. The calculator will display the positive root and it is understood that you will know that there is a negative root as well. However, there are no real roots of a negative number.

Examples 1.17

i) $\sqrt[4]{625}$ may be thought of as $625^{1/4}$. In fact it will have 2 roots, one being positive that is displayed and one being negative that is understood.
The order of key operations is

625 $x^{1/y}$ 4 =

The display is 5 whilst the other answer of -5 is understood.

ii) $\sqrt[4]{-625}$ may be thought of as $(-625)^{1/4}$ which has no real roots. If you attempt the following order of

625⌋ +/-⌋ $x^{1/y}$⌋ 4⌋ =⌋

the display will give the error message -E-.

1/x⌋ is the key that finds 1 divided by a number. It is called the **reciprocal**.

Examples 1.18

i) $\frac{1}{2}$ is the numerical way of writing 'one half'. It is also the reciprocal of 2. The usual way of finding its decimal value is to use the key operations of

1⌋ ÷⌋ 2⌋ =⌋

An alternative method is

2⌋ 1/x⌋

Both of these methods will display 0.5.

ii) $\frac{1}{13 \times 5}$ may be calculated using the key operations

13⌋ ×⌋ 5⌋ =⌋

to display 65, i.e. $\frac{1}{13 \times 5} = \frac{1}{65}$.
Pressing the key 1/x⌋ displays the decimal value of 0.0153846.
The complete order of key operations is

13⌋ ×⌋ 5⌋ =⌋ 1/x⌋

iii) $\frac{4}{13 \times 5}$ may be calculated in a few different ways. If we think of it as $4 \times \left(\frac{1}{13 \times 5}\right)$ then we may use our previously displayed value of 0.0153846 multiplied by 4 to give a final display of 0.0615384.
The complete order of key operations is

13⌋ ×⌋ 5⌋ =⌋ 1/x⌋ ×⌋ 4⌋ =⌋

π⌋ is the key representing the usual value of 3.14159... depending upon the number of decimal places displayed on your calculator. It is a useful key, saving you time and avoiding possible error inputting the value.

Examples 1.19

i) 3π, understanding a multiplication sign between 3 and π, uses the key operations

3⌋ ×⌋ π⌋ =⌋

to display 9.424778.

ii) $\dfrac{1}{3\pi}$ may be worked out using the latest two key operations in the following order because it is the reciprocal of the previous example. The key operations

$\boxed{3}$ $\boxed{\times}$ $\boxed{\pi}$ $\boxed{=}$ $\boxed{1/x}$

display an answer of 0.1061033.

iii) $\sqrt{\dfrac{1}{3\pi}}$ extends the key operations a little further to use the $\boxed{\sqrt{}}$ key. The complete order of operations is

$\boxed{3}$ $\boxed{\times}$ $\boxed{\pi}$ $\boxed{=}$ $\boxed{1/x}$ $\boxed{\sqrt{}}$

to display 0.325735.

Example 1.20

In this example we will work out $\sqrt[3]{\left(\dfrac{1}{5\pi^2}\right)^4}$ using a combination of several keys we have learned to use already. The order of key operations is important, though there are several variations of the following order that are possible. Think how you would write this down. One order might be

cube root
brackets to the power 4
one over
5
π squared.

The order of key operations is the reverse of this order of writing down, i.e. starting at the heart of the calculation and working outwards. We can take the complete order in several stages, working *up* the list above.

$\boxed{\pi}$ $\boxed{x^2}$ $\boxed{\times}$ $\boxed{5}$ $\boxed{=}$

to display 49.348022,

i.e. $\sqrt[3]{\left(\dfrac{1}{5\pi^2}\right)^4} = \sqrt[3]{\left(\dfrac{1}{49.348022}\right)^4}$

$\boxed{1/x}$ finds the reciprocal of 49.348022 to be 0.0202642.

This value can be raised to the power 4 using

$\boxed{x^y}$ $\boxed{4}$ $\boxed{=}$

The display is 1.6862463^{-07}, remembering the way the calculator represents standard form.

Finally the cube root can use

$\boxed{x^{1/y}}$ $\boxed{3}$ $\boxed{=}$

to display 5.5246784^{-03}.

EXP| is the key to input a number in standard form directly into the calculator.

Examples 1.21

i) To input 1.25×10^3 the order of key operations is

1.25| EXP| 3|

and the display becomes 1.25^{03}.

ii) To input 6.089×10^{-2} the order of key operations is

6.089| EXP| +/-| 2|

giving the display 6.089^{-02}.

The memory

Different calculators can have slightly different memory operations (functions). **Whenever you have completely finished a calculation remember to cancel the contents of the memory**. We can mention the more usual buttons common to many calculators.

MR| recalls to the screen the contents of the memory.

M in| transfers the screen display to the memory, deleting whatever was previously in the memory. When the transfer occurs the display either may go blank or remain as before.

M+| adds the screen display to the memory.

M−| subtracts the screen display from the memory.

A calculator without the key M−| presents no difficulties. +/-| M+| has the same effect because +/-| changes the display to a negative value before adding that negative value to the memory using M+|. You should recall that subtraction is just the addition of a negative number, e.g. $5-3=5+(-3)=2$. Indeed, you should test this for yourself.

EXERCISE 1.5

Use your calculator to work out the following:

1 2.479^4

2 4.3×2.96^3

3 $1.45^2+1.32^2$

4 $4.79\,(1.45^2+1.32^2)$

5 $4.79\,(1.45+1.32)^2$

6 $\sqrt[4]{74.19}$

7 $74.19^{1/4}$

8 $\sqrt[3]{33.249}$

9 $\sqrt[3]{(-33.249)}$

10 $74.19^{1/4}-\sqrt[3]{33.249}$

11 $74.19^{1/4}-\sqrt[3]{(-33.249)}$

12 $\sqrt{1.45^2+1.32^2}$

13 $\dfrac{1}{2\pi}$

14 $\dfrac{1}{2\pi + 4.9}$

15 $\sqrt{\dfrac{1}{2\pi - 4.9}}$

16 $\dfrac{\sqrt[3]{119.6}}{2.3}$

17 $\dfrac{2.3}{\sqrt[3]{119.6}}$

18 $\dfrac{1.32\pi + 2.3}{\sqrt[3]{119.6}}$

19 $1.32 + \dfrac{2.3}{\sqrt[3]{119.6}}$

20 $1.32\pi - \dfrac{2.3}{\sqrt[3]{119.6}}$

The final section of this chapter applies the calculator methods we have learned to some practical problems.

Example 1.22

Suppose we have a copper rod of length 0.20000 m which we heat up, raising its temperature by 250°C. The rod will expand so that its new length is given by the formula $l = l_0(1 + \alpha t)$. Before we attempt to calculate the newly expanded length we need to be able to understand the meaning of the formula.

l is the expanded length
l_0 is the original length
α is the coefficient of linear expansion
and t is the temperature rise.

All metals expand at different rates. The coefficient of linear expansion is a measure of that rate for every °C. In this problem we have $l_0 = 0.20000$, $\alpha = 17 \times 10^{-6}$ and $t = 250$ with all the units of measurement being consistent.

Using the methods of this chapter we can complete the calculation in a number of steps.

Step one	$\alpha t =$	4.25×10^{-3}
Step two	$1 + \alpha t =$	1.00425
Step three	$l_0(1 + \alpha t) =$	$(0.2000)(1.00425)$
	$=$	$0.20085.$

We need to interpret our answer bearing in mind the **original number of decimal places** when measuring the rod's length. This means that either we round up or down the last figure to give a new length of either 0.2009 m or 0.2008 m.

Example 1.23

In this example we will consider a body of mass 23 kg having kinetic energy of 1.8 kJ. We wish to find its speed.

The kinetic energy of a body is the energy it has by virtue of its motion, according to the formula $KE = \frac{1}{2}mv^2$.

In this formula for the body KE is the kinetic energy,
m is the mass
and v is the speed.

The energy is not given in units that are consistent with the formula. The 1.8 kJ needs to be written as 1800 J.

Firstly we will substitute the values into the formula so that

$$KE = \tfrac{1}{2}mv^2$$

becomes $$1800 = \tfrac{1}{2}(23)\,v^2$$

i.e. $$1800 = 11.5v^2.$$

Because v appears on the right we will concentrate on this side of the equation. If we had a value for v the calculation would be

input v
square
multiply by 11.5
value of 1800.

Remember that we read the list *upwards* and use the opposite operation to give a new list, i.e.

input 1800
divide by 11.5
square root
v.

Now our calculation is

$$\frac{1800}{11.5} = v^2$$ Dividing by 11.5

$$156.52 = v^2$$

$$\sqrt{156.52} = v$$ Square rooting.

$$12.5 = v.$$

The speed of the body is 12.5 ms^{-1} correct to 1 decimal place.

Example 1.24

The circuit diagram shows an A.C. emf of 240 V and frequency 50 Hz with a resistance of 550 Ω and an inductance of 8 H in series. We are going to calculate the current, i, in the circuit given that $i = \dfrac{V}{\sqrt{(R^2 + \omega^2 L^2)}}$ where $\omega = 2\pi f$.

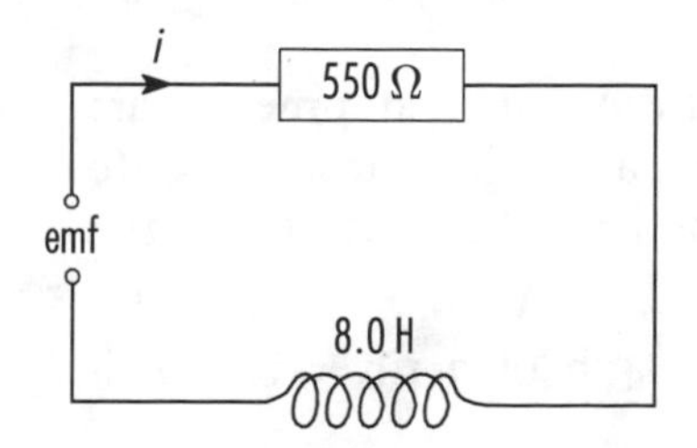

We need to check the meaning of all the letters in our formulae.

V is the voltage, $V = 240$
R is the resistance, $R = 550$
L is the inductance, $L = 8$
f is the frequency, $f = 50$
i is the current.

The first stage of the calculation is to find ω so that

$$\omega = 2\pi f$$

becomes $\omega = 2(3.14159\ldots)(50)$

$$\omega = 314.159$$

and then $\omega^2 = 98\,696.044.$

In the denominator of the more complicated formula we have terms involving R^2, ω^2 and L^2 which we can use in stages.

$$\omega^2L^2 = (98\,696.044)(64) = 6\,316\,546.8$$

and $R^2+\omega^2L^2 = 302\,500+6\,316\,546.8 = 6\,619\,046.8$

so $\sqrt{(R^2+\omega^2L^2)} = \sqrt{6\,619\,046.8} = 2\,572.751.$

Finally $i = \dfrac{V}{\sqrt{(R^2+\omega^2L^2)}}$

becomes $i = \dfrac{240}{2\,572.751} = 0.093.$

This means that there is a current of 0.093 amps or 93 mA.

With a little care you can find i using the calculator's memory to store the calculation in stages. You should try this for yourself. It will save you writing down answers and worrying about the number of decimal places to re-input into your calculator.

EXERCISE 1.6

1 The area, A, of a trapezium is given by the formula $A = \frac{1}{2}(a+b)h$. Find the area when $a=0.85$ m, $b=1.37$ m and $h=0.2$ m.

2 The surface area of a ball bearing, A, is given by $A=4\pi r^2$ where r is the radius. For $r=11$ mm calculate the surface area.

3 An increasing horizontal force is acting on a body on a rough horizontal plane. The body causes a normal reaction of $N=50$ Newtons and the coefficient of friction between the body and the plane is $\mu=0.24$. Calculate the amount of friction, F, acting if $\mu = \frac{F}{N}$. The total reaction, S, is given by $S=\sqrt{(F^2+N^2)}$. Use your value of F to calculate S.

4 The initial velocity, u, constant acceleration, a, and time, t, are connected to the displacement, s, of a vehicle according to $s=ut+\frac{1}{2}at^2$. For $u=1.56\,\mathrm{ms}^{-1}$ and $a=2.12\,\mathrm{ms}^{-2}$ find the displacement after 7.50 seconds.

5 Given $l=l_0(1+\alpha t)$ find the expanded length of a silver rod, l, if $l_0=0.3500$ m, $\alpha=19\times10^{-6}$ and $t=850°$C.

6 The rate of heat energy transfer, Q, is related by $Q = \frac{kAT}{x}$ where A is the conducting area, T is the temperature difference between the two faces, x is the thickness of the material and k is the coefficient of thermal conductivity. If $k=1.1$, $A=6.5$, $T=15$ and $x=0.1$ find the value of Q.

7 Calculate the power, P, across a resistor, R, for a current, I, related by $P = I^2R$ when $I=1.25$ A and $R=3\,\Omega$.

8 An object is placed at a distance (u) of 350 mm in front of a concave spherical mirror of radius (r) 200 mm. Find the distance of the image (v) from the mirror if $\frac{1}{v}+\frac{1}{u}=\frac{2}{r}$.

9 The effective resistance, R, of 2 resistors in parallel is $\frac{1}{R}=\frac{1}{r_1}+\frac{1}{r_2}$. Calculate R if $r_1=4.5\,\Omega$ and $r_2=3.6\,\Omega$.

10 Hooke's law relates the tension in an elastic spring, T, to the natural length of the spring, l, and its extension, x, according to $T=\frac{\lambda x}{l}$, where λ is the coefficient of elasticity. For $\lambda=28$ N and $T=9$ N find the extension from a natural length of 0.75 m.

11 A vehicle accelerates constantly from rest to $17.5\,\text{ms}^{-1}$ in 15 seconds. The displacement, s, is related to the initial velocity, u, the final velocity, v, and the time, t, by $s=\frac{1}{2}(u+v)t$. How far will it have travelled in this 15 seconds?

12 A van reaches a speed of $20.5\,\text{ms}^{-1}$ from rest over a distance of 200 m. The initial speed, u, the final speed, v, the acceleration, a, and the distance, s, are connected by the formula $v^2 = u^2 + 2as$. Calculate the acceleration of the van.

13 The area of an annulus (i.e. the area betwen two circles with the same centre and different radii), A, is given by $A = \pi(R^2 - r^2)$. Given $R=250$ mm and $r=150$ mm find the value of A.

14 For a simple pendulum the time period of an oscillation, in seconds, is given by $T = 2\pi\sqrt{(l/g)}$ where l is the length of the pendulum and g is the acceleration due to gravity. Find the value of T if $g=9.81\,\text{ms}^{-2}$ and $l=1.55$ m.

15 The greatest height, h, of a body projected vertically upwards is given by $h = \frac{v^2}{2g}$ where v is the velocity of projection and g is the acceleration due to gravity. If the maximum height is 25 m and $g=9.81\,\text{ms}^{-2}$ what is the velocity of projection?

16 The 3 sides of a triangle are a, b and c. For $a=55$ mm, $b=34$ mm and $c=46$ mm find s where $s=\frac{a+b+c}{2}$. Calculate the area of the triangle, A, where $A = \sqrt{s(s-a)(s-b)(s-c)}$.

17 The gas laws combine volume, V, pressure, P, and absolute temperature, T, according to $\frac{PV}{T} = \text{constant}$.
If $V = 19.2 \times 10^{-3}\,\text{m}^3$, $T = 273°\text{K}$ and $P = 1.013 \times 10^5\,\text{Nm}^{-2}$ find the value of the constant.
The pressure is increased by 10% due to a change in temperature whilst the other two values remain unchanged.
Calculate the new value of T.

18 The voltage magnification factor, Q, is related to the frequency, f, inductance, L, and resistance, R, by $Q = \frac{2\pi f L}{R}$. Given $f = 50\,\text{Hz}$, $L = 5\,\text{H}$ and $R = 200\,\Omega$ work out Q. For this value of Q, L is increased to 7.5 H. What is the new value of R?

19 The quantity of flow, $Q\,\text{m}^3\text{s}^{-1}$, along a pipe of cross-sectional area $0.005\ \text{m}^2$ is related to the velocity of the flow, $v\,\text{ms}^{-1}$ according to $Q = 0.005v$. Construct a table of values relating Q and v for values of v from 0 to 12 at intervals of $2\ \text{ms}^{-1}$.

20 The final velocity, v, of a van due to a constant acceleration of $2.45\,\text{ms}^{-2}$ over a period of time, t, is related to its initial velocity, $1.00\,\text{ms}^{-1}$, by $v = 1.00 + 2.45t$. From $t = 0$ to $t = 60$ seconds at intervals of 10 seconds construct a table of values relating t and v.

2 Linear Equations and Straight Line Graphs

The objectives of this chapter are to:

1. Solve a linear equation.
2. Distinguish between independent and dependent variables.
3. Plot coordinates on a pair of labelled Cartesian axes.
4. Distinguish between different types of gradients.
5. Calculate the gradient, m, and vertical intercept, c, in order to find the equation of a straight line, $y = mx + c$.
6. Draw the graph for a straight line law.
7. Deduce the gradient, vertical intercept and equation from a straight line graph.
8. Relate a linear equation to the relevant straight line graph.
9. Use a numerical substitution in an equation to find a specific value.
10. Apply straight line law techniques to experimental data.

Introduction

The first task in this chapter is to explain the meaning of **linear** and **equation**. An **equation** is a relationship which generally contains numbers and letter(s). There are many sorts of equations, some of which can be solved and some for which there is no real solution. In this book we will aim to solve equations, and in this particular chapter each equation will have just one solution. This means that only one value will be logically correct. That value is said to **satisfy** the equation. Whenever the word **linear** is mentioned it is understood to refer to straight lines. Combining both important words we will see that it is possible to represent a **linear equation** by a straight line graph.

ASSIGNMENT

The Assignment for this chapter involves the vehicle from Chapter 1 but this time paying attention to its speed. It is travelling at a speed of $30\,\text{ms}^{-1}$ when the brakes are applied to produce a deceleration of $4\,\text{ms}^{-2}$. During the motion the speed, v, is linked to the time, t, according to the formula

$$v = 30 - 4t.$$

The first use of this problem will be for one particular numerical calculation and the second will look more generally at the motion in graphical terms.

Simple equations

We could attempt to write some general algebra that is typical of an equation. As an introduction this just makes a simple problem needlessly complicated. Now $x-12=8$ is an example of a simple equation. The = sign can be thought of as a point about which the equation balances so that the left-hand side has the same overall value as the right-hand side. It is obvious from the right-hand side that in this case the value is 8. On the left-hand side reading $x-12$ we can think of this as "some number minus twelve",

i.e. $$x-12 = 8$$

is "some number minus twelve equals eight".

Easy arithmetic tells us that the number must be 20.

Suppose we have the simple equation

$$x+5 = 11$$

i.e. "some number plus 5 equals eleven".

Again easy arithmetic tells us that the number must be 6.

At this stage we have 2 important points to remember.

1. The equation must always balance about =.
2. Eventually only x must appear on one side of the equation.

Any one equation may be written in several ways. Our original example

$$x-12 = 8$$

is equivalent to $$8 = x-12$$

and $$-12 + x = 8$$

and $$8 = -12+x$$

and yet other formats.

It does *not* matter whether x, or any letter in use, lies on the left or right side.

Examples 2.1

i) Solve $x - 12 = 8$.

We can see that we are subtracting 12 on the left. So that only x appears on the left we add 12 to that side. To keep the balance we must add 12 to the right as well,

i.e. $$x-12+12 = 8+12.$$

Now $-12+12$ works out to 0, and 0 added to anything (x) leaves it unchanged,

i.e. $$x = 8+12.$$

On the right simple addition gives 20 so that

$$x = 20.$$

ii) Solve $x+5=11$.
In this case we are adding 5 on the left. So that only x appears on the left we subtract 5 from that side. To keep the balance we must subtract 5 from the right as well,

i.e. $$x+5-5 = 11-5.$$

Now $5-5$ works out to 0, and 0 added to anything (x) leaves it unchanged,

i.e. $$x = 11-5$$

and so $$x = 6.$$

To save time and space it is usual to simplify the left and right sides on the same line rather than in several repetitive stages.

iii) If α and β represent numbers solve $x-\alpha=\beta$.
To eliminate the subtraction of α on the left we must add α and do so to both sides to keep the balance,

i.e. $$x-\alpha+\alpha = \beta+\alpha$$

i.e. $$x = \beta+\alpha.$$

We have no actual values for α and β meaning that we cannot simplify the right to just a number. Instead we have to leave the answer alone.

Whenever letters are used to represent numbers in this way we term them **constants**.

We have used x as the letter in these examples but could have used any other letter in its place. This is shown in the previous examples again, given in the same formats with different letters.

i) $y-12=8$ to give $y=20$;
ii) $t+5=11$ to give $t=6$;
iii) $m-\alpha=\beta$ to give $m=\beta+\alpha$ where α and β are constants.

EXERCISE 2.1

Solve the following simple linear equations, given that in Questions **9** and **10** α, β and γ are constants.

1 $x-14=17$

2 $x+19=3$

3 $-14+x=17$

4 $5=-2+x$

5 $y+6=0$

6 $1+t+4=7$

7 $-2+t=-5$

8 $4+z-3=13$

9 $x+\alpha=\gamma$

10 $x-\alpha+\beta=0$

In this early section we have looked at the addition and subtraction of numbers in simple equations. We have dealt with them together because these arithmetic operations are the opposites of each other. Similarly multiplication and division are opposites of each other, and so now we will deal with them together.

Examples 2.2

i) Solve the simple linear equation $4x = 3$.
We are multiplying x by 4 and the opposite of multiplication is division. This means that we should divide by 4 and do it for both sides to keep the balance of the equation,

i.e.
$$\frac{4x}{4} = \frac{3}{4}$$
$$1x = 0.75$$
$$x = 0.75.$$

4 divides by 4 meaning that the 4s cancel to 1.

1 multiplied by any number leaves that number unchanged so that $1x = x$.

ii) Solve the equation $\frac{x}{6} = 2.5$.
We are dividing x by 6 and the opposite of division is multiplication. Again, attempting to leave x alone on the left we will multiply by 6 and do it for both sides to keep the balance,

i.e.
$$\frac{6x}{6} = 6(2.5)$$
$$1x = 15$$
$$x = 15.$$

iii) We may combine both the needs of multiplication and division to solve $\frac{-2x}{5} = 3.42$.
The opposite of multiplying by -2 is dividing by -2 and the opposite of dividing by 5 is multiplying by 5. We apply these operations to both the left and right of the equation,

i.e.
$$\frac{5(-2x)}{-2(5)} = \frac{5(3.42)}{-2}$$
$$1x = -8.55$$
$$x = -8.55.$$

Both the -2s cancel to 1 and both the 5s cancel to 1 as well.

iv) If α, β and γ represent numbers solve $\frac{\alpha x}{\beta} = \gamma$.
The opposite of multiplying by α is dividing by α and the opposite of dividing by β is multiplying by β. We apply these operations to both the left and right of the equation,

i.e. $$\frac{\beta(\alpha x)}{\alpha(\beta)} = \frac{\beta(\gamma)}{\alpha}$$

$$x = \frac{\beta\gamma}{\alpha}.$$

EXERCISE 2.2

Solve the following equations.

1 $5x = 4$

2 $0.3y = 0.39$

3 $-2y = 3.6$

4 $2.14 = -0.2x$

5 $\frac{y}{4} = 7$

6 $\frac{x}{0.3} = 1.1$

7 $\frac{2x}{3} = 2.84$

8 $1.7 = \frac{3x}{4}$

9 $\frac{-12x}{5} = 6$

10 $\frac{13y}{4} = -1.09$

Now we are in a position to combine the effects of these pairs of opposite arithmetic operations. In Chapter 1 we concentrated on the heart of the calculation. When solving linear equations the heart of the problem is around x (or any other letter that is being used). We must remove other numbers from the same side as the letter (left or right) working our way towards the letter until it remains alone.

Examples 2.3

i) Solve $3x+4=20$.
Addition of 4 is less closely attached to x than anything else. This means that we will deal with 4 first.

$$\therefore \quad 3x+4-4 = 20-4$$

Subtracting 4 from both sides.

$$3x = 16$$

$$\frac{3x}{3} = \frac{16}{3}$$

Dividing both sides by 3.

$$x = 5.\overline{3}.$$

ii) Solve $2x-5=12$.
Similarly, subtraction of 5 is less closely attached to x than any other part of the equation on the left.

$$\therefore \quad 2x-5+5 = 12+5$$

Adding 5 to both sides.

$$2x = 17$$

$$\frac{2x}{2} = \frac{17}{2}$$

Dividing both sides by 2.

$$x = 8.5.$$

iii) Solve $\frac{-x}{3} + 7 = 0$.

$$\frac{-x}{3} + 7 - 7 = 0 - 7$$

Subtracting 7 from both sides.

$$\frac{-x}{3} = -7$$

$$\frac{(-3)(-x)}{3} = -3(-7)$$

Multiplying both sides by -3.

$$x = 21.$$

Remembering that $(-)(-) = +$.

iv) Solve $\frac{x}{3} + 31 = \frac{-2x}{5} + 36.5$.

Terms involving x appear on both sides of this equation. The first move is to gather these terms to one side. We will gather them together on the left, but the same final solution will be reached if we gather them on the right.

$$\therefore \quad \frac{x}{3} + \frac{2x}{5} + 31 = \frac{2x}{5} - \frac{2x}{5} + 36.5$$

Adding $\frac{2x}{5}$.

$$\frac{x}{3} + \frac{2x}{5} + 31 = 36.5$$

$$\frac{x}{3} + \frac{2x}{5} + 31 - 31 = 36.5 - 31$$

Subtracting 31.

$$\frac{x}{3} + \frac{2x}{5} = 5.5.$$

We can deal with the mixture of fractions, multiplying throughout by the lowest common multiple (LCM) of 3 and 5, i.e. multiplying throughout by 15,

$$\text{i.e.} \quad 15\left(\frac{x}{3} + \frac{2x}{5}\right) = 15(5.5)$$

Multiplying each term by 15.

$$\frac{15x}{3} + \frac{30x}{5} = 82.5$$

$$5x + 6x = 82.5$$

$$11x = 82.5$$

$$\frac{11x}{11} = \frac{82.5}{11}$$

Dividing by 11.

$$x = 7.5.$$

As equations involve more and more terms we may check that the answer is correct by separately substituting the answer into each side of the equation, e.g.

Left-hand side $= \frac{x}{3} + 31$
$= \frac{7.5}{3} + 31$
$= 2.5 + 31$
$= 33.5.$

Right-hand side $= \frac{-2x}{5} + 36.5$
$= \frac{-2(7.5)}{5} + 36.5$
$= -3 + 36.5$
$= 33.5$ also.

ASSIGNMENT

Our problem involves a relationship between a vehicle's speed, v, and time, t, according to the formula $v = 30 - 4t$.

In this first discussion we are interested in the time taken for the speed to fall to 12.4 ms^{-1}. This means that we must substitute $v = 12.4$ into our formula,

i.e.

$$12.4 = 30 - 4t$$

$$12.4 - 30 = -4t$$

$$-17.6 = -4t$$

Subtracting 30 from both sides.

$$\frac{-17.6}{-4} = t$$

Dividing both sides by -4.

$$t = 4.44$$

i.e. the speed falls to 12.4 ms^{-1} after 4.4 seconds.

EXERCISE 2.3

Solve the following equations.

1 $6x+2 = 11$

2 $2y - 3 = 7.6$

3 $x + \frac{x}{5} = 24$

4 $\frac{x}{5} = 1 + \frac{x}{4}$

5 $2y + 3 = 17 - 3y$

6 $\frac{5x}{6} - \frac{x}{2} = 3$

7 $\frac{5x}{6} + 3 = 18$

8 $\frac{3x}{5} - \frac{x}{3} = \frac{7}{2}$

9 $\frac{y-1}{4} = \frac{y+2}{3}$

10 $\frac{t}{2} + \frac{t}{3} + \frac{1}{7} = \frac{15}{7}$

Examples 2.4

i) Solve $5\left(\frac{x}{2} + 6\right) = 75$.

In this example the brackets affect the order of operations. x is at the heart of this problem on the left. Using our knowledge from Chapter 1, if x is replaced by a number the order of evaluation is

x (or the number)
divide by 2
plus 6
multiply by 5.

We use this list from the bottom working upwards with the opposite operation. This leads to the new list of operations we will carry out to find x, i.e.

divide by 5
subtract 6
multiply by 2
x (or the number).

Then

$$\frac{5}{5}\left(\frac{x}{2}+6\right) = \frac{75}{5}$$ Dividing by 5.

$$1\left(\frac{x}{2}+6\right) = 15$$ Cancelling the 5s.

$$\frac{x}{2}+6 = 15$$ Multiplication by 1 – no change.

$$\frac{x}{2}+6-6 = 15-6$$ Subtracting 6.

$$\frac{x}{2} = 9$$

$$\frac{2x}{2} = 2(9)$$ Multiplying by 2.

$$x = 18.$$

Again we may check our answer by substitution, this time on the left to see that it gives the value of 75 on the right,

i.e. $5\left(\frac{x}{2}+6\right) = 5\left(\frac{18}{2}+6\right) = 5(9+6) = 5(15) = 75.$

ii) Solve $5(x-2)+3(2x-1)=7$.
As the brackets become more complicated it is useful to multiply them out,

i.e. $$5x-10+6x-3 = 7$$ 5 times each value in the first bracket and 3 times each value in the second bracket.

$$11x-13 = 7$$
$$11x-13+13 = 7+13$$
$$11x = 20$$
$$\frac{11x}{11} = \frac{20}{11}$$
$$x = 1.\overline{8}\overline{1}.$$

iii) Solve $5(x-2)-3(2x-1)=7$.

This example is similar to the previous one, again needing the multiplication of the brackets as a first step. We need to be careful with the second bracket because each value is multiplied by -3,

i.e.
$$
\begin{aligned}
5x-10-6x+3 &= 7\\
-1x-7 &= 7\\
-x-7+7 &= 7+7\\
-x &= 14\\
(-1)(-x) &= (-1)(14)\\
x &= -14.
\end{aligned}
$$

Multiplying by -1.

iv) Solve $\dfrac{(2x+7)}{4}-\dfrac{2(x+1)}{5}=\dfrac{1}{4}$.

There are a number of possible first moves in the solution of this equation. One possibility is to remove the fractions, multiplying throughout by the LCM of 4, 5 and 4, i.e. multiplying by 20,

i.e.
$$
\begin{aligned}
20\frac{(2x+7)}{4}-20\frac{2(x+1)}{5} &= 20\frac{(1)}{4}\\
5(2x+7)-8(x+1) &= 5\\
10x+35-8x-8 &= 5\\
2x+27 &= 5\\
2x+27-27 &= 5-27\\
2x &= -22\\
\frac{2x}{2} &= \frac{-22}{2}\\
x &= -11.
\end{aligned}
$$

Cancelling down the 20s.

Multiplying brackets.

Dividing both sides by 2.

EXERCISE 2.4

Solve the following linear equations.

1 $5(x+2)+3x-3=2$

2 $4(x-1)-5(1-x)-2=3x$

3 $\frac{1}{5}(2x-4)+1=\frac{1}{4}x$

4 $2(y+1)-5(y-4)=0$

5 $\frac{2}{5}x+4=\frac{1}{4}x+7$

6 $\frac{1}{2}y+1-\frac{1}{3}(y-1)+2=0$

7 $2\left(\frac{4}{3}x-1\right)=x+9$

8 $\frac{2}{3}(x-6)=\frac{x+9}{4}$

9 $\left(\frac{3y+1}{2}\right)+\left(\frac{y+2}{3}\right)=1$

10 $\frac{2}{5}(x+5)-\frac{10}{3}(x+1)=2$

All the examples so far have included the reasons and operations needed to reach a final solution. It is usual for some of the method to be omitted as you become more experienced with the algebra. We can repeat Example 2.3i) to show a simplified version,

i.e. $3x+4=20$

i.e. $3x=20-4$

Subtracting 4 from both sides.

$3x=16$

$x=\frac{16}{3}$ or $5.\bar{3}$.

Dividing both sides by 3.

Graphs and equations

A graph is a straight line or curve that represents a relationship, often between y and x. The graph of a linear equation is always a straight line. We will draw the line with reference to a pair of **Cartesian** (i.e. **rectangular**) **axes** (plural of **axis**) and attempt to relate y to x. Fig. 2.1 shows a typical set, or pair, of axes together with some points marked O, P, Q and R.

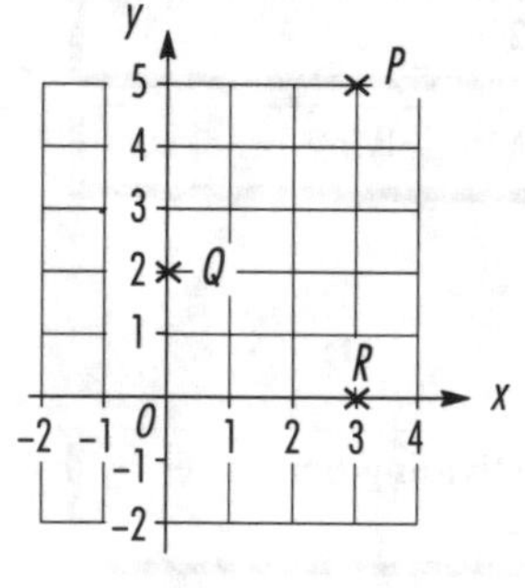

Fig. 2.1

The **horizontal** axis is the **independent** axis and is labelled in this case with x, the **independent** variable. This means that the original choice of values is a choice of values of x. The **vertical** axis is the **dependent** axis and is labelled in this case with y, the **dependent** variable. This means that the values of y depend on the choice of x values and the particular way y is related to x. y is said to be given in terms of x.

The number scale on each axis is regular. It is easier to choose scales involving 1s, 2s, 5s, 10s, 20s, 50s, 100s, . . . or 0.1s, 0.01s, ... This allows you to find parts of graph paper squares and interpret your answers accurately. Fig. 2.1 shows both axes with the same scales, but often they are different, as in Fig. 2.2.

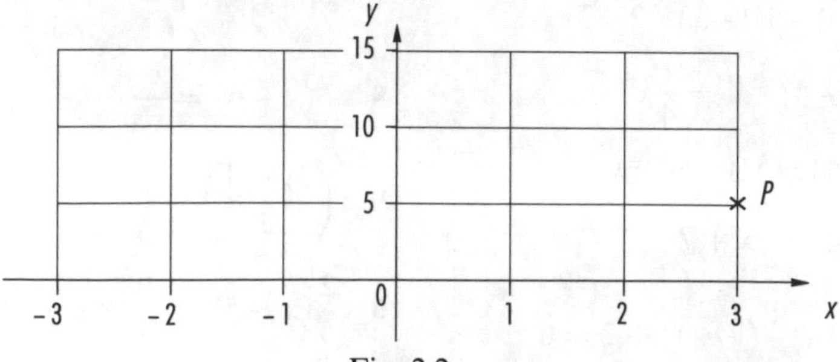

Fig. 2.2

The point O is called the **origin**. It is where the 0 values for y and for x **intersect** (i.e. **cross**). The origin is shown in Fig. 2.1. To use the graph paper efficiently it is sometimes omitted, as in Figs. 2.3.

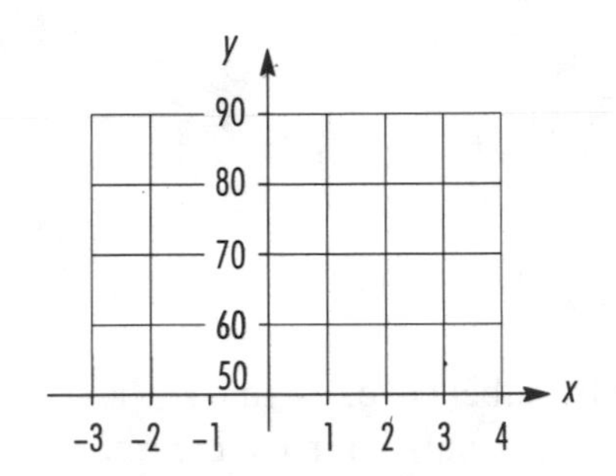

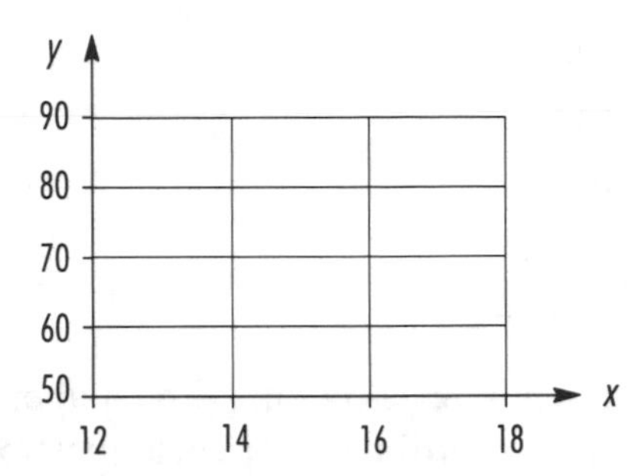

Figs. 2.3

A graph will pass through many points, some of which you will have marked on the graph paper. Each point may be marked by either a cross, x, or by a dot within a circle, ⊙. Due to error or inaccuracy you may see some points on either side of the line. We can label these points or refer to them with **coordinates**. A pair of coordinates is an ordered pair of numbers where the first is the independent variable, usually x, and the second is the dependent variable, usually y,

i.e. (independent variable, dependent variable),

i.e. (x, y).

By way of example $(3, 5)$ are the coordinates representing P in both Fig. 2.1 and Fig. 2.2 . The position of P only looks different because of the different scales in the two figures.

The origin has the coordinates $(0, 0)$, which can be written $O\,(0, 0)$.

If the first coordinate value is 0 (i.e. $x=0$) then the point lies on the **vertical axis**. Q, which may be written $Q\,(0, 2)$, is such a point in Fig. 2.1. If the second coordinate value is 0 (i.e. $y=0$) then the point lies on the **horizontal axis**. Again in Fig. 2.1 we have $R\,(3, 0)$.

Second quadrant | Positive (first) quadrant

Third quadrant | Fourth quadrant

Fig. 2.4

P is said to lie in the **positive quadrant**, i.e. the quarter of the graph paper where both x and y are positive. The "quarter" is not precise because the axes may not be in the middle of the paper to divide it equally. The **positive quadrant** is sometimes called the first quadrant, the other quadrants then being taken in the anticlockwise direction shown in Fig. 2.4.

EXERCISE 2.5

On one sheet of graph paper draw and label a pair of axes x and y. Mark on your paper the following coordinates.

1 $(0, 0)$

2 $(2, 0)$

3 $(0, 4)$

4 $(2, 2)$

5 $(-1, -1)$

6 $(-1, 6)$

7 $(7, -1)$

8 $(-\frac{1}{2}, -\frac{1}{2})$

9 $(3, 5)$

10 $(5,3)$

On a separate sheet of graph paper with fully labelled axes mark the following coordinates for each question. Is there any pattern in each case?

11 $(1, 1)$, $(6, 6)$, $(3.5, 3.5)$, $(-5, -5)$, $(-2.5, -2.5)$

12 $(-1, 1)$, $(-6, 6)$, $(-3.5, 3.5)$, $(2.5, -2.5)$, $(5, -5)$

13 $(2, 1)$, $(3, 1)$, $(-4.5, 1)$, $(-0.5, 1)$

14 $(3, 1)$, $(3, -3)$, $(3, -0.5)$, $(3, -4.5)$

15 $(1, 4)$, $(6, 9)$, $(3.5, 6.5)$, $(-5, 2)$, $(-2.5, 0.5)$

The straight line law

The 2 important features of a straight line graph are the **gradient** (i.e. **slope**) and the **intercept** on the vertical axis (the y axis) where the graph crosses that axis. In the standard equation the gradient is represented by m and the intercept by c.

There are 4 possible types of gradient.

1. Positive gradients

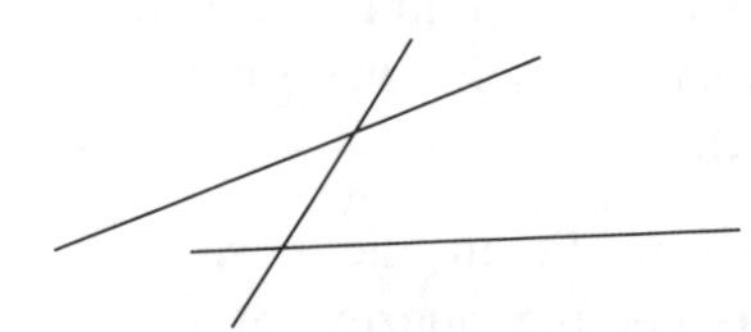

They slope from **bottom left** to **top right**.

2. Negative gradients

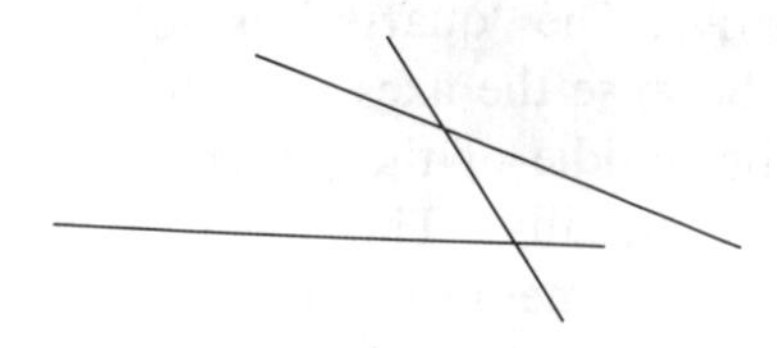

They slope from **top left** to **bottom right**.

3. Zero gradients

The lines are **horizontal**, i.e. they go neither up nor down and so have **no slope**.

4. Infinite gradients

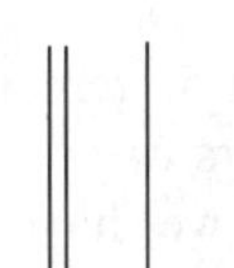

The lines are **vertical**.

Gradient, m is defined as

$$m = \frac{\text{vertical change}}{\text{horizontal change}}$$

The horizontal change is always for x increasing, i.e. moving to the right. During this movement if the vertical change is an increase then the gradient is positive, whilst if the vertical change is a decrease then the gradient is negative.

Before we look at the next set of examples we need to mention something about drawing graphs. An accurate drawing, generally on graph paper, is called a **plot**. A less accurate drawing, usually on ordinary paper, is called a **sketch**. A sketch is used as a general guide to show the basic line or curve.

Examples 2.5

i) Find the gradient of the line joining the points (2, 1) and (5, 3).
A sketch of the coordinates and this line on a pair of labelled axes is a help. As soon as it is drawn it will be obvious whether the gradient is positive or negative. In this example you see that we have a positive gradient.

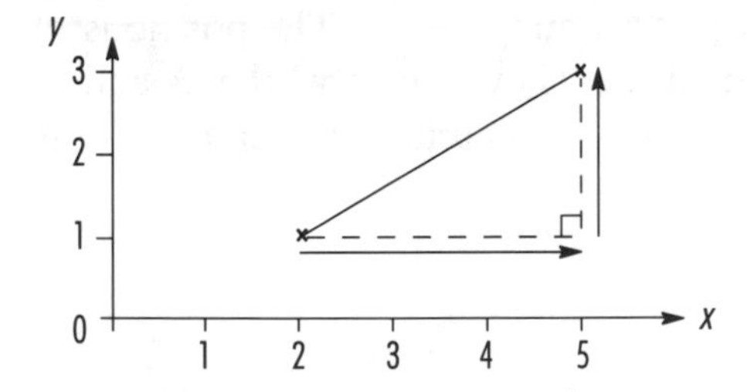

$$\begin{aligned}\text{vertical change} &= 3 - 1\\ &= 2.\end{aligned}$$

$$\begin{aligned}\text{horizontal change} &= 5 - 2\\ &= 3.\end{aligned}$$

$$\text{Gradient, } m = \frac{2}{3} \text{ or } 0.\bar{6}.$$

ii) Find the gradient of the line joining the points $(-4, -1.5)$ and $(2, -3)$.
Again we can mark the coordinates and sketch the line.

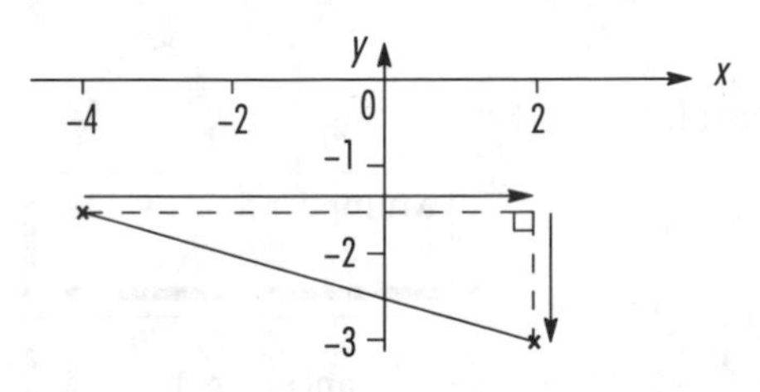

$$\begin{aligned}\text{vertical change} &= -3 - (-1.5)\\ &= -3 + 1.5\\ &= -1.5.\end{aligned}$$

$$\begin{aligned}\text{horizontal change} &= 2 - (-4)\\ &= 2 + 4\\ &= 6.\end{aligned}$$

$$\text{Gradient, } m = \frac{-1.5}{6} = -0.25.$$

iii) Find the gradient of the line joining the points $(-4, -1.5)$ and $(2, -1.5)$.

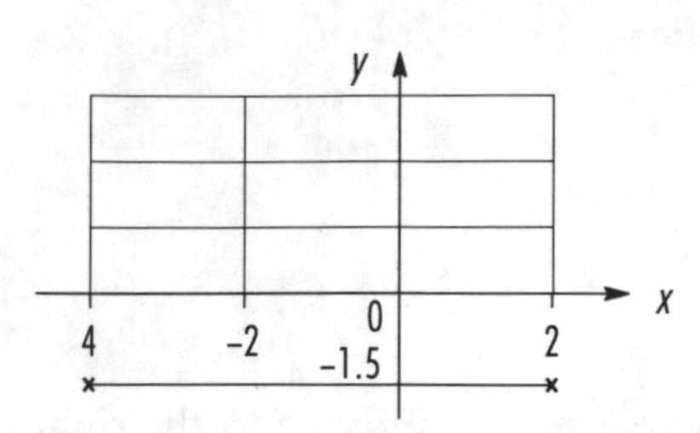

The y coordinate, -1.5, remains the same for both pairs of coordinates meaning that we have a horizontal line and so no vertical change. This means that there is no gradient,

$$\text{i.e. } m = \frac{0}{6} = 0.$$

iv) Find the gradient of the line joining the points $(1, 2)$ and $(1, 5)$.

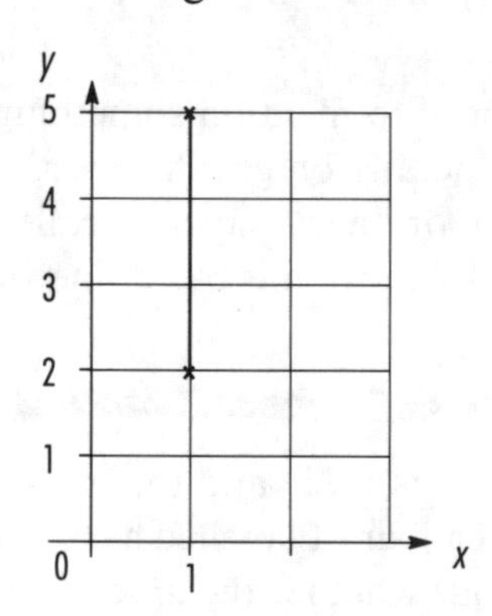

The x coordinate remains the same for both pairs of coordinates, meaning that we have a vertical line and so no horizontal change. We cannot use the gradient formula because the denominator value would be 0 and **division by 0 is not allowed in mathematics**.

There is another version of the gradient formula that saves you drawing the line. Suppose we have 2 points (x_1, y_1) and (x_2, y_2). The positions of the 1s and 2s are called **subscript** positions. They only label the xs and ys to distinguish between the points. They have no numerical meaning in the calculation.

$$\text{Vertical change} = y_2 - y_1.$$
$$\text{Horizontal change} = x_2 - x_1.$$

Using our gradient formula,

$$\text{Gradient} = \frac{\text{vertical change}}{\text{horizontal change}}$$

becomes
$$m = \frac{y_2 - y_1}{x_2 - x_1}.$$

There is another version of this formula,

i.e.
$$m = (1)\frac{(y_2 - y_1)}{(x_2 - x_1)}$$

Multiplication by 1 – no change.

$$= \frac{(-1)\,(y_2 - y_1)}{(-1)\,(x_2 - x_1)}$$

$\frac{-1}{-1}$ cancels to 1.

$$= \frac{-y_2 + y_1}{-x_2 + x_1}$$

Multiplying out the brackets.

$$m = \frac{y_1 - y_2}{x_1 - x_2}.$$

Re-writing with "–" centrally.

The formula takes into account whether the line has a positive or negative gradient. Sometimes a quick sketch can boost your confidence and confirm your answer. We will repeat Examples 2.5 i) and ii) using the formula.

Examples 2.6

i) Find the gradient of the line joining the points (2, 1) and (5, 3).
Compare these values with the general (x_1, y_1) and (x_2, y_2) so that $x_1 = 2$, $y_1 = 1$ and $x_2 = 5$, $y_2 = 3$. Now the gradient is given by

$$m = \frac{y_2 - y_1}{x_2 - x_1}$$

$$= \frac{3 - 1}{5 - 2}$$

Substituting the values.

$$= \frac{2}{3} \text{ or } 0.\bar{6}.$$

ii) Find the gradient of the line joining the points (–4,–1.5) and (2,–3).

Again we can compare these coordinates with the general (x_1, y_1) and (x_2, y_2) so that $x_1 = -4$, $y_1 = -1.5$ and $x_2 = 2$, $y_2 = -3$. The gradient is given by

$$m = \frac{y_2 - y_1}{x_2 - x_1}$$

$$= \frac{-3 - (-1.5)}{2 - (-4)}$$

Substituting the values.

$$= \frac{-3 + 1.5}{2 + 4}$$

$$= \frac{-1.5}{6}$$

$$= -0.25.$$

In both these examples the previous sketches confirm the slopes being positive in i) and negative in ii).

The second important feature of the straight line is the intercept, c, on the vertical axis (generally the y axis). At this point on the y axis the value of x is 0. Using the axes of Fig. 2.3 we may draw some straight lines.

Examples 2.7

i)

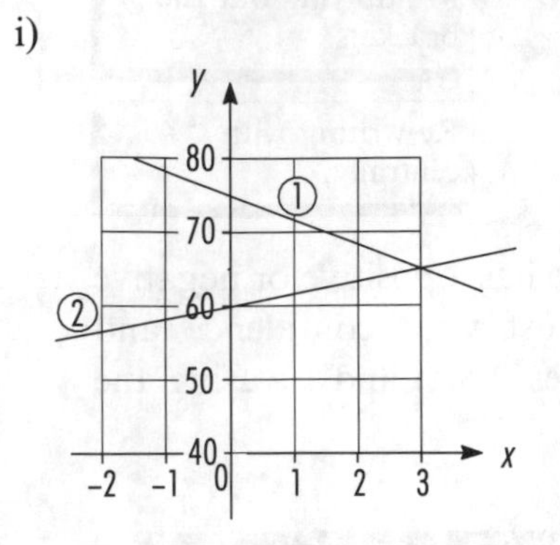

Line ① has an intercept of 75, i.e. $c=75$. Also we know from the previous section of work that the gradient is negative.
Line ② has an intercept of 60, i.e. $c=60$. Again we can see that for this line the gradient is positive.

ii)

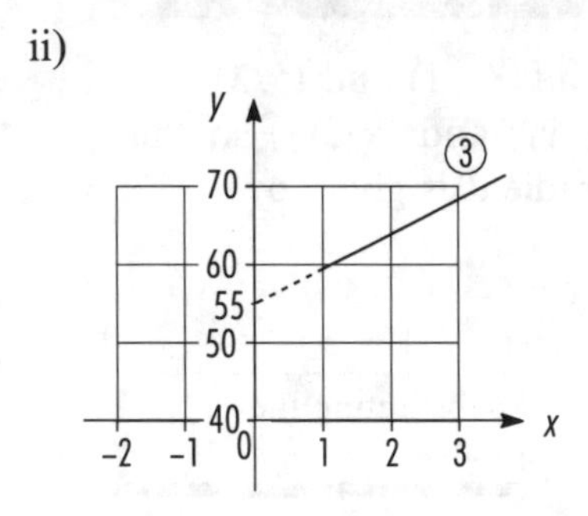

Line ③ does not actually cut the y axis, but we can extend it backwards with the same slope (-----). We see that it intercepts at $c=55$.

iii)

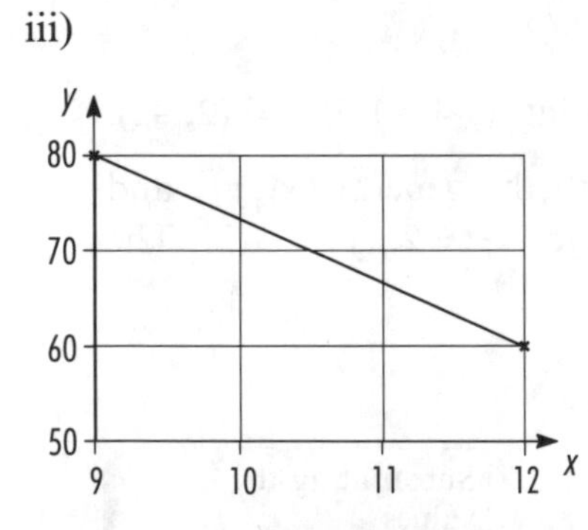

We cannot read off the intercept in this example because in our sketch the x axis does not include where $x=0$. Finding the intercept on the vertical axis needs an extra calculation which is demonstrated in Example 2.12.

So far in this section we have learned about the 2 important features of a straight line, the gradient and the intercept. Now we are able to link them together because the general equation of the straight line is given by

$$y = mx + c$$

where m is the gradient
and c is the intercept on the vertical axis.

Algebraically m, being a number just before x, is said to be the **coefficient** of x.

The format of the equation may not always be exactly the same as this, but often a little algebraic manipulation will re-create the standard format. Whatever the original form and the final, standard, format you must have both x and y to the power 1. As usual we do not write the power 1, usually treating it as being understood. In practical examples it is possible that letters other than x, y, m and c will be used. Each practical example will need to be compared with the standard $y = mx + c$.

We need to be able to interpret a known straight line to give its gradient and intercept. In addition, knowing the 2 important features we need to be able to write down the equation.

Examples 2.8

This set of examples looks to compare each one against the standard form $y = mx + c$ in order to find the values of m and c.

i) $y = 2x+1$ $\quad m = 2, \quad c = 1$

ii) $y = 2x - 1$ $\quad m = 2, \quad c = -1$

iii) $y = \frac{1}{3}x+4$ $\quad m = \frac{1}{3}, \quad c = 4$

iv) $y = 1 - 2x$ $\quad m = -2, c = 1$

Examples 2.9

Our next set of examples is just as easy when combining with some of the algebraic rearrangement we learned earlier in the chapter. This is to re-create the form of $y = mx + c$.

i) $y + 3x-5=0$ is not in the correct format because y must stand alone on one side of the equation.

i.e. $\quad y = -3x + 5$

or $\quad y = 5 - 3x$

are preferred forms.

Subtracting $3x$ from and adding 5 to both sides.

$$m = -3, c = 5.$$

ii) $2y+3x-5=0$ also needs some attention to re-create the standard format,

i.e. $\quad 2y = -3x + 5$

$$y = \frac{-3}{2}x + \frac{5}{2}$$

Dividing by 2.

$$m = \frac{-3}{2}, c = \frac{5}{2}.$$

iii) $\frac{1}{3}y - 4x+10=0$ is an equation that also needs some attention.

ie. $\quad \frac{1}{3}y = 4x - 10$

Adding $4x$ to and subtracting 10 from both sides.

$$y = 12x - 30$$

Multiplying by 3.

$$m = 12, c = -30.$$

iv) You may find the relationship $y = x$ easier to deal with in the form

$$y = x + 0$$

Coefficient of x is 1.

$$m = 1, c = 0.$$

v) $y = 2$ may be thought of as

$$y = 0x + 2$$

$$m = 0, c = 2.$$

vi) $x = 3$ does not involve any letter other than x and so it is impossible to create the standard form $y = mx + c$. In fact if you quickly sketch the graph you will see that it is a vertical line passing through the value 3 on the horizontal axis. This means that its gradient is infinite and it does not intercept the vertical axis.

Examples 2.10

In this group of examples we are going to construct the equation representing the straight line for the given gradient and intercept on the vertical axis. Each equation will be created by substituting the numerical values into the standard format $y = mx + c$.

i) A gradient of 3 and an intercept of 2,

i.e. $m = 3$ and $c = 2$.

$\therefore \quad y = 3x + 2.$

Alternative forms of this are $y - 3x = 2$ and $y - 3x - 2 = 0$.

ii) A gradient of -3 and an intercept of 2,

i.e. $m = -3$ and $c = 2$.

$\therefore \quad y = -3x + 2.$

Alternative forms of this are $y + 3x = 2$ and $y + 3x - 2 = 0$.

iii) An intercept of -2 and a zero gradient,

i.e. $m = 0$ and $c = -2$.

$\therefore \quad y = 0x - 2$

i.e. $\quad y = -2.$

An alternative form of this is $y + 2 = 0$.

iv) A gradient of $\frac{-1}{4}$ and an intercept of $\frac{3}{5}$,

i.e. $m = \frac{-1}{4}$ and $c = \frac{3}{5}$.

$\therefore \quad y = \frac{-1}{4}x + \frac{3}{5}.$

This form of the equation is perfectly all right, though alternative versions can be created to avoid the fractions,

i.e. $$20y = 20\frac{(-1x)}{4} + 20\frac{(3)}{5}$$

Multiplying by 20 and cancelling.

$$20y = -5x + 12.$$

Alternative forms of this

are $$20y + 5x = 12$$

and $$20y + 5x - 12 = 0.$$

The recent groups of examples have involved numbers given at the beginning of each one. Suppose we have no written numbers, just a graph drawn on a pair of labelled axes.

Example 2.11

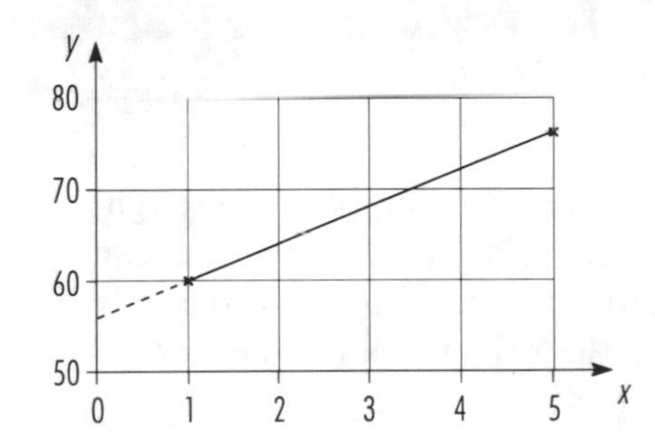

From the points marked on this graph we see that it passes through the points (1, 60) and (5, 76). We need to find the gradient and the intercept so that we may write down the equation of the line.

By comparing our coordinates with (x_1, y_1) and (x_2, y_2) we see that $x_1 = 1$, $y_1 = 60$ and $x_2 = 5$, $y_2 = 76$.

$$\therefore \qquad m = \frac{y_2 - y_1}{x_2 - x_1}$$

becomes
$$m = \frac{76 - 60}{5 - 1}$$
$$= \frac{16}{4}$$
$$= 4.$$

To find the intercept we project the line backwards and see that it intercepts at 56.

$$\therefore \qquad y = mx + c$$

becomes $y = 4x + 56$ in this case.

We can go further with the calculations and avoid drawing a graph altogether, though you will often find one very helpful. This is relevant too when we are unable to read off the intercept value because the origin does not appear on our axes.

Example 2.12

Suppose we know that a straight line passes through the points with coordinates (9, 80) and (12, 60), and that we wish to find its equation.

Comparing these coordinates with our general ones, (x_1, y_1) and (x_2, y_2) we see that $x_1 = 9$, $y_1 = 80$ and $x_2 = 12$, $y_2 = 60$.

$$\therefore \qquad m = \frac{y_2 - y_1}{x_2 - x_1}$$

$$\text{becomes} \qquad m = \frac{60 - 80}{12 - 9} = \frac{-20}{3}.$$

This value for the gradient could be given as a recurring decimal of $6.\bar{6}$, but the fraction is preferred for accuracy.

$$\therefore \qquad y = mx + c$$

$$\text{becomes} \qquad y = \frac{-20}{3}x + c.$$

Now every point that lies on the line (i.e. that the line passes through) satisifies the equation representing the line. This means that we can substitute the values for x and y from any such point to leave the only unknown, c. We can easily use either of the original points and do so below.

Using (9, 80)

$$\text{and} \qquad y = \frac{-20}{3}x + c$$

$$\text{then} \qquad 80 = \frac{-20\,(9)}{3} + c$$

$$80 = -60 + c$$

$$80 + 60 = c$$

$$\text{i.e.} \qquad c = 140.$$

or using (12, 60)

$$\text{and} \qquad y = \frac{-20}{3}x + c$$

$$\text{then} \qquad 60 = \frac{-20\,(12)}{3} + c$$

$$60 = -80 + c$$

$$60 + 80 = c$$

$$\text{i.e.} \qquad c = 140.$$

Either of these calculations may be used to complete the equation

$$y = \frac{-20}{3}x + 140.$$

We can re-write this in any format, perhaps multiplying through by 3 to remove the fraction and moving some or all terms to one side,

$$\text{i.e.} \qquad 3y = -20x + 420$$

$$\text{or} \qquad 3y + 20x = 420$$

$$\text{or} \qquad 3y + 20x - 420 = 0$$

are some of these possible alternatives, though there are others.

Example 2.13

We may extend the previous example and attempt to either sketch or plot the straight line whose equation we have found. The first piece of information we need is that we have a straight line. This means that we must have 3 points to mark – the third one being a check on the accuracy of the other 2. Already we know of our 2 original points (9, 80) and (12, 60). Also we know that the intercept on the vertical axis is 140 and that all the way along the y axis the value of x is 0. This gives us the third point of (0, 140).

As an exercise you should check for yourself with a graph on a pair of labelled axes that all the information connecting these points, the gradient and the equation is true.

EXERCISE 2.6

The first 5 questions show a straight line graph on a pair of labelled axes. In each case find the i) gradient,
ii) intercept,
iii) equation of the line.

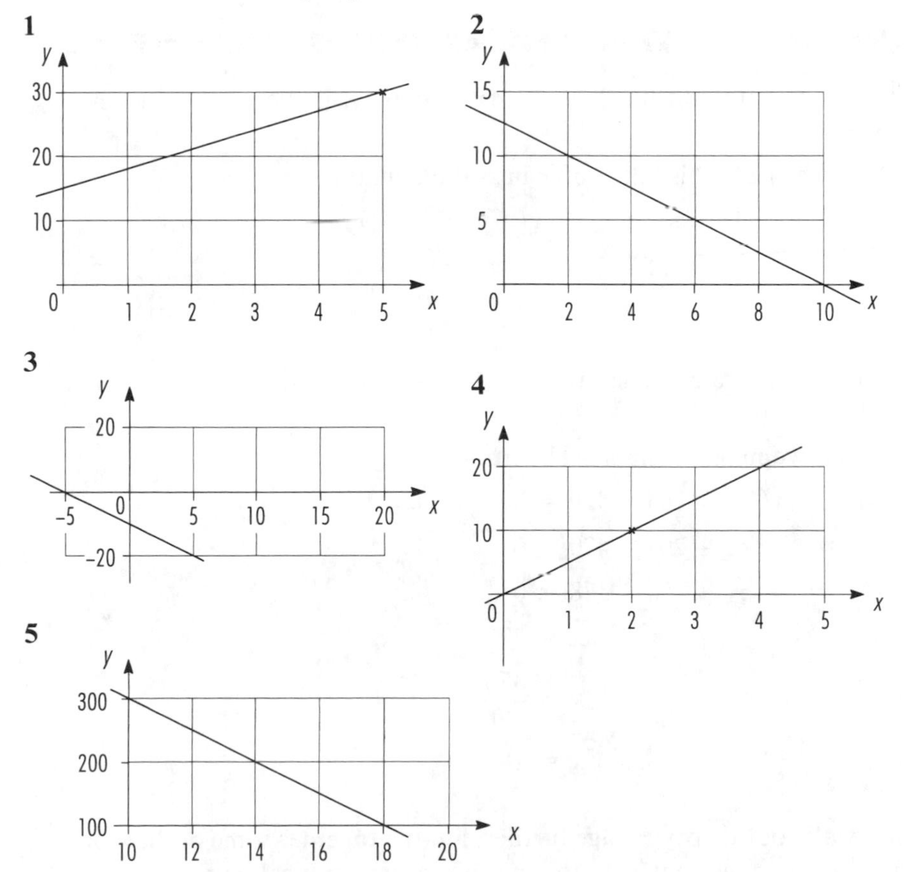

In this second set of 5 questions you have no sketch of the straight line graph. Using the pairs of coordinates in each case find

i) gradient,
ii) intercept,
iii) equation of the line.

6 $(0, 4)$ and $(2, 2)$

7 $(-1, -1)$ and $(5, 3)$

8 $(3, 5)$ and $(5, 3)$

9 $(1, 4)$ and $(6, 9)$

10 $(7, 1)$ and $(\frac{1}{2}, -\frac{1}{2})$

We have spent many pages and quite some time finding out about the straight line law. What use is an equation once we know of its features? We can use it to **predict** a value. Given a value of x we can find the associated value of y, or given a value of y we can find the value of x leading to it. The next 2 examples use equations we calculated in earlier work.

Examples 2.14

i) Given the straight line law $y = 4x + 55$ what is the value of y when $x = -7$?

We substitute this value of x into the equation,

i.e.
$$\begin{aligned} y &= 4(-7) + 55 \\ &= -28 + 55 \\ &= 27. \end{aligned}$$

ii) Given the straight line law $3y + 20x - 420 = 0$ find the value of x when $y = 14$.

Again we make a simple substitution,

i.e.
$$\begin{aligned} 3(14) + 20x - 420 &= 0 \\ 42 + 20x - 420 &= 0 \\ 20x - 378 &= 0 \\ 20x &= 378 \\ x &= \frac{378}{20} \\ x &= 18.9. \end{aligned}$$

We may take our theory a stage further, linking together some of the work we did solving equations with this more recent graphical work.

Example 2.15

We will use the simple equation $3x+4=20$ from Examples 2.3i).

i) Firstly we notice that there is no y in this equation, only x, but the left-hand side looks like an application of $mx+c$ with $m=3$ and $c=4$.

$\therefore$ let $y = 3x + 4$ which means that, linking this with the right-hand side, we have $y=20$.

We can sketch both of these lines on one pair of labelled axes.

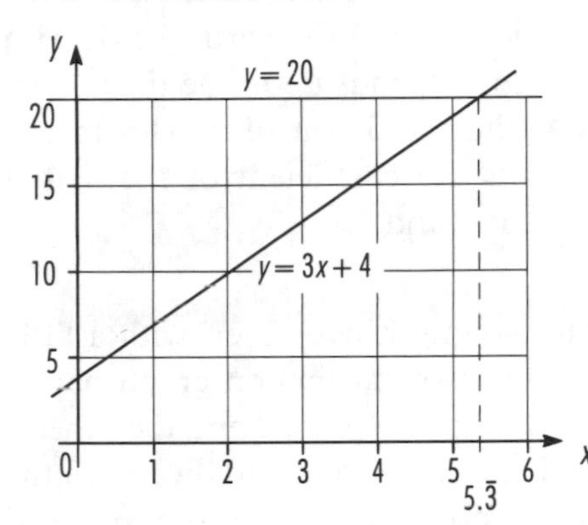

Because the original question was in terms of x our answer must give x as a number, i.e. $x=5.\bar{3}$ where the two lines cross.

ii) We may re-write our original equation

$$3x + 4 = 20$$

to become $$3x + 4 - 20 = 0$$

Subtracting 20.

$$3x - 16 = 0.$$

Again we may compare this equation with the standard format $y = mx + c$. Comparing $mx + c$ with $3x - 16$ we have $m=3$ and $c = -16$. We will sketch $y = 3x - 16$.

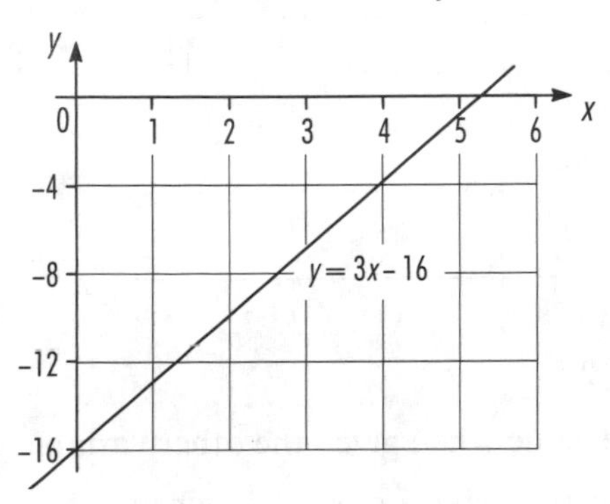

Comparing $3x - 16 = 0$ with $y = 3x - 16$ we have $y=0$. This means we will see where the graph crosses the line $y=0$, i.e. where it crosses the x axis. In fact $x=5.\bar{3}$ is that point, meaning that this is the solution to our original equation.

This method is most useful for simple linear equations. When the linear equations become complicated,

e.g. $\frac{2}{5}(x+5) - \frac{10}{3}(x+1) = 2$

with x appearing more than once then a purely algebraic solution is easier.

ASSIGNMENT

Let us return to our original Assignment. Can we draw a straight line graph of our equation $v = 30 - 4t$? The answer is yes!

We can compare our equation with the standard format,

i.e. $v = 30 - 4t$

with $y = mx + c$.

v and y look to be in similar positions and on the other side of each equation are the letters t and x. This means that we should label the horizontal axis with the independent variable t and the vertical axis with the dependent variable v, i.e. the velocity, v, depends upon the time, t, of the motion. m is the gradient, shown as the coefficient of x. This means that in our case -4 is the gradient, -4 is the coefficient of t. All that remain in the comparison are the intercept c and 30,

i.e. $c = 30$.

We have enough information to sketch the graph. However, with a little more care and accuracy we might plot the straight line on graph paper. Because it is a known straight line we need just 3 points, 2 and the other 1 as a check. The intercept on the vertical axis is 30 so already we have the first pair of coordinates (0, 30). We may choose any other values of t and calculate the values of v by substitution.

e.g. $t = 2$, $\therefore\ v = 30 - 4(2) = 22$

and $t = 5$, $\therefore\ v = 30 - 4(5) = 10$.

These give the 3 pairs of coordinates (0, 30), (2, 22) and (5, 10), through which we can draw the straight line shown in Fig. 2.5.

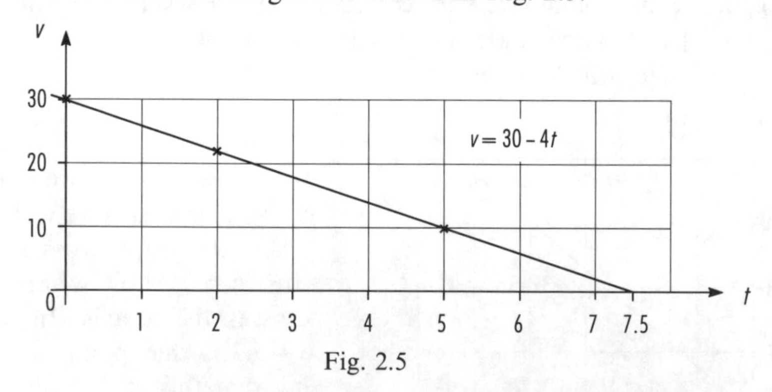

Fig. 2.5

We can use our graph to estimate one value when given the other variable.

i) Suppose we want to know when the vehicle stops, i.e. when it is at rest. This occurs when $v = 0$, the point where the graph crosses the horizontal axis. We can read off that this occurs at 7.5, i.e. after 7.5 seconds from the start of the motion the vehicle is at rest.

ii) Suppose we want to know the initial velocity of the vehicle. "Initially" means "when $t = 0$" which is where the graph crosses the vertical axis, i.e. the vertical intercept is 30 meaning that the initial velocity is $30\,\text{ms}^{-1}$.

iii) Suppose we want to know the time taken for the speed to fall to $12.4\,\text{ms}^{-1}$. Fig. 2.6 shows this line drawn on the same pair of labelled axes as our original graph $v = 30 - 4t$. Where the two lines cross we can read off the value of t as 4.4, i.e. it has taken 4.4 seconds for the speed to fall to $12.4\,\text{ms}^{-1}$.

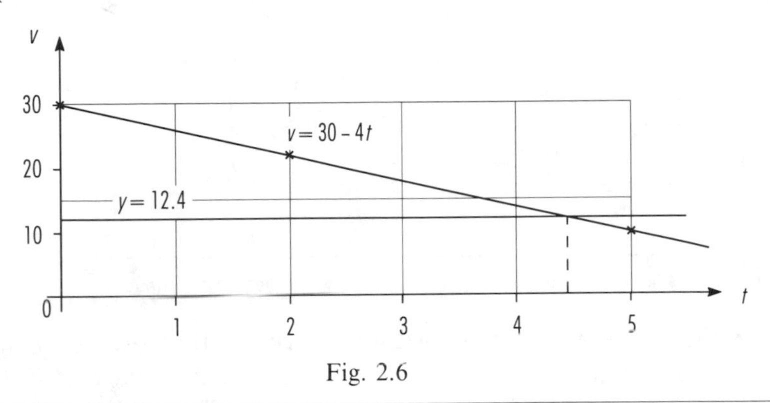

Fig. 2.6

Experimental data

Experiments and the use of test rigs often involve taking numerical readings. Even when the readings are neatly tabulated they can appear as a jumble of numbers. We need to see if there is any **pattern**. There are many ways of displaying the results in a variety of charts, diagrams and graphs. As a first step, in this chapter, we are going to see if sets of results obey a linear law, i.e. if when they are plotted accurately on a pair of labelled axes on graph paper a straight line may be drawn through them. It is highly unlikely in a practical situation that all the points will lie on the line. The aim when drawing the line is to have as many points as possible lying on the line with any others evenly scattered on either side. This will be our attempt to draw a **line of best fit**.

Example 2.16

In this example we have some mechanical test figures relating load, W Newtons, and effort, E Newtons. We wish to find out if they are related. If they are related is that relationship $E = mW + c$?

W	5	10	15	20	25	30
E	2.91	6.20	9.00	11.76	14.90	18.00

The first thing to notice is how the load figures increase by an equal amount each time. This is because they have been selected independently, meaning that the horizontal axis will be labelled with W. The effort figures are not as obviously regular because they depend on the values of W and the test rig. We will label the vertical axis E.

Because the first figures for both W and E are close to 0 we will include the origin on our graph. If this was not so we should use the graph paper more efficiently by omitting the origin.

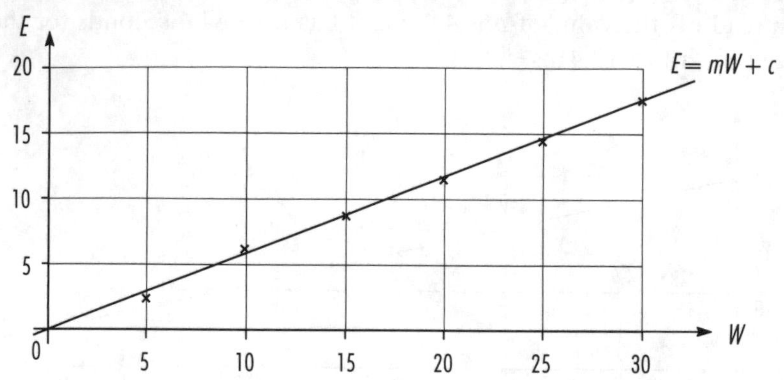

We can see from the graph that, within experimental error, the figures show that W and E are related by a straight line. In the next part of the example we can look at the form of this relationship. We need to compare our example,

$$E = mW + c$$

with the standard straight line law

$$y = mx + c.$$

In the standard format y is plotted vertically and x is plotted horizontally. Our example has E plotted vertically and W plotted horizontally. We can see that these letters have the same relative positions in their equations. Comparing further we have m as the gradient and c as the intercept in both cases.

By extending the line backwards to the vertical axis we intercept at the origin. This means that $c=0$.

We can create a right-angled triangle to find the vertical and horizontal changes so that we can calculate the gradient. If you were doing this for yourself you should aim for a large triangle. This gives a better chance of accuracy when drawing the changes and reading off the scales. In this example we have chosen the points (15, 9) and (30, 18) lying on the line and will apply the gradient formula

$$m = \frac{y_2 - y_1}{x_2 - x_1}$$

i.e.
$$m = \frac{18 - 9}{30 - 15}$$
$$= \frac{9}{15}$$
$$= 0.6.$$

At the end of this example we are in a position to bring together our results, stating that W and E are related according to the straight line law $E = 0.6W$.

We might use this relationship to estimate the effort required to move, for example, a load of 21.5 N,

i.e. $W = 21.5$ in the formula $E = 0.6W$

so that $E = (0.6)(21.5)$

$E = 12.9$

i.e. an effort of 12.9 N is needed to move 21.5 N.

Example 2.17

The table in this example relates the resistance, $R\ \Omega$, to the temperature, $t\,^\circ$C, of a length of wire.

t	20	40	60	80	100
R	5.30	5.68	6.11	6.50	7.00

We think they are related according to the straight line law $R = R_0 + kt$. This thought suggests some more questions.

i) If the relationship is true what are the values of R_0 and k?
ii) How does our equation match the usual equation $R = R_0(1 + \alpha t)$?
iii) What is the resistance at 150°C?
iv) At what temperature will the resistance be 6.25 Ω?

Firstly we will look at the table. The regularity of the values of t suggest that it is the independent variable to be plotted horizontally. Therefore R is the dependent variable to be plotted vertically.

We will omit the origin so that we can use the graph paper more efficiently. The graph shows that the table values are related according to a straight line.

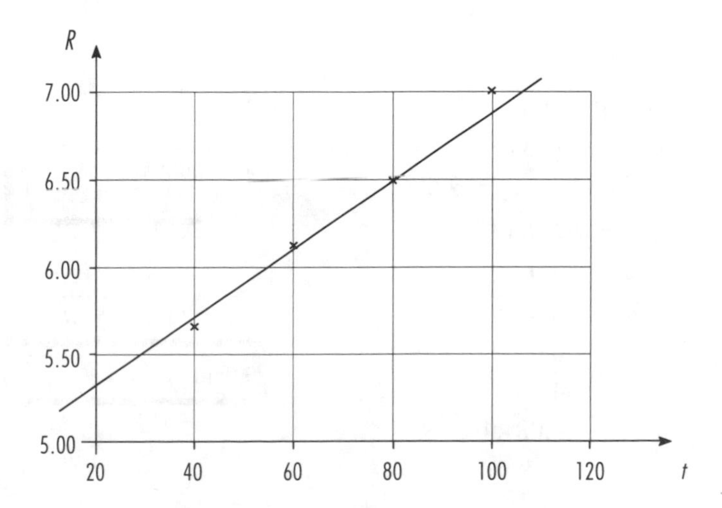

We need to see how our equation compares with the standard format,

i.e. $R = R_0 + kt$

and $y = mx + c.$

Already we have compared R and y vertically, and t and x horizontally. Earlier in the chapter we mentioned that the gradient, m, is the coefficient of x. In our equation k is the coefficient of t, meaning that it must be the gradient. This leaves R_0 to be the intercept on the vertical axis.

i) We can choose two points from the graph, (20, 5.30) and (80, 6.50), to calculate the gradient using the formula

$$m = \frac{y_2 - y_1}{x_2 - x_1}$$

i.e.

$$k = \frac{6.50 - 5.30}{80 - 20}$$

$$= \frac{1.20}{60}$$

$$= 0.02.$$

Therefore in our equation we can substitute for k so that

$$R = R_0 + 0.02t.$$

Because we are unable to use our graph to read off the intercept we must use our equation. We may choose any point that actually lies on our line, e.g. (20, 5.30), and substitute the coordinate values into our equation. This means

$$R = Ro + 0.02t$$

becomes

$$5.30 = R_0 + 0.02\,(20)$$

$$5.30 = R_0 + 0.40$$

$$5.30 - 0.40 = R_0$$

Subtracting 0.40.

$$4.90 = R_0.$$

Our completed equation may be written

$$R = 4.90 + 0.02t.$$

ii) This version may be compared to the usual one

$$R = R_0\,(1 + \alpha\,t)$$

i.e.

$$R = R_0 + R_0\,\alpha\,t.$$

Expanding the brackets.

By comparing terms we see that

$$R_0\,\alpha = 0.02$$

i.e.

$$4.90\,\alpha = 0.02$$

Substituting for R_0.

$$\alpha = \frac{0.02}{4.90}$$

Dividing by 4.90.

$$= 4.08 \times 10^{-3}.$$

iii) We will return to our version of the equation from the straight line graph and find the resistance by substituting for $t = 150$,

i.e. $$\begin{aligned} R &= 4.90 + 0.02(150) \\ &= 4.90 + 3.00 \\ &= 7.9 \end{aligned}$$

i.e. the resistance of the wire at 150 °C is 7.9 Ω.

iv) Using the same equation again, substituting for $R = 6.25$, we may calculate the temperature at which this occurs,

i.e. $6.25 = 4.90 + 0.02t$ — Subtracting 4.90.

$1.35 = 0.02t$ — Dividing by 0.02.

$67.5 = t$

i.e. the temperature of the wire is 67.5 °C when its resistance is 6.25 Ω.

EXERCISE 2.7

1 For the table of values plot y vertically against x horizontally.

x	−3	−2	−1	0	1	2	3
y	21	15	9	3	−3	−9	−15

With the aid of your graph check that the relationship is $y = 3 - 6x$.

2 y is thought to be related to x by the formula $y = mx + c$.
Using the table of values plot a graph of y vertically against x horizontally to check that it is correct. Use your graph to find the values of m and c.

x	0	1	4	9	16	25
y	−5	1	7	17	31	47

Use the equation to estimate the values of i) y when $x = 8$
and ii) x when $y = 4$.

3 A lifting machine raises loads, W kg, with effort E N according to the equation $E = mW + c$. A selection of test results are

W	10	20	40	60	100	150
E	75	102	168	225	340	500

Draw a graph to help you check that this relationship is true and use it to find values for the gradient and vertical intercept. Write down your equation using the values you have found. Estimate the effort needed to raise 130 kg.

4 The following table shows the distance travelled, d m, in time, t s, measured initially from 0 m.

d	95	190	290	385	480	575
t	30	60	90	120	150	180

By plotting a graph of d against t show that they are related by $d = kt$. Use your graph to estimate the time taken to travel 400 m. Find the value of k and use your equation to check your graphical time estimate for $d = 400$.

5 There is a temperature connection between Fahrenheit, °F, and Celsius, °C. Suppose this may be written $F = aC + b$.

i) Which letter represents the gradient and which one represents the intercept on the vertical axis?

ii) Plot F vertically against C horizontally using the 3 pairs of coordinates (0, 32), (50, 122) and (100, 212).

iii) From your graph find the values of a and b and substitute them into the equation $F = aC + b$.

iv) If your workplace temperature is 68°F what is this in °C?

6 A loaded beam has a shear force given by $S = mx + c$ where x is the distance from one end. Using the following table of values plot a graph of S against x to check that it is a straight line.

x (m)	1	2	3.5	4	5
S (N)	−800	−400	200	400	800

From your graph work out values for m and c so that you can use your new equation to find

i) the shear force when $x = 1.75$ m,

ii) the distance from the end where the shear force disappears (i.e. where the shear force is zero).

7 Hooke's law states that the tension, T N, in an elastic spring is related to the extension, x m, of the spring.

x	1.5	2.5	4.0	5.0	7.5	10.0
T	20.2	33.8	53.6	67.9	101.3	135.1

Use the table of values to decide whether this is true according to the relation $T = kx$ where k is a value to be found from your graph. Is k the gradient or the intercept?

8 The velocity of a van, $v\,\mathrm{ms}^{-1}$, is connected to time, t s, by the formula $v - mt - c = 0$ where m and c are the gradient and intercept as usual. Re-write this formula in the standard format and use the graph obtained from the following table to work out values for m and c.

t	10	14	20	26	31	35
v	3.90	5.15	6.50	7.82	9.32	10.20

Substitute your newly found values for m and c into the formula and use it to work out the velocity after a minute. Your answer will be in ms^{-1}. Convert it to kmh^{-1} and then mph to see if it is reasonable.

9 According to Ohm's law $V = IR$ where V is the voltage, I is the current (amp) and R is the resistance (ohm). The table shows some test results for a particular resistor.

V	0	3	7	10	14	20	25
I	0	0.077	0.179	0.256	0.359	0.513	0.641

Plot V vertically and I horizontally so that you can use your straight line graph to find the value of the resistor.

10 The length of a copper rod, l m, expands as its temperature, t °C is raised probably according to the formula $l = l_0 + mt$. Use the following table of values to draw a graph and check the truth of this theory.

t	0	100	250	500	750
l	0.2000	0.2017	0.2042	0.2085	0.2128

Find the values of l_0 and m from your graph. Substitute these values into the original equation and use it to estimate the temperature for a length of 0.2100 m.

3 Transposition of Formulae

The objectives of this chapter are to:

1. Transpose simple formulae involving $+/-$.
2. Transpose simple formulae involving $\times/\div$.
3. Transpose formulae in which the subject appears more than once.
4. Transpose formulae using roots.
5. Transpose formulae using powers.

Introduction

Engineers and scientists tend to remember formulae in a standard format. This may not always be convenient when attempting to solve a problem. **Transposition** of formulae tells us from the title that we will **move (trans)** the **position** of various letters and operations. **Transposition** is sometimes called **transformation**. Already we know many of the transposition rules because we used them to solve simple equations in Chapter 2. We will use some of the examples again. In fact Example 2.1 iii) was a case of simple transposition.

ASSIGNMENT

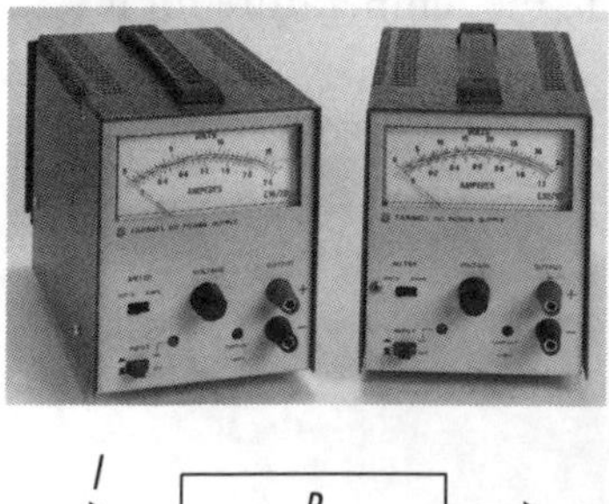

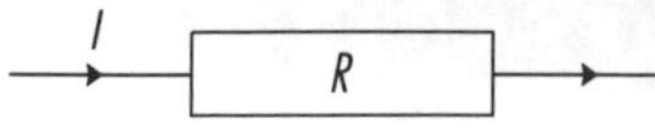

We will use an electrical problem to show how we might apply the knowledge of this chapter.

Let us consider a current, I, flowing through a resistor, R. The power, P, associated with the resistor is given by the usual formula $P = I^2R$. This format for the formula is useful if we know the current and the resistor values. However in our assignment we are going to look at finding current values for a given resistance at various powers. This will involve transposing the formula into a more useful format that will help with the calculations.

Transposition of formulae

Let us look at a basic formula involving the initial velocity of a vehicle, u, and its final velocity, v, over a time, t, due to an acceleration a. The formula is $v = u + at$. Because v is alone on one side of the formula (generally it will be the left side) v is said to be the **subject** of the formula. Alternatively v is given in terms of u, a and t. It is possible that we might not be interested in the final velocity, but in the time taken to reach this velocity. It is possible to re-arrange (transpose) the letters so that t appears on its own while all the other letters appear on the opposite side. This means that t will be given in terms of u, v and a. This re-arrangement is useful when we need to repeat a type of calculation several times for different numerical values of u, v and a.

One of the difficult things is to give all our attention to getting one particular letter in terms of all the others. Often the letters seem to be a jumble. When numbers are involved with just one letter then that letter stands out. To help in early examples we will use Greek letters in the same way as numbers so that you can concentrate on the other letter, usually x. Don't worry about understanding what the Greek letters represent; just think of them as numbers that cannot be simplified.

As a starting point we will look again at some early examples from Chapter 2.

Examples 3.1

i) Solve the equation $x - 12 = 8$.

   i.e. $x = 8 + 12$.

   > Addition and subtraction are opposites.

   It is not important in this chapter that the actual solution of the equation is $x = 20$.

ii) Make x the subject of the formula $x - \alpha = \beta$.

   i.e. $x = \beta + \alpha$

   > Adding α to both sides.

   or $x = \alpha + \beta$.

Examples 3.2

i) Solve the equation $4x = 3$.

   i.e. $x = \frac{3}{4}$.

   > Multiplication and division are opposites.

   The actual numerical answer is not important in this example. We need to pay attention to the operation used.

ii) Make x the subject of the formula $\alpha x + \beta = \gamma$.

   Addition of β is less closely attached than α to x. This means we will deal with β first.

i.e.
$$\alpha x = \gamma - \beta$$ Subtracting β.
$$x = \frac{\gamma - \beta}{\alpha}.$$ Dividing by α.

Example 3.3

Make x the subject of the formula

$$\frac{x}{\alpha} - \theta = -\beta x + \phi.$$

Terms involving x appear on both sides of this formula. The first move is to gather these terms to one side. We will gather them together on the left, but the same final solution will be reached if we gather them on the right.

$\therefore$
$$\frac{x}{\alpha} + \beta x - \theta = \phi$$ Adding βx.
$$\frac{x}{\alpha} + \beta x = \phi + \theta.$$ Adding θ.

The α in the denominator position needs some attention to ease the algebra later in the problem. We can move it from this position by multiplying throughout by α.

$$\frac{\alpha x}{\alpha} + \alpha\beta x = \alpha(\phi + \theta)$$ Multiplying by α.
$$x + \alpha\beta x = \alpha(\phi + \theta)$$ Cancelling the αs.
$$x(1 + \alpha\beta) = \alpha(\phi + \theta)$$ x is a common factor.
$$x = \frac{\alpha(\phi + \theta)}{1 + \alpha\beta}.$$ Dividing by $1 + \alpha\beta$.

Example 3.4

Make x the subject of the formula $\alpha\left(\frac{x}{\beta} + \gamma\right) = \epsilon$.

In this example the brackets affect the order of operations. x is at the heart of this formula on the left. Using our earlier knowledge, if the letters are replaced by numbers the order of evaluation is

x (or the number)
divide by β
plus γ
multiply by α.

We use this list from the bottom working upwards with the opposite operation. This leads to the new list of operations we will perform to find x, i.e.

divide by α
subtract γ
multiply by β
x (or the number).

Then $\frac{x}{\beta} + \gamma = \frac{\epsilon}{\alpha}$ — Dividing by α.

$\frac{x}{\beta} = \frac{\epsilon}{\alpha} - \gamma$ — Subtracting γ.

$x = \beta\left(\frac{\epsilon}{\alpha} - \gamma\right).$ — Multiplying by β.

Example 3.5

Make t the subject of the formula $v = u + at$.
t appears on the right. Suppose that we had actual numbers for u, a and t on the right. Using our calculator we would

input t
multiply by a
add u.

This order of operations gives the value for v. We use this list of operations from the bottom working upwards with the opposite operation. This leads to a new list

value of v
subtract u
divide by a.

This gives t as the subject of the formula,

i.e. $v = u + at$

becomes $v - u = at$ — Subtracting u.

$\frac{v - u}{a} = t.$ — Dividing by a.

It is more usual to write this as $t = \frac{v-u}{a}$ so that the subject, t, appears on the left.

Example 3.6

Make l the subject of the formula $W = \frac{\frac{1}{2}\lambda x^2}{l}$.

We can re-write the $\frac{1}{2}$ in this formula to give $W = \frac{\lambda x^2}{2l}$. Both versions of the formula have exactly the same meaning. On the right we are dividing by l. The opposite of division is multiplication. We write

$Wl = \frac{l\lambda x^2}{2l}$ — Multiplying by l.

i.e. $Wl = \frac{\lambda x^2}{2}.$ — Cancelling ls on the right.

We need l alone on the left, rather than being multiplied by W. The opposite of multiplication is division. We write

$$\frac{Wl}{W} = \frac{\lambda x^2}{2W}$$

Dividing by W.

i.e. $$l = \frac{\lambda x^2}{2W}.$$

Cancelling Ws on the left.

Example 3.7

Make r the subject of the formula $\frac{r}{1-r} = S$.

r already appears on the left in this formula; but it appears more than one. For ease we remove the fraction by multiplying by $(1 - r)$,

i.e. $$\frac{r(1-r)}{1-r} = S(1-r)$$

Multiplying by $(1 - r)$.

i.e. $$r = S(1-r)$$

Cancelling on the left.

This version of the formula is simpler than before because it has no fractional algebra. However we do need to re-gather the r terms.

$$r = S - Sr$$

Multiplying the bracket.

Then $$r + Sr = S - Sr + Sr$$

Adding Sr to both sides.

i.e. $$r + Sr = S$$

i.e. $$r(1 + S) = S$$

r is a factor on the left.

We leave r alone on the left dividing through by $(1 + S)$,

i.e. $$\frac{r(1+S)}{1+S} = \frac{S}{1+S}$$

Dividing by $(1 + S)$.

i.e. $$r = \frac{S}{1+S}$$

Cancelling on the left.

Example 3.8

Make r_1 the subject of the formula $\frac{1}{R} = \frac{1}{r_1} + \frac{1}{r_2}$ where r_1 and r_2 are resistors in parallel.

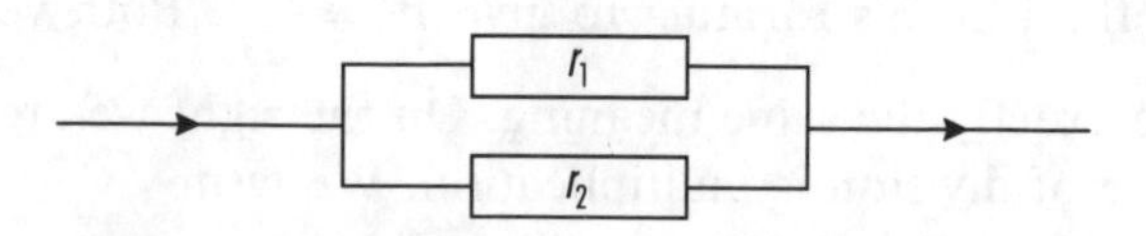

There are several ways of making r_1 the subject of the formula. In earlier examples we attempted to get rid of fractions, multiplying through by the lowest common multiple (LCM) of the denominators.

$$\frac{1}{R} - \frac{1}{r_2} = \frac{1}{r_1}.$$

Subtracting $\frac{1}{r_2}$.

The LCM of R, r_1 and r_2 is Rr_1r_2 and so

$$Rr_1r_2\left(\frac{1}{R}-\frac{1}{r_2}\right) = Rr_1r_2\left(\frac{1}{r_1}\right)$$ Multiplying by Rr_1r_2.

$$Rr_1r_2\left(\frac{1}{R}-\frac{1}{r_2}\right) = Rr_2$$ Cancelling r_1 on the right.

$$r_1\left(\frac{Rr_2}{R}-\frac{Rr_2}{r_2}\right) = Rr_2.$$ Multiplying the brackets apart from r_1.

We will keep r_1 outside the brackets because it appears only once at this point. In a few lines we will be able to move all the other terms away from r_1 to leave it alone as the subject of the equation.

$$r_1(r_2 - R) = Rr_2$$ Simplifying the left.

$$r_1 = \frac{Rr_2}{r_2 - R}.$$ Dividing by $r_2 - R$.

EXERCISE 3.1

Each question is a relationship followed by a letter in brackets. In each case transpose the relationship to make that letter the subject.

1. $x+\alpha-\beta=0$ (x)
   i.e make x the subject of this formula or give x in terms of α and β.
2. $x+\alpha-\beta=\gamma$ (x)
3. $x+\alpha=\beta+\gamma$ (x)
4. $\alpha x-\beta=\gamma$ (x)
5. $\frac{x}{\alpha}+\beta=\delta$ (x)
6. $\left(\frac{x}{\alpha}+\beta\right)\gamma=\delta$ (x)
7. $\alpha x+\gamma=\beta x$ (x)
8. $\frac{x}{a}-c=bx$ (x)
9. $\frac{x}{a}+d-c=\frac{x}{b}$ (x)
10. $\frac{1}{x}+a=b$ (x)
11. $\frac{a}{x}-b=c$ (x)
12. $ax+c=\frac{x}{b}$ (x)
13. $\frac{a}{x}-d=\frac{b}{x}$ (x)
14. $I=\frac{PRT}{100}$ (P)
15. $C=\pi d$ (d)
16. $A=\frac{1}{2}bh$ (b)
17. $C=\frac{5}{9}(F-32)$ (F)
18. $A=\frac{1}{2}(a+b)h$ (h)
19. $S=Wx-2$ (x)
20. $s=\frac{1}{2}(u+v)t$ (t)
21. $I_1+I_2+I_3+I_4=0$ (I_1)
22. $C=2\pi r$ (r)
23. $I=\frac{PRT}{100}$ (R)
24. $A=\frac{1}{2}bh$ (h)
25. $\frac{1}{C}=\frac{1}{c_1}+\frac{1}{c_2}$ (c_1)
26. $\frac{1}{v}+\frac{1}{u}=\frac{1}{f}$ (v)

27 $pV = mRT$ (T)

28 $p_1V_1 - p_2V_2 = 0$ (p_2)

29 $R = R_0(1 + \alpha t)$ (t)

30 $S = \dfrac{N - R}{N}$ (R)

31 $A = \frac{1}{2}(a + b)h$ (a)

32 $I = \dfrac{PRT}{100}$ (T)

33 $s = \frac{1}{2}(u + v)t$ (u)

34 $V = \dfrac{\pi D^2 fN}{4}$ (N)

35 $A = \frac{1}{2}(a + b)h$ (b)

36 $Ft = m(v - u)$ (v)

37 $S = \dfrac{N - R}{N}$ (N)

38 $\Omega = \omega + \alpha t$ (α)

39 $l = l_0(1 + \alpha t)$ (t)

40 $\dfrac{R_1}{R_2} = \dfrac{1 + \alpha t_1}{1 + \alpha t_2}$ (α)

Square and square roots

If we have a number which we square, this is the opposite operation to finding the square root of a number. We may try this with any number first of all using a calculator.

Suppose we choose the number 81.

Now $\quad 81^2 = 6561$

and $\quad \sqrt{6561} = 81.$

We will use the positive square root but remember that there is a negative one as well. The calculator displays only the positive root.

We may combine both these operations and write

$$\sqrt{81^2} = \sqrt{6561} = 81$$

We treat the $\sqrt{\ }$ symbol in a similar way to brackets. Usually we calculate first whatever appears in the brackets and so should calculate first whatever appears under the square root, $\sqrt{\ }$.

$\sqrt{\ }$ has another form. It is the power $\frac{1}{2}$.

e.g. $\quad (81^2)^{\frac{1}{2}} = (6561)^{\frac{1}{2}} = 81.$

We can omit the middle stage to reveal

$$(81^2)^{\frac{1}{2}} = 81^1.$$

Usually we do not write the power 1.

Looking at the powers,

$$2(\tfrac{1}{2}) = 1.$$

Simple multiplication.

Using letters makes no difference to the operations so that with x in place of 81 we have

$$\sqrt{x^2} \text{ or } (x^2)^{\frac{1}{2}} = x.$$

We can interchange the order of the square and square root operations so that

$$(\sqrt{x})^2 \text{ or } (x^{\frac{1}{2}})^2 = x.$$

Example 3.9

The volume of a cone is given by $V = \frac{1}{3}\pi r^2 h$.

We may make r the subject of the formula. The aims are to make r^2 the subject of the formula and then, by taking the square root of both sides, obtain r. On the right r^2 is multiplied by $\frac{1}{3}$, π and h. To leave r^2 on the right we must apply the opposite operation of division,

i.e. $$\frac{V}{\frac{1}{3}\pi h} = r^2.$$

This may be simplified because dividing by $\frac{1}{3}$ is the same as multiplying by 3, i.e. "invert the fraction and multiply", so that

$$r^2 = \frac{3V}{\pi h}$$

Square root of both sides.

$$r = \sqrt{\frac{3V}{\pi h}}.$$

$\sqrt{r^2} = r$.

Example 3.10

$\frac{1}{2}mu^2 = mgh + \frac{1}{2}mv^2$ represents a simple system in which mechanical energy (kinetic and potential) is conserved. We wish to make v the subject of the formula.

We may look at the formula in 3 sections (**terms**); $\frac{1}{2}mu^2$,
mgh
and $\frac{1}{2}mv^2$.

Because we need v let us give our attention to this last section. mgh is added in our formula and the opposite of addition is subtraction,

i.e. $$\tfrac{1}{2}mu^2 - mgh = \tfrac{1}{2}mv^2.$$

Subtracting mgh from both sides.

v^2 is multiplied by $\frac{1}{2}$ and m. In fact all the terms are multiplied by m so we may cancel this letter throughout the formula,

i.e. $$\tfrac{1}{2}u^2 - gh = \tfrac{1}{2}v^2.$$

Dividing each term by m.

v^2 is still multiplied by $\frac{1}{2}$, so

$$\frac{\frac{1}{2}u^2}{\frac{1}{2}} - \frac{gh}{\frac{1}{2}} = \frac{\frac{1}{2}v^2}{\frac{1}{2}}$$

Dividing each term by $\frac{1}{2}$.

$$u^2 - 2gh = v^2.$$

Division by $\frac{1}{2}$ is the same as multiplication by 2.

Finally $$\sqrt{u^2 - 2gh} = v.$$

Square root of both sides.

This is usually written as

$$v = \sqrt{u^2 - 2gh}$$

with the subject, v, on the left.

Example 3.11

Suppose that $a - 3\sqrt{2x} = b$ and that we want to make x the subject of the formula.

We need to look at the term $-3\sqrt{2x}$.

Now $$-3\sqrt{2x} = b - a$$ Subtracting a from both sides.

$$3\sqrt{2x} = -(b - a)$$ Multiplying both sides by -1.

$$3\sqrt{2x} = -b + a$$

i.e. $$3\sqrt{2x} = a - b$$

$$\sqrt{2x} = \frac{a-b}{3}$$ Dividing by 3.

$$2x = \left(\frac{a-b}{3}\right)^2$$ Squaring both sides.

$$x = \frac{1}{2}\left(\frac{a-b}{3}\right)^2.$$ Dividing by 2.

Multiplying by $\frac{1}{2}$ has the same effect as dividing by 2.

ASSIGNMENT

Our electrical problem involves the power relationship $P = I^2R$ for the current I flowing through the resistor R. We are going to make I the subject of this formula.

$$P = I^2R$$

becomes $$\frac{P}{R} = I^2$$ Dividing by R.

so that $$I = \sqrt{\frac{P}{R}}.$$ Square rooting both sides.

Our particular resistor is 470 Ω.

$\therefore$ $$I = \sqrt{\frac{P}{470}}.$$ Substituting for $R=470$.

Suppose we wish to know the current needed for a power of 5×10^{-4} W,

i.e. $$I = \sqrt{\frac{5\times10^{-4}}{470}}$$ Substituting for $P = 5 \times 10^{-4}$.

$$= \sqrt{1.06\ldots \times 10^{-6}}$$

$$= \sqrt{1.06\ldots} \times \sqrt{10^{-6}}$$

$$I = 1.03 \times 10^{-3} \text{ A or } 1.03 \text{ mA}.$$

We might go further with this Assignment taking into account a design variation allowing us to vary the power between 4.6×10^{-4} W and 5.4×10^{-4} W. The table below shows the calculations leading to the current needed for each value of P.

$P \times 10^{-4}$	4.6	4.8	5.0	5.2	5.4
$\frac{P}{470} \times 10^{-6}$	0.979	1.021	1.064	1.106	1.149
$I \times 10^{-3}$	0.99	1.01	1.03	1.05	1.07

EXERCISE 3.2

Each question is a relationship followed by a letter in brackets. In each case transpose the relationship to make that letter the subject.

1 $A = \pi r^2$ (r)
i.e. make r the subject of this formula or give r in terms of A and π.

2 $P = \frac{Wv^2}{32r}$ (v)

3 $a^2 = b^2 + c^2$ (b)

4 $V = \pi r^2 h$ (r)

5 $W = \frac{\lambda x^2}{2a}$ (x)

6 $\frac{v^2}{r} = \omega^2 r$ (ω)

7 $C = 1 - \frac{x^2}{2}$ (x)

8 $y = ax^2 + b$ (x)

9 $v = \sqrt{2gh}$ (h)

10 $y = a\sqrt{x} + b$ (x)

11 $Ch = 1 + \frac{x^2}{2}$ (x)

12 $A = P\left(1 - \frac{r}{100}\right)^2$ (r)

13 $c = \sqrt{a^2 - b^2}$ (b)

14 $A = \frac{\pi d^2}{4} + 320$ (d)

15 $V = \sqrt{V_R^2 + (V_L b - V_C)^2}$ (V_R)

16 $A = P\left(1 + \frac{r}{200}\right)^2$ (r)

17 $Ck = \frac{1}{2}m(v^2 - u^2)$ (u)

18 $v = \sqrt{\frac{gr(\upsilon + t)}{(1 - \upsilon t)}}$ (t)

19 $Lf = \frac{1}{(s - c)^2}$ (s)

20 $C = \frac{1 - t^2}{1 + t^2}$ (t)

Powers and roots

This is a more general version of what we did with squares and square roots.

Raising a number to some power n is the opposite operation to finding the nth root of a number. The nth root may be written as $\sqrt[n]{\ }$ or $(\ )^{1/n}$.

Thus $\sqrt[n]{81^n}$ or $(81^n)^{1/n} = 81^1$.

Looking at the powers only we have

$$(n)\left(\frac{1}{n}\right) = 1.$$

Simple multiplication.

Again using a letter, x, in place of 81 we have

$$\sqrt[n]{x^n} \quad \text{or} \quad (x^n)^{1/n} \quad = x^1 \quad = x.$$

Usually we do not write the power 1.

Interchanging the order of operations has no effect upon the final answer,

i.e. $\quad (\sqrt[n]{x})^n \quad$ or $(x^{1/n})^n \qquad = x.$

Example 3.12

The volume of a sphere is given by $V = \dfrac{4\pi r^3}{3}$.

Suppose we wish to make the radius, r, the subject of this formula i.e. give r in terms of V.

The style of solution is similar to the one used in Example 3.9. On the right r^3 is multiplied by $\frac{4}{3}$ and π. To leave r^3 on the right we must apply the opposite operation, division,

i.e. $$\frac{V}{\frac{4}{3}\pi} = r^3.$$

This may be simplified because dividing by $\frac{4}{3}$ is the same as multiplying by $\frac{3}{4}$, i.e. "invert the fraction and multiply", so that

$$r^3 = \frac{3V}{4\pi}$$

$$r = \left(\frac{3V}{4\pi}\right)^{\frac{1}{3}}.$$

Cube root of both sides.

$\sqrt[3]{r^3} = r.$

Example 3.13

If $a\sqrt[4]{bx} + c = d$ suppose we wish to make x the subject of the relationship.

The heart of the problem is x. If we consider numbers in place of all the letters our list of calculator operations is

input x
multiply by b
fourth root
multiply by a
add c.

This list will give a calculated value for d.

We can use this list from the bottom working upwards with the opposite operation. The new list to make x the subject of the formula is

value for d
subtract c
divide by a
raise to the power 4
divide by b.

x is now the subject of the formula.

i.e. $a\sqrt[4]{bx} + c = d$

becomes $a\sqrt[4]{bx} = d - c$ — Subtracting c.

$\sqrt[4]{bx} = \dfrac{d-c}{a}$ — Dividing by a.

$bx = \left(\dfrac{d-c}{a}\right)^4$ — Raising both sides to the power 4.

$x = \dfrac{1}{b}\left(\dfrac{d-c}{a}\right)^4.$ — Dividing by b.

Multiplying by $\frac{1}{b}$ has the same effect as dividing by b. Because of the complicated bracket, multiplying by $\frac{1}{b}$ is the preferred format.

EXERCISE 3.3

Each question is a relationship followed by a letter in brackets. In each case transpose the relationship to make that letter the subject.

1 $y = ax^3 + b$ (x)
i.e. make x the subject of this formula or give x in terms of a and b.

2 $y = mx^5 + c$ (x)

3 $y + mx^4 + c = 0$ (x)

4 $y = a\sqrt[3]{x} + b$ (x)

5 $y = m\sqrt[4]{x} + c$ (x)

6 $y + m\sqrt[5]{x} + c = 0$ (x)

7 $(c + ts)^4 = z$ (c)

8 $A = P\left(1 + \dfrac{R}{100}\right)^{10}$ (R)

9 $ST^{\frac{1}{7}} = c$ (T)

10 $z = \dfrac{\pi D^3}{32}$ (D)

11 $pV^{\gamma} = c$ (V)

12 $I = \dfrac{bd^3}{12}$ (d)

13 $ST^{0.1} = c$ (T)

14 $(c - st)^3 = z$ (s)

15 $A = P\left(1 - \dfrac{R}{100}\right)^7$ (R)

16 $T = kf^{0.75}D^{1.8}$ (D)

17 $I = \dfrac{\pi D^4}{32}$ (D)

18 $T = kf^{0.75}D^{1.8}$ (f)

19 $I = \dfrac{BD^3 - bd^3}{12}$ (D)

20 $z = \dfrac{\pi}{32}\left(\dfrac{D^4 - d^4}{D}\right)$ (d)

4 Circular Measure

The objectives of this chapter are to:

1. Define the radian.
2. Convert degrees measure to radians and vice versa.
3. Express angular rotations in multiples of π radians.
4. Use the relationship $S = r\theta$ for the arc length of a circle.
5. Use the relationship $A = \frac{1}{2}r^2\theta$ for A, the area of a sector.
6. Solve problems involving areas and angles measured in radians.

ASSIGNMENT

The Assignment for this chapter is based on teamwork between Design engineers and the Marketing Department. They are looking at packaging their new product in the shape of a cone.

Degrees

The usual way to measure an angle is by using degrees, often with a protractor. A protractor is marked off **(graduated)** in degrees.

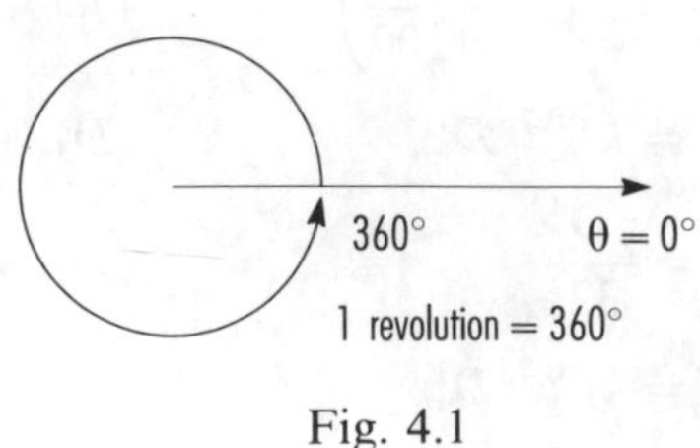

Fig. 4.1

In general we draw a horizontal line towards the right to act as a base line, or line of reference. We call this the line $\theta = 0°$ and from it rotate either clockwise or anticlockwise. In Fig. 4.1 the rotation is the more usual anti-clockwise one, with a complete revolution being 360°.

180° is said to be "the angle on a straight line". This means that from the base line $\theta = 0°$ if you turn through 180° your position is the same as if the straight line had been projected backwards. Figs. 4.2 show we can use any starting line as well as the usual horizontal $\theta = 0°$.

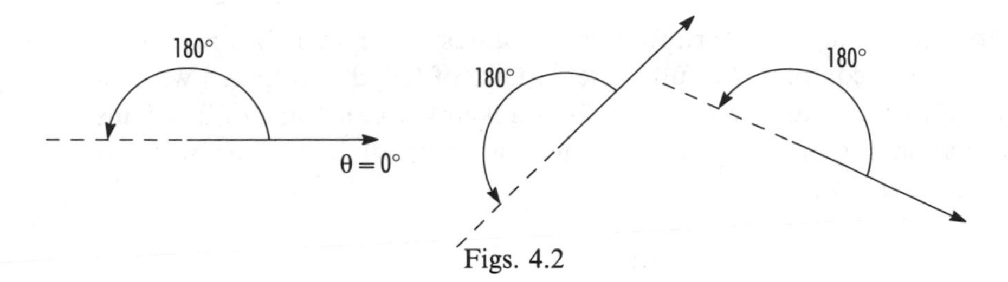

Figs. 4.2

A rotation through 90° creates a right-angle. We will pay attention to right-angled triangles (triangles containing one right-angle and 2 other acute angles) early in Chapter 5 when we look at trigonometry. Again Figs. 4.3 show that the base line does not have to be horizontally to the right.

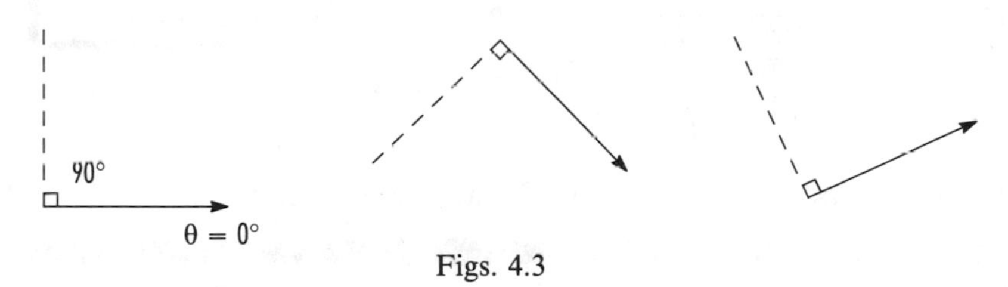

Figs. 4.3

We may relate these angles to a revolution, i.e.

$$1 \text{ rev} = 360°$$
$$\tfrac{1}{2} \text{ rev} = 180°$$
$$\tfrac{1}{4} \text{ rev} = 90°.$$

Radians

The definition of a radian is based on a circle. Draw a circle with centre O and radius r. Mark a point A on the circumference and measure an arc equal in length to the radius. If this arc length ends at B, the angle AOB, $\angle AOB$, is defined to be of size 1 radian. This is shown in Fig. 4.4.

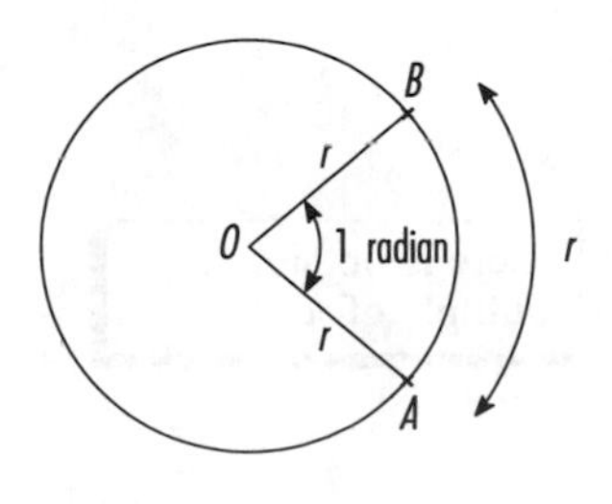

Fig. 4.4

A formal definition is that one radian is the angle **subtended** at the centre of a circle by an arc equal in length to the radius of the circle.

We can use the short form **rad** instead of radian.

There is a connection between degrees and radians. The circumference of the circle in Fig. 4.4 is given as $2\pi r$. To draw a circle we need to turn through 1 revolution (360°). When constructing (drawing) a circle we can see that the greater the angle we turn for a given radius the greater the arc formed until we have completed a full circle. An arc of length r is linked with an angle of 1 radian, an arc of length $2r$ is linked with an angle of 2 radians and so an arc of length $2\pi r$ is linked with an angle of 2π radians, which is the circumference.

We have $\boxed{1 \text{ rev} = 360° = 2\pi \text{ rad.}}$

Degrees/radians conversions

Already we have $360° = 2\pi$

so that $\dfrac{360°}{360} = \dfrac{2\pi}{360}$

Dividing by 360.

i.e. $\boxed{1° = \dfrac{\pi}{180} \text{ rad.}}$

This is the **conversion factor** to move from degrees to radians.

Examples 4.1

Convert i) 30°, ii) 90°, iii) 135°, iv) 290°, v) θ° into radians correct to 3 decimal places.

i) 30° is $30 \times 1° = \dfrac{30 \times \pi}{180}$

$= \dfrac{\pi}{6}$ or 0.524 rad.

ii) 90° is $90 \times 1° = \dfrac{90 \times \pi}{180}$

$= \dfrac{\pi}{2}$ or 1.571 rad.

iii) 135° is $135 \times 1° = \dfrac{135 \times \pi}{180}$

$= \dfrac{3\pi}{4}$ or 2.356 rad.

iv) 290° is $290 \times 1° = \dfrac{290 \times \pi}{180}$

$= 5.061$ rad

There is no simple multiple of π.

v) θ° is $\theta \times 1° = \theta \times \dfrac{\pi}{180}$ or 0.017θ rad.

Using $2\pi = 360°$

so that $\dfrac{2\pi}{2\pi} = \dfrac{360°}{2\pi}$

Dividing by 2π.

i.e. $$\boxed{1 \text{ rad} = \frac{180^\circ}{\pi}.}$$

This is the **conversion factor** to move from radians to degrees. A calculator check finds that 1 radian is slightly less than 60°, approximately 57.3° (3 sf).

Examples 4.2

Convert i) 1.5 rad, ii) 2.6 rad, iii) $\frac{7\pi}{4}$ rad, iv) θ rad to degrees.

i) 1.5 rad is $1.5 \times 1 = \frac{1.5 \times 180^\circ}{\pi}$

$= 85.94^\circ$ (4 sf).

ii) 2.6 rad is $2.6 \times 1 = \frac{2.6 \times 180^\circ}{\pi}$

$= 148.97^\circ$ (2 dp)

or 149.0° (3 sf).

iii) $\frac{7\pi}{4}$ rad is $\frac{7\pi}{4} \times 1 = \frac{7\pi}{4} \times \frac{180^\circ}{\pi}$

$= 315^\circ$

iv) θ is $\theta \times 1 = \theta \times \frac{180^\circ}{\pi}$

$= 57.3\theta^\circ$ (3 sf).

EXERCISE 4.1

Convert the following angles from degrees to radians, leaving each answer as a multiple of π.

1 10°
2 45°
3 60°
4 150°
5 180°
6 210°
7 315°
8 360°
9 540°
10 1080°

Convert the following angles from degrees to radians, giving your answer as a decimal (2 dp) using the usual calculator value of π.

11 25°
12 73.2°
13 167.4°
14 225.1°
15 255.2°
16 296.5°
17 307.75°
18 345°
19 500°
20 610°

Convert the following radian measures into degrees giving, your answer correct to 2 decimal places.

21 7.00

22 4.55

23 5.20

24 $\frac{2\pi}{3}$

25 1.25

26 13.6

27 $\frac{5\pi}{4}$

28 1.9

29 $\frac{5\pi}{3}$

30 6.15

Rotations

Most of our examples and exercises have been between 0° and 360°. There is no reason to stop there. We can keep going round and round by further revolutions using our earlier relationship that 1 rev = 360° = 2π rad.

All revolutions will be multiples of this relationship,

e.g. $2 \text{ rev} = 2 \times 2\pi = 4\pi \text{ rad}$

$5\frac{1}{2} \text{ rev} = 5\frac{1}{2} \times 2\pi = 11\pi \text{ rad}.$

Commonly used angles

Some angles in degrees or radians occur more often than others. The following lists show these common ones in both degrees and radians.

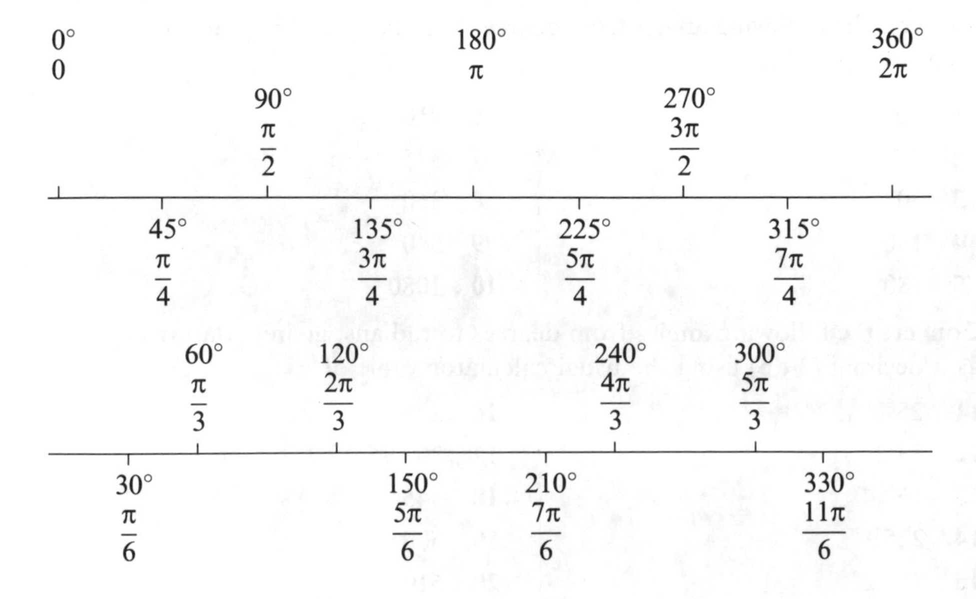

Arc length

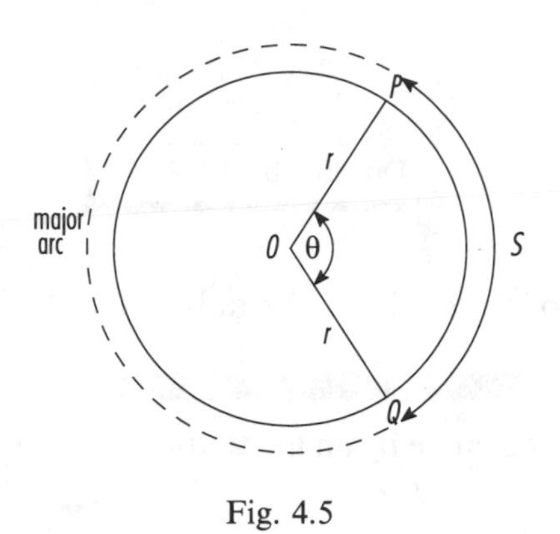

Fig. 4.5

Fig. 4.5 shows the arc PQ of length S. This is the shorter arc PQ, called the **minor arc**. The longer one on the left of this particular circle is called the major arc.

The length of arc S depends on the size of the angle θ. The greater the angle θ then the longer the arc PQ. This will continue until one complete revolution with an angle of 2π radians. Then the arc has reached its maximum length in the form of the circumference $2\pi r$. We can interpret this mathematically as

$$\frac{S}{\theta} = \frac{2\pi r}{2\pi}$$

Comparing arcs and angles.

$$\frac{S}{\theta} = r$$

Cancelling by 2π.

i.e. $$S = r\theta$$

Multiplying by θ.

i.e. the arc length is $r\theta$, where r is the radius of the circle and θ is the angle in radians subtended at the centre of the circle.

Examples 4.3

For a circle of radius 0.852 m calculate the

i) length of arc for a subtended angle of 100°,
ii) angle (radians) for an arc length of 1.55 m.

i) The formula $S = r\theta$ uses θ in radians. This means that we need to convert 100° into radians first, i.e.

$$\begin{aligned} 100^\circ &= 100 \times \frac{\pi}{180} \\ &= 1.745 \text{ rad} \end{aligned}$$

$$\begin{aligned} S &= r\theta \\ &= (0.852)(1.745) \\ &= 1.49 \text{ m (2 dp)} \end{aligned}$$

i.e. the arc length is 1.49 m.

ii) Again we use the formula and substitute $S = 1.55$ and $r = 0.852$ so that

$$S = r\theta$$

becomes $\quad 1.55 = 0.852\theta$

$$\frac{1.55}{0.852} = \theta$$

Dividing by 0.852.

$$1.82 = \theta$$

i.e. the angle subtended at the centre of the circle is 1.82 rads.

Example 4.4

Calculate the angle subtended at the centre of a circle by an arc that is two and a half times the radius of the circle.

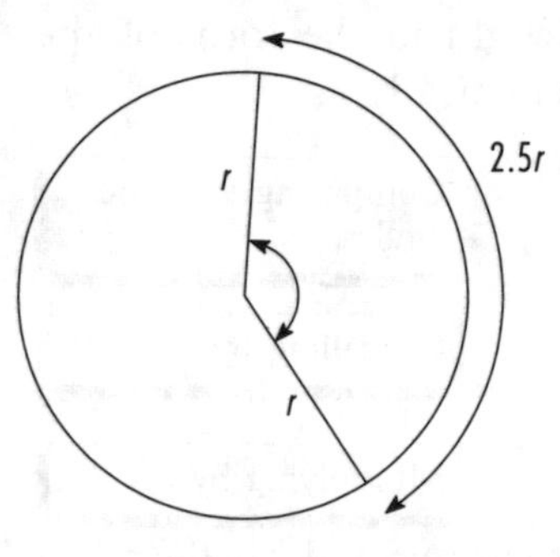

Fig. 4.6

Fig. 4.6 shows a circle. The exact length of the radius you may wish to draw does not have to be the same as this one. It will be easier to choose a length that you find convenient and then multiply by 2.5 to get the arc length. The diagram is only an aid, it does not have to be exact.

We may find the angle in 2 slightly different ways.

a) Earlier in this chapter we defined a radian to be where the arc length has the same value as the radius. If the arc length is 2.5 times the radius then we can expect the angle to be 2.5 times the original radian, i.e. 2.5 radians or

$$\frac{2.5 \times 180^\circ}{\pi} = 143.2^\circ.$$

b) Alternatively we may use the usual formula

$$S = r\theta$$

If the radius is r then the arc length is given as $2.5r$. Substituting into the formula we have

$$2.5r = r\theta$$

$$\frac{2.5r}{r} = \theta$$

Dividing by r.

$$\theta = 2.5 \text{ rad as before.}$$

Your diagram is unlikely to show exactly 143.2°, but ought to be reasonably close to that value.

Area of a sector

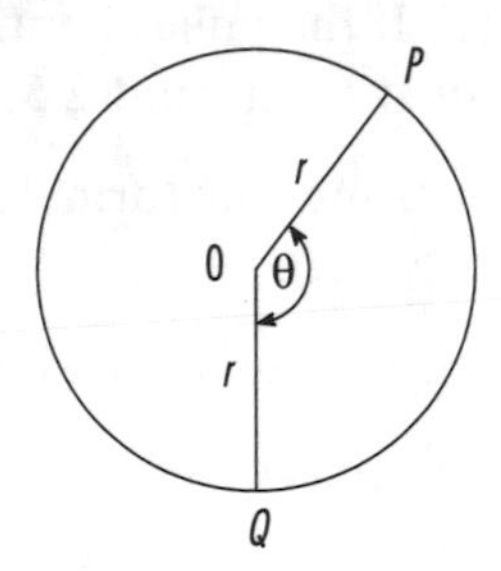

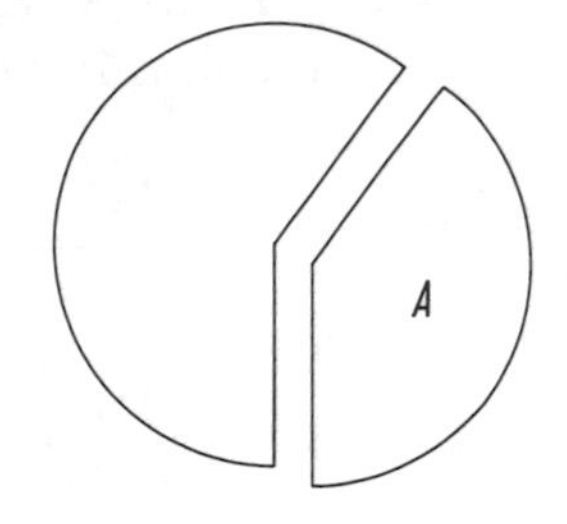

Figs. 4.7

The sector POQ is that slice of area of the original circle. The greater the angle θ then the larger the area of the sector, A. This will continue until the angle is 2π radians and the area of the circle, πr^2, has been taken.

Mathematically we may write

$$\frac{A}{\theta} = \frac{\pi r^2}{2\pi}$$

Comparing angles and areas.

$$\frac{A}{\theta} = \frac{r^2}{2}$$

Cancelling by π.

$$A = \tfrac{1}{2}r^2\theta$$

Multiplying by θ.

i.e. the area of a sector of a circle is $\frac{1}{2}r^2\theta$ where r is the radius of the circle and θ is the angle in radians subtended at the centre.

Examples 4.5

For a circle of radius 0.317 m calculate

i) the area of the sector for a subtended angle of 2.35 radians,
ii) the angle in radians for a sector of area 0.094 m^2.

i) Given that $A = \frac{1}{2}r^2\theta$ we can substitute for $r = 0.317$ and $\theta = 2.35$ so that

$$\begin{aligned} A &= \tfrac{1}{2}(0.317)^2(2.35) \\ &= 0.12\,\text{m}^2 \qquad (2\text{ dp}) \end{aligned}$$

i.e. the area of the sector is $0.12\,\text{m}^2$.

ii) Given that $A = \frac{1}{2}r^2\theta$ we can substitute for $A = 0.094$ and $r = 0.317$ so that

$$\begin{aligned} 0.094 &= \tfrac{1}{2}(0.317)^2\theta \\ 0.094 &= 0.050\theta \\ \frac{0.094}{0.050} &= \theta \\ \theta &= 1.87\text{ rad} \end{aligned}$$

i.e. the angle subtended at the centre of the circle is 1.87 rads.

Example 4.6

Rope is wound around a pulley of diameter 350 mm. If the pulley turns through 210° what length of rope will be unwound? If the pulley continues to turn how many revolutions need to be completed to unwind 2.5 m?

i) This example combines the radius and angle using our formula

$$S = r\theta.$$

Both our given values need care. The diameter is given instead of the radius, i.e. we need $r = \frac{350}{2} = 175$ mm. Also θ needs converting from degrees to radians by $210° = 210 \times \frac{\pi}{180} = 3.665$ rad.

$$\begin{aligned} \text{Then } S &= 175 \times 3.665 \\ &= 641 \text{ mm} \qquad \text{(3 sf)} \end{aligned}$$

i.e. 641 mm of rope is unwound from the pulley.

ii) The last part of the example combines 2 lengths. The 2.5 m is inconsistent with the units of mm for the radius: 2.5 m = 2.5 × 1000 = 2500 mm. Again using our formula

$$\begin{aligned} S &= r\theta \\ 2500 &= 175\theta \\ \frac{2500}{175} &= \theta \\ \theta &= 14.286 \text{ rad} \\ &= \frac{14.286}{2\pi} \text{ rev} \\ &= 2.27 \end{aligned}$$

> 1 rev = 2π rad.

i.e. the pulley will turn through 2.27 revolutions to unwind the required length of rope.

Example 4.7

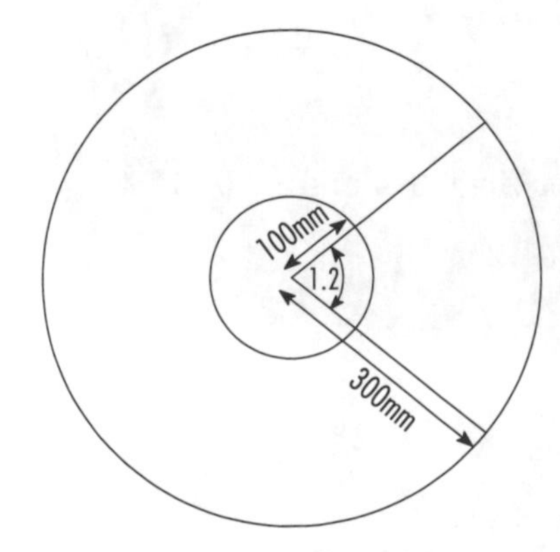

Fig. 4.8

Fig. 4.8 shows two concentric circles of radii 100 mm and 300 mm. The area between them is called an **annulus**. The angle subtended at the centre is 1.2 rad. Calculate the area of the

i) annulus,
ii) small sector,
iii) large sector.

For any angle why is the area of the large sector always 9 times the area of the small sector?

i) Area of large circle $= \pi R^2$
$= \pi(300)^2$
$= 90000\pi \text{ mm}^2.$

Area of small circle $= \pi r^2$
$= \pi(100)^2$
$= 10000\pi \text{ mm}^2.$

Area of annulus $= \pi R^2 - \pi r^2$
$= 90000\pi - 10000\pi$
$= 80000\pi$
$= 2.51 \times 10^5 \text{ mm}^2.$

ii) Area of small sector $= \frac{1}{2} r^2 \theta$
$= \frac{1}{2}(100)^2 1.2$
$= 6000 \text{ mm}^2$

iii) Area of large sector $= \frac{1}{2} R^2 \theta$
$= \frac{1}{2}(300)^2 1.2$
$= 54000 \text{ mm}^2.$

When we compare the areas of the two sectors we need to look at their formulae. $\frac{1}{2} R^2 \theta$ and $\frac{1}{2} r^2 \theta$ have $\frac{1}{2}$ and θ in common. They differ only with R^2 and r^2. Now R and r differ by a factor of 3. Because they are both squared in the formulae the differing factor is also squared, i.e. $3^2 = 9$, i.e. the large sector will always be 9 times the small sector.

ASSIGNMENT

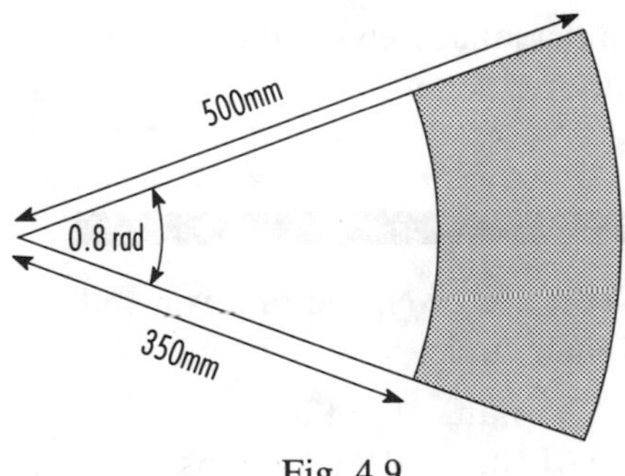

Fig. 4.9

Design engineers and the Marketing Department are looking at packaging their new product. A cone has been chosen as the preferred eye-catching shape. The diagram (Fig. 4.9) shows the cone flattened out into a sector with its probable dimensions. The shaded area is to be over-printed in red.

The team decides that full circles will be marked out and then colour printed as necessary. Finally the sectors will be cut from each circle of thin card before being formed into cones. Remember that 360° is 6.283 radians. How many sectors can be cut from each circle?

A quick calculation shows that $\frac{6.283}{0.8} = 7.854$. This means that 7 complete sectors can be cut from each circle with a wastage of 0.854. The Head of Marketing says that the waste is too great and would prefer a slightly smaller pack with less waste per circle. His aim is to get 8 slightly smaller sectors from each circle. The team agree with the suggestion of 8 sectors from a circle, each of 0.75 radians.

The area of thin card for each sector is given by

$$\begin{aligned} A &= \tfrac{1}{2}r^2\theta \\ &= \tfrac{1}{2}(500)^2 0.75 \\ &= 93750 \text{ mm}^2. \end{aligned}$$

Next the team discusses the cost related to the area of colour printing.

$$\begin{aligned} \text{Coloured area} &= \text{Area of large sector} - \text{Area of small sector} \\ &= 93750 - \tfrac{1}{2}(350)^2 0.75 \\ &= 93750 - 45937.5 \\ &= 47812.5 \text{ mm}^2. \end{aligned}$$

They think this area will be too expensive to colour print. Eventually they decide that about 20000 mm^2 would be better:

$$\begin{aligned} \text{Coloured area} &= \text{Area of large sector} - \text{Area of small sector} \\ \text{i.e.} \qquad 20000 &= 93750 - \tfrac{1}{2}r^2(0.75) \\ 20000 - 93750 &= -0.375r^2 \\ -73750 &= -0.375r^2 \\ \frac{-73750}{-0.375} &= r^2 \\ 196666.\dot{6} &= r^2 \\ \sqrt{196666.\dot{6}} &= r \\ \text{i.e.} \qquad r &= 443.5 \text{ mm}. \end{aligned}$$

Square root of both sides.

At this stage the team members are left with personal views about the final choice of 440 mm or 450 mm for the uncoloured length.

EXERCISE 4.2

1 Calculate the arc length in each case. You are given the radius r and the angle θ subtended at the centre of each circle.

a) $r = 0.5$ m, $\theta = 45°$; b) $r = 17.5$ mm, $\theta = 120°$;
c) $r = 44.4$ mm, $\theta = 150°$; d) $r = 0.61$ m, $\theta = \pi$ radians.

2 Calculate the radius of each circle. You are given the arc length S and the angle θ subtended at the centre.

a) $S = 0.97$ m, $\theta = 65°$; b) $S = 1.25$ m, $\theta = \frac{1}{2}\pi$ radians;
c) $S = 35.5$ mm, $\theta = 140°$; d) $S = 71$ mm, $\theta = 210°$.

3 Given the radius r and the arc length S calculate the angle θ subtended at the centre of each circle.

a) $r=0.85$ m, $S=1.34$ m; b) $r=25$ mm, $S=75$ mm;
c) $r=1.1$ m, $S=0.55$ m; d) $r=30$ mm, $S=105$ mm.

4 Calculate the area of the sector in each case. You are given the radius r and the angle θ subtended at the centre of each circle.

a) $r=0.65$ m, $\theta=100°$; b) $r=36$ mm $\theta=25°$;
c) $r=117$ mm, $\theta=70°$; d) $r=1.1$ m, $\theta=0.75\pi$ radians.

5 Calculate the radius of each circle. You are given the area of the sector A and the angle θ subtended at the centre.

a) $A=250\text{ mm}^2$, $\theta=40°$; b) $A=725\text{ mm}^2$, $\theta=125°$;
c) $A=950\text{ mm}^2$, $\theta=90°$; d) $A=1000\text{ mm}^2$, $\theta=\frac{1}{2}\pi$ radians.

6 Given the radius r and the area of the sector calculate the angle θ subtended at the centre of each circle.

a) $r=75$ mm, $A=750\text{ mm}^2$; b) $r=45$ mm, $A=750\text{ mm}^2$;
c) $r=25$ mm, $A=825\text{ mm}^2$; d) $r=95$ mm, $A=6000\text{ mm}^2$.

7 Wrapped around a pulley of radius 0.20 m is a rope of length 0.25 m. Calculate the angle subtended at the centre of the circle (pulley). Also find the distance between the ends of the rope around the pulley.

8 2 m of rope is wrapped around a pulley of radius 0.25 m. Calculate the i) circumference of the pulley,
ii) length of overlap of the rope,
iii) angle subtended at the centre of the pulley by the 2 ends of rope.

9 By considering the difference in areas of 2 sectors calculate the shaded area.

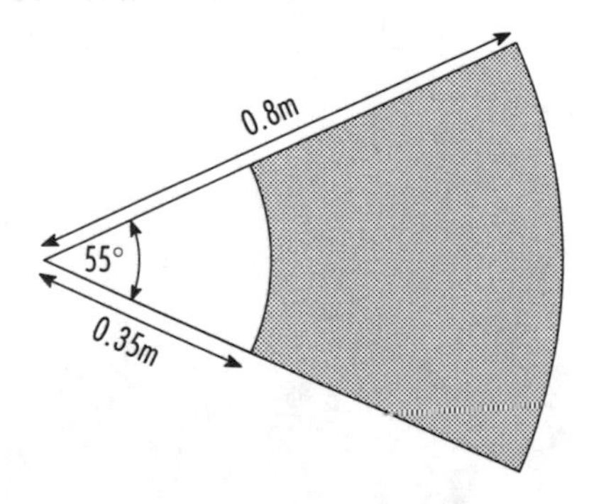

10 Two radii are inclined at 60° to each other. If the radii are of length 150 mm calculate the lengths of the major and minor arcs. Express these lengths as a ratio in its simplest form. Will that ratio be the same for similarly inclined radii of different lengths?

11 A wheel of radius 0.5 m travels a distance of 7.25 m. Through what angle has the wheel turned? How many revolutions is this?

12 A chain is wrapped around a pulley of radius 0.18 m. Through what angle must the pulley turn to unwind 0.95 m?

13 The diagram shows 2 concentric circles. Calculate the area between them. What angle must be subtended at the centre of the circles for the shaded area to be $0.1\ \text{m}^2$?

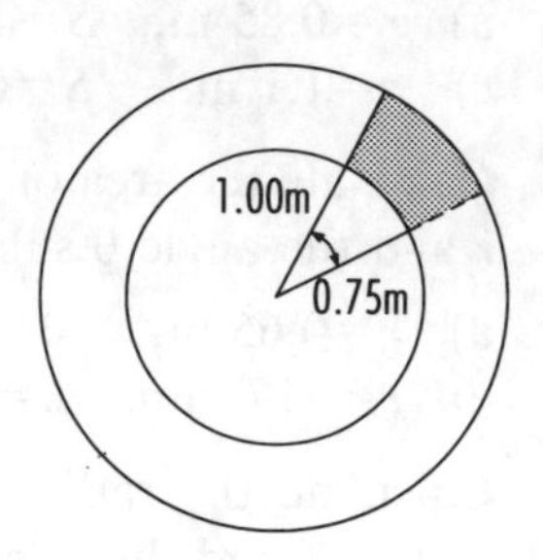

14 A sector of a circle OAB is formed into a cone so that OA and OB touch. The base of the cone is a circle of circumference 471.24 mm. What is the radius of that circular base? If the lengths OA and OB are each 200 mm what is the angle between them?

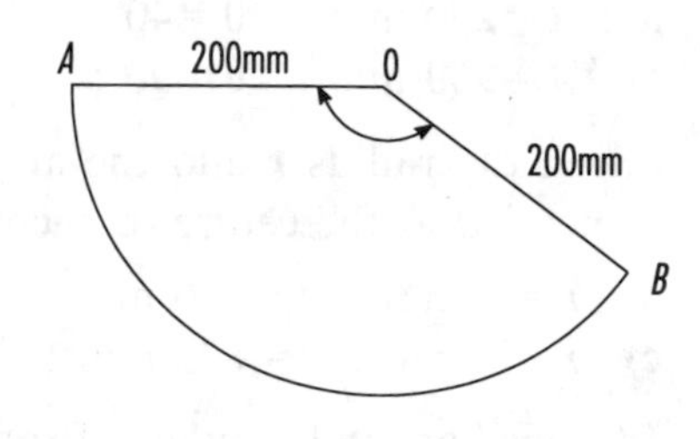

15 The volume of a cone is $\frac{1}{3}\pi r^2 h$ where r is the radius of the base and h is the vertical height. l is the slant height. What is the relationship between r, h and l? The cone may be created from the sector of a circle. How is l related to that sector?

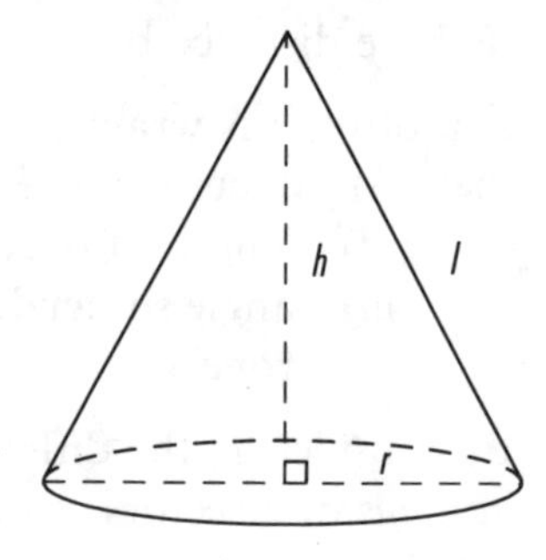

5 Trigonometry

The objectives of this chapter are to:

1. Determine values of the trigonometric ratios for angles between 0° and 360°.
2. Sketch a sine wave over one complete cycle by relating the angle of a rotating unit radius to the vertical projection.
3. Sketch a cosine wave over one complete cycle by relating the angle of a rotating unit radius to the horizontal projection.
4. Determine trigonometric ratios for angles greater than 360° and for negative angles.
5. Derive the relationship $\sin^2\theta + \cos^2\theta - 1$.
6. Define $\tan\theta = \dfrac{\sin\theta}{\cos\theta}$ and sketch the graph of $\tan\theta$.

Introduction

There are a number of different **trigonometric ratios**. They may be called **trigonometric operations** or **trigonometric functions**. In this chapter we will consider 3 of them: **sine**, **cosine** and **tangent**. The sine and cosine ratios are the most important ones in engineering and science.

ASSIGNMENT

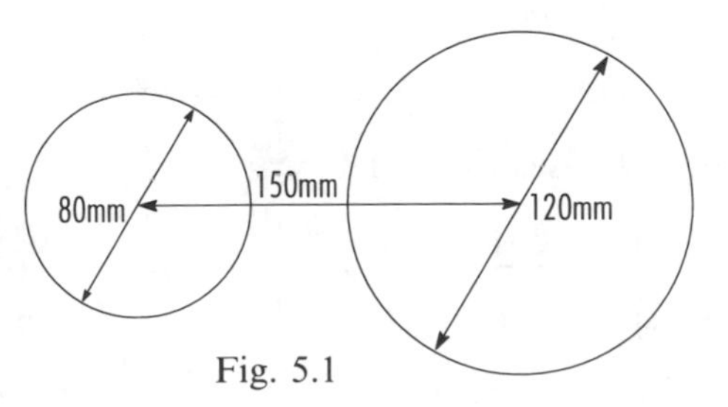

Fig. 5.1

The Assignment for this chapter looks at the 2 pulleys in Fig. 5.1. One of them has a diameter of 120 mm and the other of 80 mm. Their centres are 150 mm apart. Firstly we will find the length of the belt around them.

Using the calculator

The shortened forms of sine, cosine and tangent appear on your scientific calculator as **sin**, **cos** and **tan**. We may use either degrees or radians if there is a facility to change the angular measurement.

Examples 5.1

Use your calculator to find the values of

i) sin 53°, ii) sin 147°, iii) −sin 147°.

i) The trigonometric operation of sin acts on the angle 53°. This affects how we use the calculator as the following order of operations shows:

53 | sin |

to display 0.7986 correct to 4 decimal places.

ii) sin 147° uses the order

147 | sin |

to display 0.5446.

iii) −sin 147° just extends the previous list of operations with a final use of the +/− | button to give

147 | sin | +/− |

to display −0.5446.

Examples 5.2

Use your calculator to find the values of

i) cos 31°, ii) cos 209°, iii) cos 335.5°.

i) For cos 31° the order of operations is

31 | cos |

to display 0.8572 correct to 4 decimal places.

ii) cos 209° uses the order

209 | cos |

to display −0.8746.

iii) cos 335.5° uses the order

335.5 | cos |

to display 0.9100.

Examples 5.3

Use your calculator to find the values of
i) tan 82.75°, ii) tan 115°, iii) tan 297°.

i) For tan 82.75° the order of operations is

82.75 | tan |

to display 7.8606 correct to 4 decimal places.

ii) tan 115° uses the order

115 | tan |

to display −2.1445.

iii) tan 297° uses the order

297 | tan |

to display −1.9626.

If you use your calculator to find tan 90° or tan 270° you will find that it gives an error display. We will come back to these strange values when we look more closely at the tangent function.

Examples 5.4

Use your calculator to find the values of
i) cos (−127°), ii) −cos 127°, iii) −cos (−127°).

The orders of operations for all these examples are similar, only differing by the positions of − signs.

i) For cos (−127°) the order of operations is

127 | +/− | cos |

to display −0.6018 correct to 4 decimal places.

ii) − cos 127° puts the − sign in a different position after the cosine operation using the order

127 | cos | +/− |

to display 0.6018.

iii) −cos (−127°) extends the previous example, using the order

127 | +/− | cos | +/− |

to display 0.6018.

Examples 5.5

Use your calculator to find the values of

i) sin(−94°), ii) −sin(−94°), iii) tan(−315°).

i) For sin(−94°) the order of operations is

94 +/− sin

to display −0.9976.

ii) −sin (−94°) extends the previous example, using the order

94 +/− sin +/−

to display 0.9976.

iii) tan (−315°) uses the order

315 +/− tan

to display 1.

EXERCISE 5.1

Use your calculator to find the values of

1 sin 33°
2 sin 127°
3 cos 149°
4 cos 37°
5 tan 98°
6 tan 65°
7 sin(−59°)
8 cos 173°
9 tan 115°
10 sin 222°
11 sin 180°
12 cos 293.75°
13 tan 59.5°
14 tan 302°
15 cos 341°
16 sin 295°
17 tan 26°
18 cos(−151°)
19 cos 246°
20 tan 46°
21 tan 103°
22 sin 310.25°
23 tan(−213°)
24 cos 217°
25 sin 199°
26 tan 415°
27 cos(−340°)
28 sin 161°
29 sin 393°
30 cos 509°

Examples 5.6

Use your calculator to find values of

i) sin 3.025, ii) cos $\frac{5\pi}{12}$, iii) tan(−0.975)

where the angles are in radians.

Before you start working out these trigonometric ratios make sure that your calculator is in radian mode. If your calculator only uses degrees you must use the conversion from Chapter 4.

i) For sin 3.025 the order of operations is

3.025 sin

to display 0.1163 correct to 4 decimal places.

ii) $\cos \frac{5\pi}{12}$ uses the order

5 × π ÷ 12 = cos

to display 0.2588.

iii) tan(−0.975) uses the order

0.975 +/− tan

to display −1.475.

EXERCISE 5.2

This set of questions use radians in place of degrees.

Use your calculator to find the values of

1 sin 2

2 cos 0.76

3 sin 1.23

4 tan 6.50

5 $\tan \frac{3\pi}{4}$

6 sin (−1.54)

7 cos 4.25

8 $\sin \frac{5\pi}{3}$

9 $\cos\left(\frac{-5\pi}{6}\right)$

10 tan (−0.9)

Sine and cosine curves

We may use the calculator to construct a table of values of sine from 0° to 360°. The chosen angles are specimen ones because they generate a particular pattern of values for the sine.

θ	0°	30°	60°	90°	120°	150°
sin θ	0	0.500	0.866	1.000	0.866	0.500

θ	180°	210°	240°	270°	300°	330°	360°
sin θ	0	−0.500	−0.866	−1.000	−0.866	−0.500	0

There is an alternative method for finding sine that uses a circle in place of a calculator (see Fig. 5.2). We draw a circle with centre O and a unit radius. If you do this for yourself draw it accurately to a large scale so that it is easy to read the measurements.

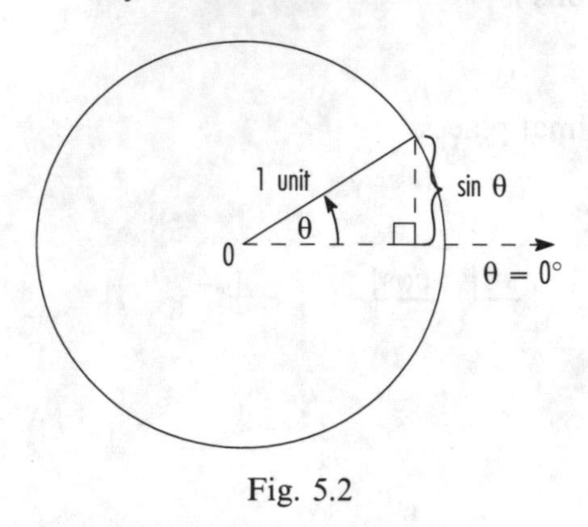

Fig. 5.2

The usual axis is $\theta=0°$ horizontally to the right, in the same place as the x-axis. Values of θ are measured from this axis in an anticlockwise direction. For each value of θ the unit radius touches the circumference of the circle at a different point. The distance of each point above or below the axis gives the value of $\sin\theta$.

A circle has 4 **quadrants** (i.e. quarters) shown in Fig. 5.3.

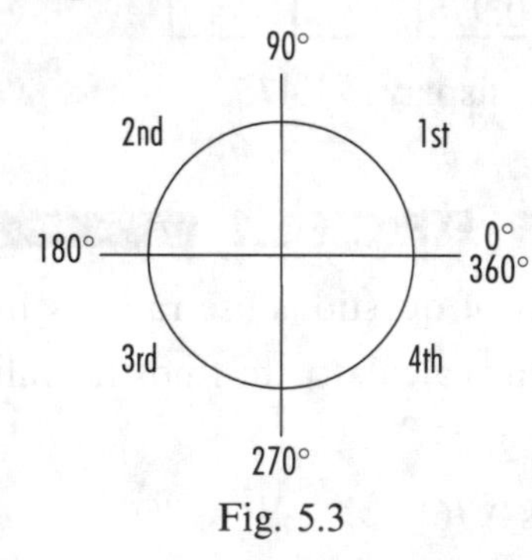

Fig. 5.3

Figs. 5.4 shows examples of sine, one from each quadrant in turn.

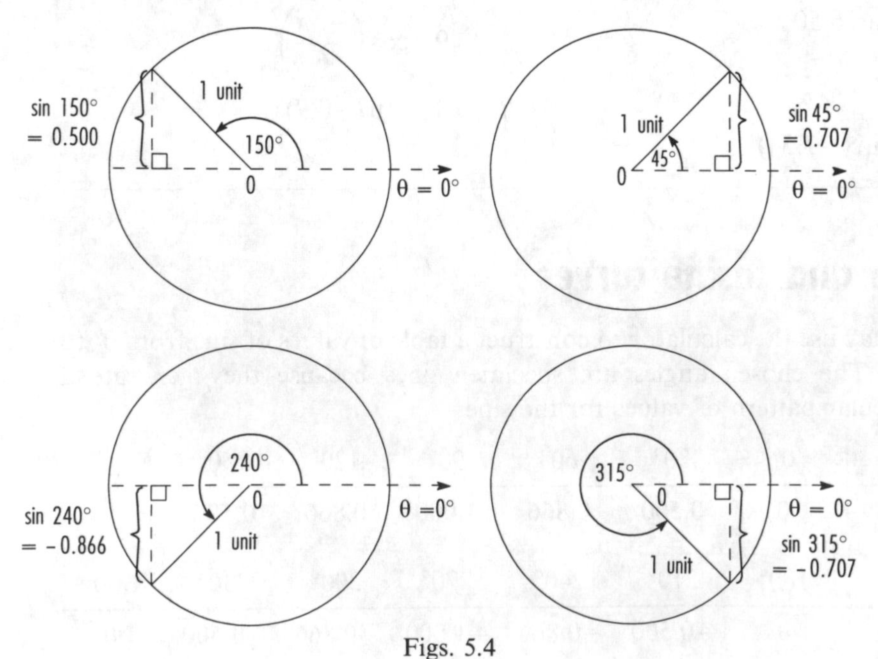

Figs. 5.4

The calculator method is quicker than the circle method. Whichever method you choose the values can be used to draw the graph of $y = \sin\theta$ shown in Fig. 5.5.

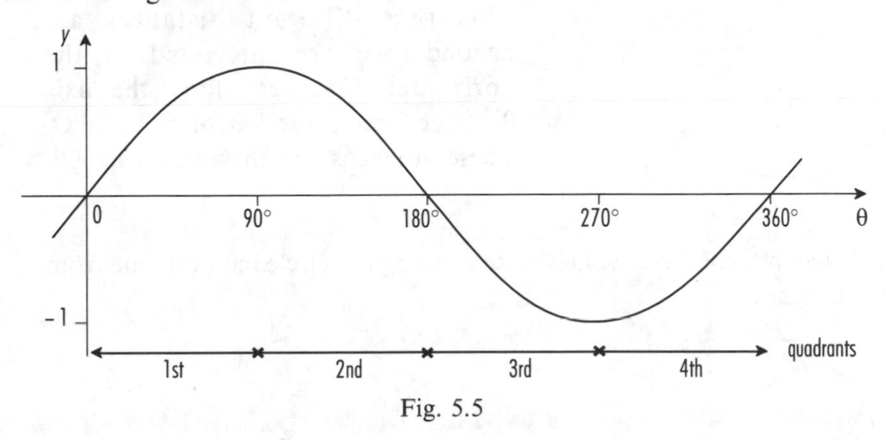

Fig. 5.5

The graph continues this pattern to the left and to the right of where we have drawn it, repeating itself every 360°. Each repetition is a **cycle**. The sine graph is important. You should know how to sketch it over one cycle.

In the first quadrant as the values of θ increase from 0° to 90° the values of $\sin\theta$ increase from 0 to a maximum of 1. Moving into the second quadrant (90° to 180°) the values of $\sin\theta$ decrease from 1 to 0. In the third quadrant (180° to 270°) that change continues from 0 down to -1. Finally in the fourth quadrant (270° to 360°) $\sin\theta$ turns, increasing from -1 back to its starting value of 0. The shape of the curve in each quadrant is just a twisted repetition of the one from the first quadrant.

You can check that a tangent drawn to the curve at these maximum and minimum values is horizontal. Tangents in the first and fourth quadrants have positive gradients. Those in the second and third quadrants have negative ones.

We may repeat these techniques with small alterations to draw the cosine curve. Firstly we can construct a table of values of cosine from 0° to 360° using a calculator. The table below again shows some specimen values at intervals of 30°. They were chosen to create a pattern.

θ	0°	30°	60°	90°	120°	150°
$\cos\theta$	1.000	0.866	0.500	0	−0.500	−0.866

θ	180°	210°	240°	270°	300°	330°	360°
$\cos\theta$	−1.000	−0.866	−0.500	0	0.500	0.866	1.000

Again we can consider an alternative method for finding the cosine. It uses a circle with centre O and a unit radius.

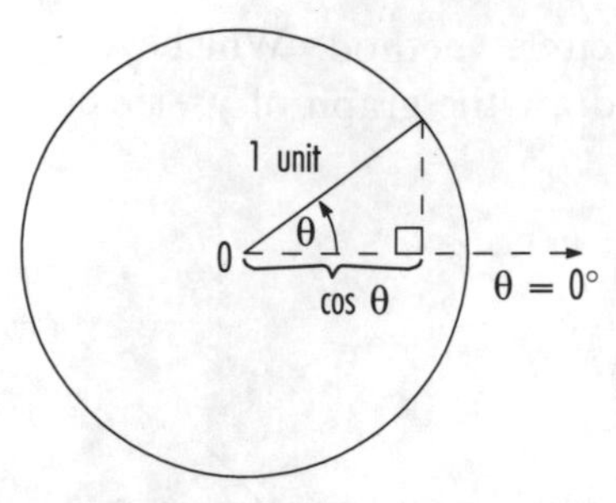

Fig. 5.6

The same axis $\theta = 0°$ is used and the points where the radius touches the circumference are important again. This time different distances are needed. We are interested in the horizontal distances along the axis $\theta = 0°$, either to the left or right of O. These distances give the values of $\cos\theta$.

Figs. 5.7 shows specimen values of $\cos\theta$ using the circle method, one from each quadrant.

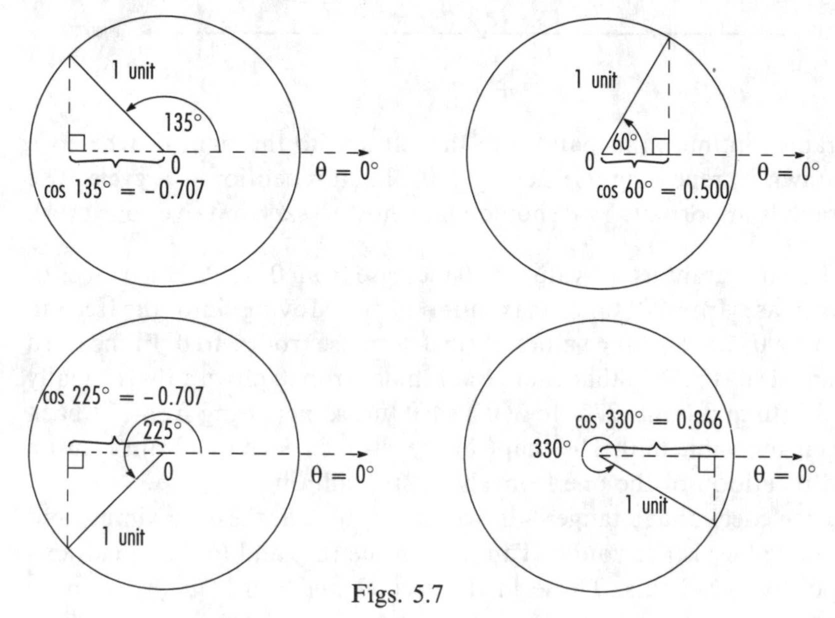

Figs. 5.7

You can choose either the calculator or circle method to work out the cosines. Again, the calculator will work out the cosine values quickly. Now we are able to draw the graph of $y = \cos\theta$ in Fig. 5.8.

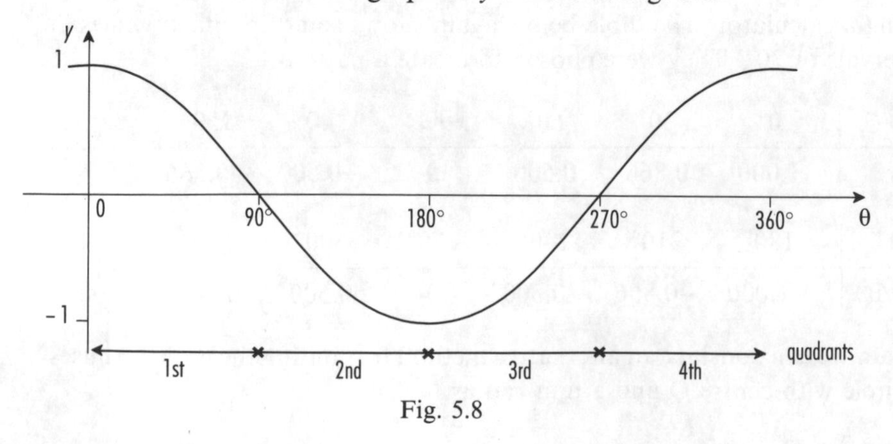

Fig. 5.8

The graph continues this pattern to the left and to the right of where we have drawn it, repeating itself every 360°. Each repetition is a **cycle**. The cosine graph is important. You should know how to sketch it over one cycle.

In the first quadrant as the values of θ increase from 0° to 90° the values of $\cos\theta$ decrease from a maximum of 1 to 0. Moving into the second quadrant (90° to 180°) the values of $\cos\theta$ continue to decrease from 0 to a minimum of -1. In the third quadrant (180° to 270°) the curve turns, increasing from -1 to 0. Finally in the fourth quadrant (270° to 360°) $\cos\theta$ returns to its starting value of 1. The shape of the curve in each quadrant is just a twisted repetition of the one from the first quadrant.

You can check that a tangent drawn to the curve at these maximum and minimum values is horizontal. Tangents in the first and second quadrants have negative gradients. Those in the third and fourth quadrants have positive ones.

If you study the graphs of $\sin\theta$ and $\cos\theta$ and attempt to sketch them you will get a "feel" for their shapes. There are a number of connections between them. We can mention a simple relationship here. **cos** stands for ***co*mplementary *s*ine**. Two angles are complementary if they add up to 90°. Thus there is a connection between sine, cosine and 90°. We may write the relationship as

$$\sin\theta = \cos(90° - \theta).$$

The next set of examples uses numerical substitution to demonstrate our new relation.

Examples 5.7

i) If $\theta = 30°$ then $\sin 30° = 0.5000$
and $\cos(90° - 30°) = \cos 60° = 0.5000.$

ii) If $\theta = 165°$ then $\sin 165° = 0.259$
and $\cos(90° - 165°) = \cos(-75°) = 0.259.$

Also sine and cosine are related by

$$\cos\theta = \sin(90° - \theta).$$

The next set of examples uses substitution to demonstrate this variation of our new relation.

Examples 5.8

i) If $\theta = 55°$ then $\cos 55° = 0.5736$
and $\sin(90° - 55°) = \sin 35° = 0.5736.$

ii) If $\theta = 201°$ then $\cos 201° = -0.9336$
and $\sin(90° - 201°) = \sin(-111°) = -0.9336.$

iii) If $\theta = -24°$ then $\cos(-24°) = 0.9135$

$$\begin{aligned} \text{and } \sin(90° - -24°) &= \sin(90° + 24°) \\ &= \sin 114° \\ &= 0.9135. \end{aligned}$$

We have mentioned already that each graph repeats itself every 360°. This means that the sine curve has the same shape from −360° to 0° as from 0° to 360°. Then that shape is repeated from 360° to 720°, and from 720° to 1080° and so on. The same rule applies to the cosine curve. In fact Figs. 5.9 show each graph over 3 cycles. A cycle can start at any point on the curve. It stops when the next similar position on the curve is reached. Two examples of cycles are drawn in Figs. 5.9.

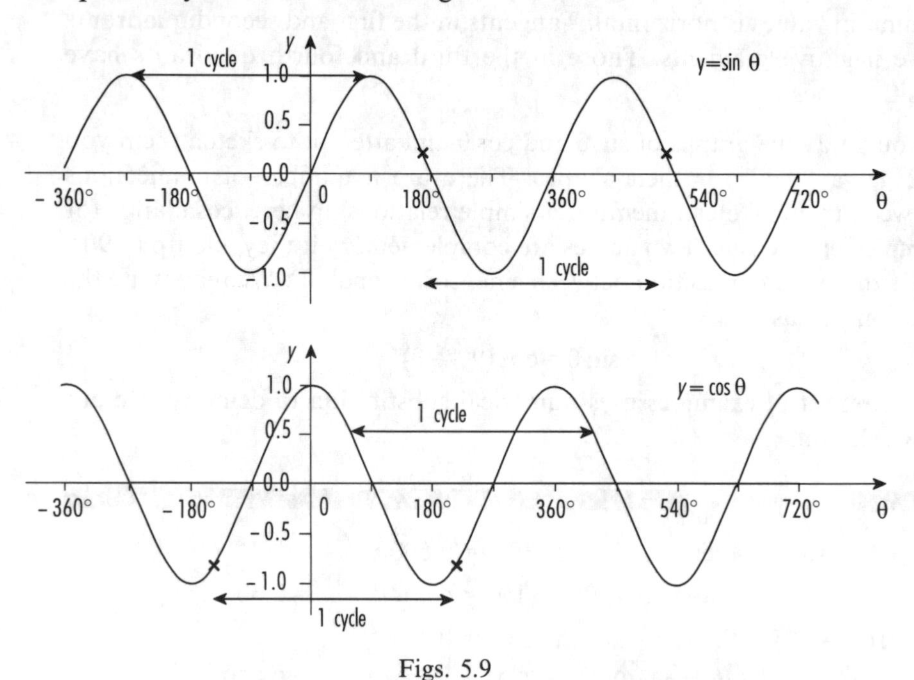

Figs. 5.9

Inverse sine and cosine

So far in this chapter we have always worked from the angle and found either the sine or cosine. What happens if we start with $\sin\theta$ and attempt to find the angle θ? What happens if we start with $\cos\theta$ and attempt to find θ? In other words we are using the opposite or inverse operations to those we have used already. $\mathbf{sin^{-1}}$ represents **inverse sin** and $\mathbf{cos^{-1}}$ represents **inverse cos**. $\sin^{-1}$ is the opposite operation to sin and $\cos^{-1}$ is the opposite operation to cos. Alternatives to $\mathbf{sin^{-1}}$ and $\mathbf{cos^{-1}}$ are **arcsin** and **arccos**.

Firstly we will return to the calculator. Most calculators use the keys inv and sin or inv and cos with the shortened form $\sin^{-1}$ or $\cos^{-1}$ above the sin and cos buttons.

With sin we input an **angle** and found the value of the **sine** of that angle.

With inv sin we input the **sine value** and find the **angle**. Similarly with inv cos we input the **cosine value** and find the **angle**. The next set of examples show the techniques.

Examples 5.9

Use your calculator to find the values (angles) of

i) $\sin^{-1}0.5$, ii) $\sin^{-1}(-0.6)$,
iii) $\cos^{-1}0.75$, iv) $\cos^{-1}(-0.866)$.

i) For $\sin^{-1}0.5$ the order of operations is

.5 inv sin

to display 30°.

ii) For $\sin^{-1}(-0.6)$ the order is

.6 +/– inv sin

to display −36.87°.

iii) For $\cos^{-1}0.75$ the order is

.75 inv cos

to display 41.41°.

iv) For $\cos^{-1}(-0.866)$ the order is

.866 +/– inv cos

to display 150°.

We can check these answers by looking at the graphs of sine and cosine. An accurate check depends upon the accuracy and scale of your drawing. When we wanted to find the sine of 30° we started at 30° on the horizontal axis. We moved directly to the curve and then towards the vertical axis. The reading of 0.5000 from that vertical allowed us to write $\sin 30° = 0.5000$. The next set of examples, for inverse sine, reverses the process. It starts with the value from the vertical axis and ends with angles on the horizontal axis.

Examples 5.10

Use your calculator and the sine graph between 0° and 360° to find the values (angles) of

i) $\sin^{-1}0.5$, ii) $\sin^{-1}(-0.6)$.

i) We can look at the sine curve and draw a horizontal line through 0.5 on the vertical axis (Fig. 5.10). The line cuts the curve in 2 places, the first one being at $\theta = 30°$. You can see from the graph that it cuts the curve at 30° before the curve crosses the horizontal axis again,

i.e. $180° - 30° = 150°$.

This means that there are two angles with a sine value of 0.5, i.e. $\sin^{-1}0.5 = 30°$ and $150°$.

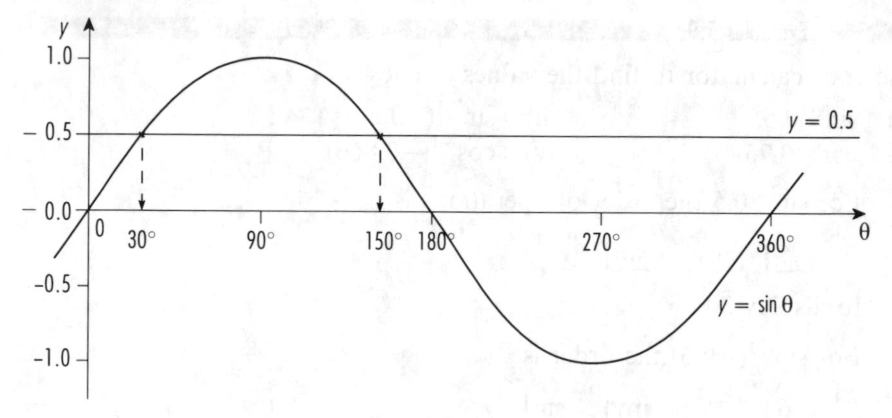

Fig. 5.10

ii) This time we draw a horizontal line through −0.6 on the vertical axis. Fig. 5.11 shows that it cuts the curve in 2 places between 0° and 360°.

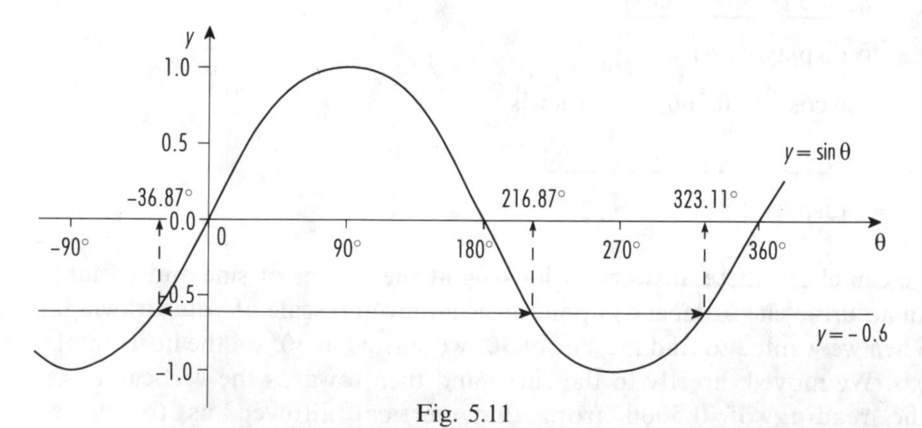

Fig. 5.11

The calculator's answer of −36.87° is before our cycle of interest. By comparing positions with our cycle we see these are 36.87° after (addition) 180° and before (subtraction) 360°,

i.e. $180° + 36.87° = 216.87°$

and $360° - 36.87° = 323.13°$

so that $\sin^{-1}(-0.6) = 216.87°$ and $323.13°$.

We can turn these into 2 rules. You should always use the calculator **and** a sketch of the sine curve rather than blindly following them.

$\sin^{-1}$(POSITIVE VALUE) = CALCULATOR DISPLAY and
180° – CALCULATOR DISPLAY

$\sin^{-1}$(NEGATIVE VALUE) gives a negative angle.

Press the +/– calculator button to get a new positive angle on the CALCULATOR DISPLAY so that the answers are

180° + CALCULATOR DISPLAY and
360° – CALCULATOR DISPLAY.

These ideas can be extended to other cycles. Because each cycle is 360° you just add 360° or 720° or . . . for further cycles. For earlier cycles you subtract the multiples of 360°.

Examples 5.11

Use your calculator and the cosine graph between 0° and 360° to find the values (angles) of

i) $\cos^{-1}0.75$, ii) $\cos^{-1}(-0.866)$.

i) We can look at the cosine curve and draw a horizontal line through 0.75 on the vertical axis (Fig. 5.12). The line cuts the curve in 2 places, the first one being at $\theta = 41.41°$. You can see from the graph that it cuts the curve at 41.41° before it reaches another maximum value, i.e. 360° – 41.41° = 318.59°. This means that there are 2 angles with a cosine value of 0.75, i.e. $\cos^{-1}0.75 = 41.41°$ and 318.59°.

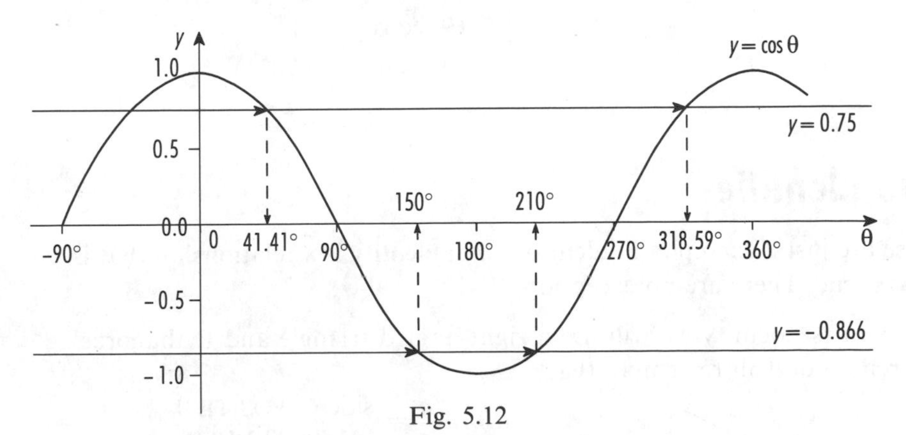

Fig. 5.12

ii) This time we draw the horizontal line through –0.866 on the vertical axis. Again it cuts the curve in 2 places, the first one being at $\theta = 150°$ after the first maximum value. You can see that it cuts the curve again at 150° before the next maximum value, i.e. at $360° - 150° = 210°$,

so that $\cos^{-1}(-0.866) = 150°$ and $210°$.

We can turn these into 2 rules. You should always use the calculator **and** a sketch of the cosine curve rather than blindly following them.

$\cos^{-1}$(POSITIVE VALUE) = CALCULATOR DISPLAY and
360° − CALCULATOR DISPLAY

$\cos^{-1}$(NEGATIVE VALUE) = CALCULATOR DISPLAY and
360° − CALCULATOR DISPLAY.

These ideas can be extended to other cycles. Because each cycle is 360° you just add 360° or 720° or . . . for further cycles. For earlier cycles you subtract the multiples of 360°.

EXERCISE 5.3

Find the angles, in the range 0° to 360°, for the following inverse trigonometric functions. (Use your graphs to check that your answers lie in the correct quadrants.)

1 $\sin^{-1}0.866$
2 $\sin^{-1}0.7071$
3 $\cos^{-1}0.5$
4 $\cos^{-1}0.236$
5 $\cos^{-1}0.7071$
6 $\sin^{-1}(-0.5)$
7 $\sin^{-1}(-0.866)$
8 $\sin^{-1}0.5936$
9 $\cos^{-1}1$
10 $\cos^{-1}(-0.893)$
11 $\cos^{-1}0.1326$
12 $\sin^{-1}1$
13 $\sin^{-1}0$
14 $\cos^{-1}(-1.372)$
15 $\cos^{-1}(-0.5)$
16 $\sin^{-1}(-1)$
17 $\sin^{-1}(4.09)$
18 $\cos^{-1}0$
19 $\cos^{-1}(-1)$
20 $\sin^{-1}(-0.7071)$

Two identities

These are just 2 examples of identities. An **identity** is a relationship that is **always true**. There are no exceptions.

For the first identity we will use a right-angled triangle and Pythagoras' theorem. You will remember that

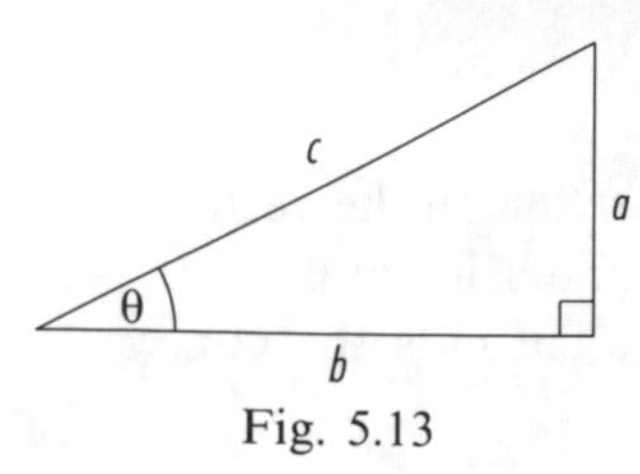

Fig. 5.13

$$\sin\theta = \frac{a}{c} = \frac{\text{side OPPOSITE } \theta}{\text{HYPOTENUSE}}$$

$$\cos\theta = \frac{b}{c} = \frac{\text{side ADJACENT to } \theta}{\text{HYPOTENUSE}}$$

$$\tan\theta = \frac{a}{b} = \frac{\text{side OPPOSITE } \theta}{\text{side ADJACENT to } \theta}$$

Also the **hypotenuse** is the longest side of a right-angled triangle. It is opposite the right-angle, the largest angle.

Applying Pythagoras' theorem we have

$$a^2 + b^2 = c^2$$

i.e. $$\frac{a^2}{c^2} + \frac{b^2}{c^2} = \frac{c^2}{c^2}$$

Dividing by c^2.

i.e. $$\left(\frac{a}{c}\right)^2 + \left(\frac{b}{c}\right)^2 = 1$$

then $$(\sin\theta)^2 + (\cos\theta)^2 = 1$$

i.e. $$\sin^2\theta + \cos^2\theta = 1.$$

Because this is an identity it is true for all angles and strictly we should use $\equiv$ instead of $=$.

You will see the slight re-arrangement from $(\sin\theta)^2$ to $\sin^2\theta$. These forms mean the same. $\sin^2\theta$ is the neater way. The same applies to $(\cos\theta)^2$ and $\cos^2\theta$. We can check out the order of operations using a calculator in the next set of examples. To find $\sin^2\theta$ we input θ, press the sin key and then square the display to give the answer.

We have worked out (**derived**) the trigonometric identity using an acute angle in a right-angled triangle. Examples 5.12 show the identity working for a selection of other angles.

Examples 5.12

Check (**verify**) that the identity $\sin^2\theta + \cos^2\theta = 1$ is true for

i) $\theta = 150°$, ii) $\theta = 231°$, iii) $\theta = 327.2°$, iv) $\theta = 450°$, v) $\theta = 4.6$ radians.

We start with the left-hand side (LHS) of the identity and aim to finish with the value of 1 on the right.

i) $\theta = 150°$

$$\sin^2 150° + \cos^2 150° = (0.5)^2 + (-0.866)^2$$
$$= 0.25 + 0.75$$
$$= 1.$$

ii) $\theta = 231°$

$$\sin^2 231° + \cos^2 231° = (-0.7771)^2 + (-0.6293)^2$$
$$= 0.6040 + 0.3960$$
$$= 1.$$

iii) $\theta = 327.2°$

$$\sin^2 327.2° + \cos^2 327.2° = (-0.5417)^2 + (0.8406)^2$$
$$= 0.2934 + 0.7066$$
$$= 1.$$

iv) $\theta = 450°$

$$\sin^2 450° + \cos^2 450° = 1^2 + 0^2 = 1.$$

v) $\theta = 4.6$ radians,

$$\sin^2 4.6 + \cos^2 4.6 = (-0.9937)^2 + (-0.1122)^2 = 0.9874 + 0.0126 = 1.$$

For the second trigonometric identity we will combine all 3 of our trigonometric ratios.

Now $\dfrac{\sin\theta}{\cos\theta} = \dfrac{a/c}{b/c}$

$= \dfrac{a}{c} \times \dfrac{c}{b}$ — Rule for dividing fractions.

$= \dfrac{a}{b}$ — Cancelling the *c*s.

i.e. $\dfrac{\sin\theta}{\cos\theta} = \tan\theta.$

Again this is an identity true for all angles and strictly we should use $\equiv$ instead of $=$. The next set of examples shows a selection of angles combined with this identity.

Examples 5.13

Verify that the identity $\dfrac{\sin\theta}{\cos\theta} = \tan\theta$ is true for

i) $\theta = 135°$, ii) $\theta = 262°$, iii) $\theta = 309.3°$, iv) $\theta = 540°$,
v) $\theta = 1.25$ radians.

Separately we will work out the left- and right-hand sides of the identity. In each case we expect the same value for our answers.

i) $\theta = 135°$

$$\frac{\sin 135°}{\cos 135°} = \frac{0.7071}{-0.7071} = -1$$

and $\tan 135° = -1.$

ii) $\theta = 262°$

$$\frac{\sin 262°}{\cos 262°} = \frac{-0.9903}{-0.1392} = 7.115$$

and $\tan 262° = 7.115.$

iii) $\theta = 309.3°$

$$\frac{\sin 309.3°}{\cos 309.3°} = \frac{-0.7738}{0.6334} = -1.222$$

and $\tan 309.3° = -1.222.$

iv) $\theta = 540°$

$$\frac{\sin 540°}{\cos 540°} = \frac{0}{-1} = 0$$

and $\tan 540° = 0.$

v) $\theta = 1.25$ radians

$$\frac{\sin 1.25}{\cos 1.25} = \frac{0.9490}{0.3153} = 3.01$$

and $\tan 1.25° = 3.01.$

The tangent curve

We may use the calculator to construct a table of values for tangent from 0° to 360°. Strange things happen to the tangent curve around 90° and 270°. Because of this more table values are included.

θ	0°	15°	30°	45°	60°	75°	90°
$\tan\theta$	0	0.268	0.577	1.000	1.732	3.732	∞

θ		105°	120°	135°	150°	165°	180°
$\tan\theta$		−3.732	−1.732	−1.000	−0.577	−0.268	0

θ		195°	210°	225°	240°	255°	270°
$\tan\theta$		0.268	0.577	1.000	1.732	3.732	∞

θ		285°	300°	315°	330°	345°	360°
$\tan\theta$		−3.732	−1.732	−1.000	−0.577	−0.268	0

∞ is the symbol for **infinity**, i.e. a very large number. It is possible that your calculator may display -E- instead of the symbol for tan 90° and tan 270°. This is because it cannot cope with the calculation. The problem arises because $\tan 90° = \frac{\sin 90°}{\cos 90°}$ and $\tan 270° = \frac{\sin 270°}{\cos 270°}$. Remember that both $\cos 90°$ and $\cos 270°$ are 0, and division by 0 is not allowed in Mathematics.

The graph of $y = \tan\theta$ is drawn in Fig. 5.14. It continues this pattern to the left and to the right of where we have drawn it. It repeats itself every 180°. You can see the same basic shape from the first quadrant (0° to 90°)

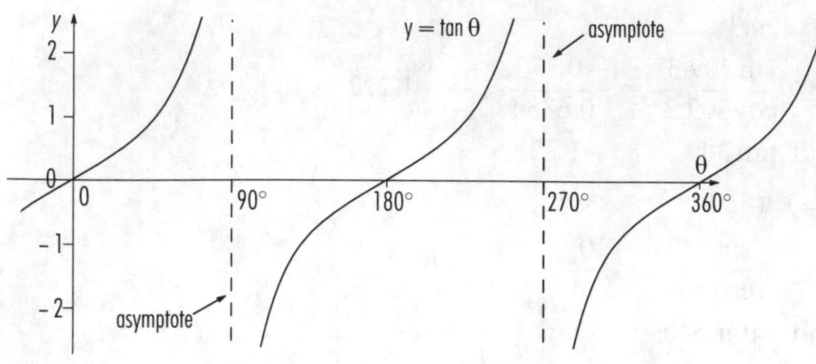

Fig. 5.14

is re-drawn in a different position for the second quadrant. The pattern is then repeated. The ¦ lines are not part of the tangent graph, but are included as a guide. They are **asymptotes**. A graph approaches an asymptote but never quite touches it. Unlike the graphs of sine and cosine, tangent is discontinuous at 90° and 270°, shown by the breaks in the curve.

Inverse tangent

We start with $\tan\theta$ and attempt to find the angle θ using the inv and tan calculator keys. The shortened form $\tan^{-1}$ appears above the tan key. The alternative to $\mathbf{tan^{-1}}$ is **arctan**. Finding the angle from the tangent value is similar to the method used for $\sin^{-1}$ and $\cos^{-1}$.

Examples 5.14

Use your calculator and the graph of $\tan\theta$ between 0° and 360° to find the values of i) $\tan^{-1}0.65$, ii) $\tan^{-1}(-1.29)$.

i) We can look at the tangent curve and draw a horizontal line through 0.65 on the vertical axis. The line cuts the curve in 2 places, the first one being at 33° (2 sf). You can see from Fig 5.15 that it cuts the curve at 33° after the curve next crosses the horizontal axis, i.e. at $180° + 33° = 213°$.

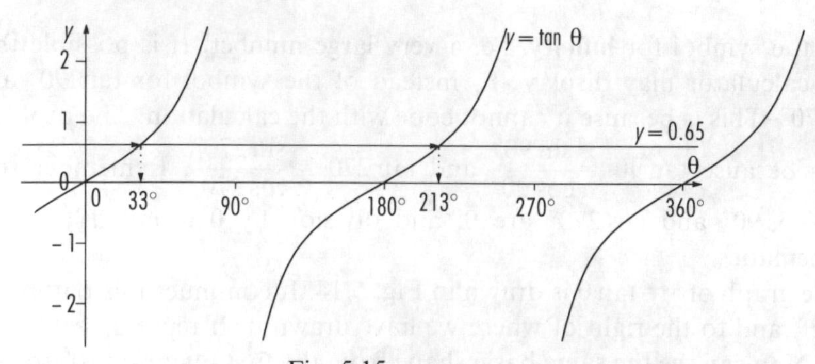

Fig. 5.15

This means there are 2 angles with a tangent value of 0.65,
i.e. $\tan^{-1}0.65 = 33°$ and $213°$.

ii) This time we draw a horizontal line through -1.29 on the vertical axis. It cuts the curve in 2 places between 0° and 360°. The calculator's angle of $-52.22°$ is not in this range. By comparing the position of $-52.22°$, in Fig. 5.16, with those of our cycle we have 52.22° before 180° and before 360°,

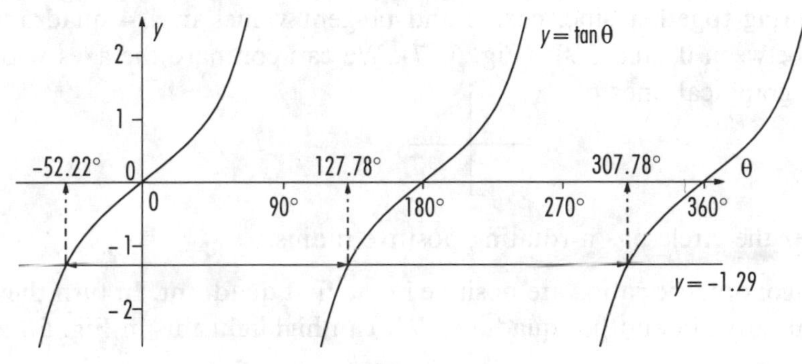

Fig. 5.16

i.e. $\quad 180° - 52.22° = 127.78°$

and $\quad 360° - 52.22° = 307.78°$

i.e. $\quad \tan^{-1}(-1.29) = 127.78°$ and $307.78°$.

We can turn these into 2 rules. You should always use the calculator **and** a sketch of the tangent curve rather than blindly following them.

$\tan^{-1}$(POSITIVE VALUE) = CALCULATOR DISPLAY and
180° + CALCULATOR DISPLAY

$\tan^{-1}$(NEGATIVE VALUE) gives a negative angle on the display.

Press +/– to get a new, positive angle, on the CALCULATOR DISPLAY. Now the answers are

180° – CALCULATOR DISPLAY and
360° – CALCULATOR DISPLAY.

These ideas can be extended to other cycles. Because each cycle is 360° you just add 360° or 720° or . . . for further cycles. For earlier cycles you subtract the multiples of 360°.

EXERCISE 5.4

Find the angles for the following inverse tangents. Use your calculator and graph in the range 0° to 360°.

1 $\tan^{-1}0.5$

2 $\tan^{-1}1.5$

3 $\tan^{-1}(-2.45)$

4 $\tan^{-1}6.28$

5 $\tan^{-1}1$

6 $\tan^{-1}(-1.732)$

7 $\tan^{-1}(-1.5)$	**9** $\tan^{-1}(-3.25)$
8 $\tan^{-1}(3.25)$	**10** $\tan^{-1}(-8.35)$

Summary of sine, cosine and tangent values

We can bring together sine, cosine and tangent values in a 4 quadrant diagram between 0° and 360° (Fig. 5.17). We can compare the axes with the usual graphical ones of

Remember the circle has a rotating positive radius.

All the trigonometric ratios are positive in the first quadrant. In turn they are each positive in another quadrant. We can highlight this in Fig. 5.18.

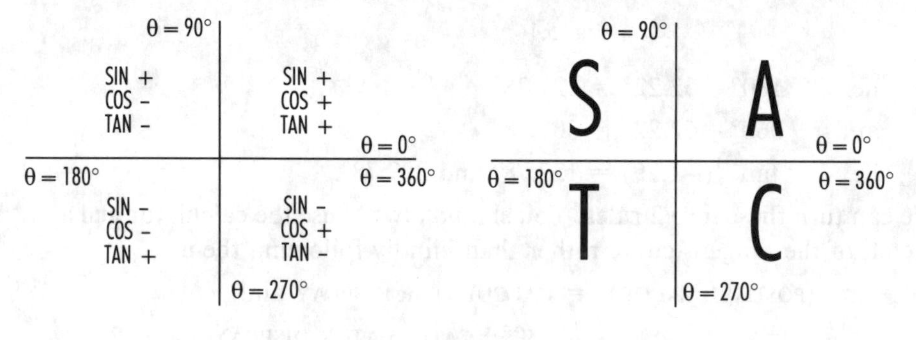

Fig. 5.17

Fig. 5.18

ASSIGNMENT

We can return to our pulley problem and the belt's length around them. Fig. 5.19 shows both pulleys with 4 important points labelled. There is a line of symmetry joining their centres. Remember the benefits of a line of symmetry. In this case the top half is a mirror image, about the line of

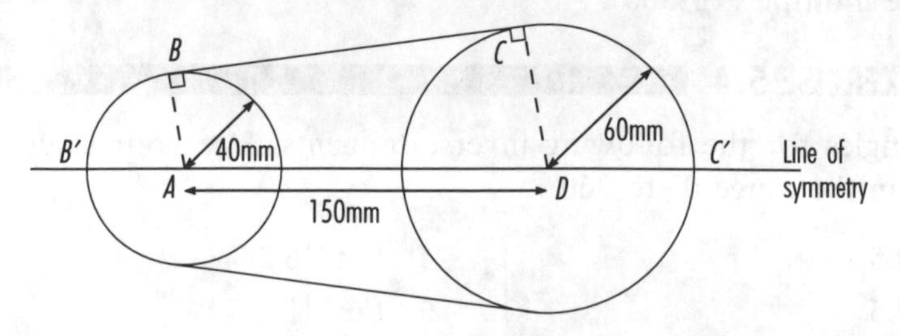

Fig. 5.19

symmetry, of the bottom half. It means we can look simply at the upper half and just double our answers. ABCD is an awkward shape. We can simplify it (Fig. 5.20) into a rectangle and a right-angled triangle.

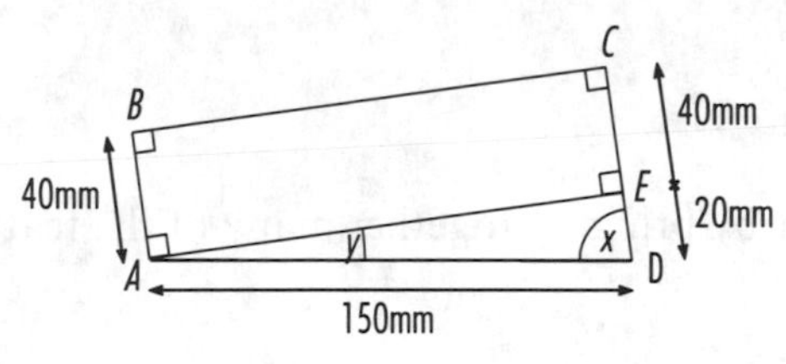

Fig. 5.20

We can now start to calculate some lengths and angles.

$AE = BC$, part of the belt.

In triangle ADE,

$$\cos x = \frac{20}{150} = 0.1333$$

$$\therefore \quad x = 82.34^\circ$$

$$\text{and} \quad y = 90^\circ - 82.34^\circ = 7.66^\circ.$$

Now we use our skills learned in Chapter 4.

For the larger pulley the upper belt section turns around C to C'. This is an angle of $180^\circ - 82.34^\circ = 97.66^\circ$. It converts to approximately 1.70 radians. Arc length $= 60 \times 1.70 = 102.0$ mm.

For the smaller pulley the upper belt section turns around B to B'. This is an angle of $180^\circ - (90^\circ + 7.66^\circ) = 82.34^\circ$. It converts to approximately 1.44 radians. Arc length $= 40 \times 1.44 = 57.6$ mm.

Returning to triangle ADE we can calculate the length AE in one of many ways. By Pythagoras' Theorem

$$AE^2 = 150^2 - 20^2$$

$$= 22500 - 400$$

$$AE = \sqrt{22100}$$

$$\text{i.e.} \quad AE = 148.7 \text{ mm}.$$

Now we have the 3 upper sections of the pulley belt which we can add together.

Upper section $= BB' + BC + CC'$.

The complete length is found by doubling this value.

$$\text{Belt length} = 2(57.6 + 148.7 + 102.0)$$

$$= 2 \times 308.3$$

$$= 616.6 \text{ mm}$$

$$= 617 \text{ mm} \quad (3 \text{ sf}).$$

You might like to extend this Assignment. Calculate the area of metal required to make a guard for the belt and pulley system. You have to decide on the design, including the necessary overlap for safety.

Periodic properties

We complete this chapter by bringing together many of the features of the 3 trigonometric functions.

The sine curve

i) is continuous;
ii) has a period of 360°;
iii) has a value of 0 at 0°, 180°, 360°, . . . ;
iv) has a maximum value of 1 at 90°, 450°, . . . ;
v) has a minimum value of −1 at 270°, 630°, . . .

The cosine curve

i) is continuous;
ii) has a period of 360°;
iii) has a value of 0 at 90°, 270°, . . . ;
iv) has a maximum value of 1 at 0°, 360°, . . . ;
v) has a minimum value of −1 at 180°, 540°, . . .

The tangent curve

i) is discontinuous at 90°, 270°, 450°, . . . ;
ii) has a period of 180°;
iii) has a value of 0 at 0°, 180°, 360°, . . . ;
iv) tends to infinity as it approaches 90°, 270°, 450°, . . .

6 Trigonometry II and Areas

The objectives of this chapter are to:

1. State and use the sine rule for a labelled triangle in the form of $\dfrac{a}{\sin A}=\dfrac{b}{\sin B}=\dfrac{c}{\sin C}$.
2. State and use the cosine rule for a labelled triangle in the form $a^2=b^2+c^2-2bc\cos A$.
3. Recognise the conditions under which the sine and cosine rules can be used.
4. Calculate the area of any triangle using any of the formulae $\frac{1}{2}bh$, $\frac{1}{2}ab\sin C$ and $\sqrt{s(s-a)(s-b)(s-c)}$.
5. Solve problems on triangles and quadrilaterals involving the use of the sine rule, cosine rule and formulae for areas of triangles.
6. Define the angles of elevation and depression.
7. Define the angle between a linc and a plane.
8. Define the angle between two intersecting plancs.
9. Identify relevant planes in a given three-dimensional problem.
10. Solve three-dimensional triangulation problems capable of being specified within a rectangular prism.

Introduction

We start this chapter with non-right-angled triangles. For such triangles we are *unable* to use our simple ratios for sine, cosine and tangent. Also we are *unable* to use Pythagoras' theorem.

Remember that the **sizes of angles correspond to the sizes of sides**. The smallest angle is opposite the smallest side. The largest angle is opposite the largest side. Also the 3 angles of a triangle add up to 180°.

ASSIGNMENT

A construction company has bought a plot of land adjacent to a road junction. The plan shows the plot as triangular together with its labelled dimensions. As you might expect the site is not level. We will look at that aspect later in the chapter. The company has outline

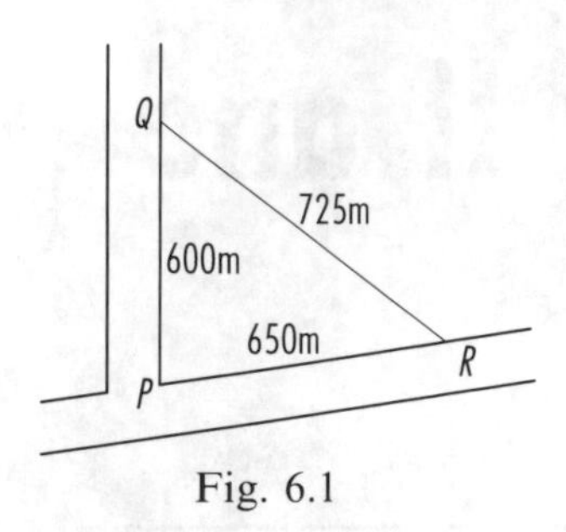

Fig. 6.1

planning permission for new offices. Before drawing a site plan it wishes to calculate the angles of this triangle. Also it wishes to find the overall site area.

The sine rule

The **sine rule** applies to any triangle. However, for right-angled triangles we use the simple sine ratio. In this section we look at non-right-angled triangles.

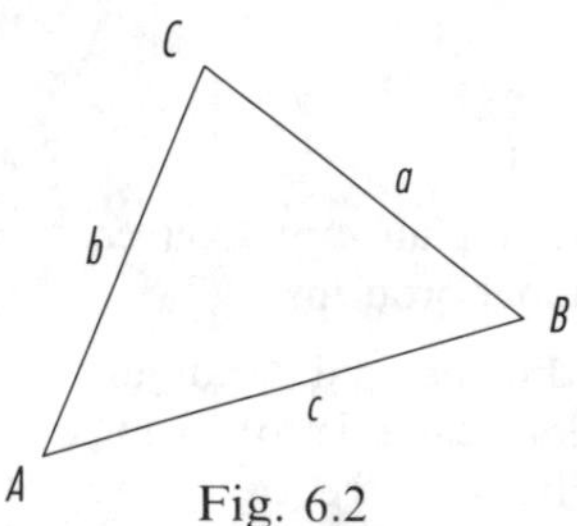

Fig. 6.2

The **sine rule** states $\dfrac{a}{\sin A} = \dfrac{b}{\sin B} = \dfrac{c}{\sin C}$. We do *not* always need all of these 3 sections. It is usual to use just 2 of them at a time, i.e. $\dfrac{a}{\sin A} = \dfrac{b}{\sin B}$ or $\dfrac{a}{\sin A} = \dfrac{c}{\sin C}$ or $\dfrac{b}{\sin B} = \dfrac{c}{\sin C}$

We may use the sine rule in any triangle given

i) 2 angles and a side,
ii) 2 sides and a *non-included* angle (i.e. an angle that is *not* between the 2 sides).

For these cases, by using the sine rule repeatedly we can **solve a triangle**, i.e. we can find all its angles and sides.

Example 6.1

In triangle ABC (Fig. 6.3) if $\angle A = 105°$, $\angle B = 40°$ and $c = 0.475$ m solve the triangle.

This example applies the sine rule twice, given 2 angles and a side.

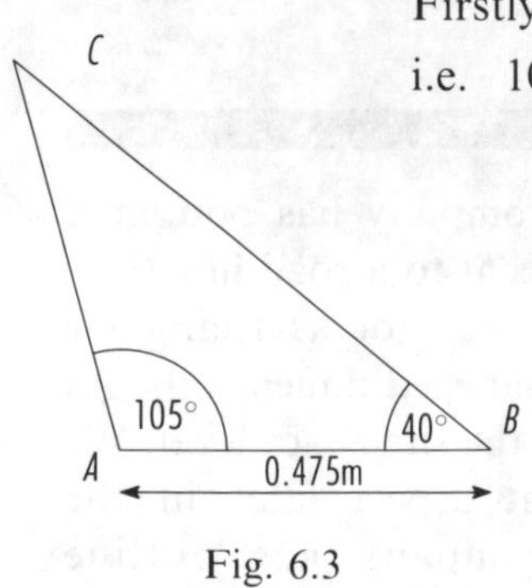

Fig. 6.3

$$\begin{aligned} \text{Firstly} \quad A + B + C &= 180°, \\ \text{i.e.} \quad 105° + 40° + C &= 180° \\ \angle C &= 180° - 145° \\ \angle C &= 35° \end{aligned}$$

Sum of angles of a triangle.

This first simple calculation means we have values for all 3 angles of our triangle *ABC*. Now we know $\angle C$ and side *c*, meaning we must use $\dfrac{c}{\sin C}$ in our application of the sine rule. We have to find the sides *a* and *b*.

Using $\dfrac{a}{\sin A} = \dfrac{c}{\sin C}$ we can substitute our values

to get $\dfrac{a}{\sin 105°} = \dfrac{0.475}{\sin 35°}$

i.e. $\dfrac{a}{0.9659} = \dfrac{0.475}{0.5736}$

$$\frac{a}{0.9659} \times 0.9659 = \frac{0.475}{0.5736} \times 0.9659$$

This multiplication leaves a alone on left.

i.e. $a = \dfrac{0.475}{0.5736} \times 0.9659$

i.e. $a = 0.800$ m.

Now using $\dfrac{b}{\sin B} = \dfrac{c}{\sin C}$ and substituting

we get $\dfrac{b}{\sin 40°} = \dfrac{0.475}{\sin 35°}$

i.e. $\dfrac{b}{0.6428} = \dfrac{0.475}{0.5736}$

i.e. $\dfrac{b}{0.6428} \times 0.6428 = \dfrac{0.475}{0.5736} \times 0.6428$

This multiplication leaves b alone on left.

$$b = \frac{0.475}{0.5736} \times 0.6428$$

i.e. $b = 0.532$ m.

Knowing all angles and sides we can fully label our completely solved triangle ABC in Fig. 6.4.

We see the figure confirms that the smallest side is opposite the smallest angle. Also the largest side is opposite the largest angle.

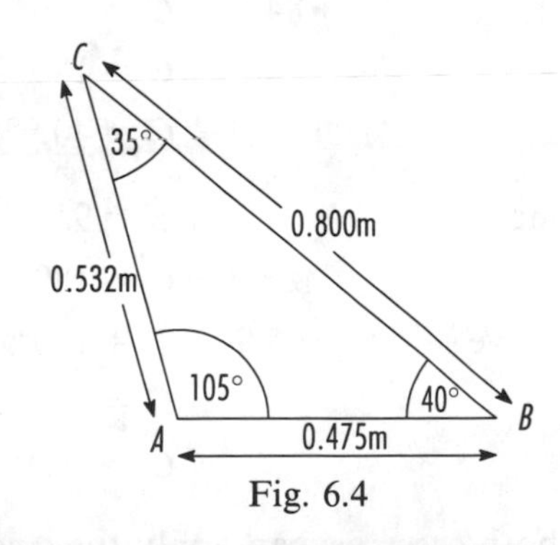

Fig. 6.4

Example 6.2

In triangle ABC (Fig. 6.5) if $\angle A = 55°$, $b = 0.85$ m and $a = 0.70$ m solve the triangle.

This example applies the sine rule given 2 sides and a non-included angle.

We have $\angle A$ and side a, meaning we must use $\dfrac{a}{\sin A}$ in the sine rule. We have no information about $\angle C$ or side c and so will look at these later.

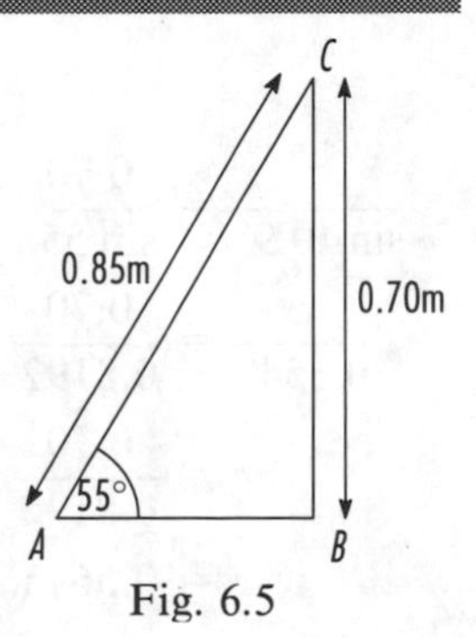

Fig. 6.5

Using $\quad \dfrac{b}{\sin B} = \dfrac{a}{\sin A}$ we can substitute values

to get $\quad \dfrac{0.85}{\sin B} = \dfrac{0.70}{\sin 55^\circ}$

i.e. $\quad \dfrac{0.85}{\sin B} = \dfrac{0.70}{0.8192}$

$$\sin B \times \frac{0.85}{\sin B} = \sin B \times \frac{0.70}{0.8192}$$

This move puts $\sin B$ in the numerator.

$$0.85 \times 0.8192 = \sin B \times \frac{0.70}{0.8192} \times 0.8192$$

$$\frac{0.85 \times 0.8192}{0.70} = \frac{\sin B \times 0.70}{0.70}$$

The moves leave $\sin B$ alone.
Left side is purely numbers.

i.e. $\quad 0.9947 = \sin B$

$\therefore \quad \angle B = 84.1^\circ,\ 180^\circ - 84.1^\circ$

i.e. $\quad \angle B = 84.1^\circ,\ 95.9^\circ.$

Sine is positive in the 1st and 2nd quadrants.

The mathematics has given us 2 possible values for $\angle B$. We need to consider them both.

Using $\quad \angle B = 84.1^\circ$

and $\quad A + B + C = 180^\circ$

we have $\quad 55^\circ + 84.1^\circ + C = 180^\circ$

i.e. $\quad \angle C = 180^\circ - 139.1^\circ$

i.e. $\quad \angle C = 40.9^\circ.$

Using $\quad \angle B = 95.9^\circ$

and $\quad A + B + C = 180^\circ$

we have $\quad 55^\circ + 95.9^\circ + C = 180^\circ$

i.e. $\quad \angle C = 180^\circ - 150.9^\circ$

i.e. $\quad \angle C = 29.1^\circ.$

In both cases we can apply the sine rule again:

$$\frac{c}{\sin C} = \frac{a}{\sin A}$$

i.e. $\quad \dfrac{c}{\sin 40.9^\circ} = \dfrac{0.70}{\sin 55^\circ} \quad$ and $\quad \dfrac{c}{\sin 29.1^\circ} = \dfrac{0.70}{\sin 55^\circ}$

$\dfrac{c}{0.6547} = \dfrac{0.70}{0.8192} \qquad \dfrac{c}{0.4863} = \dfrac{0.70}{0.8192}$

$c = \dfrac{0.70}{0.8192} \times 0.6547 \qquad c = \dfrac{0.70}{0.8192} \times 0.4863$

$c = 0.56\,\text{m}. \qquad c = 0.42\,\text{m}.$

You can see there are 2 possible triangles for ABC. This is because the positive sine has solutions in the range 0° to 90° and 90° to 180°. It is called the **ambiguous case**. This possibility can arise when the first calculation finds an angle. Both answers are correct mathematically. In a practical situation you would have to choose which solution matched the physical constraints. Figs. 6.6 show both triangles.

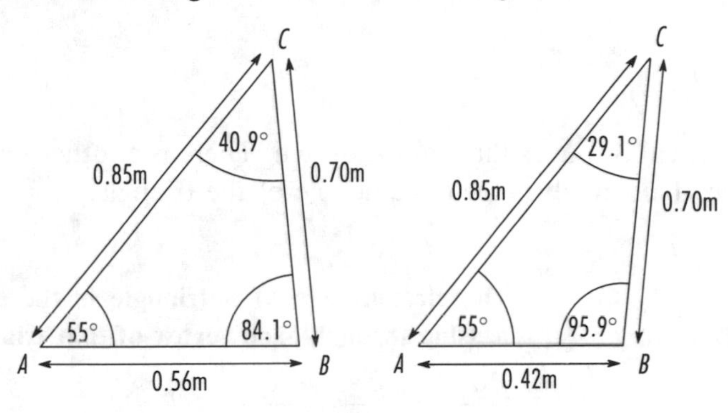

Figs. 6.6

So far our examples have correctly matched the diagrams and the mathematics. Not all triangle information is genuine. In this next example we see the solution quickly shows that a triangle cannot exist based on the original information.

Example 6.3

In triangle ABC if $\angle A = 75°$, $b = 0.85$ m and $a = 0.70$ m solve the triangle. Notice that the format and numbers are similar to those of Example 6.2, with only a slight change. At this stage we will look at the mathematics and leave the diagram until later.

Let us use
$$\frac{b}{\sin B} = \frac{a}{\sin A}$$
so that
$$\frac{0.85}{\sin B} = \frac{0.70}{\sin 75°}.$$
Eventually this gives
$$\sin B = \frac{0.85 \times 0.9659}{0.70}$$
i.e.
$$\sin B = 1.1729.$$
Remember that the maximum value of sine is 1. The value $\sin B = 1.1729$ shows that it is *not* possible to draw this particular triangle. You might like to check this out for yourself.

Draw a horizontal line and label the left end A. With a protractor centred on A measure an angle of 75° above the line. From A along this inclined line measure 0.85 m (to your chosen scale) to C. You will find that C is greater than the necessary 0.70 m (to your chosen scale) from the

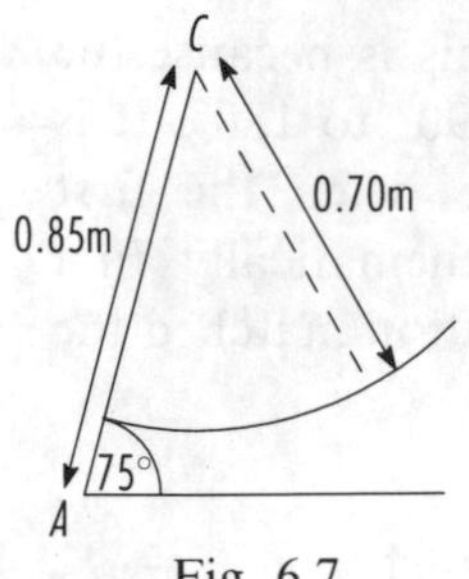

Fig. 6.7

original horizontal line. This is shown by an arc of radius 0.70 m with centre C. It does not reach the horizontal line. The calculation and Fig. 6.7 confirm that the original information will *not* create a triangle.

Let us return to real triangles that can be drawn. There is another section to the sine rule. This involves the circumcircle of the triangle.

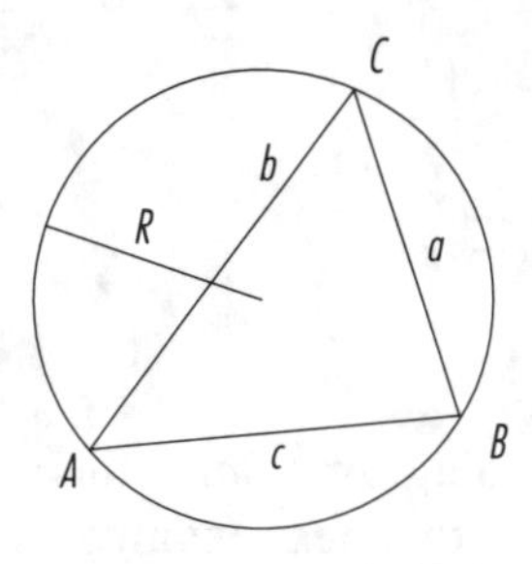

Fig. 6.8

The circumcircle of a triangle is the circle passing through each vertex of that triangle.

$$\frac{a}{\sin A} = \frac{b}{\sin B} = \frac{c}{\sin C} = 2R$$

where R is the radius of the circumcircle.

ASSIGNMENT

The only triangular information we have for our building site is for the 3 sides. We know that the sine rule cannot be applied in this situation. This means we need some more theory if we are to solve our triangle successfully.

EXERCISE 6.1

Use the sine rule to solve the triangles ABC in each question.

1 $\angle A = 40°$, $\angle B = 70°$, $a = 0.25$ m.

2 $\angle A = 55°$, $\angle C = 73°$, $c = 0.145$ m.

3 $\angle B = 34°$, $\angle C = 28°$, $a = 2.75$ m.

4 $c = 9.45$ m, $\angle B = 32°$, $\angle A = 68°$.

5 $a = 0.75$ m, $\angle C = 48°$, $\angle B = 46°$.

6 $\angle A = 50°$, $a = 0.29$ m, $b = 0.36$ m.

7 $\angle B = 25.5°$, $b = 0.42$ m, $c = 0.36$ m.

8 $\angle C = 120°$, $c = 1.7$ m, $a = 0.75$ m.

9 $\angle A = 41°$, $a = 0.24$ m, $c = 0.31$ m.

10 $\angle B = 48°$, $a = 0.49$ m, $b = 0.81$ m.

The cosine rule

The **cosine rule** applies to any triangle. However, for right-angled triangles we use the simple cosine ratio. In this section we look at non-right-angled triangles.

The **cosine rule** states

$$a^2 = b^2 + c^2 - 2bc \cos A.$$

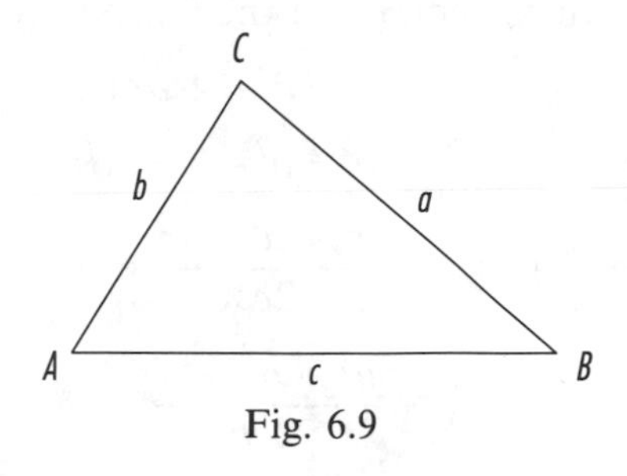

Fig. 6.9

Let us discuss some patterns of letters in our new formula. In $a^2 = b^2 + c^2 - 2bc \cos A$ notice how the side a is on the left of the formula as a^2. Its opposite angle, A, is on the opposite, right, side of the formula as $\cos A$. The other sides, b and c appear on the right. In separate appearances they are each squared and in the other, third appearance they are multiplied together.

We can re-arrange this formula to get an alternative version:

$$a^2 + 2bc \cos A = b^2 + c^2 - 2bc \cos A + 2bc \cos A$$

Adding $2bc \cos A$ to both sides.

i.e. $$a^2 + 2bc \cos A = b^2 + c^2$$

$$a^2 + 2bc \cos A - a^2 = b^2 + c^2 - a^2$$

Subtracting a^2.

i.e. $$2bc \cos A = b^2 + c^2 - a^2$$

$$\frac{2bc \cos A}{2bc} = \frac{b^2 + c^2 - a^2}{2bc}$$

Dividing by $2bc$.

i.e. $$\cos A = \frac{b^2 + c^2 - a^2}{2bc}.$$

There are similar patterns in this formula too. Notice how $\angle A$ is separate as $\cos A$. Its opposite side, a, is squared and subtracted on the right. Again b and c appear twice. Each is squared. Also they are multiplied together in the denominator.

We may use the cosine rule in any triangle given

i) 2 sides and an included angle (i.e. angle between the 2 given sides) (version 1 of the formula),

ii) 3 sides (version 2 of the formula).

Already, in the sine rule section, we have seen information that is not genuine. Whenever we have 3 sides we can test to see if they will actually create a triangle. The test is that the sum of the 2 shorter lengths must exceed the third length.

Now let us return to the formulae. The first formula finds side a and the second formula finds $\angle A$. There are similar versions for the other sides and angles. The letters in each formula move around on a cyclical basis, i.e. b moves to a, c moves to b and a moves to c. This means you do *not* need to remember all the various versions. We have

$$b^2 = c^2+a^2-2ca\cos B$$

and $$c^2 = a^2+b^2-2ab\cos C.$$

Also $$\cos B = \frac{c^2+a^2-b^2}{2ca}$$

and $$\cos C = \frac{a^2+b^2-c^2}{2ab}.$$

The patterns we described for the early versions of our formulae apply to these versions. The same ideas are true with the letters moving around cyclically.

Examples 6.4

In triangle ABC find side a if

i) $\angle A=75°$, $b=0.375$ m and $c=0.415$ m;
ii) $\angle A=125°$, $b=0.375$ m and $c=0.415$ m.

For both triangles we are finding the side a opposite the given $\angle A$. In each case 2 sides and an included angle are given, allowing us to use the cosine rule.

i) Fig. 6.10 shows the triangle labelled with our given information. We use the original version of our formula,

$$a^2 = b^2+c^2-2bc\cos A.$$

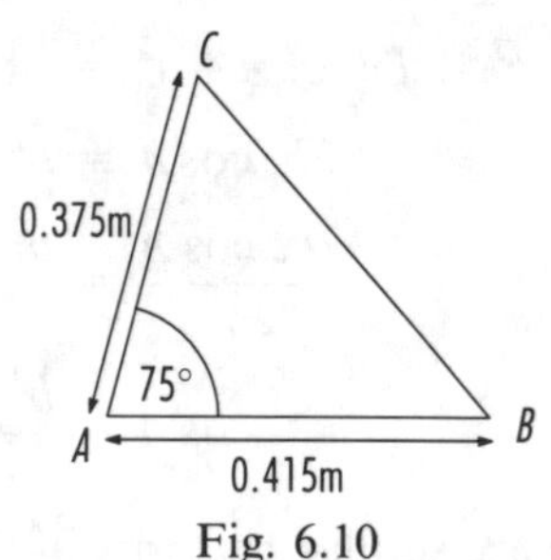

Fig. 6.10

We substitute our given values to get

$$\begin{aligned} a^2 &= 0.375^2+0.415^2-2(0.375)(0.415)(\cos 75°) \\ &= 0.375^2+0.415^2-2(0.375)(0.415)(0.2588) \\ &= 0.1406+0.1722-0.0805 \end{aligned}$$

i.e. $a^2 = 0.2323$

$\therefore \quad a = 0.48$ m.

ii) The given information in this case is very similar to that in the previous case. Fig. 6.11 shows the labelled triangle. Whilst b and c have the same values as before, $\angle A$ has been increased.

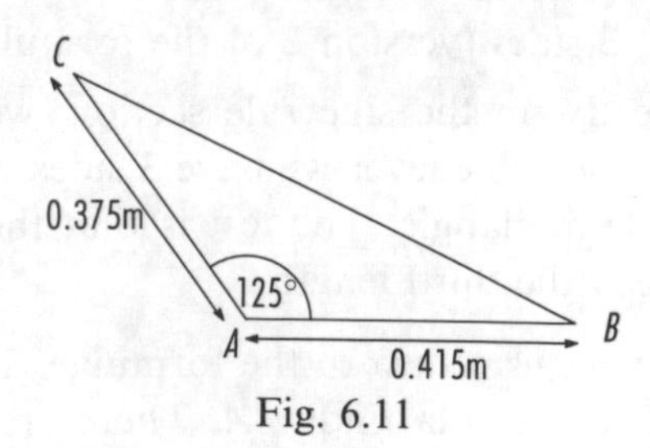

Fig. 6.11

Because of this we would expect the length of side a to increase as well. Also $\angle A$ is obtuse which means the cosine value will be negative. Using the same formula,

$$a^2 = b^2 + c^2 - 2bc \cos A$$

we substitute our given values to get

$$\begin{aligned} a^2 &= 0.375^2 + 0.415^2 - 2(0.375)(0.415)(\cos 125°) \\ &= 0.375^2 + 0.415^2 - 2(0.375)(0.415)(-0.5736) \\ &= 0.1406 + 0.1722 + 0.1785 \end{aligned}$$

i.e. $a^2 = 0.491$ $\qquad$ $(-)(-) = +$.

$\therefore$ $a = 0.70$ m.

As we expected side a is indeed larger than in the previous case.

Examples 6.5

In triangle ABC we are given the lengths of the sides, $a = 1.15$ m, $b = 1.30$ m and $c = 2.25$ m.

Find i) the smallest angle,
ii) the largest angle
and iii) the third angle of the triangle.

Before using the cosine rule we ought to check that the 3 lengths do create a triangle. We need the sum of the 2 shorter lengths to be greater than the third length. Now $1.15 + 1.30 = 2.45$ which is greater than 2.25. The test works and so we do have a genuine triangle.

i) The smallest angle is opposite the smallest side, i.e. $\angle A$ is opposite a. We use version 2 of the cosine rule,

$$\cos A = \frac{b^2 + c^2 - a^2}{2bc}.$$

Substituting our given values we get

$$\begin{aligned} \cos A &= \frac{1.30^2 + 2.25^2 - 1.15^2}{2(1.30)(2.25)} \\ &= \frac{1.6900 + 5.0625 - 1.3225}{5.85} \\ &= \frac{5.43}{5.85} \end{aligned}$$

i.e. $\cos A = 0.9282$

$\therefore$ $\angle A = 21.8°$ is the smallest angle.

ii) The greatest angle is opposite the greatest side, i.e. $\angle C$ is opposite c. We use a similar version of the cosine rule, with the letters moved around cyclically,

$$\cos C = \frac{a^2 + b^2 - c^2}{2ab}.$$

Substituting our given values we get

$$\cos C = \frac{1.15^2 + 1.30^2 - 2.25^2}{2(1.15)(1.30)}$$

$$= \frac{1.3225 + 1.6900 - 5.0625}{2.99}$$

$$= \frac{-2.05}{2.99}$$

i.e. $\cos C = -0.6856$

$\therefore$ $\angle C = 133.3°$ is the largest angle.

iii) The third angle of the triangle is $\angle B$. We know that the sum of the angles of a triangle is 180°,

i.e. $A + B + C = 180°.$

Substituting our calculated angles we get

$$21.8° + B + 133.3° = 180°$$

i.e. $\angle B = 180° - 155.1°$

i.e. $\angle B = 24.9°.$

Alternatively, and longer, we could have found $\angle B$ using

$$\cos B = \frac{c^2 + a^2 - b^2}{2ca}.$$

Fig 6.12 shows our completely solved and fully labelled triangle ABC.

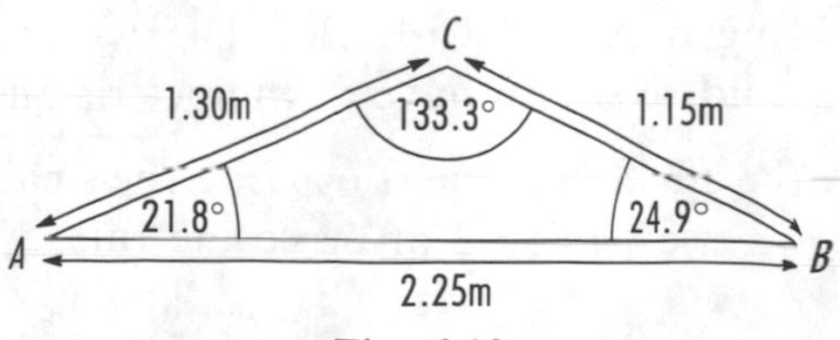

Fig. 6.12

Example 6.6

Given $a = 2.72$ m, $b = 5.56$ m and $c = 5.35$ m solve the triangle ABC.

In the previous example we saw how to test whether the 3 lengths created a triangle. We checked that the sum of the 2 shorter lengths was greater than the third length.

We can apply the test again. Now $5.35 + 2.72 = 8.07$ which is greater than 5.56. The test works and so we do have a genuine triangle.

In the solution of our triangle we may start by finding any angle. Suppose we start with $\angle A$. Using

$$\cos A = \frac{b^2 + c^2 - a^2}{2bc}$$

we substitute our values to get

$$\cos A = \frac{5.56^2 + 5.35^2 - 2.72^2}{2(5.56)(5.35)}$$

$$= \frac{52.14}{59.49}$$

i.e. $\cos A = 0.8764$

$\therefore \quad \angle A = 28.8°.$

Let us draw our triangle with the information we have so far. This is shown in Fig. 6.13.

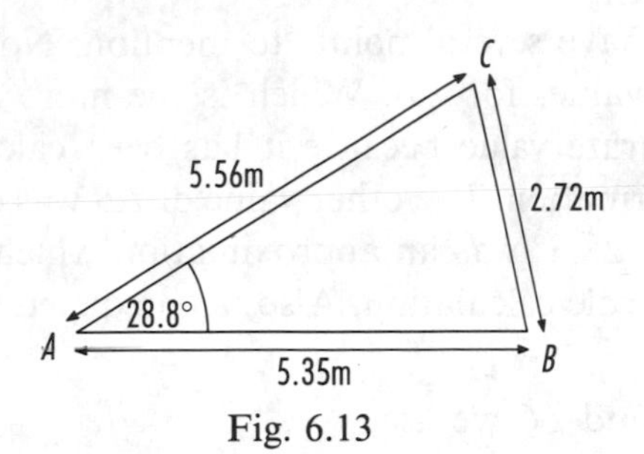

Fig. 6.13

Before we continue let us consider the angles in our triangle.

We know that $A + B + C = 180°$

i.e $28.8° + B + C = 180°$

i.e. $B + C = 180° - 28.8°$

i.e. $B + C = 151.2°.$

Sum of angles of a triangle.

Now sides b and c are of very similar lengths, b being slightly longer than c. This means $\angle B$ and $\angle C$ will be very similar angles, $\angle B$ being slightly larger than $\angle C$, i.e. $\angle B$ being slightly more than half of 151.2° (75.6°) and $\angle C$ being slightly less than half of 151.2° (75.6°).

Next we can find either $\angle B$ or $\angle C$. Also we have enough triangle information to use either the sine rule or the cosine rule. Suppose we find $\angle B$.

Using the sine rule we have

$$\frac{b}{\sin B} = \frac{a}{\sin A}.$$

Substituting our values we get

$$\frac{5.56}{\sin B} = \frac{2.72}{\sin 28.8°}.$$

We can re-arrange this equation to get

$$\sin B = \frac{5.56 \times 0.4818}{2.72}$$

i.e. $\sin B = 0.9848$

$\therefore \quad \angle B = 80.0°.$

Alternatively using the cosine rule we have

$$\cos B = \frac{c^2 + a^2 - b^2}{2ca}$$

i.e. $$\cos B = \frac{5.35^2 + 2.72^2 - 5.56^2}{2(5.35)(2.72)}$$

$$= \frac{5.107}{29.10}$$

i.e. $\cos B = 0.1755$

$\therefore \quad \angle B = 79.9°.$

We have several points to mention. Notice the slight difference between the values for $\angle B$. Which is the more accurate? $\angle B = 79.9°$ is the more accurate value because it has been calculated directly from the original information. The other value of $\angle B$ was calculated from 28.8° and the sine rule. 28.8° was an approximation which was compounded further in the sine rule calculation. Also, as predicted $\angle B = 79.9°$ is slightly greater than 75.6°.

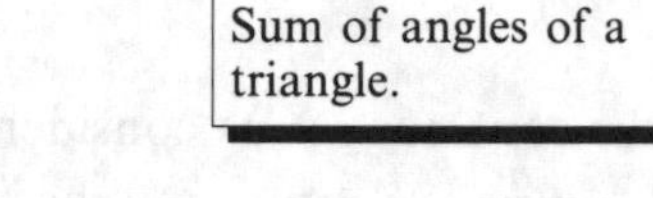

To find $\angle C$ we use

$$A + B + C = 180°$$

and substitute our calculated angles to get

$$28.8° + 79.9° + C = 180°$$

i.e. $\angle C = 180° - 108.7°$

i.e. $\angle C = 71.3°.$

To complete our predictions $\angle C = 71.3°$ is slightly less than 75.6°.

In Fig. 6.14 we have the complete and fully labelled triangle ABC.

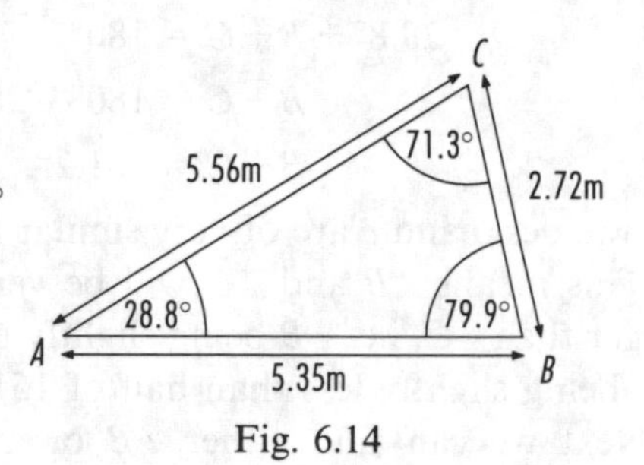

Fig. 6.14

ASSIGNMENT

Our basic triangle PQR is drawn in Fig. 6.15 showing our 3 known sides.

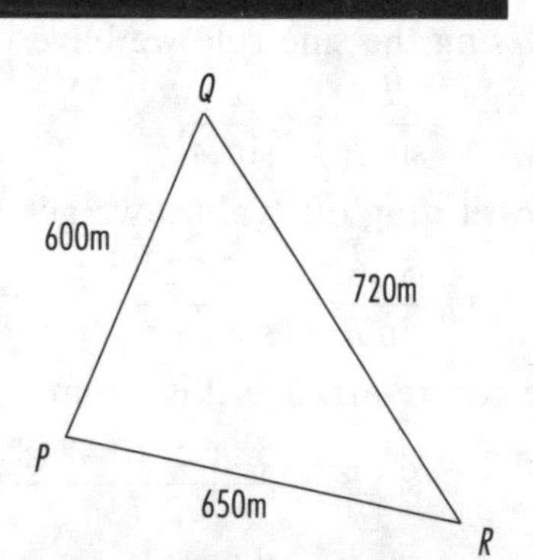

Fig. 6.15

We start our calculation by finding any angle. Suppose we find $\angle P$. We need to interpret our original formula and the patterns of letters so that

$$\cos P = \frac{q^2 + r^2 - p^2}{2qr}.$$

We substitute our values to get

$$\cos P = \frac{650^2 + 600^2 - 725^2}{2(650)(600)}$$

$$= \frac{256875}{780000}$$

i.e. $\cos P = 0.3293$

$\therefore \quad \angle P = 70.8°.$

Our experience in Example 6.6 suggests, for accuracy, that we apply the cosine rule again rather than the sine rule.

$$\text{Using} \quad \cos Q = \frac{600^2 + 725^2 - 650^2}{2(600)(725)}$$

$$= \frac{463125}{870000}$$

i.e. $\cos Q = 0.5323$

$\therefore$ $\angle Q = 57.8°$.

The sum of the 3 angles of a triangle is 180°,

i.e. $P + Q + R = 180°$

so that $70.8° + 57.8° + R = 180°$

i.e. $\angle R = 180° - 128.6°$

$\therefore$ $\angle R = 51.4°$.

In Fig. 6.16 we have a complete and fully labelled triangle PQR.

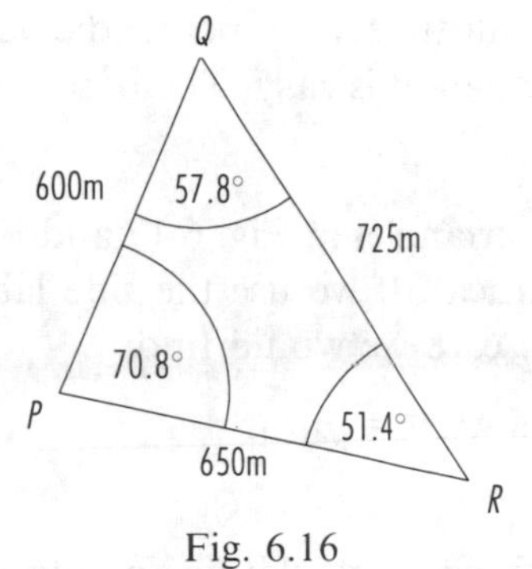

Fig. 6.16

EXERCISE 6.2

Use the cosine rule to solve the triangles ABC in each question.

1 $a = 0.75$ m, $b = 0.85$ m, $\angle C = 62°$.

2 $b = 0.84$ m, $c = 0.34$ m, $\angle A = 43°$.

3 $c = 0.56$ m, $a = 0.14$ m, $\angle B = 71°$.

4 $b = 1.12$ m, $a = 1.43$ m, $\angle C = 25°$.

5 $a = 0.32$ m, $c = 0.53$ m, $\angle B = 109°$.

6 $a = 2$ m, $b = 7$ m, $c = 6$ m.

7 $a = 2.4$ m, $b = 7.1$ m, $c = 6.4$ m.

8 $a = 0.76$ m, $b = 0.86$ m, $c = 1.32$ m.

9 $a = 1.14$ m, $b = 1.42$ m, $c = 2.61$ m.

10 $a = 7.26$ m, $b = 3.31$ m, $c = 4.81$ m.

Area of a triangle

We are going to look at 3 different formulae for finding the area of a triangle. Which one you use depends on the original information of each particular problem.

Remember the units for area. If we have distances in metres (m) then the area is in metres squared (m^2).

1.

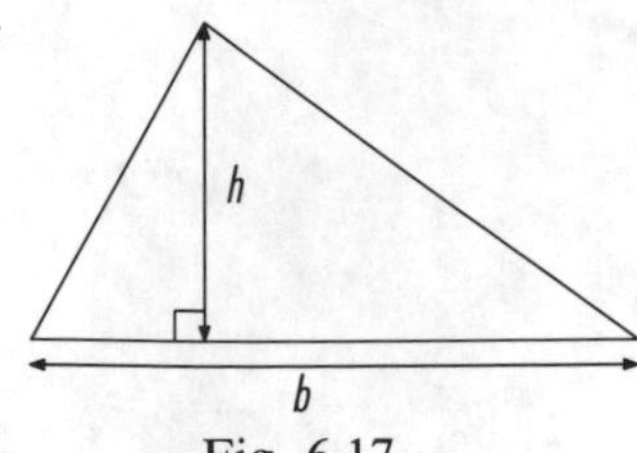

Fig. 6.17

The area of a triangle is half the product of the base and the altitude,

i.e. $\text{Area} = \frac{1}{2}bh.$

The **altitude** is the perpendicular distance between the vertex and its opposite side.

This formula is the simplest of the triangle area formulae. We use it when the altitude (the perpendicular height from the base) is given or to calculate it is easy. The base does *not* have to be horizontal. Any side will do.

The triangles in Fig. 6.17 and Fig. 6.18 are identical. If we use the side labelled b_1 in Fig. 6.18 we would find

$$\text{Area} = \tfrac{1}{2}b_1h_1.$$

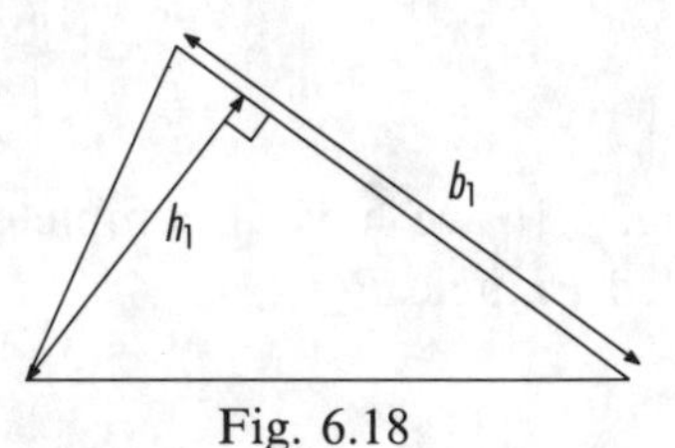

Fig. 6.18

Both area calculations would give exactly the same answers because the triangles are exactly the same. A variation from b to b_1 is compensated by a change from h to h_1 so the products $\frac{1}{2}bh$ and $\frac{1}{2}b_1h_1$ remain the same,

i.e. $\frac{1}{2}bh = \frac{1}{2}b_1h_1$

i.e. $bh = b_1h_1$

i.e. $\dfrac{b}{b_1} = \dfrac{h_1}{h}.$

Dividing by b_1 and h.

This ratio of distances will occur in a triangle to keep the area calculations consistent.

Example 6.7

The triangle in Fig. 6.19 has a base of 0.68 m and an altitude of 0.35 m. Find the area of the triangle.

We use $b = 0.68$ and $h = 0.35$ in our formula

$$\text{Area} = \tfrac{1}{2}bh$$

to get

$$\text{Area} = \tfrac{1}{2}(0.68)(0.35) = 0.12\,\text{m}^2$$

as the area of the triangle.

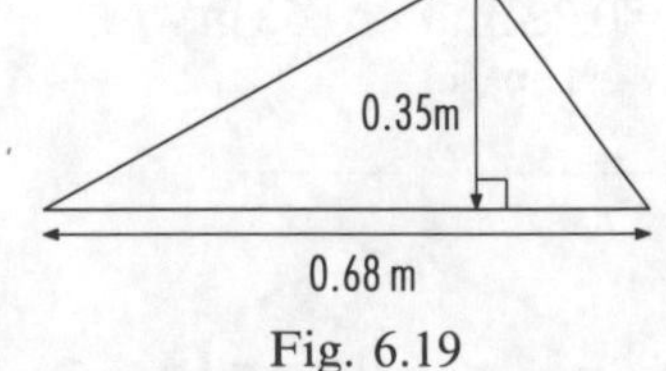

Fig. 6.19

2. The area of a triangle is half the product of any 2 sides and the sine of the included angle,

i.e. $\text{Area} = \frac{1}{2}ab\sin C$

where $\angle C$ is included between sides a and b.

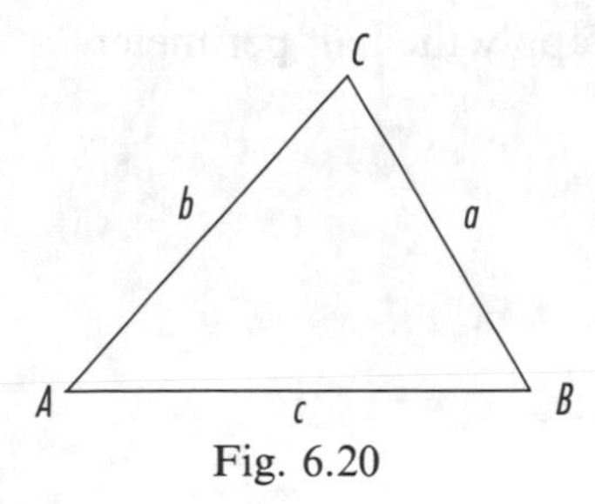

Fig. 6.20

In every case we might not have these particular sides and angle. Other versions of the formula are

$$\text{Area} = \tfrac{1}{2}bc\sin A$$

and

$$\text{Area} = \tfrac{1}{2}ca\sin B.$$

When you compare all 3 versions of this area formula you can see the letters move around cyclically. Remember the cyclical movement of letters in the cosine rule?

We can apply the formula using the values from an earlier example, Example 6.4i).

Example 6.8

In triangle ABC $\angle A = 75°$, $b = 0.375$ m and $c = 0.415$ m shown in Fig. 6.21. Find the area of the triangle.

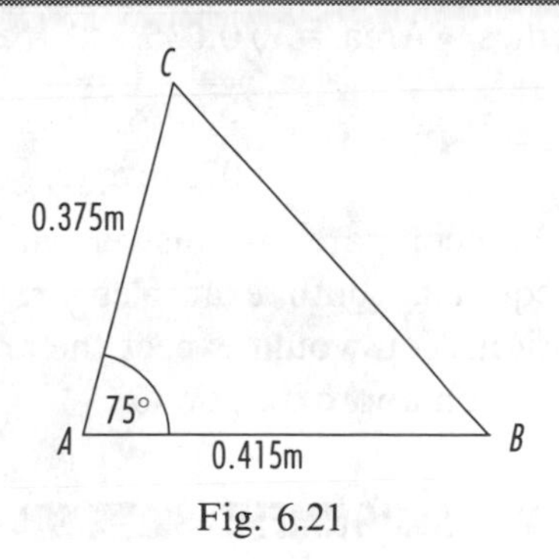

Fig. 6.21

Immediately you can see we have 2 sides and an included angle. Using the second version of our area formula we have

$$\text{Area} = \tfrac{1}{2}bc\sin A.$$

We substitute our given values to get

$$\begin{aligned}\text{Area} &= \tfrac{1}{2}(0.375)(0.415)(\sin 75°)\\ &= \tfrac{1}{2}(0.375)(0.415)(0.9659)\\ &= 0.075\,\text{m}^2.\end{aligned}$$

3.

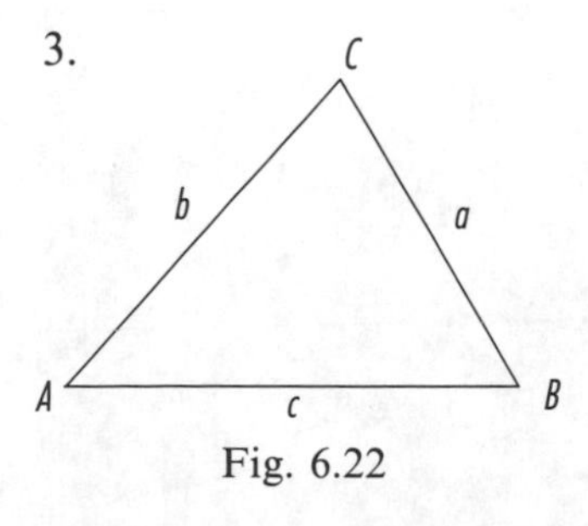

Fig. 6.22

The final of our trio of area formulae is

$$\text{Area} = \sqrt{s(s-a)(s-b)(s-c)}$$

where $\quad s = \tfrac{1}{2}(a+b+c)$

i.e. s is half the perimeter of the triangle.

As you can see this is the formula we choose when we know only the sides of the triangle.

Example 6.9

Calculate the area of triangle ABC given $a = 0.480$ m, $b = 0.375$ m and $c = 0.415$ m.

You might like to look again at Example 6.4i). These are the lengths of the sides of the triangle in that example.

Before we can use our area formula we need to apply the half perimeter formula,

i.e. $s = \frac{1}{2}(a+b+c)$.

We substitute our given values to get

$$\begin{aligned} s &= \tfrac{1}{2}(0.480+0.375+0.415) \\ &= \tfrac{1}{2}(1.27) \end{aligned}$$

i.e. $s = 0.635\,\text{m}$.

Using $s = 0.635$

we have $s-a = 0.635-0.480 = 0.155$,

$s-b = 0.635-0.375 = 0.260$

and $s-c = 0.635-0.415 = 0.220$.

Then $\text{Area} = \sqrt{s(s-a)(s-b)(s-c)}$

becomes

$$\begin{aligned} \text{Area} &= \sqrt{0.635 \times 0.155 \times 0.260 \times 0.220} \\ &= \sqrt{5.63 \times 10^{-3}} \\ &= 0.075\,\text{m}^2. \end{aligned}$$

You can compare this answer with the area answer in Example 6.8. They are equal as both examples are based on a triangle from an earlier question. You would expect the answers to be the same, allowing for any calculation approximations.

ASSIGNMENT

We can return to our triangular plot of land and its area. Remember we have the lengths of the 3 sides; $p=725\,\text{m}$, $q=650\,\text{m}$ and $r=600\,\text{m}$. This means we use the third of our formulae, first finding half the perimeter.

Using $s = \frac{1}{2}(p+q+r)$

we substitute to get

$$\begin{aligned} s &= \tfrac{1}{2}(725+650+600) \\ &= \tfrac{1}{2}(1975) \\ &= 987.5\,\text{m}. \end{aligned}$$

Applying $s = 987.5$

we have $s-p = 987.5-725 = 262.5$,

$s-q = 987.5-650 = 337.5$

and $s-r = 987.5-600 = 387.5$.

Then $\text{Area} = \sqrt{s(s-p)(s-q)(s-r)}$

becomes

$$\begin{aligned} \text{Area} &= \sqrt{987.5 \times 262.5 \times 337.5 \times 387.5} \\ &= \sqrt{3.39 \times 10^{10}} \\ &= 1.84\times 10^5\,\text{m}^2 \end{aligned}$$

is the area of the site.

EXERCISE 6.3

In each question find the area of the triangle ABC.

1 Base = 0.245 m, altitude = 0.780 m.

2 Base = 2.90 m, altitude = 1.35 m.

3 Base = 0.55 m, altitude = 0.55 m.

4 Base = 0.425 m, altitude = 0.150 m.

5 Base = 0.150 m, altitude = 0.425 m.

6 $a = 0.75$ m, $b = 0.85$ m, $\angle C = 62°$.

7 $b = 0.84$ m, $c = 0.34$ m, $\angle A = 43°$.

8 $c = 0.56$ m, $a = 0.14$ m, $\angle B = 71°$.

9 $b = 1.12$ m, $a = 1.43$ m, $\angle C = 25°$.

10 $a = 0.32$ m, $c = 0.53$ m, $\angle B = 109°$.

11 $a = 2$ m, $b = 7$ m, $c = 6$ m.

12 $a = 2.4$ m, $b = 7.1$ m, $c = 6.4$ m.

13 $a = 0.76$ m, $b = 0.86$ m, $c = 1.32$ m.

14 $a = 1.14$ m, $b = 1.42$ m, $c = 2.61$ m.

15 $a = 7.26$ m, $b = 3.31$ m, $c = 4.81$ m.

Area of a quadrilateral

A quadrilateral is a 4 sided plane figure. The square and rectangle are well known standard types with simple area formulae (Fig. 6.23 and Fig. 6.24).

1. The square

Perimeter $= 4a$

Area $= a^2$

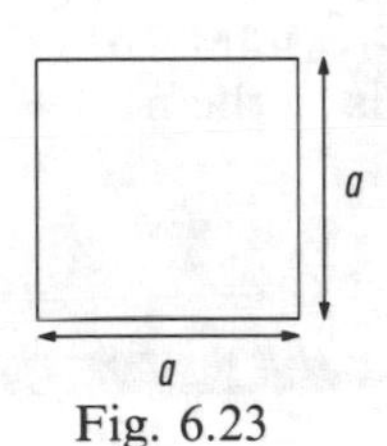

Fig. 6.23

2. The rectangle

Perimeter $= 2a + 2b = 2(a + b)$

Area $= ab$

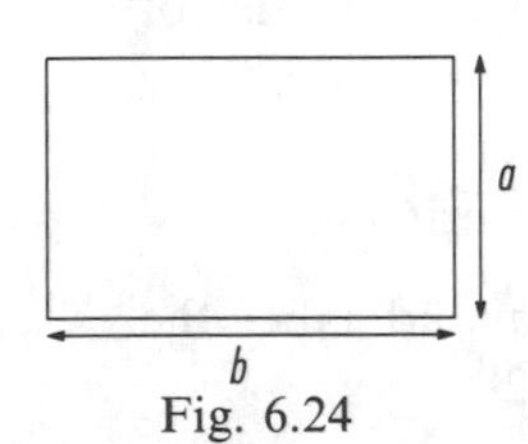

Fig. 6.24

3. The parallelogram

Perimeter $= 2a + 2b = 2(a + b)$

We can split the parallelogram into 2 triangles as in Figs. 6.25.

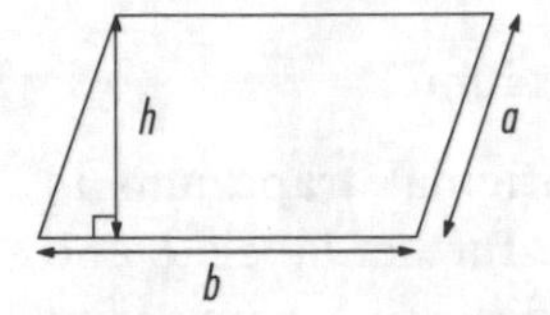

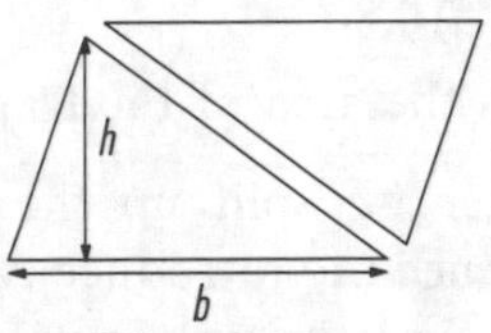

or

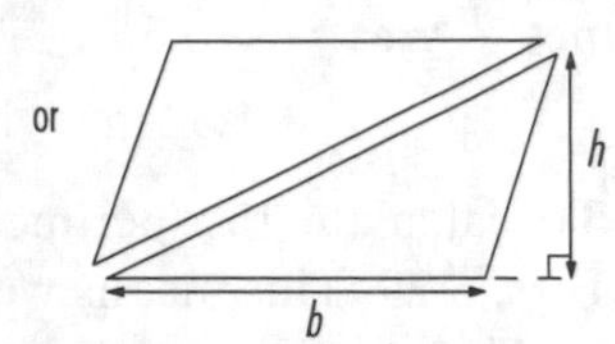

Figs. 6.25

Using these triangles and their area formula we have

$$\text{Area of triangle} = \tfrac{1}{2}bh.$$

$$\begin{aligned}\text{Area of parallelogram} &= 2\times\text{Area of triangle}\\ &= 2\times\tfrac{1}{2}bh\\ &= bh.\end{aligned}$$

We can apply this same principle using another of the triangle area formulae, i.e.

$$\text{Area of Triangle} = \tfrac{1}{2}ab\sin C$$

so that

$$\text{Area of parallelogram} = ab\sin C.$$

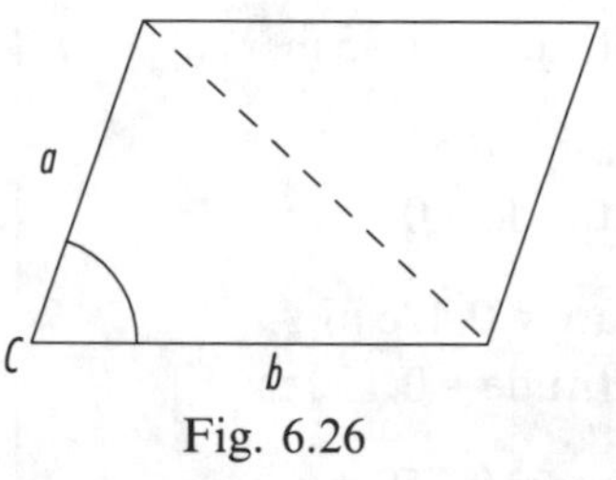

Fig. 6.26

The area of a triangle formula that uses the perimeter is less useful. For the area of a parallelogram we would not have a simple factor of 2 as for the other formulae.

4. The trapezium

$$\text{Area} = \tfrac{1}{2}(a+b)h.$$

To find the perimeter we need to include the lengths of the 2 unknown sides (Fig 6.27). We do this in the next example.

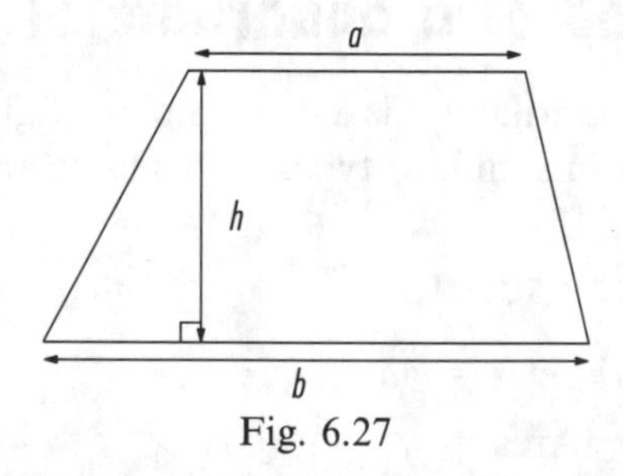

Fig. 6.27

Example 6.10

This trapezium has parallel sides of 0.55 m and 2.75 m separated by 0.80 m. Find its i) area, and ii) perimeter.

Firstly let us sketch and label the trapezium in Fig. 6.28.

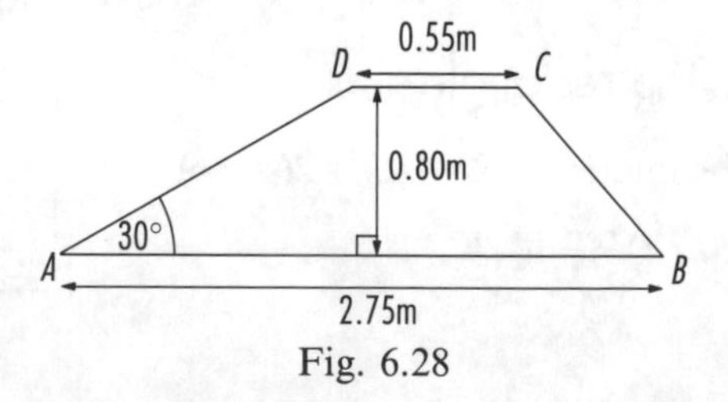

Fig. 6.28

i) In our area formula we substitute for $a=0.55$ m, $b=2.75$ m and $h=0.80$ m.

Then $\quad \text{Area} = \tfrac{1}{2}(a+b)h$

becomes $\quad \text{Area} = \tfrac{1}{2}(0.55+2.75)\,0.80$

$\qquad\qquad = 1.32\,\text{m}^2$ is the area of the trapezium.

ii) To calculate the perimeter we split up the original trapezium in Figs. 6.29. This means we need to introduce two further labels E and F. The split allows us to find the unknown sides using convenient right-angled triangles. We apply simple trigonometric ratios for sine, cosine and tangent as necessary.

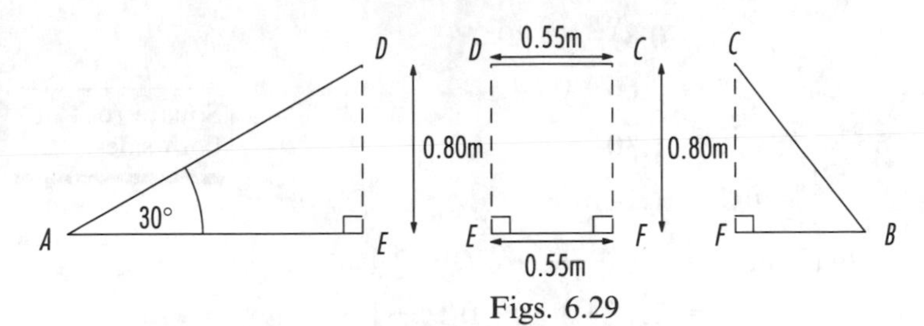

Figs. 6.29

In triangle AED using

$$\sin\theta = \frac{\text{opposite}}{\text{hypotenuse}}$$

we get $$\sin 30^\circ = \frac{0.80}{AD}$$

i.e. $$0.50 = \frac{0.80}{AD}.$$

Then $$0.50 \times AD = \frac{80}{AD} \times AD$$ Multiplying by AD.

i.e. $$0.50 \times AD = 0.80.$$

Also $$\frac{0.50 \times AD}{0.50} = \frac{0.80}{0.50}$$ Dividing by 0.50.

i.e. $$AD = \frac{0.80}{0.50} = 1.60\,\text{m}.$$

Also $$\tan\theta = \frac{\text{opposite}}{\text{adjacent}}$$

becomes $$\tan 30^\circ = \frac{0.80}{AE}$$

i.e. $$0.5774 = \frac{0.80}{AE}.$$

Using the same steps as before, eventually we get

$$AE = \frac{0.80}{0.5774} = 1.39\,\text{m}.$$

In the memory we retain the calculator value to avoid too much inaccuracy later in triangle BCF (i.e. $AE = 1.3856\ldots$).

In the trapezium we have

$$AE + EF + FB = AB$$

i.e. $$1.39 + 0.55 + FB = 2.75$$

i.e. $$FB = 2.75 - 1.94$$

$$FB = 0.81.$$ Calculator value is 0.814...

In triangle BCF, by Pythagoras' theorem

$$BC^2 = FB^2 + CF^2$$
$$= 0.81^2 + 0.80^2$$
$$= 0.66 + 0.64$$

i.e. $\quad BC^2 = 1.30\ldots$

Square root of both sides.

$\therefore \quad BC = 1.14\,\text{m}.$

Finally, $\quad \text{Perimeter} = AB + BC + CD + DA$
$$= 2.75 + 1.14 + 0.55 + 1.60$$
$$= 6.04\,\text{m}.$$

This technique of splitting up a figure can be useful when finding the area of a quadrilateral. We apply it to an area in the next example.

Example 6.11

The diagram, Fig. 6.30, shows a rectangular piece of metal measuring 0.75 m by 0.25 m. We are interested in the shaded area once the 2 triangles and quadrant have been removed. Find the shaded area.

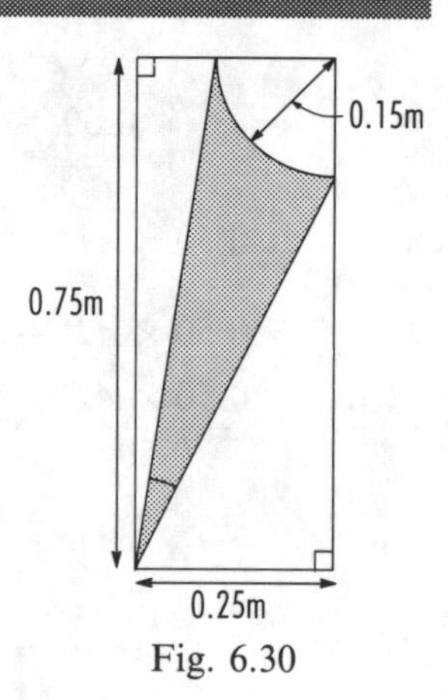

Fig. 6.30

We think of the shaded area as a rectangle minus the triangles and quadrant. For ease we re-draw these components in Figs. 6.31.

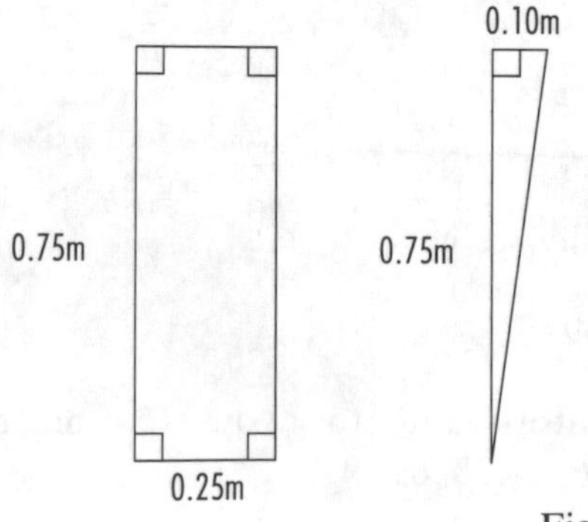

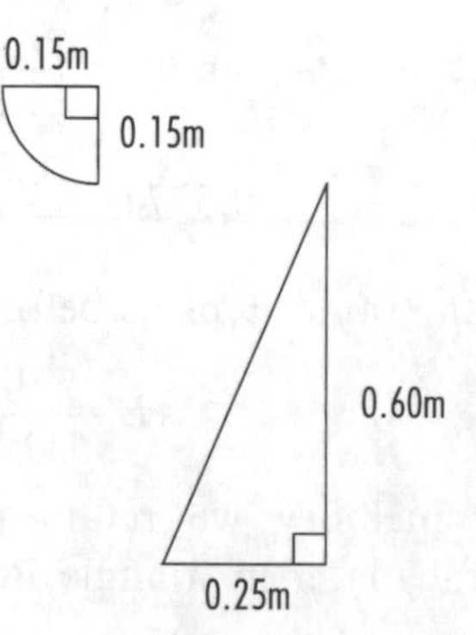

Fig. 6.31

For accuracy in our final answers we quote part answers but retain more decimal places in the calculator's memory.

Rectangle: $\quad \text{Area} = 0.25 \times 0.75 \quad = 0.1875\,\text{m}^2$

Top triangle: Area $= \frac{1}{2} \times 0.75 \times 0.10 = 0.0375\,\text{m}^2$

Quadrant: Area $= \frac{1}{4}\pi 0.15^2 = 0.0177\,\text{m}^2$

Lower triangle: Area $= \frac{1}{2} \times 0.25 \times 0.60 = 0.075\,\text{m}^2$

Shaded area $= 0.1875 - 0.0375 - 0.177 - 0.075$
$= 0.057\,\text{m}^2$.

The next set of exercises is based on areas, using formulae from this and the circular measure chapters.

EXERCISE 6.4

1 The diagram shows the cross-section of a V-block. What cross-sectional area has been removed to create the V? What area of metal remains?

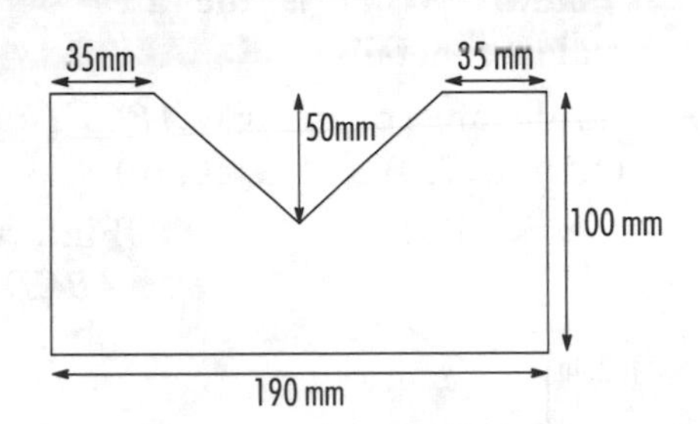

2 A small swimming pool is 1.25 m deep at the shallow end and 2.15 m deep at the deep end. The slope of the bottom is uniform. The length and width of the pool are 10.00 m and 6.50 m respectively. Make a sketch of the pool. What area of tiling is required to line the pool completely? The new owner decides to have a 0.50 m width of tiling to surround the pool. What is this extra area of tiling? Express the extra tiling as a percentage of the original tiling correct to 1 decimal place.

3 In the diagram we have a circle of radius 125.0 mm and a sector of angle 75°. Find the area of the shaded segment.

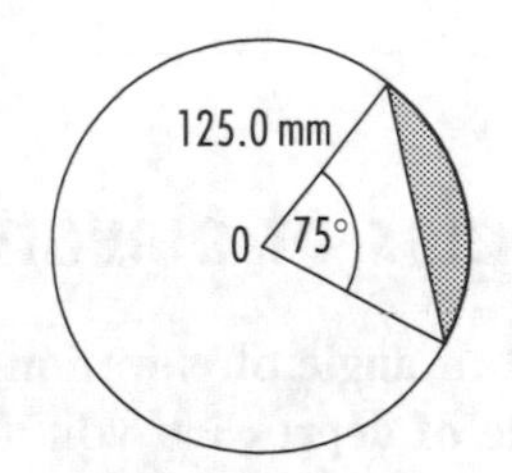

4 The diagram shows a large 30°, 60° and 90° triangular set square used for demonstrations. The length of the larger hypotenuse is 1.00 m and of the shorter is 0.50 m. Calculate the area of material in this set square.

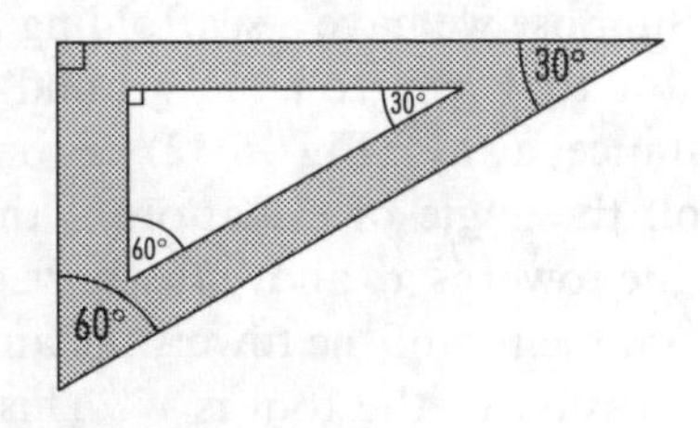

5 From a piece of high density polymer in the form of a trapezium we have removed a circle of radius 0.20 m. Some other dimensions are shown on the diagram. What is the area remaining?

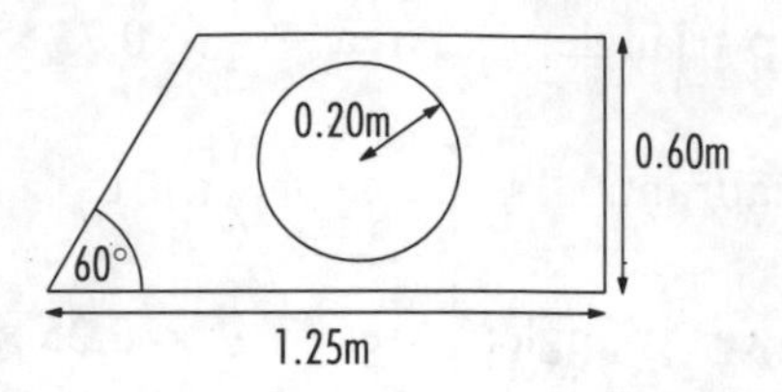

6 A parallelogram has sides of 0.725 m and 0.895 m. One diagonal measures 0.820 m. Draw this parallelogram to aid your calculation of its area.

7 The diagram shows a trapezium shaped piece of sheet metal. A semi-circle of diameter 0.60 m has been removed from the trapezium as shown. What is the area of remaining metal?

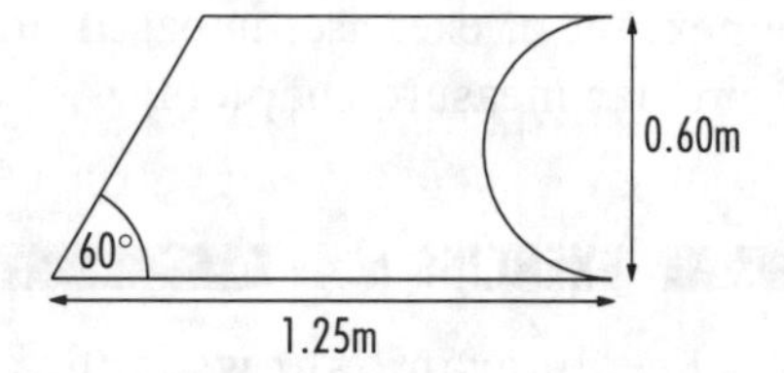

8 Find the area of triangle ABC. You are given the coordinates to be $A(1,1)$, $B(7,4)$ and $C(10,10)$.

9 Find the area of the field labelled $ABCD$.

D 108.6m C
35°
102.4m B
30°
A

10 Find the area of triangle XYZ using the information on the accompanying diagram.

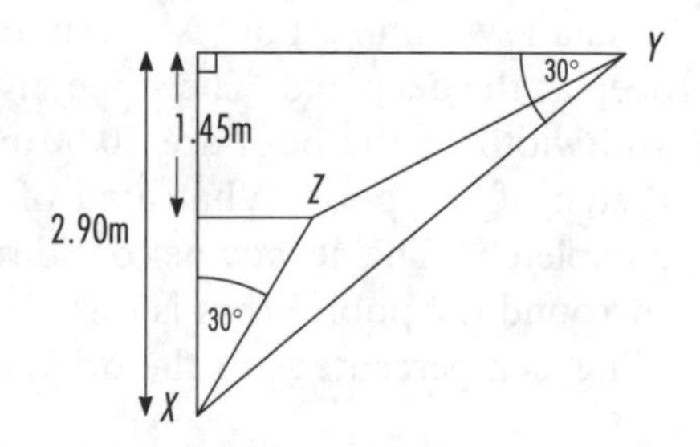

Angles of elevation and depression

For an **angle of elevation** you are looking up from the horizontal. For an **angle of depression** you are looking down from the horizontal.

Suppose we have a scaffolding tower and a tool lying on the ground some distance away (Fig. 6.32). From the tool, the angle of elevation of the top of the tower is x° above the horizontal. From the top of the tower, the angle of depression of the tool is y°. This is y°

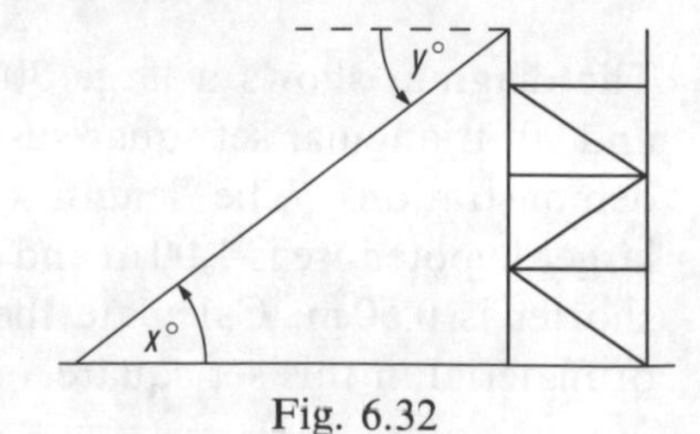

Fig. 6.32

below the horizontal. Simple geometry tells us that these are the same angles,

i.e. $x = y$.

Alternate angles.

We can put some numbers to this general idea in the next example.

Example 6.12

Our scaffolding tower is 7.4 m high and an angle grinder is 12 m away from its base on the horizontal ground.

i) What is the angle of elevation of the top of the tower from the grinder?

A man is three-quarters of the way up the tower when he sees the tool he has forgotten.

ii) What is the angle of depression of the grinder?

In each case we can draw a diagram for our information and apply some trigonometry (Fig. 6.33 and Fig. 6.34).

i)

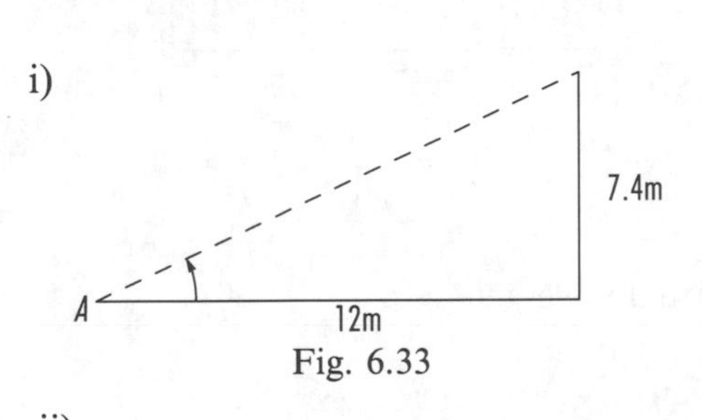

Fig. 6.33

$$\tan A = \frac{7.4}{12} = 0.61\bar{6}$$

i.e. $\angle A = 31.7°$ is the angle of elevation of the top of the tower from the grinder.

ii)

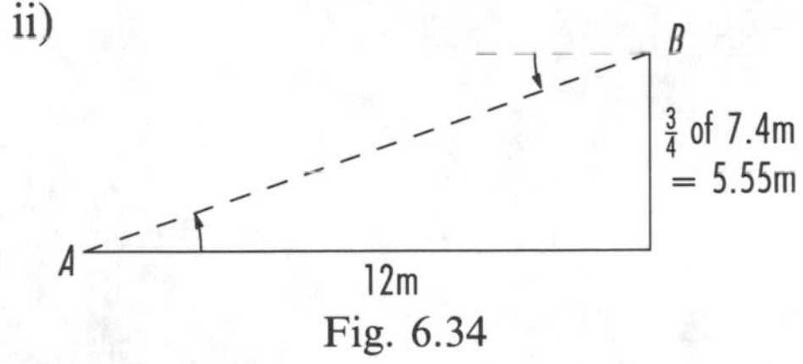

Fig. 6.34

$$\tan B = \frac{12}{5.55} = 2.16$$

i.e. $\angle B = 65.2°$.

$\therefore\ 90° - 65.2° = 24.8°$ is the angle of depression of the forgotten tool.

ASSIGNMENT

We can take another look at our triangular building plot. Suppose the plot, as expected, is *not* level. Hence our dimensions follow the lie of the land rather than the horizontal.

From P the angle of elevation of Q is found to be 2.25°. From P the angle of depression of R is found to be 0.80°. Using our theory we can find the height of Q above the level of P, and the height of R below P. We can follow this with simple arithmetic to find the height difference between Q and R.

First we deal with P and Q, drawing a supplementary diagram (Fig. 6.35) from our extra information. Q' is vertically below Q along the level of P.

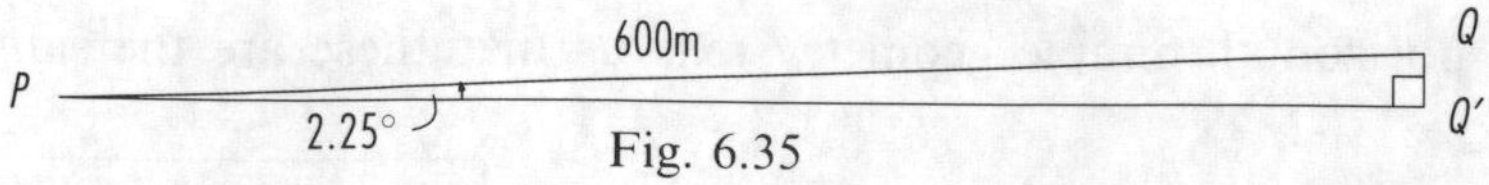

Fig. 6.35

In triangle $PQ'Q$, $\sin 2.25° = \dfrac{QQ'}{600}$

i.e. $0.0393 = \dfrac{QQ'}{600}$

i.e. $0.0393 \times 600 = \dfrac{QQ'}{600} \times 600$ — Multiplying by 600.

so that $0.0393 \times 600 = QQ'$.

We may complete the multiplication to get

$$QQ' = 23.6\,\text{m}$$

i.e. Q is 23.6 m above the level of P.

Now we look at P and R, again drawing another separate diagram, Fig. 6.36, for our information.

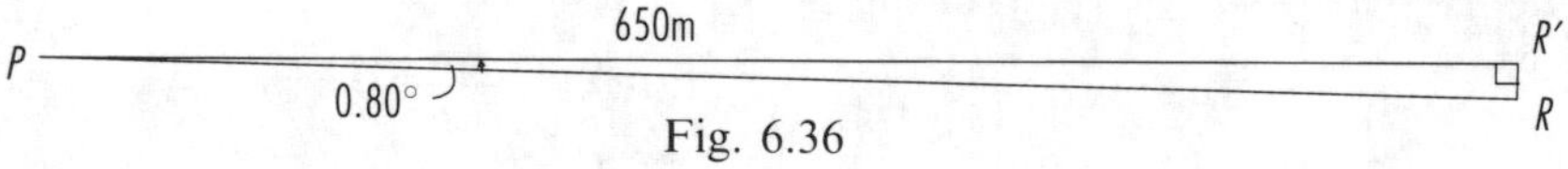

Fig. 6.36

In triangle PRR', $\tan 0.80° = \dfrac{RR'}{650}$.

Using the same method as before eventually we get

$$RR' = 0.0140 \times 650$$
$$= 9.1\,\text{m},$$

i.e. R is 9.1 m below the level of P.

Finally, linking together our results (Fig. 6.37) we can find the difference in heights of Q and R.

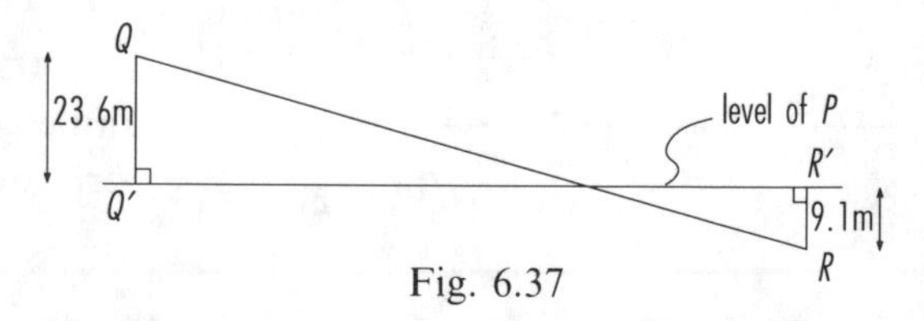

Fig. 6.37

$$\text{Difference in heights} = 23.6 + 9.1$$
$$= 32.7\,\text{m}.$$

The next short exercise concentrates on angles of elevation and depression.

EXERCISE 6.5

1 A ladder of length 4 m has its foot on horizontal ground. It leans against a vertical wall. If its angle of elevation is 76° how far is its foot from the wall? What is the acute angle between the top of the ladder and the wall?

2 A railway line has a gradient of 1 in 125 measured along the line. What is its angle of elevation?

3 The eyes of a man 1.82 m tall are 1.725 m above horizontal ground. Unknowingly he drops his pay slip and then something catches his attention. If he sees the pay slip 10 m away what is its angle of depression? Before he can run towards it the wind blows it a further 2 m away. What is the new angle of depression?

4 On a small hillock stands a tree. Along the line of sight the base of the tree is 150 m away at an angle of elevation of 4.05°. From the same point the angle of elevation of the top of the tree is 7.84°. Estimate the height of the tree.

5 A church tower needs a new weather vane. The height of the vane is 1.35 m. The blacksmith places the vane on top of the tower at the front. From 100 m away the angle of elevation of the top of the vane is 14.4°. How high is the tower? It is placed at the back on top of the tower. From the same place the angle of elevation is 13.6°. What is the depth of the tower, front to back?

Angle between a line and a plane

In Fig. 6.38 $WXYZ$ is any plane and AC is a line. A lies on the plane and B, vertically below C, also lies on the plane. This means we have a right-angled triangle ABC. $\angle CAB$ is the angle between the line and the plane.

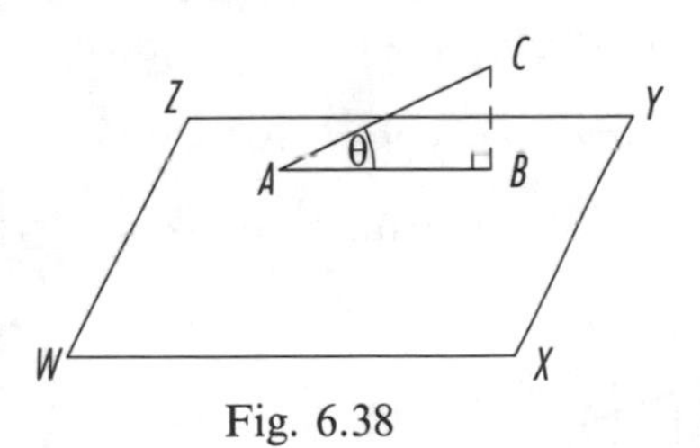

Fig. 6.38

The **projection** of AC onto the plane is AB, $\frac{AB}{AC} = \cos\theta$

i.e. $AB = AC \times \cos\theta$.

Suppose A does *not* lie on the plane $WXYZ$ as in Fig. 6.39.

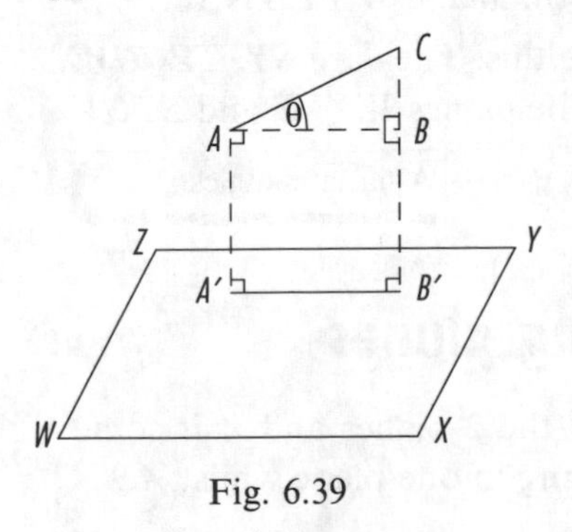

Fig. 6.39

Again B is vertically below C. Also A and B are at the same height above the plane. The projections of A and B onto the plane are A' and B', i.e. A' and B' lie on the plane, vertically below A and B respectively. Hence AB and $A'B'$ are equal in length and parallel. Also we have parallel lines AA' and CBB'. Notice also there are quite a few right-angles marked in the diagram.

Again $\angle CAB$ is the angle between the line and the plane.
Also AB or $A'B' = AC \times \cos\theta$.

Example 6.13

The diagram shows a rectangular metal block measuring 0.25 m by 0.15 m by 0.10 m. We have labelled it for reference.
What is the angle between the line SY and the plane $WXYZ$?

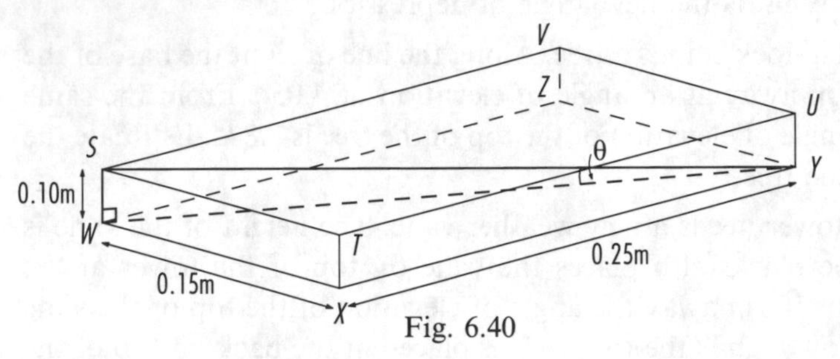

Fig. 6.40

We start with the projection of SY onto the plane $WXYZ$, WY. To show this more clearly we draw a supplementary diagram, Fig. 6.41.

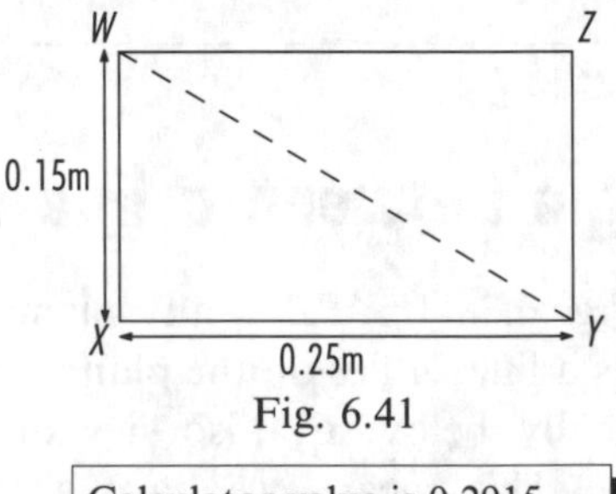

Fig. 6.41

In triangle WXY, by Pythagoras' theorem

$$\begin{aligned} WY^2 &= WX^2 + XY^2 \\ &= 0.15^2 + 0.25^2 \\ &= 0.0225 + 0.0625 \\ \text{i.e.} \quad WY^2 &= 0.085 \\ \therefore \quad WY &= \sqrt{0.085} \\ &= 0.29\,\text{m.} \end{aligned}$$

Calculator value is 0.2915...

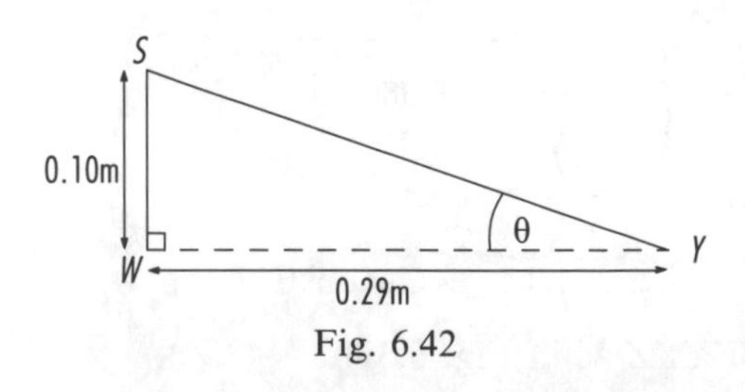

Fig. 6.42

Picking out $\angle\theta$ from the block we have

$$\tan\theta = \frac{0.10}{0.29..} = 0.3430$$

$$\therefore \quad \theta = 18.9°$$

i.e. the angle between the line SY and the plane $WXYZ$ is 18.9°.

There are 4 lines joining opposite corners like this. They are SY, TZ, UW and VX. Each of them is inclined at 18.9° to the planes $WXYZ$ and $STUV$ (Fig. 6.42).

Alternate angles.

Angle between two intersecting planes

In Fig. 6.43 XY is the line of intersection of the 2 planes and A is some point along this common line. AC is a line lying in one plane whilst AB is lying in the other plane. Both AC and AB are perpendicular to XY. Then $\angle CAB$ is the angle between the 2 planes.

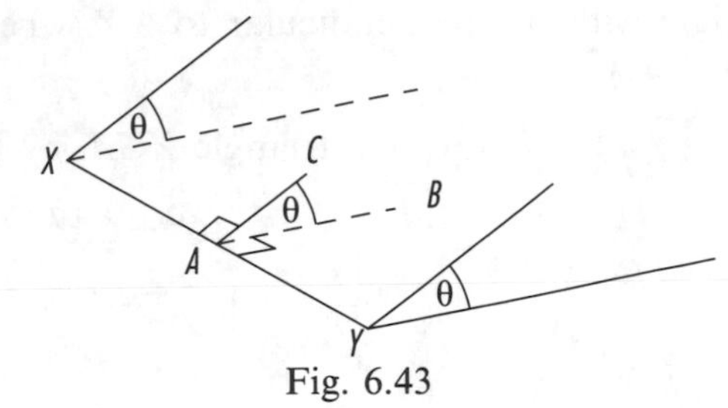

Fig. 6.43

If C is vertically above B (Fig. 6.44) then the projection of AC onto the other plane is AB.

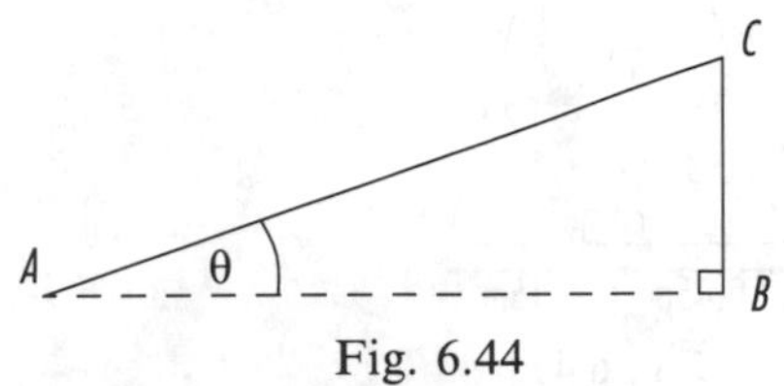

Fig. 6.44

Again $\quad \dfrac{AB}{AC} = \cos\theta$

i.e. $\quad AB = AC \times \cos\theta.$

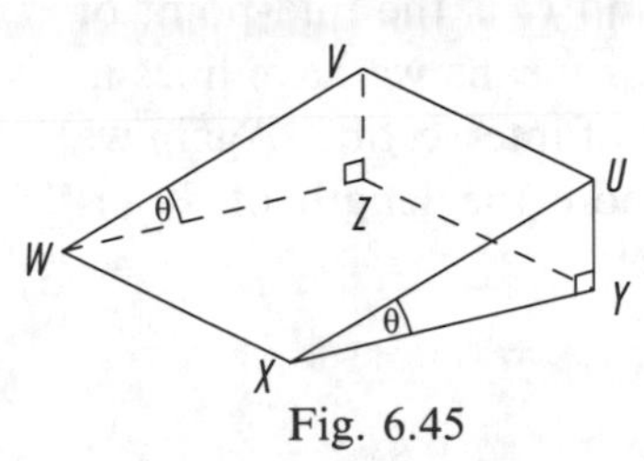

Fig. 6.45

Similarly we can project areas. In Fig. 6.45 U is vertically above Y and V is vertically above Z. WX is common to both planes as their line of intersection. The projection of area $WXUV$ onto the other plane is $WXYZ$, i.e.

Area $WXYZ$ = Area $WXUV \times \cos\theta$.

Suppose $WXYZ$ is a horizontal plane and plane $WXUV$ is inclined at the angle θ. On $WXUV$ there are many lines inclined at many angles to various other lines of reference. An important line is a **line of greatest slope**. A line of greatest slope is inclined at θ to the horizontal, just like its plane.

Example 6.14

The diagram in Fig. 6.46 shows a pyramid with a horizontal rectangular base $WXYZ$ measuring 0.42 m by 0.30 m. The apex of the pyramid, A, is vertically above the centre of the base. The length of each sloping edge is 0.54 m. What are the angles of inclination of the sloping sides to the base?

The centre of the rectangular base $WXYZ$ is B, vertically below A. The opposite sides of a rectangle are equal. Hence for our pyramid the opposite sloping sides are equally inclined to the base.

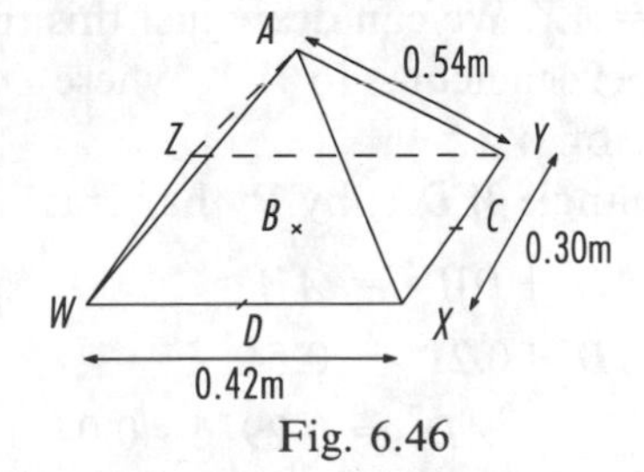

Fig. 6.46

Let us look at the sloping side AXY. XY is the common line of intersection between the plane AXY and the base plane. The symmetry of the pyramid means that triangle AXY is isosceles with $AX = AY$. We can draw just this triangle with AC perpendicular to XY, where C is the mid-point of XY (Fig. 6.47).

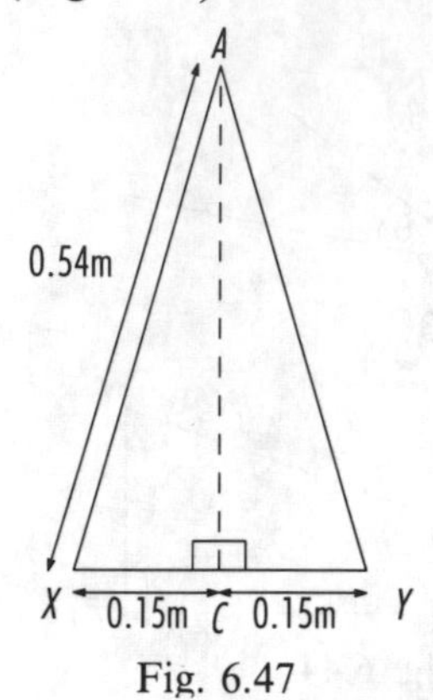

Fig. 6.47

In triangle XCA, by Pythagoras' theorem

$$AC^2 + CX^2 = AX^2$$

i.e. $$AC^2 + 0.15^2 = 0.54^2$$

i.e. $$AC^2 = 0.2916 - 0.0225$$

$$AC^2 = 0.2691$$

$$\therefore \quad AC = \sqrt{0.2691}$$

$$= 0.52\,\text{m}.$$

Calculator value is 0.5187...

Remember B is at the centre of the rectangle and C is the mid-point of side, XY. Hence BC is perpendicular to XY. This means we have lines in each plane both perpendicular to the planes' line of intersection. Again we use a supplementary diagram, Fig. 6.48. We note the length of BC is 0.21 m ($\frac{1}{2} \times 0.42$ m), half the side of the rectangle.

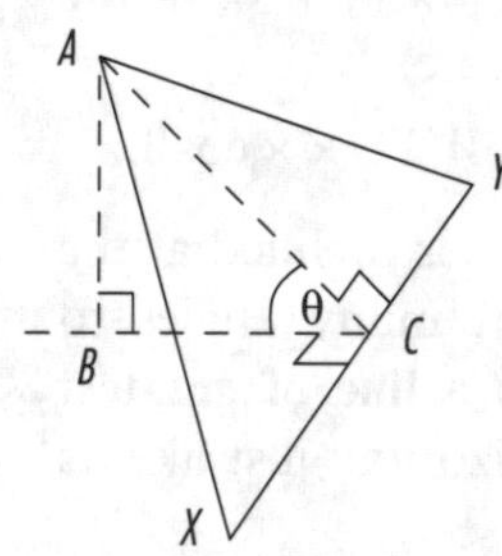

Fig. 6.48

In triangle ABC,

$$\cos\theta = \frac{BC}{AC}$$

i.e. $$\cos\theta = \frac{0.21}{0.51\ldots} = 0.4048.$$

$$\therefore \quad \angle\theta = 66.1^\circ$$

is the angle of inclination of both sloping sides AXY and AWZ to the rectangular base.

We apply the same technique to the other 2 opposite sides (Fig. 6.49). For the sloping side AWX, WX is the common line of intersection with the base plane. The symmetry of the pyramid means triangle AWX is isosceles with $AW = AX$. We can draw just this triangle with AD perpendicular to WX, where D is the mid-point of WX.

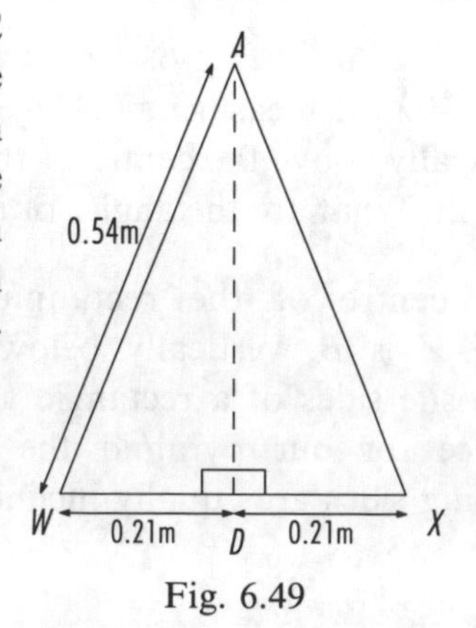

Fig. 6.49

In triangle WDA, by Pythagoras' theorem

$$AD^2 + DW^2 = AW^2$$

i.e. $$AD^2 + 0.21^2 = 0.54^2$$

i.e. $$AD^2 = 0.2916 - 0.0441$$

$$AD^2 = 0.2475$$

$$\therefore \quad AD = \sqrt{0.2475}$$

$$= 0.50\,\text{m}.$$

Calculator value is 0.497...

Remember B is the centre of the rectangle and D is the mid-point of side WX. Hence BD is perpendicular to WX. This means we have lines in each plane perpendicular to the planes' line of intersection. Again we use a supplementary diagram, Fig. 6.50. We note the length of BD is 0.15 m ($\frac{1}{2} \times 0.30$ m), half the other side of the rectangle.

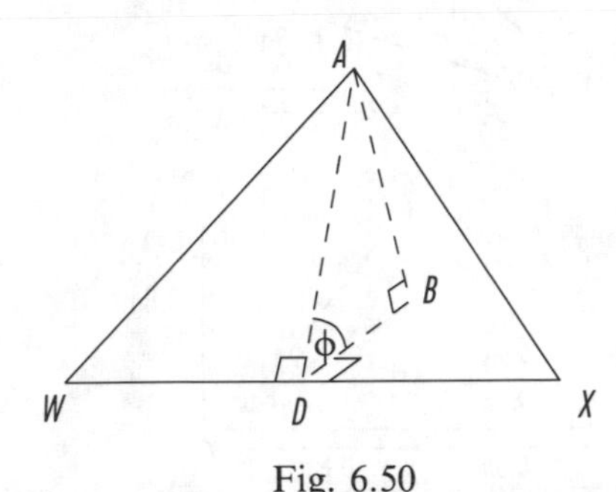

Fig. 6.50

In triangle ABD,

$$\cos\phi = \frac{BD}{AD}$$

i.e. $$\cos\phi = \frac{0.15}{0.497\ldots} = 0.3015$$

$$\therefore \quad \angle\phi = 72.5^\circ$$

is the angle of inclination of both sloping sides AWX and AYZ to the rectangular base.

Our final example in this chapter looks at both these angle techniques. We distinguish between the 2 angles by looking at the inclinations of an edge and a sloping face to the horizontal. Much of the numerical work is omitted and left as an exercise for yourself.

Example 6.15

We have a porch to the front of a house with rectangular floor dimensions of 1.6 m and 2.0 m, drawn and labelled in Fig. 6.51. The height of the porch to the internal ceiling is 2.20 m and the overall height is 3.00 m. Calculate the angle of inclination to the horizontal of AX (and hence AY) and of the sloping hipped roof sections.

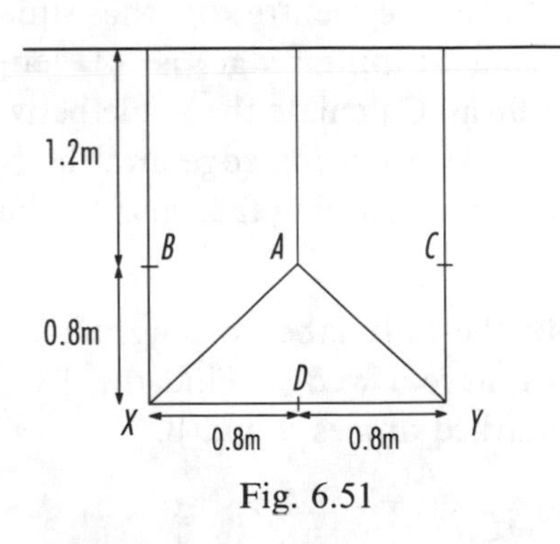

Fig. 6.51

Firstly the height of the roof alone is $3.00 - 2.20 = 0.80$ m. We can project vertically down 0.80 m from A to A' so that A' lies on the horizontal plane of the internal ceiling.

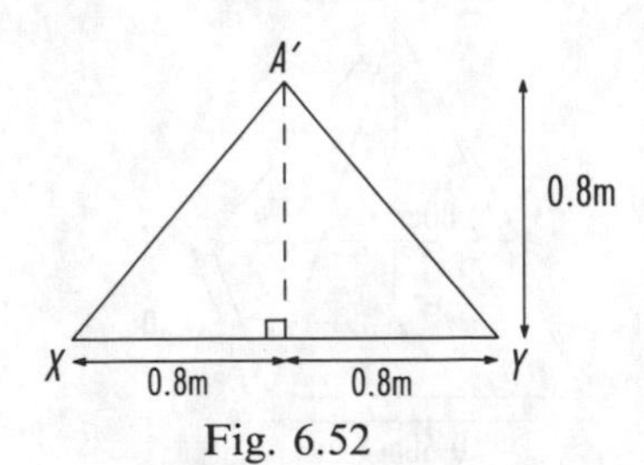

Fig. 6.52

Fig. 6.52 shows $A'XY$ to be an isosceles triangle.

Using Pythagoras' theorem

$$A'X = 1.13\text{ m}.$$

The projection of the sloping edge AX onto the horizontal is $A'X$. With the aid of Fig. 6.53 we have used the tangent ratio to get $\theta = 35.3°$, i.e. to the horizontal the angle of inclination of AX, and AY, is 35.3°.

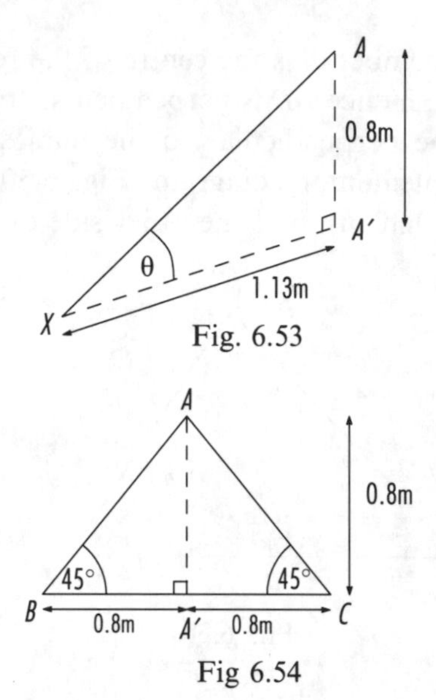

Fig. 6.53

Fig 6.54

Fig. 6.54 shows a cross-sectional triangle through $A'A$ in the form of triangle ABC. To the horizontal the angle of inclination of the similar sloping sides, using tangent, is 45°.

Similarly we use tangent in right-angled triangle $AA'D$. To the horizontal the angle of inclination of side AXY is 45°.

EXERCISE 6.6

1 A pyramid has a horizontal square base $PQRS$ of side 0.50 m. The apex of the pyramid, A, is vertically above the centre of the square. Each sloping edge is of length 1.00 m. Calculate the angle between
i) a sloping edge and the base
and ii) a sloping face and the base.

2 In the fully labelled diagram we have a wedge. Find the 2 marked angles α and β.

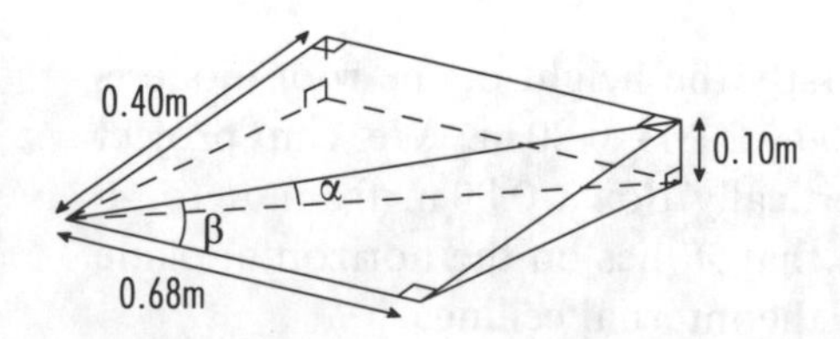

3 The pyramid in the diagram has a square base of side 0.60 m. Its apex is at an altitude of 0.80 m above the centre of the base. Calculate the length of a sloping edge. What are the angles of inclination of each sloping edge and of each sloping face to the horizontal base?

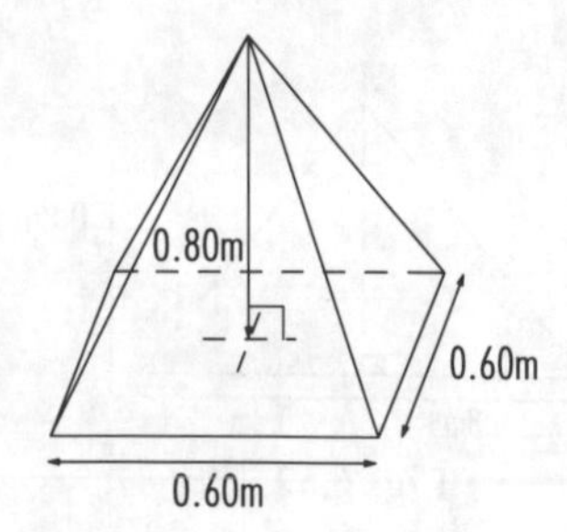

4 In this diagram of a triangular wedge A is vertically above D and B is vertically above E. Both vertical heights are 0.44 m. In triangle ABC $AC = 1.24$ m, $BC = 1.50$ m and $AB = 0.68$ m. What is the area of this triangle? What is the area of triangle CDE? Find $\angle DCE$.

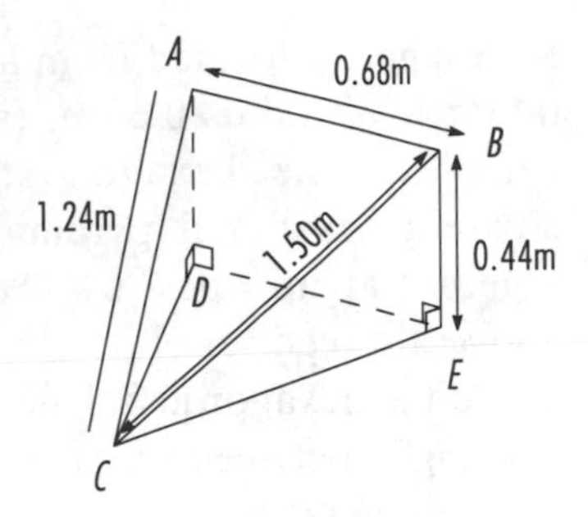

5 A weight is supported by 4 chains of equal length, 1.4 m. All the chains meet at a point to carry the weight. The other ends of the chains form a horizontal square bolted to a network of steel beams. The weight hangs 0.8 m below the square. What is the angle of inclination between a chain and the vertical? What is the length of a side of the square?

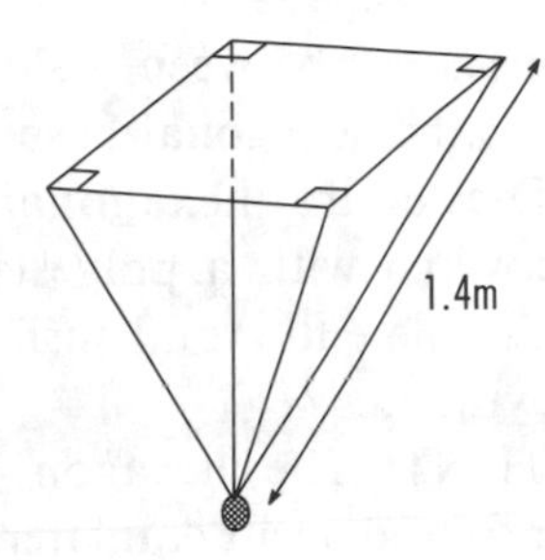

6 The diagram shows a wedge $ABCD$ of four faces each in the shape of a triangle. A is vertically above D. $AB = 0.98$ m, $AC = 1.06$ m, $BC = 0.56$ m and $AD = 0.30$ m. Calculate the total surface area of the wedge.

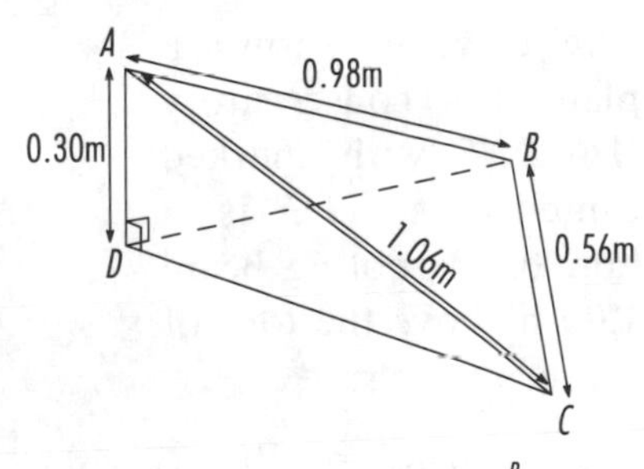

7 The figure shows a hipped roof with a horizontal rectangular base $WXYZ$ measuring 8 m by 14 m. Each roof section is inclined at 39° to the horizontal. Estimate the area of roof to be tiled. Ridge tiles will be used along AB, AW, AX, BY and BZ. Calculate the total length requiring ridge tiles.

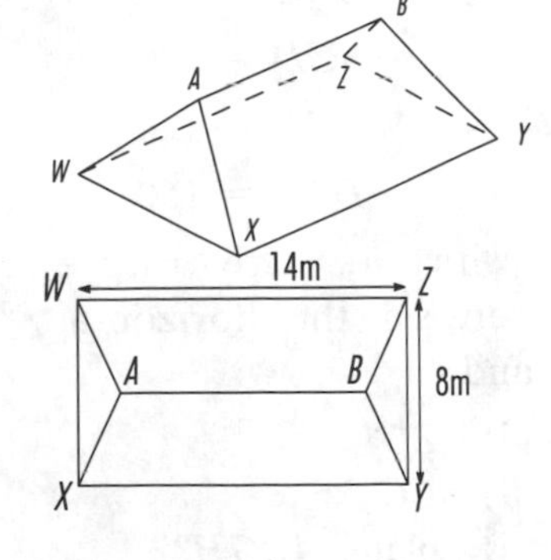

8 $ABCDEF$ is a 1.50 m length of moulding of triangular cross-section. That cross-section forms an equilateral triangle of side 250 mm. Calculate the angle between BE and the plane $ABCD$. G is the mid-point of BC. What is the angle between EG and the plane $ABCD$?

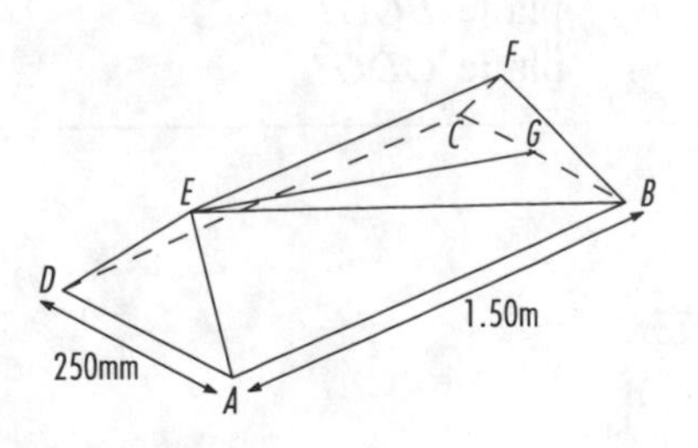

9 A hexagon is formed from a horizontal circle of radius 0.60 m. From each vertex of the hexagon stiff wires create a hexagonal pyramid, joining together at an apex O. O is 0.80 m above the centre of the circle (and hence the hexagon). Calculate the

i) angle between a wire and the hexagonal base,

ii) angle between a sloping face and the hexagonal base.

The entire hexagonal pyramid is covered with a polyester membrane. Calculate the total surface area of the pyramid.

(HINT: The hexagon may be split into 6 adjacent equilateral triangles of side 0.60 m.)

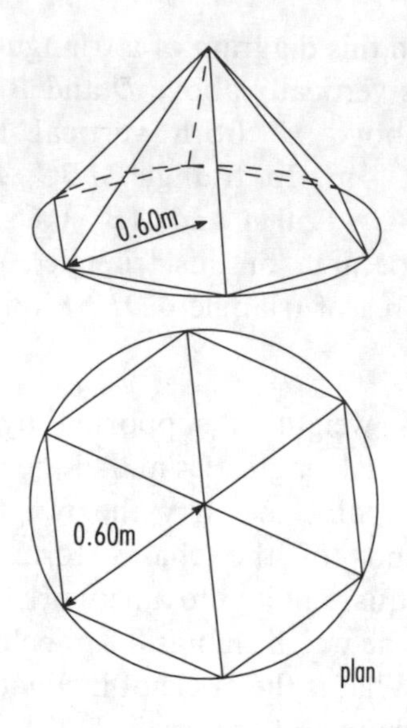

10 The diagram shows a plan of a conservatory $ABCDEF$ with marked dimensions. GH is a horizontal roofing beam 1.00 m above the tops of the side panels.

$$\angle ABC = 150°$$
$$\angle BCD = 120°$$
$$\angle CDE = 120°$$
$$\angle DEF = 150°$$

What is the angle between the horizontal and

i) BG?
ii) CG?
iii) plane $ABGH$?
iv) plane BCG?
v) plane CDG?

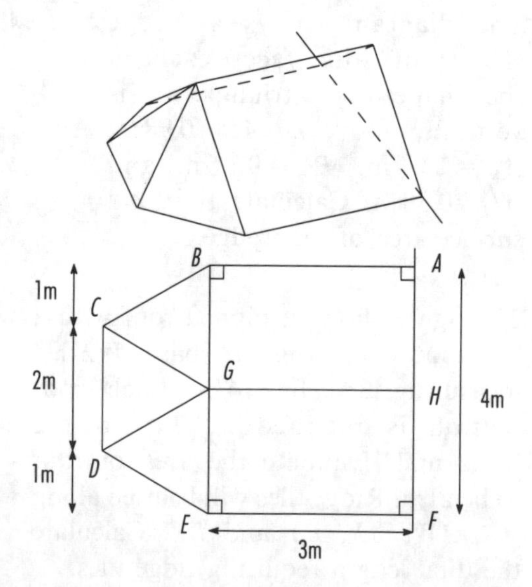

7 Quadratic Equations

The objectives of this chapter are to:

1. Distinguish between a quadratic expression and a quadratic equation.
2. Plot the graph of a quadratic expression.
3. Determine the roots of a quadratic equation by the intersection of the graph with the horizontal axis.
4. Recognise factors of quadratic expressions including $(a+b)^2$, $(a-b)^2$ and $(a+b)(a-b)$.
5. Determine the equation which is satisfied by a given pair of roots.
6. Solve quadratic equations with real roots by factorisation.
7. Recognise that some simple quadratic expressions do not factorise, e.g. a^2+b^2.
8. Solve quadratic equations by the method of completing the square.
9. Derive a formula for solving quadratic equations.
10. Solve quadratic equations using the formula.

Introduction

In Chapter 2 we looked at linear equations and straight line graphs. Example 2.10i) used $y=3x+2$, the highest power of x being 1, i.e. written as x but understood to be x^1. Quadratics move one stage further with the highest power of x being 2, written as x^2, i.e. x squared. There may/may not be other terms involving x and/or pure numbers.

We need to distinguish between slightly different ideas connected with the word **quadratic**. These are **quadratic expressions** and **quadratic equations**.

ASSIGNMENT

The Assignment for this chapter looks at a conservatory to the rear of a house. Its external measurements are 6 m by 4 m. We are interested in putting a concrete path around the 3 sides and calculating the best width for that path.

Quadratic expressions

There is a difference between a **quadratic expression** and a **quadratic equation**. An expression is a list of algebraic terms, preferably written in some order as simply as possible. You cannot do anything really useful with a quadratic expression on its own. On the other hand a quadratic equation, because it is an equation, may be solved.

Example 7.1

This table distinguishes between quadratic expressions and equations. = is the clue to deciding between them.

Quadratic expression	Quadratic equation
x^2+5x+6	$x^2+5x+6=0$
$12x^2-7x-14$	$12x^2-7x-14=0$
$3x^2+8$	$3x^2+8=0$
$3x^2-8$	$3x^2=-8$
$5x^2-10x$	$5x^2-10x=0$
$2x^2$	$2x^2=0$

At first glance it may not always be obvious that an expression is a quadratic expression. For example $(5x+2)(4x-7)$ is a quadratic expression but no term in x^2 is shown. Firstly let us check the meaning of each bracket.

$5x+2$ has two terms, $5x$ and $+2$.

Also $4x-7$ has two terms, $4x$ and -7.

Notice how the sign immediately before the pure number is attached to it. Because there are no signs written before $5x$ and $4x$ these terms are understood to be positive.

Now we will attempt to multiply out the brackets. Each term in the first bracket multiplies each term in the second bracket to give the pattern

$$(5x+2)(4x-7)$$

i.e. $5x$ multiplies $4x$ as $(+)(+) = +$

$$(5)(4) = 20$$

$$(x)(x) = x^2, \text{ all written as } 20x^2.$$

Where possible think of multiplying

SIGNS
NUMBERS
LETTERS

Also $5x$ multiplies -7 to give $-35x$,
2 multiplies $4x$ to give $8x$
and 2 multiplies -7 to give -14.
Thus $(5x+2)(4x-7) = 20x^2-35x+8x-14$
$= 20x^2-27x-14.$

Simplifying like terms.

Examples 7.2

Multiply out the brackets and fully simplify the algebra for

i) $(2+x)(7+3x)$, ii) $(3x-7)(9x-5)$,
iii) $x(7+3x)$, iv) $8x(7+3x)$,
v) $-8x(7+3x)$.

i) For $(2+x)(7+3x)$

2 multiplies 7 to give 14
2 multiplies $3x$ to give $6x$
x multiplies 7 to give $7x$
x multiplies $3x$ to give $3x^2$
i.e. $(2+x)(7+3x) = 14+13x+3x^2$.

Simplifying like terms.

ii) For $(3x-7)(9x-5)$

$3x$ multiplies $9x$ to give $27x^2$
$3x$ multiplies -5 to give $-15x$
-7 multiplies $9x$ to give $-63x$
-7 multiplies -5 to give 35
i.e. $(3x-7)(9x-5) = 27x^2-78x+35$.

Simplifying like terms.

iii) For $x(7+3x)$ the multiplication is easier because the first bracket has been reduced to a simple x term. Then x multiplies each term in the bracket to give

$$x(7+3x)=7x+3x^2.$$

iv) $8x(7+3x)$ follows a pattern similar to the previous example, being increased by a factor of 8 so that

$$8x(7+3x)=56x+24x^2.$$

v) $-8x(7+3x)$ shows the pattern continuing, with the inclusion of a minus sign, so that

$$-8x(7+3x)=-56x-24x^2.$$

The order of terms **within** the brackets can be important if there are some minus signs. However, which bracket is written first does not matter. Exactly the same result of $14-13x+3x^2$ will come from both $(2-x)(7-3x)$ and $(7-3x)(2-x)$.

Examples 7.3

Multiply out the brackets and fully simplify the algebra for

i) $(2+3x)^2$, ii) $(2-3x)^2$,
iii) $(2+3x)(2-3x)$.

i) The whole bracket is squared. This means that the bracket is multiplied by itself to give

$$\begin{aligned}(2+3x)^2 &= (2+3x)(2+3x)\\ &= 4+6x+6x+9x^2\\ &= 4+12x+9x^2.\end{aligned}$$

ii) This example follows a style similar to the previous one,

i.e. $$\begin{aligned}(2-3x)^2 &= (2-3x)(2-3x)\\ &= 4-6x-6x+9x^2\\ &= 4-12x+9x^2.\end{aligned}$$

The answers to these two examples are very much alike, differing only by the $+12x$ and $-12x$ terms. You can expect this type of result again, but do not attempt to learn it according to any complicated rule.

iii) $$\begin{aligned}(2+3x)(2-3x) &= 4-6x+6x-9x^2\\ &= 4-9x^2.\end{aligned}$$

Again you can expect to see this type of result again. Where the brackets look similar, differing only by the middle +/– sign, the x terms disappear.

Examples 7.3 can be generalised by

$$(a+b)^2 = a^2+2ab+b^2$$

and $$(a-b)^2 = a^2-2ab+b^2.$$

You may find it helpful to remember:

"square the first term (a^2), square the second term (b^2), and twice the product of the terms ($+2ab$ or $-2ab$), the results being added".

This is often shortened to:

"square the first, square the second, twice the product".

Also $(a+b)(a-b)=a^2-b^2$

or $(a-b)(a+b)=a^2-b^2$.

Read from the right this is the difference ($-$) of two squares (a^2 and b^2).

ASSIGNMENT

Let us have an initial look at our Assignment. This is not a solution, more a formulation of the problem. Fig. 7.1 is a plan of our conservatory with the external dimensions. The shaded section is our path. We have no known dimension for the path so let it be of width w. There are several methods of obtaining an expression for the area of the path. One is to look at the overall area and the area of the conservatory alone. The difference in these 2 values is the area of the path.

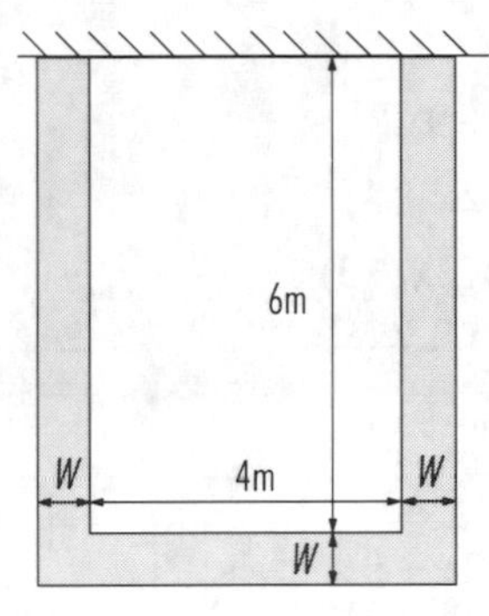

Fig. 7.1

$$\text{Overall area} = (6+w)(4+2w).$$

$$\begin{aligned}\text{Area of conservatory} &= 6\times 4\\ &= 24.\end{aligned}$$

$$\begin{aligned}\text{Path area} &= (6+w)(4+2w)-24\\ &= 24+16w+2w^2-24\\ &= 16w+2w^2.\end{aligned}$$

The next important question to answer is how much concrete? From our limited budget we can afford $1.5\,\text{m}^3$ and expect a wastage of 10%. This leaves us with 90% of $1.5\,\text{m}^3$,

i.e. $\dfrac{90}{100}\times 1.5 = 1.35\,\text{m}^3$.

For this type of path we expect to lay the concrete to a depth of 75 mm (0.075 m) over the hardcore. This means the total path area is to be $\dfrac{1.35}{0.075} = 18\,\text{m}^2$.

We can link together our 2 calculations for path area so that

$$2w^2+16w = 18$$

i.e. $2w^2+16w-18 = 0$

i.e. $w^2+8w-9 = 0$.

Dividing each term by 2.

EXERCISE 7.1

Multiply out the following brackets, giving fully simplified answers.

1 $(x+2)(x+5)$

2 $(5x+1)(2x+1)$

3 $(x+5)(x+2)$

4 $(5x+3)(2x+4)$

5 $(x+5)(3x+2)$

6 $(3x+5)(4x+2)$

7 $(x-2)(x-5)$

8 $(4-x)(3-x)$

9 $(x-4)(x-10)$
10 $(4x-3)(3x-2)$
11 $(3x-2)(4x-7)$
12 $(2x-1)(3x-1)$
13 $(x+2)(x-5)$
14 $(x+4)(x-9)$
15 $(2x+3)(2x-4)$
16 $(2x+4)(x-\frac{1}{2})$
17 $(7x+1)(7x-3)$
18 $(6x+5)(11x-9)$
19 $(x-2)(x+5)$
20 $(x-1)(9x+10)$
21 $(2x-2)(3x+13)$
22 $(11x-1)(2x+8)$
23 $(x-2)(2x+5)$
24 $(3x-2)(2x+3)$
25 $(x+1)^2$
26 $(4x-5)(3x+2)$
27 $(4-x)^2$
28 $(x-1)^2$
29 $(x-4)^2$
30 $(x+5)^2$
31 $(2x+7)^2$
32 $(7-2x)^2$
33 $(x+2)(x-2)$
34 $(2x-13)(2x+13)$
35 $(2x-2)(3x+3)$
36 $x(7+5x)$
37 $4x(7-x)$
38 $-7x(2x+9)$
39 $(5x+4)(2x+3)$
40 $(4x-11)(2x+3)$
41 $(3+2x)(1-6x)$
42 $(9-x)(1-9x)$
43 $(2+x)x$
44 $(1-7x)(2x+9)$
45 $(2x+\frac{1}{2})(3x-6)$
46 $(3+t)(5-2t)$
47 $(2t-7)(3t+7)$
48 $(2-3x)4x$
49 $(2t-7)(7+3t)$
50 $(7t-1)(9+2t)$

Quadratic graphs

Let us introduce these graphs by looking at a particular example of a quadratic expression, x^2+3x+2. At this stage we do not know its likely shape. We may write the three terms x^2+3x+2 in the form

$$y=x^2+3x+2.$$

Just one letter, y, represents the sum of the three terms. y depends on their values for any given values of x.

Before attempting to plot a graph we construct a table of values, using some specimen values of x. In this case the specimen values range from $x=-4$ to $x=2$. The body of the table between the horizontal lines has three rows of working, one for each term. The x^2 row shows each specimen value of x squared, e.g. $(-4)^2=16$. The $3x$ row shows each value of x multiplied by 3, e.g. $3(-4)=-12$. The row labelled 2 has the same constant value all the way along.

x	-4	-3	-2	-1	0	1	2	⟷
x^2	16	9	4	1	0	1	4	
$3x$	-12	-9	-6	-3	0	3	6	
2	2	2	2	2	2	2	2	
y	6	2	0	0	2	6	12	↕

Finally for each x value, the column is added to give the last value, the value of y below the second horizontal line. With the table complete we can plot a graph of y against x. Fig. 7.2 shows the points connected by a smooth curve which dips slightly below the y values of 0. This is to maintain the smoothness of the curve.

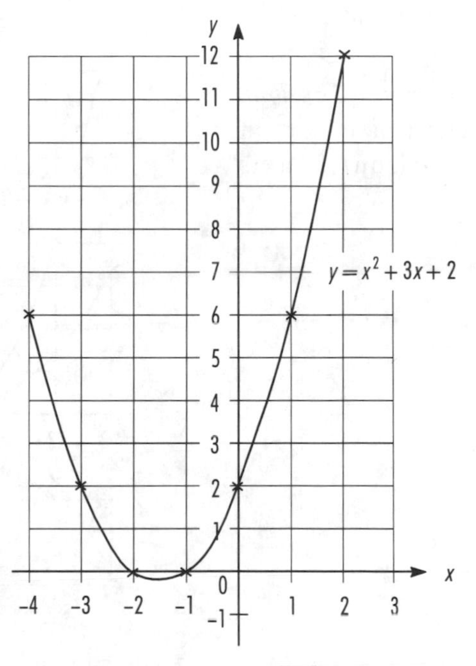

Fig. 7.2

We can draw a smooth curve more easily with more points plotted closer together. In later examples we will use intervals of 0.5 rather than 1.0 for x.

Let us look at a few features of the graph. We are interested in where it crosses the axes, i.e. at $x = -2$, $x = -1$ and $y = 2$. The x values will be more important to us later in this chapter. Remember that the curve crosses the y-axis when $x = 0$

so that $\quad y = x^2 + 3x + 2$

becomes $\quad y = 0^2 + 3(0) + 2 = 2.$

The shape is important. The graph of any quadratic expression is called a **parabola** (Figs. 7.3) and, usually, is either

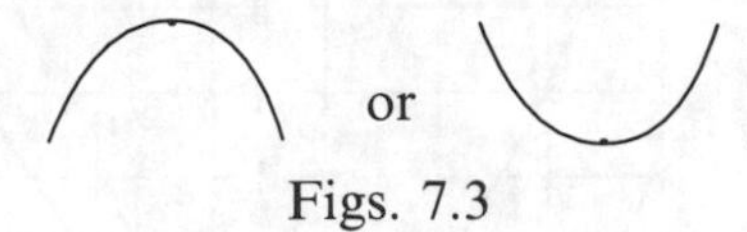

Figs. 7.3

Where the graph turns, at either the peak or trough, is called the **vertex**.

In our example the curve crosses the x-axis but this might not always happen. Other curves may only touch the x-axis or may not cross it at all.

The general form of a quadratic expression can be written as

$$y = ax^2 + bx + c.$$

The letters a, b and c are standard ones, each representing a number that may be positive, negative or perhaps 0. We will look at them in turn.

Firstly let us look at $y=ax^2$ (a being the coefficient of x^2) and let $a=2$, then $a=5$ and finally $a=-2$. Fig. 7.4 shows that as the size of a gets larger so the curve rises more steeply, i.e. $y=5x^2$ rises more steeply than $y=2x^2$. A negative value changes the shape from ∪ to ∩.

In fact $y=-2x^2$ is a reflection of $y=2x^2$ in the horizontal axis (Fig. 7.4).

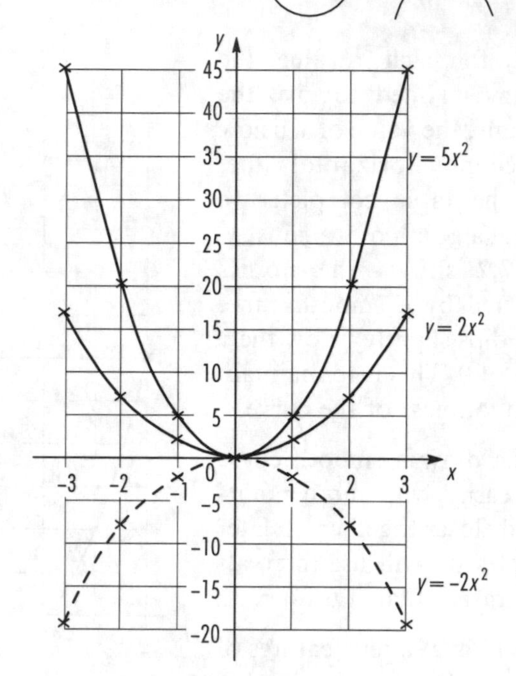

Fig. 7.4

Now let us see what happens as we change the value of b (the coefficient of x) using $y=x^2+bx$. In turn let $b=1$, then $b=3$ and finally $b=-3$. As b increases the vertex moves to the left. When b is negative the vertex is to the right of the vertical axis. This is shown by the 3 examples in Fig. 7.5.

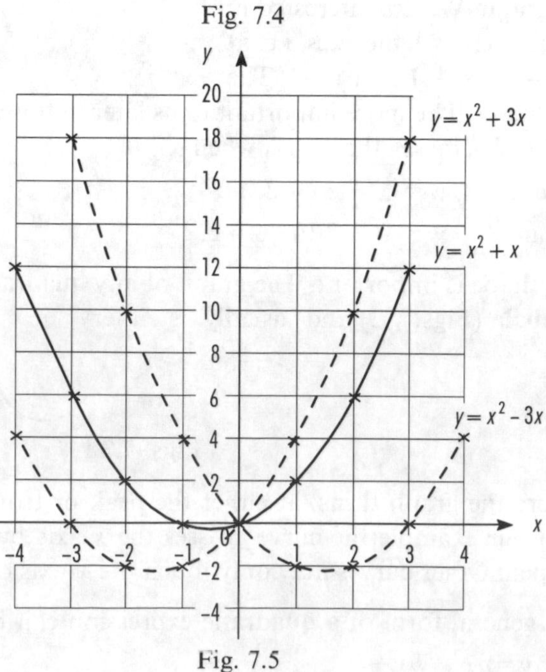

Fig. 7.5

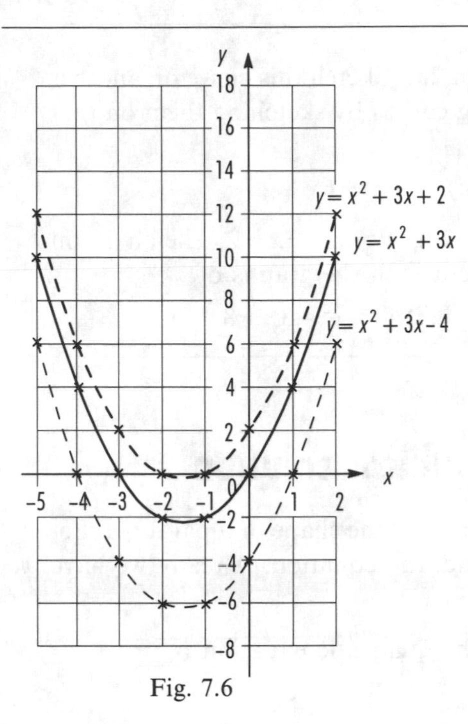

Fig. 7.6

Finally we can see that a change in c (the constant) causes the graph to shift on the vertical axis. Fig. 7.6 shows that as c increases the graph shifts vertically upwards, whilst a decrease shifts it vertically downwards. The value of c corresponds to where the curve cuts the vertical axis. Thus $y=x^2+3x$ passes through the origin, with $c=0$.

$$y = x^2+3x+2$$

$c=2$.

$$y = x^2+3x$$

$c=0$.

$$y = x^2+3x-4$$

$c=-4$.

EXERCISE 7.2

1 Complete the following table and then plot the graph of $y=2x^2+4x-5$.

x	-3.5	-3.0	-2.5	-2.0	-1.5	-1.0	-0.5	0	0.5	1.0
$2x^2$	24.5					2.0			0.5	
$4x$	-14.0						-2.0		2.0	
-5	-5.0		-5.0						-5.0	
y	5.5								-2.5	

2 For each quadratic expression construct a table of values and plot a graph of y against x.

i) $y=x^2$ from $x=-4$ to $x=3$ at intervals of 1.0.

ii) $y=x^2-2x$ from $x=-2$ to $x=5$ at intervals of 1.0.

iii) $y=3x^2+5x$ from $x=-2$ to $x=2$ at intervals of 0.5.

iv) $y=3x^2-5x$ from $x=-2$ to $x=2$ at intervals of 0.5.

v) $y=x^2-3x+2$ from $x=-3$ to $x=3$ at intervals of 0.5.

3 In Question **2i**) you plotted $y=x^2$. Sketch this curve on another pair of axes. Compare the following curves by sketching them on that same set of axes

i) $y=\frac{1}{2}x^2$, ii) $y=-\frac{1}{2}x^2$, iii) $y=4x^2$, iv) $y=-3x^2$.

4 You plotted $y=x^2-2x$ in Question **2ii**). Sketch this curve on another set of axes. Compare the following curves by sketching them on that same set of axes

i) $y=x^2+2x$, ii) $y=x^2-5x$, iii) $y=x^2+6x$.

5 Using the graph from Question **2v**), $y=x^2-3x+2$, sketch it on another set of axes. Now compare it with the graphs of

i) $y=x^2-3x-2$, ii) $y=x^2-3x$, iii) $y=x^2-3x+8$.

Quadratic equations – graphical solution

The previous section helped us to understand the shape of a parabola. Let us link together the graph and the quadratic equation. Already we have drawn the graph (Fig. 7.2) of $y=x^2+3x+2$.

Suppose we wish to use this to solve the quadratic equation

$$x^2+3x+2=0.$$

Patterns within Mathematics encourage us to compare these equations. Both of them have in common x^2, $3x$ and 2. That leaves y from the first and 0 from the second one. The only way for the two relationships to be the same is for $y=0$, which occurs on the x-axis. Therefore if we plot the parabola, $y=x^2+3x+2$, we can use it to solve the quadratic equation, $x^2+3x+2=0$. The solutions are where the graph cuts the horizontal axis. The **solutions** of the quadratic equation are called the **roots** of the equation. Fig. 7.7 shows the roots at $x=-2$ and $x=-1$.

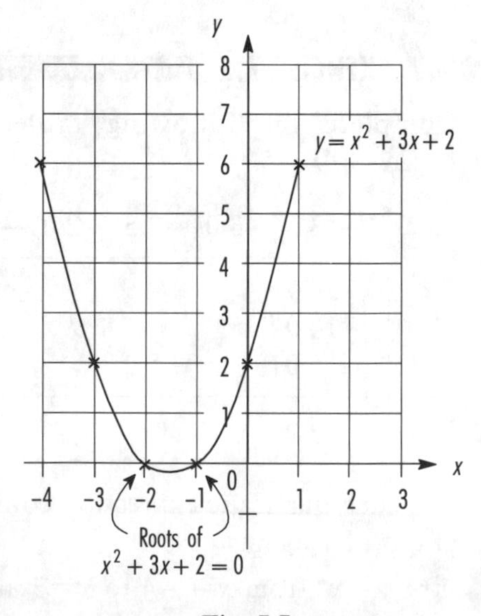

Fig. 7.7

Example 7.4

Graphically solve the quadratic equation $x^2-5x+3.5=0$.

We start by looking at the terms of $y=x^2-5x+3.5$. The coefficient of x^2 is positive which means that the basic shape of the graph is $\cup$. The constant being 3.5 means that the graph cuts the vertical axis at 3.5. At this stage this is as much as we know. At least we have some idea against which to check our plot.

We suggest that we use values of x from 0 to 3.5 at intervals of 0.5. If the curve doesn't cut the horizontal axis twice we can re-assess the situation. If necessary we might then plot some more coordinates. Now let us construct a table of values before attempting the plot.

x	0.0	0.5	1.0	1.5	2.0	2.5	3.0	3.5 ⟷
x^2	0.00	0.25	1.00	2.25	4.00	6.25	9.00	12.25
$-5x$	0.00	−2.50	−5.00	−7.50	−10.00	−12.50	−15.00	−17.50
$+3.5$	3.50	3.50	3.50	3.50	3.50	3.50	3.50	3.50
y	3.50	1.25	−0.50	−1.75	−2.50	−2.75	−2.50	−1.75 ↕

At this stage we might plot y (vertically) against x (horizontally).

We know that the roots of the quadratic equation occur as the curve crosses the horizontal axis, i.e. as $y=0$. As this happens the value of y changes from being positive to being negative, or from being negative to being positive. In the table we see that y changes from 1.25 to −0.50, meaning that there is a root somewhere in this region. The root lies between the corresponding x values of 0.5 and 1.0.

There is no other change in the sign of y so perhaps we should extend our table.

x	4.0	4.5
x^2	16.00	20.25
$-5x$	−20.00	−22.50
$+3.5$	3.50	3.50
y	−0.5	1.25

These extra values show that the sign of y does change again. Hence there is another root somewhere between thc x values of 4.0 and 4.5.

From our graph in Fig. 7.8 we can read off the roots of our quadratic equation. The accuracy of these values depends on the quality of our graph. A high quality plot should get close to roots of $x=0.84$, 4.16.

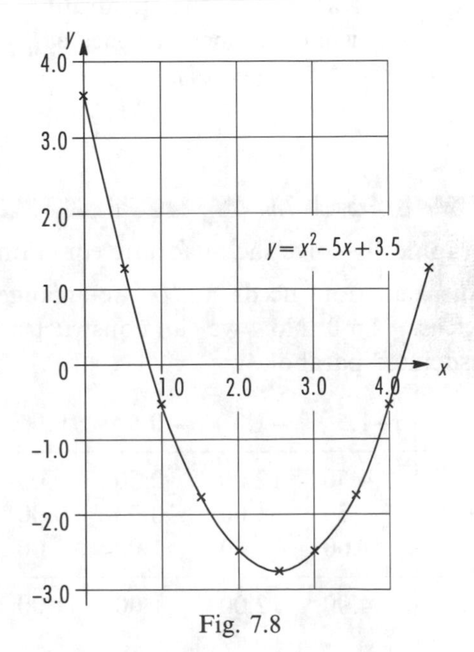

Fig. 7.8

Example 7.5

Graphically solve the quadratic equation $x^2+3x+2.25=0$.

Using the method of Example 7.4 we can plot the graph of $y=x^2+3x+2.25$. In this case the specimen x values are to range from $x=-2.5$ to $x=1.5$ at intervals of 0.5. The table shows the usual method for working out the coordinates.

x	−2.5	−2.0	−1.5	−1.0	−0.5	0.0	0.5	1.0	1.5	⟷
x^2	6.25	4.00	2.25	1.00	0.25	0.00	0.25	1.00	2.25	
$+3x$	−7.50	−6.00	−4.50	−3.00	−1.50	0.00	1.50	3.00	4.50	
$+2.25$	2.25	2.25	2.25	2.25	2.25	2.25	2.25	2.25	2.25	
y	1.00	0.25	0.00	0.25	1.00	2.25	4.00	6.25	9.00	↕

A glance at the table shows no change of sign for y, but a root is obvious immediately. We see $y=0$ corresponds to the root $x=-1.5$.

Let us plot the coordinates in Fig 7.9 and draw a smooth curve through them.

This time, following through the points smoothly, the curve does not actually cross the x-axis. It touches the axis at the point where $x=1.5$, i.e. the horizontal axis is a tangent to the curve at this point. This means that $x=-1.5$ is a root to the quadratic equation, usually termed a **repeated root** or $\boldsymbol{x=-1.5}$ **(repeated)**.

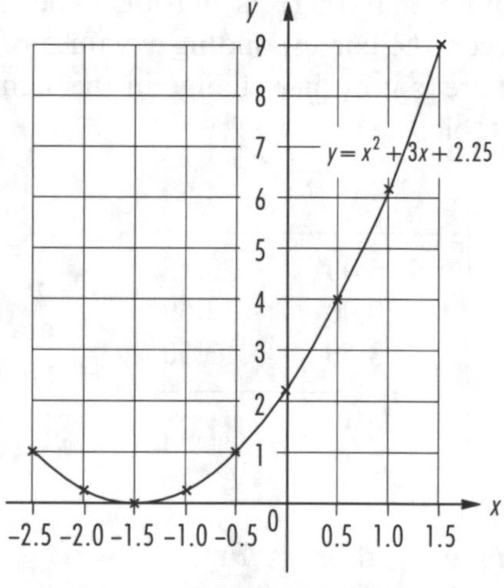

Fig. 7.9

Example 7.6

Graphically solve the quadratic equation $2x^2=-x-1$.

This equation needs to be re-arranged into the correct format of $2x^2+x+1=0$. Now we can construct a table of specimen values for the associated parabola $y=2x^2+x+1$.

x	−1.5	−1.0	−0.5	0.0	0.5	1.0	1.5	2.0	⟷
$2x^2$	4.50	2.00	0.50	0.00	0.50	2.00	4.50	8.00	
$+x$	−1.50	−1.00	−0.50	0.00	0.50	1.00	1.50	2.00	
$+1$	1.00	1.00	1.00	1.00	1.00	1.00	1.00	1.00	
y	4.00	2.00	1.00	1.00	2.00	4.00	7.00	11.00	↕

The y values in the table show no change of sign and no hint of any change. Therefore it is unlikely that plotting more coordinates will be useful. Perhaps the curve doesn't cross the horizontal axis. In fact Fig. 7.10 shows our graph always above the horizontal axis. The $\cup$ shape has its lowest y value a little less than 1.00 with increasing values on both sides of this. Because the graph does **not** cross the horizontal axis there are **no real roots** to the quadratic equation $2x^2+x+1=0$.

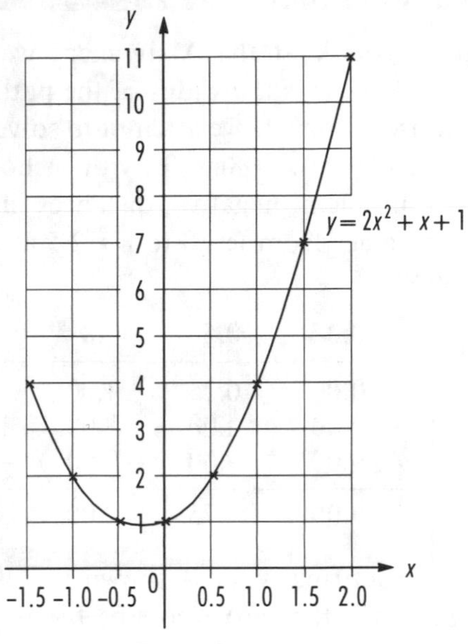

Fig. 7.10

We have looked at 3 possible types of quadratic equations.

i) Curve cutting the x-axis twice, i.e. 2 different roots;
ii) curve touching the x-axis, i.e. 1 repeated root;
iii) curve neither cutting nor touching the x-axis, i.e. no real roots.

For consistency all our curves have been $\cup$ shaped, but exactly the same ideas apply to $\cap$ shaped curves.

Let us look at some features of quadratic equations for a variety of curves in Figs. 7.11 to 7.13.

The general equation is $ax^2+bx+c=0$.

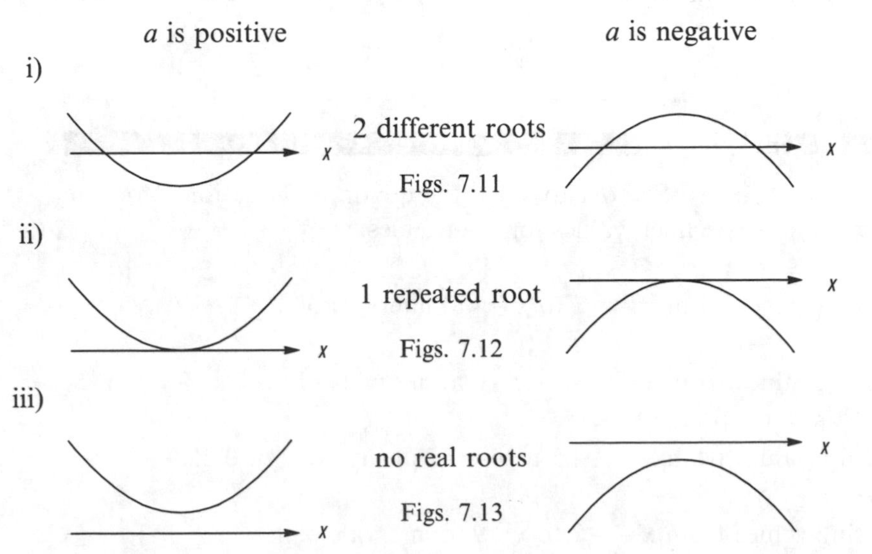

Figs. 7.11

Figs. 7.12

Figs. 7.13

ASSIGNMENT

In our first look at the Assignment we deduced a quadratic equation $w^2+8w-9=0$. w is the width of the path to be concreted around 3 sides of the conservatory. If we attempt to solve this graphically we need to plot the graph of $y=w^2+8w-9$, with w horizontal and y vertical. In this practical problem negative distances have no meaning. We can try specimen values from $w=0$ to $w=2.5$ at intervals of 0.5 according to the table below.

w	0.0	0.5	1.0	1.5	2.0	2.5	⟷
w^2	0.00	0.25	1.00	2.25	4.00	6.25	
$+8w$	0.00	4.00	8.00	12.00	16.00	20.00	
-9	-9.00	-9.00	-9.00	-9.00	-9.00	-9.00	
y	-9.00	-4.75	0.00	5.25	11.00	17.25	↓

Fig. 7.14 confirms what is shown in the table, that the curve crosses the horizontal axis at $w=1.0$.

You can see that our graph shows only 1 crossing of the horizontal axis when we might have expected to get 2 crossings. Also you can appreciate that plotting the curve for negative values of w would give that other crossing point. Remember that negative dimensions have no practical meaning. This leaves our only sensible solution to be a path around the conservatory of width 1 m.

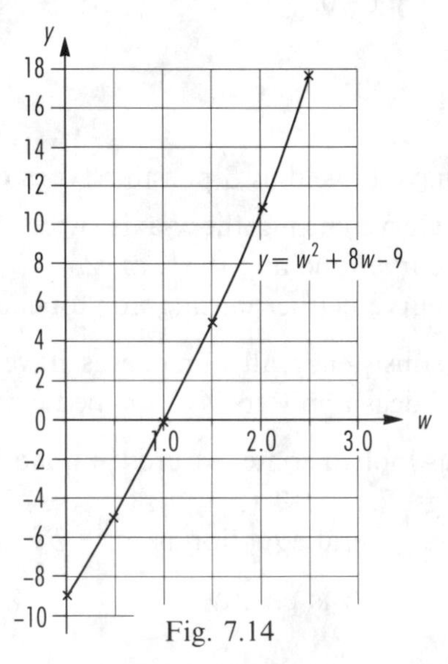

Fig. 7.14

EXERCISE 7.3

In each case plot a parabola to solve the quadratic equation. You are given ranges of specimen values for your tables.

1 $x^2+4x-5=0$
using values from $x=-7$ to $x=4$ at intervals of 1.

2 $x^2-x-6=0$
using values from $x=-4$ to $x=6$ at intervals of 1.

3 $x^2+6x+9=0$
using values from $x=-4.5$ to $x=-2$ at intervals of 0.25.

4 $x^2-7x=0$
using values from $x=-2$ to $x=9$ at intervals of 1.

5 $x^2-25=0$
using values from $x=-6$ to $x=6$ at intervals of 1.

6 $x^2-3x-1=0$
using values from $x=-1$ to $x=4$ at intervals of 0.5.

7 $5x^2-9x-6=0$
using values from $x=-1$ to $x=4$ at intervals of 0.5.

8 $2x^2+3x-3=0$
using values from $x=-3$ to $x=1.5$ at intervals of 0.5.

9 $3x^2+5=8x$
using values from $x=0.5$ to $x=2.0$ at intervals of 0.1.

10 $6x^2-2x+1=0$
using values from $x=-2$ to $x=2$ at intervals of 0.5.

Quadratic equations – solution by factors

The first section of this chapter saw us multiplying out pairs of brackets. By way of example we started with $(x+2)(x+5)$ and reached $x^2+7x+10$. Both $(x+2)$ and $(x+5)$ are factors of the quadratic expression $x^2+7x+10$. Suppose we start with the quadratic equation

$$x^2+7x+10 = 0$$

aiming to factorise it so that

$$(x+2)(x+5) = 0.$$

Always check back – multiply out the brackets.

These two brackets are multiplied together and their overall result is zero. Remember that any value multiplied by zero gives zero as a result. This means that one or other of the brackets is zero,

i.e. $x+2 = 0$ or $x+5 = 0$

i.e. $x = -2$ or $x = -5$.

Often this is shortened to $x=-2, -5$.

Both $x=-2$ and $x=-5$ are roots of the quadratic equation. Reversing the method of factorisation we can create that quadratic equation.

If $x = -2$ or $x = -5$

i.e. $x+2 = 0$ or $x+5 = 0$

and multiplying these results together we get

$$(x+2)(x+5) = 0\times 0$$

i.e. $x^2+7x+10 = 0$.

Whenever we get roots we can check that they are correct. Just substitute them into the original equation. Separately if we substitute -2 and -5 for x in $x^2+7x+10$ we should get a result of 0,

i.e. when $x=-2$, $(-2)^2+7(-2)+10 = 4-14+10 = 0$
and when $x=-5$, $(-5)^2+7(-5)+10 = 25-35+10 = 0$.

When factorising a quadratic equation (or expression) we think about the **products** we will get. They are from the first terms in the brackets and the second terms,

i.e. $(x+2)(x+5)$.

These are important because $x \times x = x^2$ and $2 \times 5 = 10$ are the first and last terms of $x^2+7x+10$. Also $2+5=7$ which is the coefficient of x in the middle term.

There are other factors of 10. Because $10 \times 1 = 10$ we have 10 and 1 as those factors. However $10+1=11$ and we know that we need $7x$, not $11x$, as the middle term. There is some trial and error in discovering which pair of factors are the correct ones. The +/– signs in a quadratic equation (or expression) are also important. Let us demonstrate the factors and signs using the next set of examples.

Examples 7.7

By factorisation solve the quadratic equations

i) $x^2+10x+21=0$, ii) $x^2-10x+21=0$,
iii) $x^2+4x-21=0$, iv) $x^2-4x-21=0$.

These examples are similar to show various combinations of factors. Firstly they all start with x^2, and we know that $x \times x = x^2$, i.e. x and x are factors of x^2. They all finish on the left-hand side with 21. The factors of 21 are 1, 3, 7, and 21 combined as $1 \times 21 = 21$
and $3 \times 7 = 21$.

i) In $x^2+10x+21=0$ we have $+21$. The + sign means both factor brackets have the **same** sign. The + sign of $+10x$ means both factor signs are +. So we have

$$(x+\ \ )(x+\ \ ) = 0.$$

The next step is to look at the addition (+) of factors of 21,

i.e. $1+21 = 22$

and $3+7 = 10$.

These results and the $10x$ mean we choose 3 and 7 to give

$$(x+3)(x+7) = 0.$$

Then $x+3 = 0$ or $x+7 = 0$

i.e. $x = -3$ or $x = -7$

i.e. $x = -3, -7$.

ii) In $x^2-10x+21=0$ the + of $+21$ means both factor brackets have the **same** sign. The – of $-10x$ means the signs are –. So we have

$$(x-\ \ )(x-\ \ ) = 0.$$

Looking at the factors of 21 again we choose 3 and 7 because $-3-7=-10$ to give

$$(x-3)(x-7) = 0.$$

Then $x-3 = 0$ or $x-7 = 0$

i.e. $x = 3$ or $x = 7$

i.e. $x = 3, 7.$

iii) In $x^2+4x-21=0$ the $-$ of -21 means the factor brackets have **different** signs, i.e.

$$(x+\ \ )(x-\ \ ) = 0.$$

We get the same result using $(x-\ \ )(x+\ \ )$.

For the next step with factors of 21 we look at their **difference**:

$$1-21 = -20 \quad \text{and} \quad 21-1 = 20,$$

$$3-7 = -4 \quad \text{and} \quad 7-3 = 4.$$

The last option, giving 4, agrees with the middle term's coefficient so that

$$(x+7)(x-3) = 0.$$

Then $x+7 = 0$ or $x-3 = 0$

i.e. $x = -7$ or $x = 3$

i.e. $x = -7, 3.$

iv) In $x^2-4x-21=0$ we make decisions similar to those of the previous example. We choose the third option of $3-7=4$ to agree with the middle term's coefficient so that

$$(x+3)(x-7) = 0.$$

Then $x+3 = 0$ or $x-7 = 0$

i.e. $x = -3$ or $x = 7$

i.e. $x = -3, 7.$

For each example remember to substitute each root into it's original quadratic equation. This check should give an answer of 0.

Examples 7.8

By factorisation solve the quadratic equations

i) $x^2+14x+24=0$, ii) $x^2-11x+24=0$,
iii) $x^2+5x-24=0$, iv) $x^2-2x-24=0$.

All these examples have 24 on the left-hand side. The factors of 24 combine as 1×24, 2×12, 3×8 and 4×6.

i) In $x^2+14x+24=0$ the $+$ sign of $+24$ means the same signs for the factor brackets. The $+$ sign of $+14x$ means both the factor signs are $+$.

Now $1+24 = 25$,

$2+12 = 14$,

$3+8 = 11$

and $4+6 = 10$.

The second option agrees with the middle term's coefficient. Combining all these decisions we get

$(x+2)(x+12) = 0$

i.e. $x+2 = 0$ or $x+12 = 0$

i.e. $x = -2$ or $x = -12$

i.e. $x = -2, -12$.

ii) In $x^2-11x+24=0$ we have the same four pairs of factor brackets based on x^2 and 24. The + sign of $+24$ and the − sign of $-11x$ mean the factor bracket signs are the same and negative.

Now $-1-24 = -25$

$-2-12 = -14$

$-3-8 = -11$

and $-4-6 = -10$.

This time the third option agrees with the middle term's coefficient. Combining all these decisions we get

$(x-3)(x-8)=0$

i.e. $x-3 = 0$ or $x-8 = 0$

i.e. $x = 3$ or $x = 8$

i.e. $x = 3, 8$.

iii) In $x^2+5x-24=0$ the − sign of -24 indicates different factor bracket signs. Hence we are looking at the differences of the numbers,

$1-24 = -23$, $24-1 = 23$,

$2-12 = -10$, $12-2 = 10$,

$3-8 = -5$, $8-3 = 5$, *

** $4-6 = -2$, $6-4 = 2$.

The * option agrees with the middle term's coefficient. Combining all these decisions we get

$(x+8)(x-3) = 0$

i.e. $x+8 = 0$ or $x-3 = 0$

i.e. $x = -8$ or $x = 3$

i.e. $x = -8, 3$.

iv) For the solution of $x^2-2x-24=0$ we use the previous example as a basis and choose the ** option to give

$$(x+4)(x-6) = 0$$

i.e. $x+4 = 0$ or $x-6 = 0$

i.e. $x = -4$ or $x = 6$

i.e. $x = -4, 6.$

For each example remember to substitute each root into its original quadratic equation. This check should give an answer of 0.

ASSIGNMENT

Now we have enough factorisation skill to look at our assignment problem again. Our assignment is based on the quadratic equation $w^2+8w-9=0$. Remember that w is the width of our concrete path around 3 sides of the conservatory. The $-$ sign of -9 means there are different signs in the factor brackets. Now the difference of those numbers is to be $+8$. Factors of 9 are 9×1 and 3×3. We choose the first option and the difference because $9-1=8$;

i.e. $(w+9)(w-1) = 0$

i.e. $w+9 = 0$ or $w-1 = 0$

i.e. $w = -9$ or $w = 1.$

Because this is a practical problem we need to interpret our answers. Path widths can only be positive and so we need the value $w=1$ m.

EXERCISE 7.4

By factorisation solve the following quadratic equations.

1 $x^2+6x+5=0$

2 $x^2+5x+6=0$

3 $x^2+7x+12=0$

4 $x^2+3x+2=0$

5 $x^2+14x+49=0$

6 $x^2-5x+6=0$

7 $x^2-12x+27=0$

8 $x^2-11x+30=0$

9 $x^2-10x+25=0$

10 $x^2-8x+12=0$

11 $x^2+4x-5=0$

12 $x^2+4x-12=0$

13 $x^2+2x-8=0$

14 $x^2+7x-18=0$

15 $x^2+13x-48=0$

16 $x^2-2x-15=0$

17 $x^2-3x-10=0$

18 $x^2-2x-48=0$

19 $x^2-x-56=0$

20 $x^2-2x-63=0$

21 $x^2+6x-40=0$

22 $x^2+3x-70=0$

23 $x^2+20x+100=0$

24 $x^2-6x+9=0$

25 $x^2-x-72=0$

26 $x^2-x-12=0$

27 $x^2+6x-72=0$

28 $x^2+5x-36=0$

29 $x^2-18x-40=0$

30 $x^2-11x=60$

So far in this section the coefficient of x^2 has always been 1. This might not always happen but the basic ideas about +/− signs and factors continue to apply. Any coefficient change just means we have to check more combinations of factors.

Examples 7.9

By factorisation solve the quadratic equations
i) $2x^2+9x+10=0$, ii) $3x^2-14x+8=0$,
iii) $4x^2-5x-6=0$.

i) All our signs are + which means both the factor bracket signs are +. Now we need to look at the factors of $2x^2$ and 10. Firstly, with x in each factor, the factors of $2x^2$ are x and $2x$. This gives us

$$(x+\quad)(2x+\quad) = 0.$$

Previously at this stage the brackets were the same so order was not important. Different brackets now mean we must be more careful. The factors of 10 are

$$10\times 1, \quad 1\times 10,$$
$$5\times 2, \quad 2\times 5.$$

By trial and error we test these options in turn, i.e.

$$(x+10)(2x+1), \qquad (x+1)(2x+10),$$
$$(x+5)(2x+2), \qquad (x+2)(2x+5).$$

In each case we multiply out the brackets and collect together the terms. The first term is $2x^2$ and the last term is 10. Only the middle term varies. You should check for yourself that those middle terms are $21x$, $12x$, $12x$ and $9x$. Therefore we need the fourth option, i.e.

$$2x^2+9x+10 = 0$$

factorises to $(x+2)(2x+5) = 0$

i.e. $x+2 = 0$ or $2x+5 = 0$

i.e. $x = -2$ or $2x = -5$

i.e. $x = -2, -\dfrac{5}{2}$

(or -2, -2.5).

ii) The signs in $3x^2-14x+8=0$ mean both factor brackets contain −. The factors of $3x^2$ are x and $3x$ so we have

$$(x-\quad)(3x-\quad) = 0.$$

The factors of 8 are 8×1, 1×8, 2×4 and 4×2. By trial and error we test these options in turn, i.e.

$$(x-8)(3x-1), \qquad (x-1)(3x-8),$$
$$(x-2)(3x-4), \qquad (x-4)(3x-2).$$

In each case, when we multiply out the brackets and collect the terms the first term is $3x^2$ and the last term is 8. You should check for yourself that the middle terms of these options in turn are $-25x$, $-11x$, $-10x$ and $-14x$.

Thus we need the fourth option, i.e.

$$3x^2-14x+8 = 0$$

factorises to $\quad (x-4)(3x-2) = 0$

i.e. $\quad x-4 = 0 \quad$ or $\quad 3x-2 = 0$

i.e. $\quad x = 4 \quad$ or $\quad 3x = 2$

i.e. $\quad x = 4, \frac{2}{3}.$

iii) In $4x^2-5x-6=0$ the $-$ of -6 means the factor brackets have different signs. The possible factors of $4x^2$ are either $4x$ and x or $2x$ and $2x$. The possible factors of 6 are either 1 and 6 or 2 and 3. By trial and error we test all these options in turn, i.e.

$$(2x+1)(2x-6), \qquad (2x-1)(2x+6),$$
$$(2x+2)(2x-3), \qquad (2x-2)(2x+3),$$
$$(4x+1)(x-6), \qquad (4x-1)(x+6),$$
$$(4x+2)(x-3), \qquad (4x-2)(x+3),$$
$$* \ (4x+3)(x-2), \qquad (4x-3)(x+2),$$
$$(4x+6)(x-1), \qquad (4x-6)(x+1).$$

The * option, when multiplied out, gives the correct middle term of $-5x$ so that

$$4x^2-5x-6 = 0$$

factorises to $\quad (4x+3)(x-2) = 0$

i.e. $\quad 4x+3 = 0 \quad$ or $\quad x-2 = 0$

i.e. $\quad 4x = -3 \quad$ or $\quad x = 2$

i.e. $\quad x = -\frac{3}{4}, 2$

(or -0.75, 2).

For each example remember to substitute each root into its original quadratic equation. This check should give an answer of 0.

Not all quadratics factorise easily. Quadratic expressions (and equations) of the type a^2+b^2 will not factorise. The hint is the 2 square terms both being positive, (though both being negative would also fail). For example $x^2+9=0$ has no real roots as we will see in Examples 7.11.

We will look at other cases in the next section.

EXERCISE 7.5

By factorisation solve the quadratic equations.

1 $3x^2+x-2=0$

2 $4x^2-4x+1=0$

3 $6x^2-x-2=0$

4 $3x^2-x-2=0$

5 $2x^2+7x+6=0$

6 $2x^2+13x+15=0$

7 $3x^2+8x-3=0$

8 $2x^2-7x+6=0$

9 $2x^2+11x+5=0$

10 $6x^2+7x-20=0$

11 $6x^2-11x-7=0$

12 $6x^2-13x+6=0$

13 $15x^2-x-2=0$

14 $3x^2+13x+4=0$

15 $4x^2-12x+9=0$

16 $25x^2-30x+9=0$

17 $8x^2+2x-15=0$

18 $6x^2+13x+6=0$

19 $9x^2+12x+4=0$

20 $8x^2+45x-18=0$

During solution by factorisation we have looked at quadratic equations with all three terms, i.e. a term in x^2, a term in x and a constant. They combine to equal zero. There may be cases where either the x term or the constant is missing. These types have easy solutions.

Examples 7.10

By factorisation solve the quadratic equations

i) $2x^2+5x=0$, ii) $5x^2-4x=0$,

iii) $9x^2-6x=0$.

All these examples have the same feature: only two terms. We are looking for a factor that is common to both terms. The method does **not** use division through by x. You will recall that division by 0 is **not allowed** in Mathematics. Each equation has a root $x=0$ and so care is needed.

i)

$$2x^2+5x = 0$$

factorises to $x(2x+5) = 0$.

x is the factor common to both terms.

Either $x = 0$ or $2x+5 = 0$

i.e. $x = 0$ or $2x = -5$

i.e. $x = 0, -2.5$.

ii)

$$5x^2-4x = 0$$

factorises to $x(5x-4) = 0$.

Either $x = 0$ or $5x-4 = 0$

i.e. $x = 0$ or $5x = 4$

i.e. $x = 0, 0.8$.

iii) $9x^2 - 6x = 0$

i.e. $3x(3x-2) = 0.$

3x is the factor common to both terms.

Either $3x = 0$ or $3x - 2 = 0$

i.e. $x = 0$ or $3x = 2$

i.e. $x = 0, \frac{2}{3}.$

For each example remember to substitute each root into its original quadratic equation. This check should give an answer of 0.

Examples 7.11

Solve the quadratic equations

i) $x^2 - 25 = 0,$ ii) $2x^2 - 36 = 0,$

iii) $x^2 + 9 = 0.$

You will see this is an alternative method to factorisation in the first and second examples.

i) $x^2 - 25 = 0$

becomes $x^2 = 25$

i.e. $x = \pm\sqrt{25}.$

Square root of both sides.

Remember we need to include the negative solution. A negative value squared, e.g. $(-5)^2 = 25$, gives a positive result,

i.e. $x = \pm 5,$

or $x = -5, 5.$

ii) $2x^2 - 36 = 0$

becomes $2x^2 = 36$

i.e. $x^2 = 18$

so that $x = \pm\sqrt{18}$

i.e. $x = \pm 4.243$ (3 decimal places)

or $x = -4.243, 4.243.$

iii) $x^2 + 9 = 0$

starts off its solution in the same way with

$x^2 = -9$

i.e. $x = \pm\sqrt{(-9)}$

This is beyond our skill at the moment because we cannot find the square root of a negative number. There are no real solutions to this quadratic equation, i.e. there are no real roots.

EXERCISE 7.6

Solve the quadratic equations.

1 $x^2+4x=0$

2 $3x^2-8x=0$

3 $\frac{1}{2}x^2-2x=0$

4 $9x^2-100=0$

5 $4x^2-4x=0$

6 $9x^2+30x=0$

7 $6x^2=18x$

8 $2x^2+7=0$

9 $(x-3)^2=0$

10 $(2x+1)^2=25$

Quadratic equations – solution using the formula

We know that not all quadratic equations (and expressions) factorise. Also some might not factorise easily so this alternative method uses a formula. It is easy to use, simply a matter of substituting the correct values and using a calculator.

We need to remind ourselves of a particular relation

$$(x+\alpha)^2 = x^2+2\alpha x+\alpha^2.$$

We will deduce the formula from the general quadratic equation

$$ax^2+bx+c = 0.$$

Alongside we will work a numerical example so that you may compare the letters and numbers. Remember that a, b, and c represent numbers in the general equation. It is possible that a may be 1. Also if b or c is zero then an easier solution follows like Examples 7.10 or 7.11.

The method we use is called "**completing the square**".

$$ax^2+bx+c = 0 \qquad\qquad 2x^2-7x-2 = 0$$

$$ax^2+bx = -c \qquad\qquad 2x^2-7x = 2$$

$$x^2+\frac{b}{a}x = \frac{-c}{a} \qquad\qquad x^2-\frac{7}{2}x = 1$$

We compare $x^2+\frac{b}{a}x$ and $x^2-\frac{7}{2}x$ with part of the relation $x^2+2\alpha x+\alpha^2$. Both cases have x^2 and a term in x. However the constant term, represented by α^2, is missing. We can create this term and add it to both sides of each equation so maintaining their balance.

Now $\qquad 2\alpha = \frac{b}{a} \qquad$ or $\qquad 2\alpha = -\frac{7}{2}$

i.e $\qquad \alpha = \frac{b}{2a} \qquad\qquad \alpha = -\frac{7}{4}$

Hence $\qquad \alpha^2 = \left(\frac{b}{2a}\right)^2 \qquad\qquad \alpha^2 = \left(-\frac{7}{4}\right)^2$

To create the correct format we add this to both sides of each equation. Remember that by adding to both sides we maintain the balance of each equation.

$$x^2+\frac{bx}{a}+\left(\frac{b}{2a}\right)^2=\left(\frac{b}{2a}\right)^2-\frac{c}{a} \qquad x^2-\frac{7}{2}x+\left(-\frac{7}{4}\right)^2=\left(-\frac{7}{4}\right)^2+1.$$

In each case we have completed the square on the left-hand side. It will factorise into one bracket all squared to give

$$\left(x+\frac{b}{2a}\right)^2=\frac{b^2}{4a^2}-\frac{c}{a} \qquad \left(x-\frac{7}{4}\right)^2=\frac{49}{16}+1$$

$$\left(x+\frac{b}{2a}\right)^2=\frac{b^2-4ac}{4a^2} \qquad \left(x-\frac{7}{4}\right)^2=\frac{49+16}{16}$$

$$x+\frac{b}{2a}=\pm\sqrt{\frac{b^2-4ac}{4a^2}} \qquad x-\frac{7}{4}=\pm\sqrt{\frac{65}{16}}$$

$$x+\frac{b}{2a}=\frac{\pm\sqrt{b^2-4ac}}{2a} \qquad x-\frac{7}{4}=\frac{\pm\sqrt{65}}{4}$$

There is always a pattern of 2 denominators that are the same ($2a$ in the general one and 4 in this particular example) in each equation.

$$x=\frac{-b}{2a}\frac{\pm\sqrt{b^2-4ac}}{2a} \qquad x=\frac{7}{4}\frac{\pm\sqrt{65}}{4}$$

$$x=\frac{-b\pm\sqrt{b^2-4ac}}{2a} \qquad x=\frac{7\pm\sqrt{65}}{4}$$

$$x=\frac{7+8.0623}{4},\ \frac{7-8.0623}{4}$$

$$x=\frac{15.0623}{4},\ \frac{-1.0623}{4}$$

$$x=3.766,\ -0.266.$$

The next set of examples shows how to use the formula. Just remember to substitute correctly for a, b and c with any relevant $-$ signs.

Examples 7.12

Use the formula to solve the quadratic equations

i) $3x^2-2x-12=0$, ii) $x^2+5x-19=0$.

i) We compare $3x^2-2x-12=0$ with the general quadratic equation $ax^2+bx+c=0$.

This gives $a=3$, $b=-2$, $c=-12$.

Now $$x=\frac{-b\pm\sqrt{b^2-4ac}}{2a}$$

becomes $$x = \frac{-(-2) \pm \sqrt{(-2)^2 - 4(3)(-12)}}{2(3)}$$

$$x = \frac{2 \pm \sqrt{4 + 144}}{6}$$

$$x = \frac{2 \pm \sqrt{148}}{6}$$

$$x = \frac{2 + 12.166}{6}, \frac{2 - 12.166}{6}$$

$$x = 2.36, \ -1.69.$$

ii) We compare $x^2 + 5x - 19 = 0$ with the general quadratic equation $ax^2 + bx + c = 0$.

This gives $a = 1$, $b = 5$, $c = -19$.

Now $$x = \frac{-b \pm \sqrt{b^2 - 4ac}}{2a}$$

becomes $$x = \frac{-5 \pm \sqrt{5^2 - 4(1)(-19)}}{2(1)}$$

$$x = \frac{-5 \pm \sqrt{25 + 76}}{2}$$

$$x = \frac{-5 \pm \sqrt{101}}{2}$$

$$x = \frac{-5 + 10.0499}{2}, \quad \frac{-5 - 10.0499}{2}$$

$$x = 2.52, \ -7.52.$$

For each example remember to substitute each root into its original quadratic equation. This check should give an answer of 0. More decimal places used means a more accurate check.

Examples 7.13

Use the formula to solve the quadratic equations

i) $x^2 + 3x + 2.25 = 0$, ii) $x^2 + 2x + 5.5 = 0$.

i) The values for substitution are $a = 1$, $b = 3$, $c = 2.25$,

so that $$x = \frac{-b \pm \sqrt{b^2 - 4ac}}{2a}$$

becomes $$x = \frac{-3 \pm \sqrt{3^2 - 4(1)(2.25)}}{2(1)}$$

$$x = \frac{-9 \pm \sqrt{9-9}}{2}$$

$$x = \frac{-3}{2}$$

i.e. $x = -1.5$.

You can see this solution leads to $\sqrt{0}$ and then only gives one solution for x. This is the case of a repeated root, i.e. $x = -1.5$ repeated.

ii) The values for substitution are $a = 1$, $b = 2$, $c = 5.5$

so that $$x = \frac{-b \pm \sqrt{b^2 - 4ac}}{2a}$$

becomes $$x = \frac{-2 \pm \sqrt{2^2 - 4(1)(5.5)}}{2(1)}$$

$$x = \frac{-2 \pm \sqrt{-18}}{2}.$$

We can go no further with this solution because there are no real answers to the square root of a negative number. This means this quadratic equation has no real roots.

We can look a little more closely under the $\sqrt{}$ in the general formula

$$x = \frac{-b \pm \sqrt{b^2 - 4ac}}{2a}.$$

It is possible to look at 3 solution types which we can link with our earlier graph work.

i) $b^2 - 4ac > 0$, i.e. $b^2 > 4ac$

is the usual situation leading to 2 different roots. The graph crosses the horizontal axis twice (Figs. 7.15).

Figs. 7.15

ii) $b^2 - 4ac = 0$, i.e. $b^2 = 4ac$

leads to one repeated root. The graph touches the horizontal axis, i.e. that axis acts as a tangent (Figs. 7.16).

Figs. 7.16

iii) $b^2 - 4ac < 0$, i.e. $b^2 < 4ac$

leads to no real roots. The graph does not cross the horizontal axis, i.e. the graph and axis do not intersect (Figs. 7.17).

Figs. 7.17

ASSIGNMENT

We return to our conservatory project and the concrete path going around the 3 sides. Suppose the hardcore is rather uneven meaning that in total we waste slightly more concrete than before. If now we lose 15% this leaves us with 85% of our $1.5\,\text{m}^3$ delivery, i.e. $1.5 \times \frac{85}{100} = 1.275\,\text{m}^3$.

Remember that the depth of concrete is to be 75 mm which gives us a total area of $\frac{1.275\,\text{m}^3}{75\,\text{mm}} = 17\,\text{m}^2$.

In the quadratic equation we replace our original 18 with 17

$$\begin{aligned} \text{so that} \quad 2w^2 + 16w &= 17 \\ \text{i.e} \quad 2w^2 + 16w - 17 &= 0 \\ w &= \frac{-16 \pm \sqrt{16^2 - 4(2)(-17)}}{2(2)} \\ w &= \frac{-16 \pm \sqrt{392}}{4} \\ w &= \frac{-16 + 19.799}{4}, \quad \frac{-16 - 19.799}{4} \\ w &= 0.95 \text{ only.} \end{aligned}$$

The second answer, being negative, has no physical meaning relevant to our concrete path. We can have only positive widths, in this case 0.95 m.

EXERCISE 7.7

Use the formula to solve the quadratic equations.

1 $x^2 + 20x - 34 = 0$

2 $x^2 - x - 1 = 0$

3 $x^2 - 5x - 12 = 0$

4 $x^2 - 28x - 18 = 0$

5 $x^2 + 3x - 2 = 0$

6 $x^2 - 6x - 3 = 0$

7 $x^2 - 12x + 36 = 0$

8 $3x^2 + 14x + 2 = 0$

9 $2x^2 + 4x + 5 = 0$

10 $15x^2 - 9x + 1 = 0$

Our final exercise contains problems that need solving, rather than simply quadratic equations. If you can spot the factors of the quadratic equation in a question the solution will be quick. However, it is more likely you will need to use the formula and your calculator.

EXERCISE 7.8

1 The power dissipated by a circuit component is 6W. It is connected in series with a resistor of 15Ω and an emf of 25V, as shown in the diagram.
Calculate the current, I, if $15I^2 - 25I + 6 = 0$.

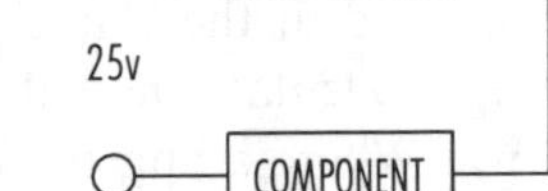

2 $s = ut + \frac{1}{2}at^2$ refers to the motion of a vehicle.
s is the displacement,
u is the initial velocity,
a is the acceleration
and t is the time for the motion.
Calculate the time for the motion over a displacement of 70 m if the initial velocity is 1.56 ms^{-1} and the acceleration is 2.12 ms^{-2}.

3 Apart from the units, the area of a circle ($A = \pi r^2$) and its circumference ($C = 2\pi r$) are equal. What is the length of the radius, r? The area of the circle is now increased so that, apart from the units again, the area is 3 times the circumference; i.e. $A = 3C$. Calculate the length of the new radius.

4 Your company needs some new equipment and you decide to borrow £75 000 repayable over 2 years. The bank tells you the total repayments will be £105 000. The relevant compound interest formula is

$$105\,000 = 75\,000\left(1 + \frac{R}{100}\right)^2$$

where R is the rate of interest you will be paying.
Calculate this value of R.

5 The voltage (V volts) and current (I amps) of a non-linear resistor are related according to $I = 0.025V + 0.005V^2$. If $I = 0.91$ amp calculate the voltage, V.

6 The acceleration, f, of a vehicle over a distance, s, changes its velocity from u to v. These are related by $v^2 = u^2 + 2fs$. You are given $f = 2\,ms^{-2}$ and $s = 150\,m$. Find the initial velocity, u, if the final velocity is 4 times larger than it.

7 The diagram shows a floor area of $120\,\text{m}^2$ to be tiled with new quarry tiles. Calculate the value of x.

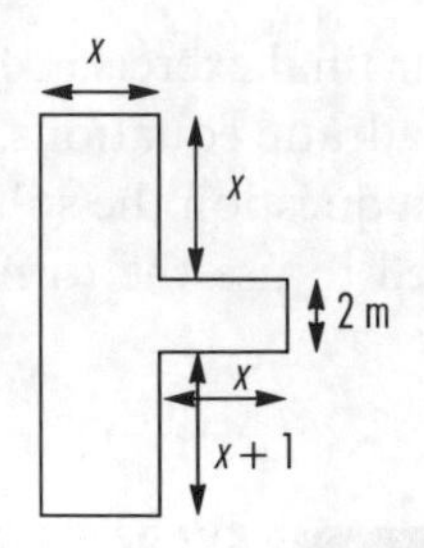

8 A rigid beam of length 10 m carries a uniformly distributed total load of 10 000 N. The bending moment, M, is related to its distance from one end of the beam, x, according to $M = -10\,000 + 5000x - 500x^2$

i) Calculate the value of x when $M = 2000$ Nm.

ii) Where along the beam is the bending moment 0?

9 The diagram shows a cone of height h, radius r and slant height l. They are related according to Pythagoras' Theorem. You are given that $l = 0.4$ m.

i) If the height and radius are equal calculate their values.

ii) The curved surface area is πrl and the area of the circular base is πr^2. Their total area is $10\,\text{m}^2$, i.e. $\pi rl + \pi r^2 = 10$. Calculate the radius of the cone.

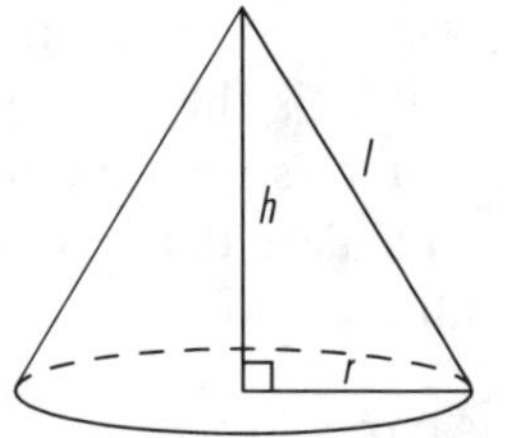

10 2 resistors, R_1 and R_2, are in parallel with a combined resistance of 4Ω according to $\frac{1}{4} = \frac{1}{R_1} + \frac{1}{R_2}$.

When they are in series we are given their combined resistance to be 21Ω according to $21 = R_1 + R_2$.

We may make R_2 the subject of the second formula so that

$$21 - R_1 = R_2.$$

Substituting into the first equation we get

$$\frac{1}{4} = \frac{1}{R_1} + \frac{1}{21 - R_1}.$$

Re-arrange this relation into a more usual quadratic equation. Solve your equation for R_1.

8 Cubic Equations

The objectives of this chapter are to:

1 Plot the graph of a cubic equation for a specified interval and range.

2 Graphically solve a cubic equation.

Introduction

We have learned about linear equations (e.g. $3x+2=0$) where the highest power of x is 1, i.e. x^1 written as x. We have learned about quadratic equations (e.g. $x^2+3x+2=0$) where the highest power of x is 2, i.e. x^2. In a cubic equation the highest power of x is 3, i.e. x cubed written as x^3. For example $x^3+x^2+3x+2=0$ is a cubic equation. We need not always have terms in x^2, x or a constant.

ASSIGNMENT

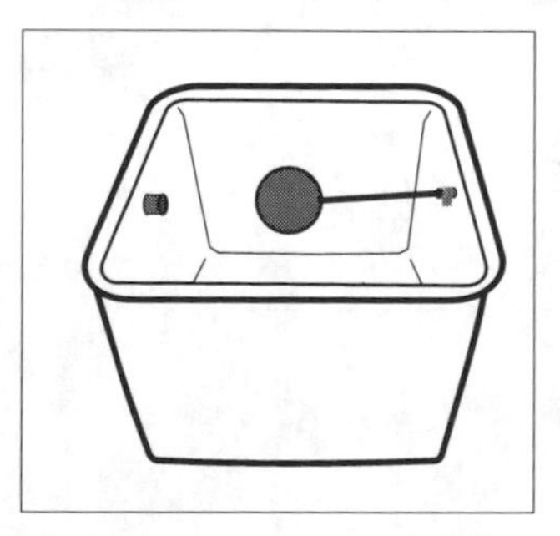

Our Assignment for this chapter looks at a particular open cistern. It might be used as a heading tank for a domestic hot water system. We have a cistern with a square base and total surface area (a base and 4 sides) of 12m^2. Later in the chapter we will look at the area of material used in its construction and the effect this has upon the volume.

Cubic equations – graphical solution

Firstly we can look at some simple examples of cubic equations. Then we can see how these may look graphically before attempting to solve them.

Examples 8.1

The following are a few examples of cubic equations:

i) $x^3=0$,

ii) $2x^3=x^2+x-7$,

iii) $4x^3+3x^2+2=0$,

iv) $4x^3+x-3=0$,

v) $x^3=8$.

In Chapter 7 there was a neat formula for solving quadratic equations. Unfortunately we have no formula for cubic equations, so we must rely on our graphical efforts. The graphs of cubics are quite varied as the diagrams in Figs. 8.1 show. Here are some examples.

i) 3 solutions (roots)

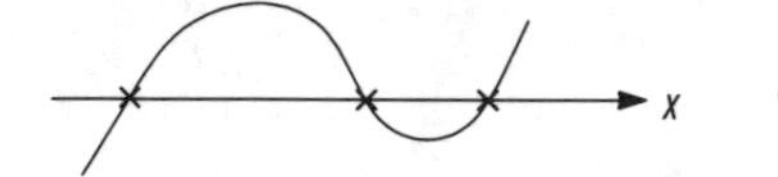

or

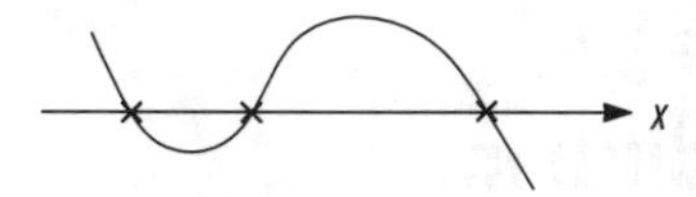

ii) 2 solutions (roots), 1 being repeated

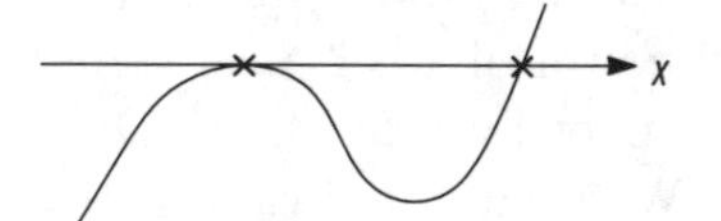

or

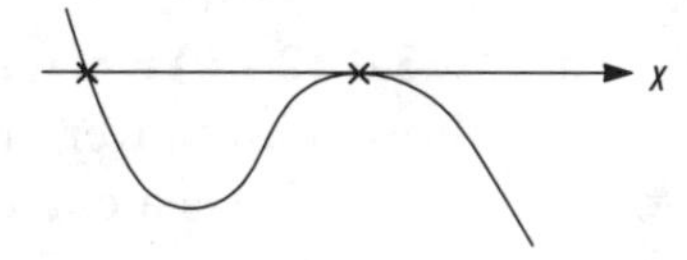

iii) 1 solution (root)

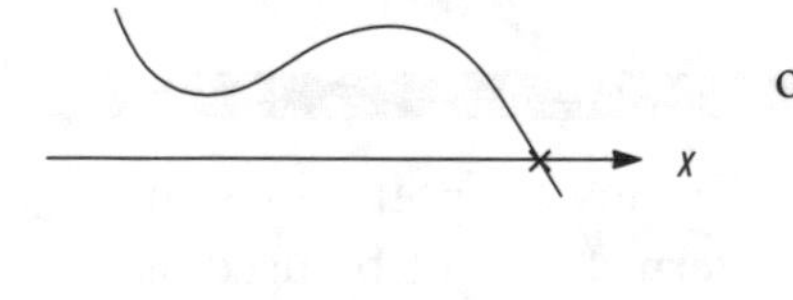

or

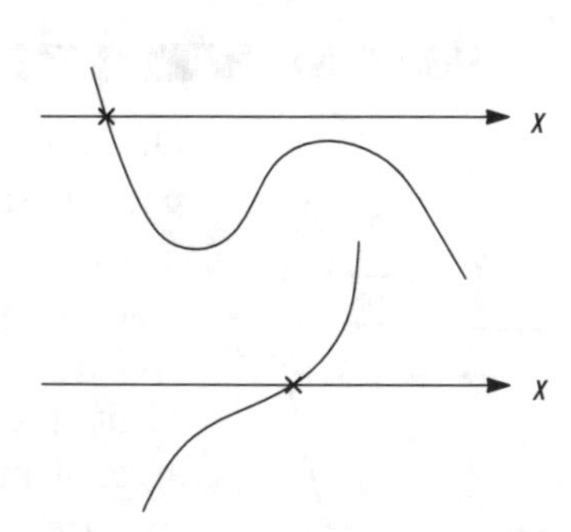

Figs. 8.1

Our aim is to plot the graph of $y=ax^3+bx^2+cx+d$ (a, b, c and d are numbers that may be positive or negative or zero). Remember that the solution(s) to the cubic equation $ax^3+bx^2+cx+d=0$ occur where the graph crosses or touches the horizontal axis. We could look at the graph of $y=ax^3+bx^2+cx+d$ generally and change values of a or b or c or d. This would give us some idea about shape. However the task of looking at all possible changes is too long for us. Instead we will only have a brief look at changes in a and d.

Examples 8.2

Plot and compare the graphs of

i) $y=x^3$, ii) $y=2x^3$, iii) $y=-2x^3$.

Firstly we will construct our tables in the usual way. The specimen values of x are to range from $x=-2.5$ to $x=2.5$ at intervals of 0.5 . We will plot the graphs on one set of axes for easy comparison.

i) For $y=x^3$ we have a very simple table just cubing each specimen value to give

x	−2.500	−2.000	−1.500	−1.000	−0.500	0.000
y	−15.625	−8.000	−3.375	−1.000	−0.125	0.000

	0.500	1.000	1.500	2.000	2.500
	0.125	1.000	3.375	8.000	15.625

ii) For $y=2x^3$ each specimen value is cubed and that result multiplied by 2. You will see the table builds on the previous one to give

x	−2.500	−2.000	−1.500	−1.000	−0.500	0.000
x^3	−15.625	−8.000	−3.375	−1.000	−0.125	0.000
y	−31.25	−16.00	−6.75	−2.00	−0.25	0.00

	0.500	1.000	1.500	2.000	2.500
	0.125	1.000	3.375	8.000	15.625
	0.25	2.00	6.75	16.00	31.25

iii) For $y=-2x^3$ we continue with this building process. We just apply the − sign to each value of y in the previous table so that

x	−2.500	−2.000	−1.500	−1.000	−0.500	0.000
x^3	−15.625	−8.000	−3.375	−1.000	−0.125	0.000
y	31.25	16.00	6.75	2.00	0.25	0.00

	0.500	1.000	1.500	2.000	2.500
	0.125	1.000	3.375	8.000	15.625
	−0.25	−2.00	−6.75	−16.00	−31.25

Fig. 8.2 shows these 3 graphs on one set of axes. An increase in the value of a steepens the slope. When a is multiplied by a – sign the original graph is reflected in the horizontal axis.

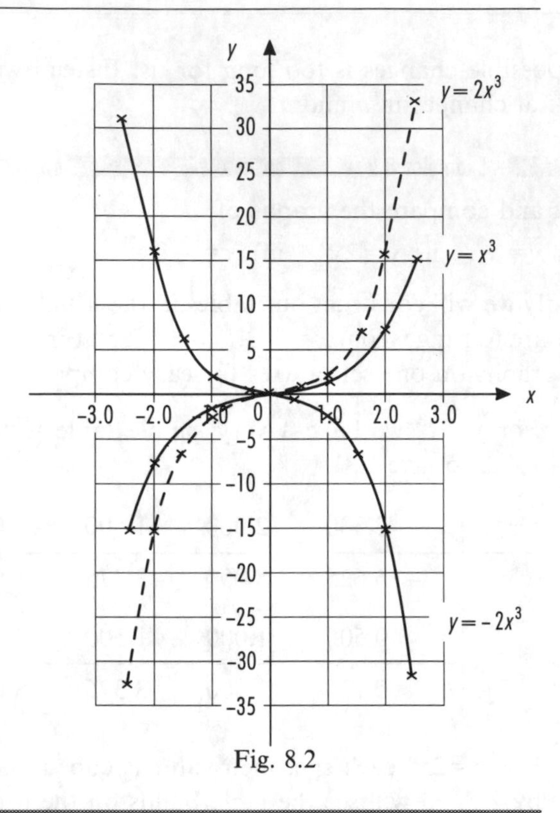

Fig. 8.2

Examples 8.3

Plot and compare the graphs of

i) $y=x^3$, ii) $y=x^3+8$, iii) $y=x^3-8$.

Firstly we will construct our tables in the usual way.

The specimen values of x are to range from $x=-2.5$ to $x=2.5$ at intervals of 0.5. We will plot the graphs on one set of axes for easy comparison.

i) For $y=x^3$ we have a very simple table just cubing each specimen value to give

x	-2.500	-2.000	-1.500	-1.000	-0.500	0.000
y	-15.625	-8.000	-3.375	-1.000	-0.125	0.000

	0.500	1.000	1.500	2.000	2.500
	0.125	1.000	3.375	8.000	15.625

ii) For $y=x^3+8$ we build on the previous table simply adding 8 to each value of x^3 to give

x	-2.500	-2.000	-1.500	-1.000	-0.500	0.000
x^3	-15.625	-8.000	-3.375	-1.000	-0.125	0.000
y	-7.625	0.000	4.625	7.000	7.875	8.000

	0.500	1.000	1.500	2.000	2.500
	0.125	1.000	3.375	8.000	15.625
	8.125	9.000	11.375	16.000	23.625

iii) For $y=x^3-8$ again we build on the first table. As might be expected we simply subtract 8 in each case to give

x	-2.500	-2.000	-1.500	-1.000	-0.500	0.000
x^3	-15.625	-8.000	-3.375	-1.000	-0.125	0.000
y	-23.625	-16.000	-11.375	-9.000	-8.125	-8.000

	0.500	1.000	1.500	2.000	2.500
	0.125	1.000	3.375	8.000	15.625
	-7.875	-7.000	-4.625	0.000	7.625

Fig. 8.3 shows these 3 graphs on one set of axes. A change in the value of d shifts a graph vertically. An increase in d causes a vertical upward shift. A decrease in d causes a downward vertical shift.

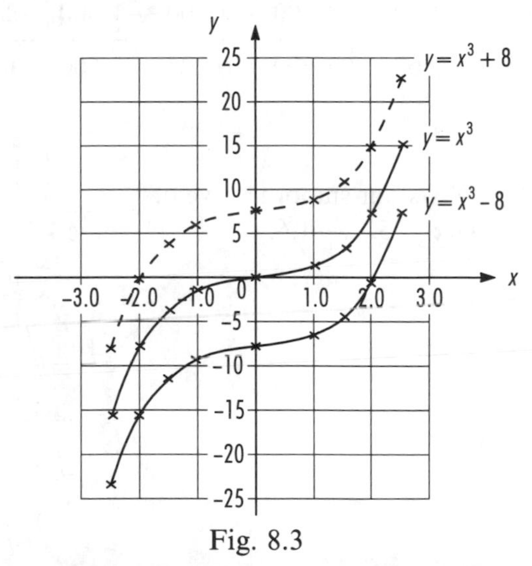

Fig. 8.3

Examples 8.4

Graphically solve the equations

i) $x^3=0$, ii) $2x^3=0$, iii) $-2x^3=0$, iv) $x^3+8=0$, v) $x^3-8=0$.

In Examples 8.2 and 8.3 we plotted all the relevant cubic graphs. Comparing each graph with its corresponding equation we need to look where $y=0$. This means the roots of a cubic equation, (just like a quadratic equation of Chapter 7) occur as the graph crosses the horizontal axis. In each case we can refer to Fig. 8.2 and Fig. 8.3 and read off the roots from the horizontal axis.

i) $x^3=0$ has one solution, $x=0$.
ii) $2x^3=0$ has one solution, $x=0$.
iii) $-2x^3=0$ has one solution, $x=0$.

Really all these 3 equations are very similar. In $2x^3=0$ if we divide through by 2 we get our first equation of $x^3=0$. Similarly in $-2x^3=0$ when we divide by -2 we get $x^3=0$.

iv) $x^3+8=0$ has one solution, $x=-2$.

v) $x^3-8=0$ has one solution, $x=2$.

Examples 8.5

Graphically solve the cubic equation $x^3-4.4x^2+5.6x-1.6=0$ using specimen values of x from $x=-0.5$ to $x=2.5$ at intervals of 0.5.

As usual our first step is to construct a table.

x	-0.500	0.000	0.500	1.000	1.500	2.000	2.500 ↔
x^3	-0.125	0.000	0.125	1.000	3.375	8.000	15.625
$-4.4x^2$	-1.100	0.000	-1.100	-4.400	-9.900	-17.600	-27.500
$5.6x$	-2.800	0.000	2.800	5.600	8.400	11.200	14.000
-1.6	-1.600	-1.600	-1.600	-1.600	-1.600	-1.600	-1.600
y	-5.625	-1.600	0.225	0.600	0.275	0.000	0.525 ↕

Fig. 8.4 shows our smooth curve of $y=x^3-4.4x^2+5.6x-1.6$.

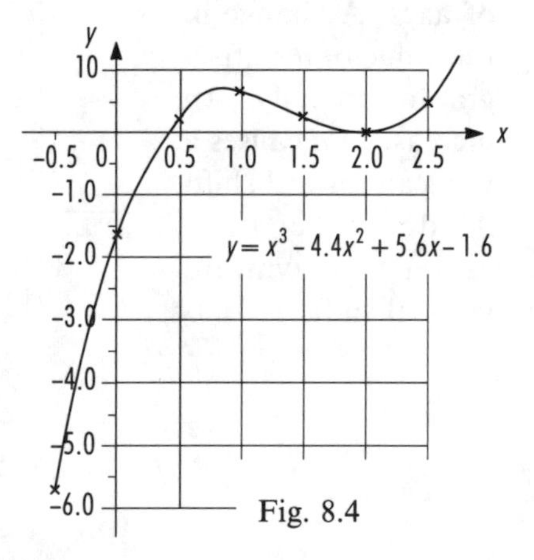

Fig. 8.4

The solutions (roots) of $x^3-4.4x^2+5.6x-1.6=0$ occur where the graph crosses the horizontal axis (i.e. where $y=0$). From our graph these are $x=0.4$ and $x=2.0$ (repeated). For a repeated root the graph touches rather than crosses the horizontal axis; i.e. the axis is a tangent to the graph at this point.

We can look at the table for some hint of where there may be any solutions. Remember that for a solution at $y=0$ we look for a change of sign in the y values. The table shows a change as y goes from -1.600 to 0.225. This means there is a root somewhere between the corresponding x values of 0.000 and 0.500. Our root of $x=0.4$ confirms this. The table shows an exact y value of 0.000 and hence a root of $x=2.0$. However there is no way of telling from the table that this is a repeated root. We really do need to plot the graph.

Example 8.6

Graphically solve the cubic equation $4x^3-3x^2-11x+2.5=0$ using specimen values of x from $x=-2.0$ to $x=2.5$ at intervals of 0.5.

We need to plot $y=4x^3-3x^2-11x+2.5$ and so start with a table of values

x	-2.00	-1.50	-1.00	-0.50	0.00	0.50	1.00	1.50	2.00	2.50
$4x^3$	-32.00	-13.50	-4.00	-0.50	0.00	0.50	4.00	13.50	32.00	62.50
$-3x^2$	-12.00	-6.75	-3.00	-0.75	0.00	-0.75	-3.00	-6.75	-12.00	-18.75
$-11x$	22.00	16.50	11.00	5.50	0.00	-5.50	-11.00	-16.50	-22.00	-27.50
$+2.5$	2.50	2.50	2.50	2.50	2.50	2.50	2.50	2.50	2.50	2.50
y	-19.50	-1.25	6.50	6.75	2.50	-3.25	-7.50	-7.25	0.50	18.75

Looking at the sign changes in y we can spot whereabouts roots may lie. The roots lie between the corresponding x values.

y changes sign from -1.25 to 6.50;
$\therefore$ a root lies between $x=-1.50$ and $x=-1.00$.

y changes sign from 2.50 to -3.25;
$\therefore$ a root lies between $x=0.00$ and $x=0.50$.

y changes sign from -7.25 to 0.50;
$\therefore$ a root lies between $x=1.50$ and $x=2.00$.

Our accurate plot of $y=4x^3-3x^2-11x+2.5$ is given in Fig. 8.5.

The roots of the cubic equation are approximately $x=-1.45$, 0.20 and 1.95.

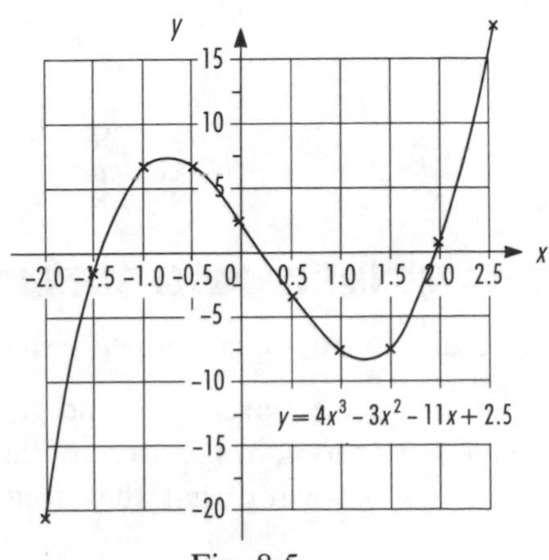

Fig. 8.5

Example 8.7

Graphically solve the equation $2x^3-7x^2+3x+8=0$ using values of x from $x=-1.0$ to $x=3.0$ at intervals of 0.5 .

Our first step in the plot of $y=2x^3-7x^2+3x+8$ is to construct the table below.

x	−1.00	−0.50	0.00	0.50	1.00	1.50
$2x^3$	−2.00	−0.25	0.00	0.25	2.00	6.75
$-7x^2$	−7.00	−1.75	0.00	−1.75	−7.00	−15.75
$+3x$	−3.00	−1.50	0.00	1.50	3.00	4.50
$+8$	8.00	8.00	8.00	8.00	8.00	8.00
y	−4.00	4.50	8.00	8.00	6.00	3.50

	2.00	2.50	3.00	⟷
	16.00	31.25	54.00	
	−28.00	−43.75	−63.00	
	6.00	7.50	9.00	
	8.00	8.00	8.00	
	2.00	3.00	8.00	↕

When we look over the table we see only one sign change for y. Perhaps there is only 1 root somewhere between $x=-1.00$ and $x=-0.50$. At this stage we do not know whether we may need to plot more values. However, as Fig. 8.6 shows this cubic equation has only 1 root at $x=-0.8$.

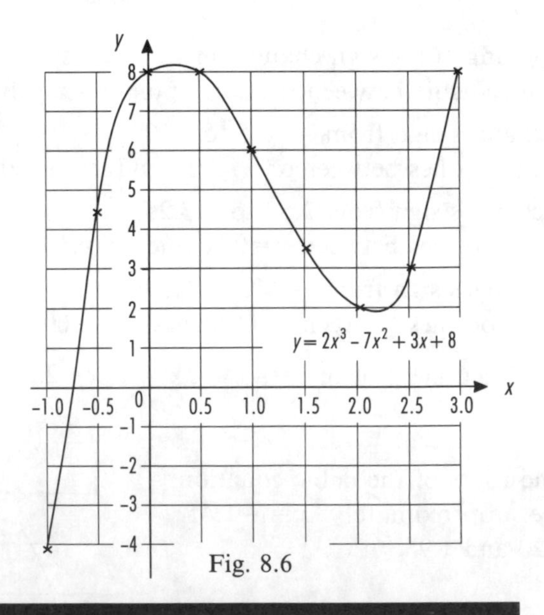

Fig. 8.6

ASSIGNMENT

We can return to our open topped cistern.

Let the sides of the base be x and the height be h. Let us look at the areas of the base and sides, knowing that they total $12\,\text{m}^2$.

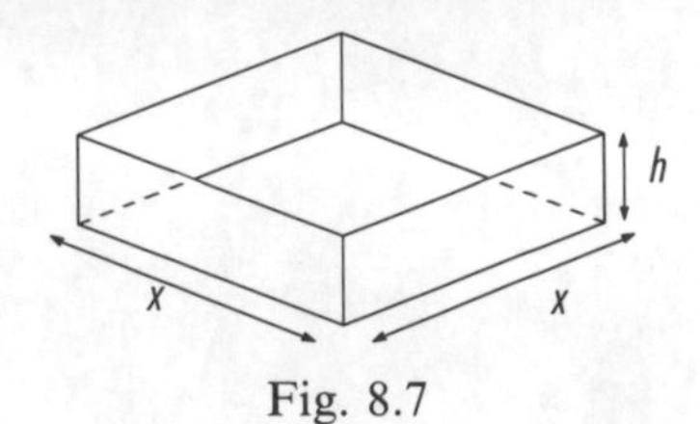

Fig. 8.7

Area of base	$= x \times x$	$= x^2$
Area of a side	$= x \times h$	
Area of all 4 sides	$= 4 \times x \times h$	$= 4xh$
Total area		$= x^2 + 4xh$

From the original information we can link together our area values to give $12 = x^2 + 4xh$.

We have not met 2 variables multiplied together (xh) before. Also we still have to discuss the volume of the cistern.

Volume $= x \times x \times h$

i.e. $V = x^2 h$

Now we have 3 variables, V, x and h. We only know how to deal with 2 variables on our graphs. In our 2 formulae x appears 3 times but h appears only twice. Because h appears less often we will get rid of it by substitution. From the first equation

$$h = \frac{12 - x^2}{4x}$$

so that $$V = x^2 \frac{(12 - x^2)}{4x}$$

i.e. $$V = \frac{x(12 - x^2)}{4}$$

Cancelling x.

$$\text{or } 3x - \frac{x^3}{4}.$$

We can look at the graph of V (vertically) against x (horizontally). Because x is a distance it is always positive. We have constructed a table using values of x up to 3.5. This is a bit further than we really need: a square base of side 3.5 m would exceed the available $12\,\text{m}^2$ for the whole job.

x	0	0.50	1.00	1.50	2.00	2.50	3.00	3.50	$\longleftrightarrow$
$-x^2$	0	−0.25	−1.00	−2.25	−4.00	−6.25	−9.00	−12.25	
$12-x^2$	12.00	11.75	11.00	9.75	8.00	5.75	3.00	−0.25	*
$\frac{x}{4}$	0	0.125	0.250	0.375	0.500	0.625	0.750	0.875	**
V	0	1.47	2.75	3.66	4.00	3.59	2.25	−0.22	$\updownarrow$

We multiply together rows * and **, i.e. $(12 - x^2)$ and $\frac{x}{4}$, to complete our formula and the table values for V.

Fig. 8.8 shows the graph of the table. Because of the problem we are interested in positive values for both x and V.

It is interesting to see how the volume, V, increases as x increases. It reaches a maximum value of $4\,m^3$ when $x=2\,m$. Using the formula for volume, $V=x^2h$, we can substitute these values for V and x to calculate $h=1$. Thus the height of the cistern is 1 m and its square base is of side 2 m. The Mathematics may need some adjustment, perhaps by other design constraints. For example one might be the weight of water and the support necessary. However it makes economic sense to get the largest possible volume from the material being used.

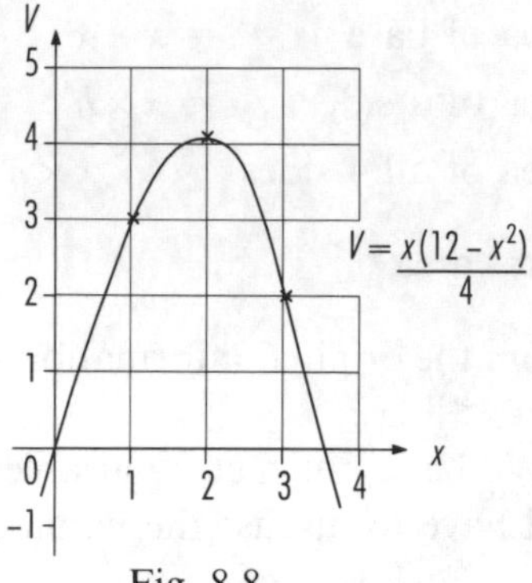

Fig. 8.8

The Assignment can be extended. Areas of metal other than $12\,m^2$ may be available. If, for example there were 10 or 15 or $16\,m^2$ or . . ., just substitute that number in place of 12 in the volume formula. Now repeat the table and graph for the new value.

EXERCISE 8.1

1 Plot the graph of $y=x^3-2x^2-5x+6$ using values of x from $x=-2.5$ to $x=4.0$ at intervals of 0.5. Use your graph to solve the cubic equation $x^3-2x^2-5x+6=0$.

2 Using values of x from $x=-1.5$ to $x=4.5$ at intervals of 0.5 plot the graph of $y=x^3-3x^2+2$. Hence solve the cubic equation $x^3-3x^2+2=0$.

3 Graphically solve the cubic equation $x^3+x^2-1.75x+0.5=0$. You should plot the necessary graph for values of x from $x=-3.0$ to $x=2.0$ at intervals of 0.5.

4 Solve the cubic equation $3x^3+6x+8=0$. It is suggested that you plot the necessary graph for values of x from $x=-3$ to $x=3$.

5 Plot the graph of $y=1-3x+\frac{1}{2}x^3$ using values of x from $x=-3$ to $x=3$ at intervals of 0.5. Hence use your graph to solve the cubic equation $x^3-6x+2=0$.
(*Hint*: Amend one of these relationships by a factor of 2.)

6 The second moment of area, I_{xx}, for this section is given by $I_{xx}=\dfrac{bd^3}{12}$.
For a section where $b=0.10\,m$ plot a graph of I_{xx} against d to investigate the change in this second moment. Use values of d from 0.05 m to 0.50 m at intervals of 0.05 m.

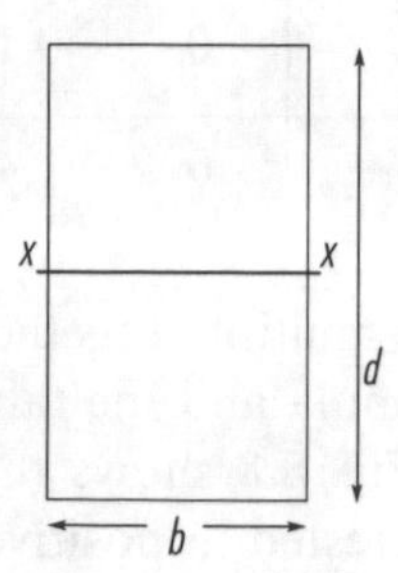

7 A disc is spun from rest. It spins through $y°$ in t seconds according to $y = 100t - 5t^3$. Investigate this relation by plotting a graph of y against t. Use values of t from 0 to 5.0 seconds at intervals of 0.5.
From your graph estimate the maximum value of y and the time at which this occurs.

8 A hollow cylinder with closed ends is made from 6×10^4 mm^2 of sheet metal. For a cylinder of radius, r, and volume, V, this means they are related by the formula $V = 3 \times 10^4 r - \pi r^3$.
Plot V against r using values of r from 0 to 120 mm at intervals of 10 mm. From your graph read off the greatest volume. What is the radius for this volume?

9 One particular machine in an engineering workshop costs £C to lease each week according to the formula $C = 200 + \frac{t^3}{20}$.
t is the number of hours/week worked by the machine. Plot a graph of C against t. Use values of t from 0 to 48 hours at intervals of 4 hours.

i) What is the minimum weekly lease cost?
ii) If one particular week's lease was approximately £4000, for how long was the machine worked?
Give your answer to the nearest half-hour.
iii) The usual week without overtime was $37\frac{1}{2}$ hours. What was that usual lease cost?
Give your answer to the nearest hundred pounds.

10 A chemical process plant can produce up to 5 tonnes each week. Its costs, y_1, are related to the tonnage produced, x, by $y_1 = 10 + 3x - x^2$. Its sales income, y_2, is also related to tonnage by $y_2 = 0.2x^3$.
The units for y_1 and y_2 are £000 (thousand pounds). Using intervals of 0.5 tonne plot both of these graphs on one set of axes. From your graph read off the **break even point**, i.e. where costs and income are **the same**.

9 Simultaneous Equations

The objectives of this chapter are to:

1 Solve a pair of simultaneous linear equations
  i) graphically,
  ii) by elimination.

2 Graphically solve simultaneous linear and quadratic equations.

3 Graphically solve simultaneous linear and cubic equations.

4 Graphically solve simultaneous quadratic and cubic equations.

Introduction

We know equations are for solving in Mathematics. In fact we have already attempted to solve separately linear, quadratic and cubic equations.

Firstly let us consider our chapter title. Things that occur **simultaneously** are things that occur at the **same time**. For example, if 2 people join the end of a queue simultaneously then they join it at the same place and time. When we solve equations simultaneously we solve them together: the solution for one of them must apply to them all. Most of our examples and exercises will consider a pair of simultaneous equations. The same principles in fact apply to more than just 2 equations.

ASSIGNMENT

This Assignment looks at a production company. It produces 3 types of interior car trim for the domestic automotive industry. These are Standard, Deluxe and Prestige styles. The company has a split site operation over two factories. Later in this chapter we will look at its attempts to meet a production target.

The company's sales force has been particularly active. Many of these new smaller orders have to be squeezed into the production schedules.

However, there is a very important order for a longstanding customer. It is for 700 Standards, 900 Deluxes and 500 Prestiges.

The 2 factories are differently equipped. This means their daily production capacities are different. The following table shows the capacity figures for each type of trim.

Daily Production	Standard	Deluxe	Prestige
First Site	100	300	100
Second Site	300	100	100

We want to know how to schedule this order using both sites. Suppose the order will use x days of production at the first site and y days at the second site. Combining the daily production capacities with these days gives

Order Production	Standard	Deluxe	Prestige
First Site	$100x$	$300x$	$100x$
Second Site	$300y$	$100y$	$100y$
Total	700	900	500

Having set out the problem we will return to it once we have looked at some relevant Mathematics.

Coordinates and graphs

Before we attempt to solve any equations let us look at points lying on graphs. Whenever we plot a graph we plot only a **selection of points**, usually (x, y). Then we draw either a straight line or a smooth curve through them. The graph passes through many more points than our selected ones. All such points satisfy the particular relationship between x and y. Our first set of examples demonstrates whether points lie or do not lie on a graph.

Examples 9.1

For the quadratic function $y = x^2 + 3x + 2$ decide if the following points lie on the graph

i) $(0, 2)$, ii) $(3, 20)$, iii) $(2, 0)$, iv) $(0.50, 3.75)$, v) $(-0.4, 0.9)$.

The right-hand side of $y = x^2 + 3x + 2$ is the more complicated side. Hence we will start with that side, substituting the values of x in each case.

i) Using $(0, 2)$ we have $x=0$ to substitute in x^2+3x+2 to give

$$0^2+3(0)+2=0+0+2=2.$$

This answer agrees with the y value of 2 in $(0, 2)$. Hence $(0, 2)$ lies on the curve.

ii) Using $(3, 20)$ we have $x=3$ to substitute in x^2+3x+2 to give

$$3^2+3(3)+2=9+9+2=20.$$

This answer agrees with the y value of 20 in $(3, 20)$. Hence $(3, 20)$ lies on the curve.

iii) Using $(2, 0)$ we have $x=2$ to substitute in x^2+3x+2 to give

$$2^2+3(2)+2=4+6+2=12.$$

This answer *differs* from the y value in $(2, 0)$. Hence $(2, 0)$ does *not* lie on the curve.

iv) Using $(0.50, 3.75)$ we have $x=0.50$ to substitute in x^2+3x+2 to give

$$(0.5)^2+3(0.5)+2=0.25+1.50+2=3.75.$$

This answer agrees with the y value in $(0.50, 3.75)$. Hence $(0.50, 3.75)$ lies on the curve.

v) Using $(-0.4, 0.9)$ we have $x=-0.4$ to substitute in x^2+3x+2 to give

$$(-0.4)^2+3(-0.4)+2=0.16-1.20+2=0.96.$$

This answer *differs* from the y value in $(-0.4, 0.9)$. Hence $(-0.4, 0.9)$ does *not* lie on the curve.

Fig. 9.1 shows the curve together with these 5 specimen pairs of coordinates. It confirms our "lies" and "not lie" decisions.

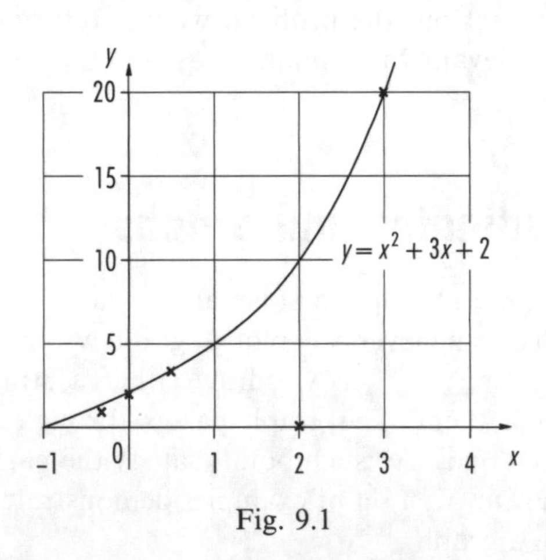

Fig. 9.1

Simultaneous linear equations – graphical solution

Let us start with 2 straight line (linear) graphs. We know that every point lying on a straight line satisfies the x and y relation. If possible, we need to find a point that satisfies **both relations**. This can only happen if the point lies on both lines, i.e. if those lines **cross (intersect)**.

Example 9.2

Graphically solve the pair of simultaneous equations

$$2x+3y = 24$$

and $\quad y-4x = 1.$

We know the general equation of a straight line is $y=mx+c$. It might be easier if we re-arrange our equations into this form.

$$2x+3y = 24$$

becomes $\quad 3y = 24-2x \qquad$ Subtracting $2x$ from both sides.

i.e. $\quad y = 8-\frac{2}{3}x. \qquad$ Dividing by 3.

We know this is a straight line of gradient $-\frac{2}{3}$ and vertical intercept 8.

Also $\quad y-4x = 1$

becomes $\quad y = 4x+1.$

We know this is a straight line of gradient 4 and vertical intercept 1.

Because these are known straight lines our tables need only 3 values for x: 2 and 1 as a check. We can choose those values of x at random.

$y=8-\frac{2x}{3}$

x	-6	0	3
8	8	8	8
$-\frac{2x}{3}$	4	0	-2
y	12	8	6

$y=4x+1$

x	-5	0	4
$4x$	-20	0	16
$+1$	1	1	1
y	-19	1	17

The graphs are plotted in Fig. 9.2. We see that the graphs do intersect and can read off their point of intersection as (1.5, 7). This means the solution to the pair of simultaneous equations is $x=1.5$ and $y=7$.

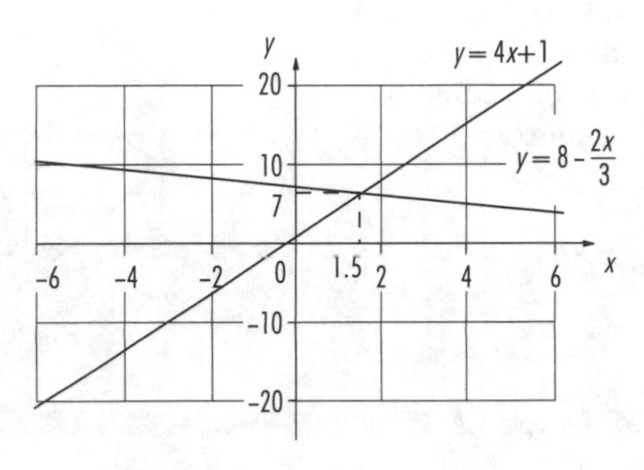

Fig. 9.2

The pair of equations $2x+3y=24$ and $y-4x=1$ of Example 9.2 have 2 **unknowns (variables)**. Whenever we attempt to solve any simultaneous equations we need at least as many equations as unknowns.

Example 9.3

Graphically solve the simultaneous equations

$$2x+y+3 = 0$$
$$2y+x = 0$$
$$3y-4x = 11.$$

Like our method in Example 9.2 we can re-arrange these 3 equations so y is the subject,

i.e. $y = -2x-3,$

$$y = -\frac{x}{2}$$

$$y = \frac{4x}{3}+\frac{11}{3}.$$

As an exercise for yourself you might like to check these re-arrangements are correct.

We recognise these equations represent straight lines and so need to plot just 3 points. The method of table construction in the previous example will have refreshed your memory. So the ones for this example show just the first (x) and last (y) rows.

$y=-2x-3$

x	-2	0	2
y	1	-3	-7

$y=-\dfrac{x}{2}$

x	-4	2	6
y	2	-1	-3

$y=\dfrac{4x}{3}+\dfrac{11}{3}$

x	-2	1	4
y	1	5	9

The graphs are plotted in Fig. 9.3. We see all 3 graphs do intersect at one point, $(-2, 1)$. This means the solution to our original set of 3 equations is $x=-2$ and $y=1$.

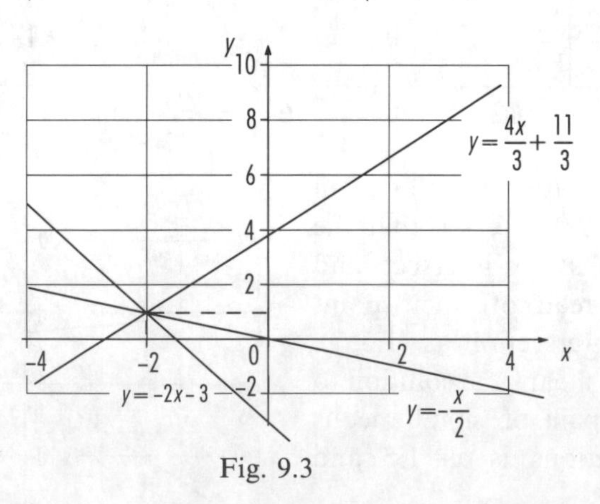

Fig. 9.3

Example 9.4

Graphically solve the simultaneous equations

$$2x+y+3 = 0$$
$$2y+x = 0$$
$$3y-2x = 12.$$

Like our method in Example 9.2 we can re-arrange these 3 equations so y is the subject,

i.e.
$$y = -2x-3$$
$$y = -\frac{x}{2}$$
$$y = \frac{2x}{3}+4.$$

As an exercise for yourself you might like to check these re-arrangements are correct.

We recognise these equations represent straight lines and so will plot just 3 points. Again the tables show just the first (x) and last (y) rows.

$y=-2x-3$

x	-2	0	2
y	1	-3	-7

$y=-\frac{x}{2}$

x	-4	2	6
y	2	-1	-3

$y=\frac{2x}{3}+4$

x	-3	3	6
y	2	6	8

The graphs are plotted in Fig. 9.4. We see all 3 graphs do *not* pass through one point simultaneously. This means there is no one solution. We do have solutions to pairs of equations.

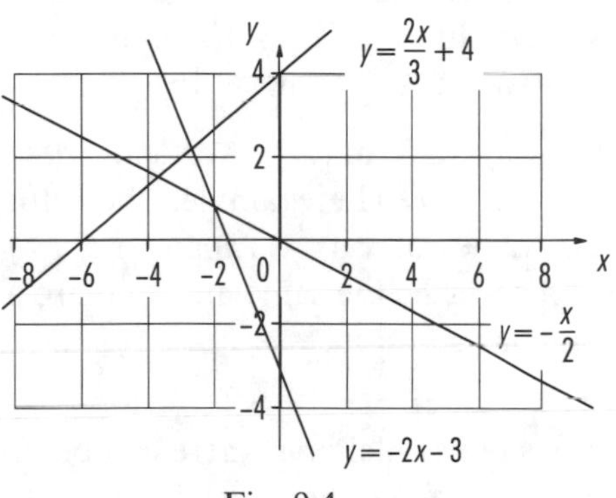

Fig. 9.4

$$2x+y+3 = 0$$
and $\quad 2y+x = 0 \quad$ both pass through $(-2, 1)$

$$2x+y+3 = 0$$
and $\quad 3y-2x = 12 \quad$ both pass through $(-2.625, 2.25)$

$$2y+x = 0$$
and $\quad 3y-2x = 12 \quad$ both pass through $(-3.43, 1.71)$.

Only as pairs, but not as all 3 together, can we solve them simultaneously.

Example 9.5

Graphically solve the simultaneous equations

$$y = 2x+3$$
and $\quad y = 2x-4$

We can plot these straight lines as usual (Fig. 9.5).

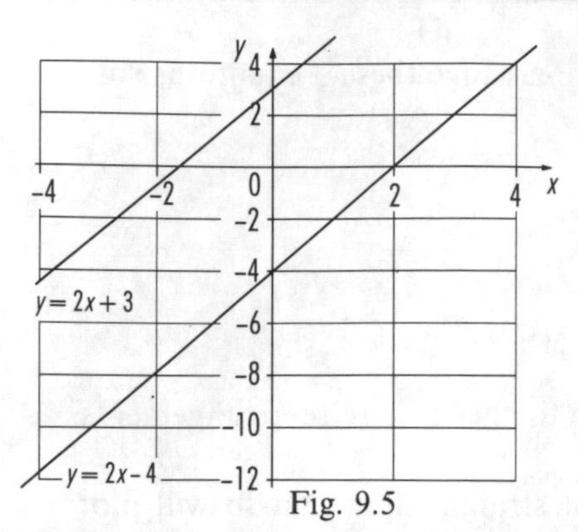

Fig. 9.5

Comparing the original equations with the general equation $y = mx + c$ we see they both have a gradient of 2. This means they are **parallel** and so **cannot intersect**, i.e. there is **no solution**.

There are 2 other types of parallel lines that you may meet. They may be horizontal lines like those in Fig. 9.6. The examples show totally different values of y as $\frac{1}{2}$ and 3. Whatever the values of x separately these values of y never change. Never can the same value of y satisfy both lines and so there is no simultaneous solution.

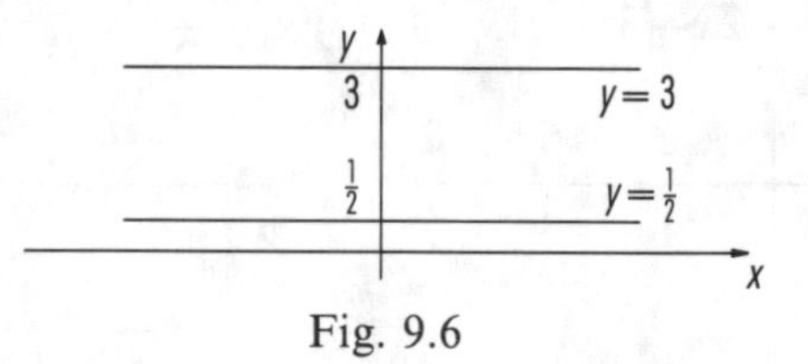

Fig. 9.6

Alternatively they may be vertical lines like those in Fig. 9.7. The examples show totally different values of x as -1.0 and 1.5 . Again these can never be the same and so there is no simultaneous solution.

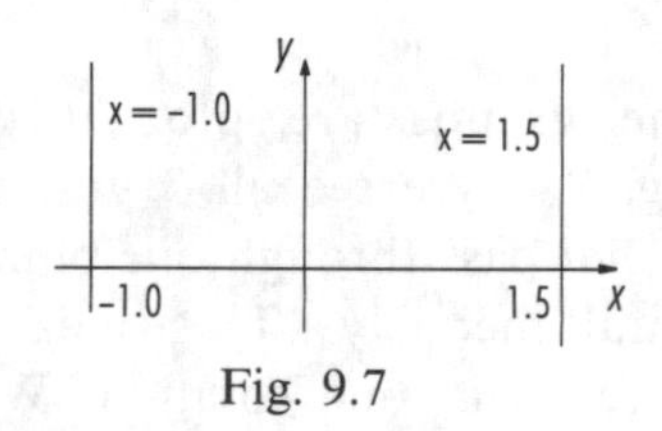

Fig. 9.7

Thus for any parallel lines there can be no simultaneous solution. Do not be disappointed. Mathematics is a tool to aid your decision making. We must move away from early ideas of absolute truth into the real world.

There is one more case where you will get no solution. Sometimes you *think* you have a pair of equations. Closer inspection reveals that one is just a multiple of the other.

Examples 9.6

We look at pairs of equations in this set of examples.

i) $y - 2x = 3$
and $2y - 4x = 6$.

The second equation is twice the first equation, i.e. in the second equation we can divide through by a common factor of 2 to create the first equation.

ii) $2y - 4x = 6$
and $-3y + 6x + 9 = 0$.

In this case the second equation is -1.5 times the first equation. Do not be misled by the slightly different forms. In the second equation, if you prefer, move the 9 to the right-hand side. This gives $-3y+6x=-9$. Now you will be able to compare them both more easily and spot the factor of -1.5 .

ASSIGNMENT

Earlier in this chapter we set out the order production figures in total and for each site. We may now combine them to create 3 linear equations

$$100x+300y = 700$$
$$300x+100y = 900$$
$$100x+100y = 500.$$

Each term in each equation is a multiple of 100. We can cancel through by this figure and still preserve the balances. Also we may re-arrange each one to make y the subject,

i.e. $$y = \frac{7}{3}-\frac{x}{3}$$
$$y = 9-3x$$
$$y = 5-x.$$

The next step is to construct tables of values. Each equation represents a straight line so we need just 3 specimen values. Because production has to be positive the values of x and y have to be positive too.

$y=\frac{7}{3}-\frac{x}{3}$

x	1	2.5	4
y	2	1.5	1

$y=9-3x$

x	0	2	3
y	9	3	0

$y=5-x$

x	0	1	3
y	5	4	2

The graphs are plotted in Fig. 9.8. We see that all 3 straight lines do not pass through the same point. This means there is no one solution. There is no optimum solution. Instead we have 3 solutions because we have 3 intersections. These are $x=2$, $y=3$; $x=2.5$, $y=1.5$ and $x=4$, $y=1$. The points are marked as A, B and C on Fig. 9.8. We will look at the meaning of these results at the end of the next section.

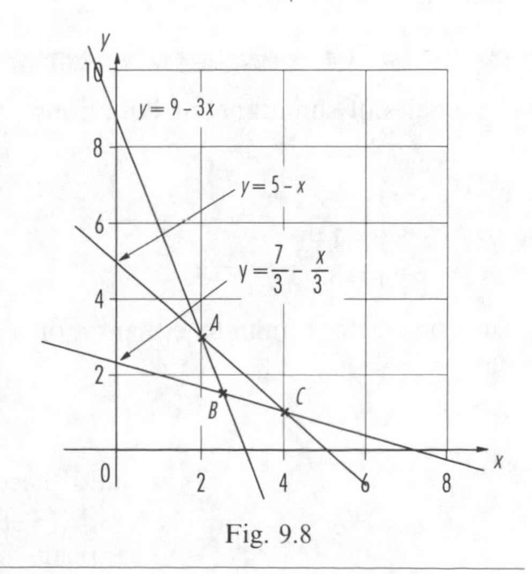

Fig. 9.8

EXERCISE 9.1

Graphically solve the following simultaneous equations. In each case choose 3 specimen values of x and plot the straight lines.

1 $x+y=11$ and $x-y=1$

2 $x+3y=8$ and $x-2y=3$

3 $3y=x+1$ and $5y=20-2x$

4 $2x-y=0$ and $4x-5y=-3$

5 $3y-5x=4$ and $y=1$

6 $x+2y=5$ and $2x-y=7$

7 $2x+3y=3$ and $x=0$

8 $7x-6y=18$ and $6y=7x-12$

9 $3x+2y-4=0$ and $x+3y-11=0$

10 $2x-5y-23=0$ and $15y=69-6x$

Simultaneous linear equations – solution by elimination

For our second method we can look again at some of our earlier examples. We will have 2 equations with 2 unknowns, x and y. The aim is to make both x terms the same, or both y terms the same. Often we can achieve this by multiplying one or both equations by selected values. Remember that any multiplication must be **consistent throughout an equation**. Also, equations must balance themselves about the "=" sign. We will maintain this balance if we act consistently with the left-hand sides and right-hand sides.

Examples 9.7

Solve the pairs of simultaneous equations

i) $x+y=9$
and $x-y=3$;

ii) $4x+y=14$
and $x+y=8$.

We can choose to eliminate either x or y in each case. Firstly let us eliminate y.

i) In $x+y=9$
and $x-y=3+9$

we add these equations knowing that $(+y)+(-y)=y-y=0$ so the y terms disappear,

$$\text{i.e.} \quad x+x = 9+3$$
$$2x = 12$$
$$\frac{2x}{2} = \frac{12}{2}$$
$$x = 6.$$

To find y we substitute this value into one or other of our original equations. Let us substitute into the first equation so that

$$6+y = 9$$
$$\text{i.e.} \quad y = 9-6$$
$$y = 3.$$

Our complete solution is $x=6$, $y=3$.

We should check our solution with the other equation. Substitute both values into the left-hand side to get

$$x-y = 6-3 = 3$$

This is consistent with our second equation so confirming our solution.

ii) In $4x+y = 14$

and $x+y = 8$ — we subtract these equations knowing that $(+y)-(+y)=y-y=0$ so the y terms disappear again,

$$\text{i.e} \quad 4x-x = 14-8$$
$$3x = 6$$
$$\frac{3x}{3} = \frac{6}{3}$$
$$x = 2.$$

To find y we substitute this value into one or other of our original equations. Let us substitute into the first equation to get

$$4(2)+y = 14$$
$$8+y = 14$$
$$y = 14-8$$
$$y = 6.$$

Our complete solution is $x=2$, $y=6$.

We should check our solution with the other equation. Substitute both values into the left-hand side to get

$$x+y = 2+6 = 8.$$

Again this is consistent with our second equation so confirming our solution.

ii) Again. Let us repeat the second example, this time eliminating x.

In $\quad 4x+y = 14$

and $\quad x+y = 8 \quad$ we need both x terms the same.

It is easier numerically if they become $4x$. We achieve this by multiplying throughout the second equation by 4,

i.e. $\quad 4x+y = 14$

and $\quad 4x+4y = 32.$

We subtract the equations since the signs for both $4x$ terms are the same (i.e. $4x-4x=0$).

Then $\quad y-4y = 14-32$

i.e. $\quad -3y = -18$

$$\frac{-3y}{-3} = \frac{-18}{-3}$$

$$y = 6.$$

To find x we substitute this value into one or other of our original equations. Let us substitute into the first equation to get

$$4x+6 = 14$$

i.e. $\quad 4x = 14-6$

$$4x = 8$$

$$\frac{4x}{4} = \frac{8}{4}$$

$$x = 2.$$

Just as before we can check our complete solution of $x=2$, $y=6$ by substitution.

Examples 9.8

Solve the pairs of simultaneous equations

i) $\quad 5x-3y=21$
and $\quad 7x+8y=5;$

ii) $\quad 2x+3y=7$
and $\quad x+4y=11.$

We can choose to eliminate either x or y in each case. Our choice is to eliminate y.

i) In $\quad 5x-3y = 21$

and $\quad 7x+8y = 5$

the y terms are not the same. To make them the same we multiply the first equation by 8 and the second equation by 3. Then they will be $24y$, 24 being the lowest common multiple (LCM) of 3 and 8.

$$\begin{aligned}
& & 5x-3y &= 21 & \times 8 \\
&\text{and} & 7x+8y &= 5 & \times 3 \\
&\text{become} & 40x-24y &= 168 & \\
&\text{and} & 21x+24y &= 15 & \\
&\text{Add} & 40x+21x &= 168+15 & \\
&\text{i.e.} & 61x &= 183 & \\
& & \frac{61x}{61} &= \frac{183}{61} & \\
& & x &= 3. &
\end{aligned}$$

To find y we substitute this value into one or other of our original equations. Let us substitute into the second equation for a change to get

$$\begin{aligned}
& & 7(3)+8y &= 5 \\
&\text{i.e.} & 21+8y &= 5 \\
& & 8y &= 5-21 \\
& & 8y &= -16 \\
& & \frac{8y}{8} &= \frac{-16}{8} \\
& & y &= -2.
\end{aligned}$$

Our complete solution is $x=3$, $y=-2$.

We check our solution with the other equation. Substitute both values into the left-hand side so that

$$5x-3y = 5(3)-3(-2) = 15+6 = 21$$

This is consistent with our second equation so confirming our solution.

ii) In $2x+3y = 7$

and $x+4y = 11$

the y terms are not the same. To make them the same we multiply the first equation by 4 and the second equation by 3. Then they will be $12y$, 12 being the lowest common multiple (LCM) of 4 and 3.

$$\begin{aligned}
& & 2x+3y &= 7 & \times 4 \\
&\text{and} & x+4y &= 11 & \times 3 \\
&\text{become} & 8x+12y &= 28 & \\
&\text{and} & 3x+12y &= 33 & \\
&\text{Subtract} & 8x-3x &= 28-33 & \\
&\text{i.e.} & 5x &= -5 & \\
& & \frac{5x}{5} &= \frac{-5}{5} & \\
& & x &= -1. &
\end{aligned}$$

To find y we substitute this value into one or other of our original equations. Let us substitute into the first equation to get

$$2(-1)+3y = 7$$

i.e.

$$-2+3y = 7$$
$$3y = 7+2$$
$$3y = 9$$
$$y = 3.$$

We check our solution with the other equation. Substitute both values into the left-hand side so that

$$x+4y = -1+4(3) = -1+12 = 11.$$

Examples 9.9

Solve the pairs of simultaneous equations

i) $4x+y=14$
and $4x+y=9$;

ii) $4x+\ y=14$
and $8x+2y=28$.

i) This first pair of equations has the same terms on the left-hand sides yet different ones on the right. Remember that the solutions for x and y must apply to **both** equations simultaneously. How can $4x+y$ give different answers on the right? This cannot happen. These equations are inconsistent. they have no simultaneous solution. As we saw earlier in the chapter graphs show them to be a pair of parallel lines.

ii) Look closely at the second pair of equations. Really we have only one equation because $8x+2y=28$ is just twice $4x+y=14$. As we have only one equation repeated and not a pair of them we are unable to solve them simultaneously. Each equation is **dependent** on the other. Remember because we have 2 unknowns, x and y, we need at least 2 different equations.

ASSIGNMENT

Let us return to our Assignment. We will attempt to solve our 3 equations simultaneously. Using the first pair

$$y = \frac{7}{3}-\frac{x}{3}$$

and

$$y = 9-3x$$

we subtract them, so eliminating y,

i.e. $$0 = \frac{7}{3} - 9 - \frac{x}{3} + 3x$$

$-(-3x) = +3x.$

$$0 = -\frac{20}{3} + \frac{8x}{3}$$

$$\frac{20}{3} = \frac{8x}{3}$$

$$2.5 = x.$$

Dividing by $\frac{8}{3}$.

To find y we can substitute for x in the second equation,

i.e. $$y = 9 - 3(2.5)$$

$$y = 9 - 7.5$$

$$y = 1.5.$$

As an exercise you should check this solution, $x=2.5$ and $y=1.5$, by substituting into the first equation.

We can check to see whether it satisfies the third equation, $y=5-x$. In fact $x-2.5$ gives $y=5-2.5=2.5$. We need $y=1.5$, meaning $y=5-x$ does *not* pass through our solution point of (2.5, 1.5).

Now we know the 3 equations do not pass through the same point. Our method continues by taking the equations in pairs.

Using the first and last equations

$$y = \frac{7}{3} - \frac{x}{3}$$

and $$y = 5 - x$$

we subtract them, so eliminating y,

i.e. $$0 = \frac{7}{3} - 5 - \frac{x}{3} + x$$

$$0 = -\frac{8}{3} + \frac{2x}{3}$$

$$\frac{8}{3} = \frac{2x}{3}$$

$$4 = x.$$

Dividing by $\frac{2}{3}$.

To find y we can substitute for x in the last equation,

i.e. $$y = 5 - 4$$

$$y = 1.$$

As an exercise you should check this solution, $x=4$ and $y=1$, by substitution in the usual way.

To find the final point of intersection we use the remaining equation pairing of

$$y = 9 - 3x$$

and $$y = 5 - x.$$

We subtract them to eliminate y,

i.e.
$$0 = 9-5-3x+x$$
$$0 = 4-2x$$
$$2x = 4$$
$$x = 2.$$

To find y we can substitute for x in the last equation,

i.e.
$$y = 5-2$$
$$y = 3.$$

Having solved our problem by 2 different methods let us interpret our answers. x is the number of days' production at the First Site and y is the number of days' production at the Second Site. We have the following table from earlier in the chapter

Order Production	Standard	Deluxe	Prestige
First Site	$100x$	$300x$	$100x$
Second Site	$300y$	$100y$	$100y$
Total	700	900	500

and can substitute for x and y in turn.

$x=2.5$ and $y=1.5$

Actual Production	Standard	Deluxe	Prestige
First Site	250	750	250
Second Site	450	150	150
Total	700	900	400

These production totals show that the 500 Prestige trim part of the order would not be met. Hence we must reject this solution. However, it is possible that, unknown to us, there may be stock to make up the shortfall.

$x=4$ and $y=1$

Actual Production	Standard	Deluxe	Prestige
First Site	400	1200	400
Second Site	300	100	100
Total	700	1300	500

These production totals show all the order being met. We would have a surplus of 400 Deluxe trims. If we used this solution we would need to be able to sell this surplus to another customer.

$x=2$ and $y=3$

Actual Production	Standard	Deluxe	Prestige
First Site	200	600	200
Second Site	900	300	300
Total	1100	900	500

All the order is met with a surplus of 400 Standard trims. Again we ask whether this surplus might be sold to another customer.

The Mathematics has helped us with some possible production plans. However there are many queries yet to be resolved. For example, here are a few of them.

i) Do we hold stock to meet any production shortfalls?
ii) Can we easily sell production surpluses?
iii) With different numbers of production days can we balance up the schedule with other orders?
iv) What are the production costs between plants?

EXERCISE 9.2

Solve the following pairs of simultaneous equations.

1 $x-y=19$ and $x+y=1$

2 $3x+y=7$ and $x+y=11$

3 $2x+y=9$ and $x-y=0$

4 $x-2y=-4$ and $x+2y=36$

5 $3x-y=2$ and $5x+2y=40$

6 $-3x+8y=5$ and $x-5y=3$

7 $3x+2y=2$ and $x+3y=6.5$

8 $x+\frac{1}{2}y=26$ and $\frac{1}{3}x-y=4$

9 $2x-5y=9$ and $7x-y=15$

10 $\frac{1}{2}x-2y=2.5$ and $\frac{1}{3}x+y=0.5$

Linear and quadratic equations – graphical solution

In Chapter 7 we looked at the various shapes associated with quadratics. Now we aim to link together our knowledge and skills concerning straight lines and quadratic curves. The next comment applies to all curves: the accuracy of your plot will affect your answers. Do not expect accuracy beyond 1 or 2 decimal places.

Example 9.10

Solve the simultaneous equations $y=x^2+3x+2$
and $y=x+2$.

We construct 2 tables. Because the first equation represents a curve we need to use quite a few values of x. The specimen values of x are from $x=-3.0$ to $x=1.0$ at intervals of 0.5. If we need more values we just extend our table.

$y=x^2+3x+2$

x	-3.0	-2.5	-2.0	-1.5	-1.0	-0.5	0.0	0.5	1.0 ↔
x^2	9.00	6.25	4.00	2.25	1.00	0.25	0.00	0.25	1.00
$+3x$	-9.00	-7.50	-6.00	-4.50	-3.00	-1.50	0.00	1.50	3.00
$+2$	2.00	2.00	2.00	2.00	2.00	2.00	2.00	2.00	2.00
y	2.00	0.75	0.00	-0.25	0.00	0.75	2.00	3.75	6.00 ↕

The second equation can be compared with the general $y=mx+c$. It is a straight line of gradient 1 and intercept 2. We need to plot only 3 points for a straight line. For our table we have chosen these at random.

$y=x+2$

x	-3	0	3
y	-1	2	5

The graphs are plotted in Fig. 9.9. We can read off the solutions where the graphs intersect. The pairs of solutions are $x=-2$, $y=0$ and $x=0$, $y=2$. you will notice the solutions lie on the axes. This is neither significant nor usual.

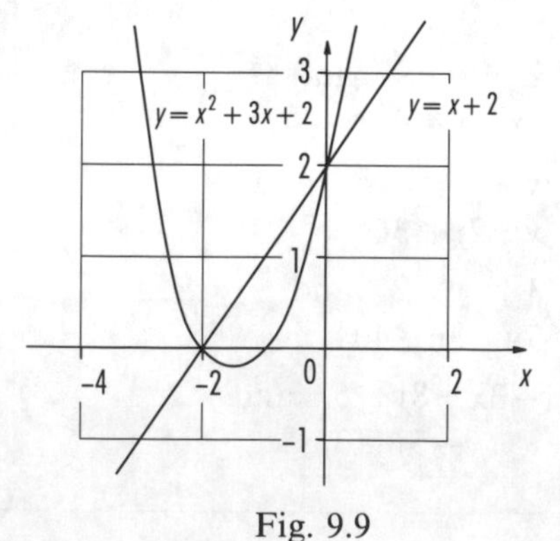

Fig. 9.9

Suppose we look again at this pair of equations

$$y = x^2+3x+2$$
and $$y = x+2.$$

We can eliminate y by subtracting the second equation from the first one,

i.e. $\quad 0 = x^2+2x.$

This gives us a quadratic equation in x which we can solve by factorisation

i.e. $\quad 0 = x(x+2)$

$\therefore \quad x = 0$ or $x+2 = 0$

i.e. $\quad x = 0, -2.$

To find y now we may substitute these values in turn into one of our original equations. We choose the second equation, $y = x + 2$, because it is the simpler one.

$$x = 0 \quad \text{gives } y = 0 + 2 = 2$$
$$\text{and} \quad x = -2 \quad \text{gives } y = -2 + 2 = 0.$$

This alternative method gives the same pair of solutions, $x = -2$, $y = 0$ and $x = 0$, $y = 2$, as the graphical method.

We can use the other equation, $y = x^2 + 3x + 2$, to check our results,

$$\text{i.e.} \quad x = 0 \quad \text{gives } y = 0^2 + 3(0) + 2 = 2$$
$$\text{and} \quad x = -2 \quad \text{gives } y = (-2)^2 + 3(-2) + 2 = 0.$$

This confirms our solutions to be correct.

Example 9.11

Solve the simultaneous equations $y = x^2 + 3x + 2$
and $y + 3x + 7 = 0$.

We have the table for $y = x^2 + 3x + 2$ in Example 9.10. $y + 3x + 7 = 0$ can be re-arranged to $y = -3x - 7$. This is known to be a straight line with a gradient of -3 and an intercept of -7. Choosing 3 specimen values for x we have the table below.

$y = -3x - 7$

x	-4	-2	-1
y	5	-1	-4

Both our graphs are plotted in Fig. 9.10.

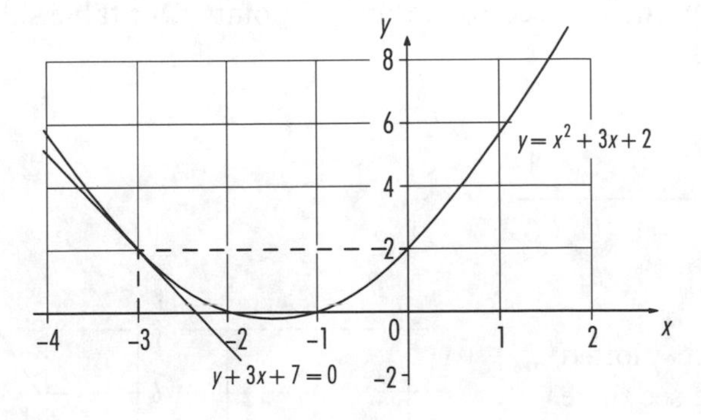

Fig. 9.10

We see the straight line is a tangent to the curve at the point where $x = -3$. The coordinates of this point are $(-3, 2)$ meaning the simultaneous solution is $x = -3$, $y = 2$. The **tangential** aspect means this is the **only solution**. you can think of it as a repeated solution, similar to the idea we saw in Chapter 7 on quadratic equations.

Suppose we look again at this pair of simultaneous equations

$$y = x^2+3x+2$$

and $\quad y = -3x-7.$

We can eliminate y by subtracting the second equation from the first one,

i.e. $\quad 0 = x^2+6x+9.$

This is a quadratic equation in x which we can solve by factorisation,

i.e. $\quad 0 = (x+3)^2$

$\therefore\ x+3 = 0$

i.e. $\quad x = -3$ repeated.

To find y we substitute this value into one of our original equations. We choose the second equation, $y+3x+7=0$, because it is the simpler one.

$x=-3$ gives $\quad y+3(-3)+7 = 0$

i.e.
$$\begin{aligned} y-9+7 &= 0 \\ y-2 &= 0 \\ y &= 2. \end{aligned}$$

The complete solution is $x=-3$, $y=2$ repeated.

We can use the other equation, $y=x^2+3x+2$, to check our result.

$x=-3$ gives $\quad y=(-3)^2+3(-3)+2 = 9-9+2=2.$

This confirms our solution to be correct.

Example 9.12

Solve the simultaneous equations $\quad y=x^2+3x+2$

and $\quad y=x-1.$

We have the table for $y=x^2+3x+2$ in Example 9.10. Because $y=x-1$ is a known straight line we need to plot only 3 points. Our table shows just 3 specimen values.

$y=x-1$

x	-1	0	$1.$
y	-2	-1	$0.$

Both graphs are plotted in Fig. 9.11. We see there is **no intersection** of the curve and the straight line. This means there is **no simultaneous solution**. No pair of coordinates satisfies both equations.

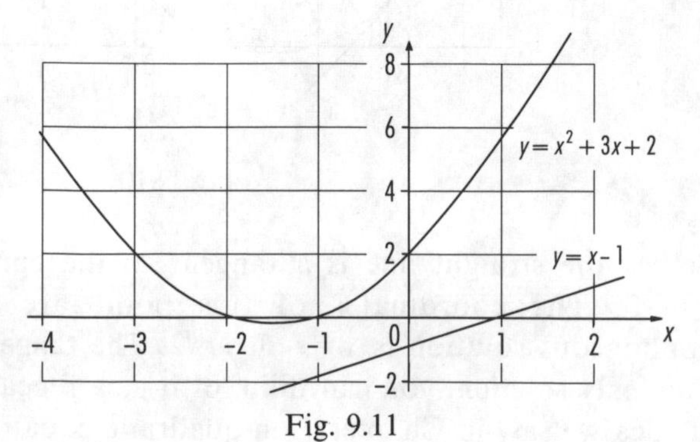

Fig. 9.11

As we have done previously we may look at another method. Again we eliminate y by subtraction,

i.e. $\quad y = x^2+3x+2$

and $\quad y = \quad\quad x-1$

give $\quad 0 = x^2+2x+3$

This quadratic equation will not factorise. The quadratic equation formula will not work because it involves the **square root of a negative number**. This is consistent with the curve not crossing the horizontal axis. Fig. 9.12 provides us with that confirmation.

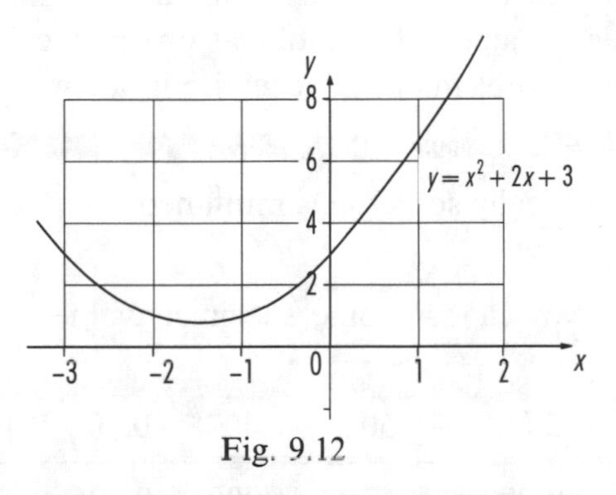

Fig. 9.12

EXERCISE 9.3

Graphically solve the following pairs of simultaneous equations. In each case you will see they are a quadratic equation and a linear equation.

1 $y=x^2+x-2$ and $y=4x+2$.
In your tables use values of x from $x=-2$ to $x=5$ at intervals of 0.5.

2 $y=3x^2+x+1$ and $y=-1.5x+6$.
In your tables use values of x from $x=-2$ to $x=2$ at intervals of 0.5.

3 $y=2x^2+11x+5$ and $y=3x-3$.
In your tables use values of x from $x=-6$ to $x=1$ at intervals of 0.5.

4 $y=4x^2-4x+1$ and $y+x+2=0$.
In your tables use values of x from $x=-1$ to $x=3$ at intervals of 0.5.

5 $y=2x^2-7x+6$ and $2y+x=6$.
In your tables use values of x from $x=0$ to $x=4$ at intervals of 0.5.

Linear and cubic equations – graphical solution

The principles we have learned already in this chapter apply here. We know from earlier work what linear and cubic graphs look like. This section just combines the graphs. There is a variety of possible solutions:

i) they may intersect 3 times, or
ii) they may intersect once and touch once, or
iii) they may intersect once only.

When the straight line touches the curve it acts as a tangent to the curve at that point. Remember we say the solution is repeated in this case.

There are many variations and combinations of curves and straight lines. Example 9.13 looks at one particular situation. Just work through other problems in the exercise in a similar way.

Example 9.13

Graphically solve the simultaneous equations $y = x^3$
and $y = 2.5x + 1$.

We can choose some specimen values for x and construct a table for $y = x^3$.

x	−2.00	−1.50	−1.00	−0.50	0	0.50	1.00	1.50	2.00	2.50
y	−8.000	−3.375	−1.000	−0.125	0	0.125	1.000	3.375	8.000	15.625

Because $y = 2.5x + 1$ is a known straight line the table is simpler.

x	−2.00	0	2.50
y	−4.00	1.00	7.25

The graphs are plotted in Fig. 9.13. There are 3 points of intersection. These give the simultaneous solutions of $x = 1.75$, $y = 5.37$; $x = -0.44$, $y = -0.10$ and $x = -1.32$, $y = -2.30$. Remember that how well the graph is plotted affects the accuracy of the solution.

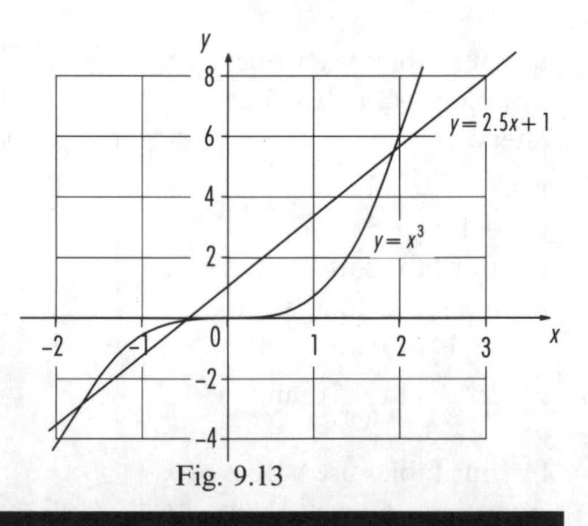

Fig. 9.13

EXERCISE 9.4

Graphically solve the following pairs of simultaneous equations. In each case you will see they are a cubic equation and a linear equation.

1 $y = x^3$ and
$y = 3x + 1$.
In your tables use values of x from $x = -2$ to $x = 3$ at intervals of 0.5.

2 $y = 2x^3 + x^2$ and
$y = 4x$.
In your tables use values of x from $x = -2$ to $x = 2$ at intervals of 0.5.

3 $y=x^3$ and
$y=-9$.
In your tables use values of x from $x=-3$ to $x=1$ at intervals of 0.5.

4 $y=(x+2)(x-1)(x-3)$ and
$y=x+1$.
In your tables use values of x from $x=-3$ to $x=4$ at intervals of 0.5.

5 $y=x^3+x^2+x+1$ and
$y=x+3.5$.
In your tables use values of x from $x=-2$ to $x=3$ at intervals of 0.5.

Quadratic and cubic equations – graphical solution

In this final section of the chapter we continue with our basic principles. Again one example is sufficient.

Example 9.14

Graphically solve the simultaneous equations $y=x^3$
and $y=2x^2+5x-6$.

We can choose some specimen values for x and construct a table for $y=x^3$.

x	−2.50	−2.00	−1.50	−1.00	−0.50	0	0.50	1.00	1.50	2.00	2.50	3.00	3.50
y	−15.625	−8.000	−3.375	−1.000	−0.125	0	0.125	1.000	3.375	8.000	15.625	27.000	42.875

The table for $y=2x^2+5x-6$ is

x	−2.50	−2.00	−1.50	−1.00	−0.50	0	0.50	1.00	1.50	2.00	2.50	3.00	3.50
$2x^2$	12.50	8.00	4.50	2.00	0.50	0	0.50	2.00	4.50	8.00	12.50	18.00	24.50
$+5x$	−12.50	−10.00	−7.50	−5.00	−2.50	0	2.50	5.00	7.50	10.00	12.50	15.00	17.50
-6	−6.00	−6.00	−6.00	−6.00	−6.00	−6.00	−6.00	−6.00	−6.00	−6.00	−6.00	−6.00	−6.00
y	−6.00	−8.00	−9.00	−9.00	−8.00	−6.00	−3.00	1.00	6.00	12.00	19.00	27.00	36.00

The graphs are plotted in Fig. 9.14. There are 3 points of intersection. These give the simultaneous solutions of $x=-2$, $y=-8$; $x=1$, $y=1$; and $x=3$, $y=27$.

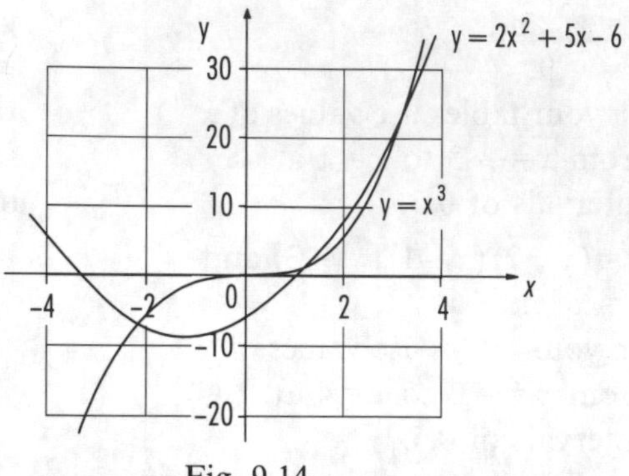

Fig. 9.14

EXERCISE 9.5

Graphically solve the following pairs of simultaneous equations. In each case you will see they are a quadratic equation and a cubic equation.

1 $y=x^3$ and
$y=4x^2+4x$.
In your tables use values of x from $x=-1$ to $x=5$ at intervals of 0.5.

2 $y=x^3$ and
$y=x^2-x-2$.
In your tables use values of x from $x=-3$ to $x=3$ at intervals of 0.5.

3 $y=\frac{1}{2}x^3$ and
$y=-5+x-x^2$.
In your tables use values of x from $x=-4$ to $x=1$ at intervals of 0.5.

4 $y=3x^3$ and
$y=10(x^2-1)$.
In your tables use values of x from $x=-1.5$ to $x=2$ at intervals of 0.5.

5 $y+2x^3=0$ and
$y=3x^2+1$.
In your tables use values of x from $x=-2.5$ to $x=0.5$ at intervals of 0.5.

The theory is complete. Now you are in a position to try a mixed set of questions in this last exercise. In each question the equations have been formed for you.

EXERCISE 9.6

1 The electrical circuit shows 2 emfs together with various resistors. The sums of the relevant voltages produce the simultaneous equations. They are in terms of the currents, I_1 and I_2 amps.

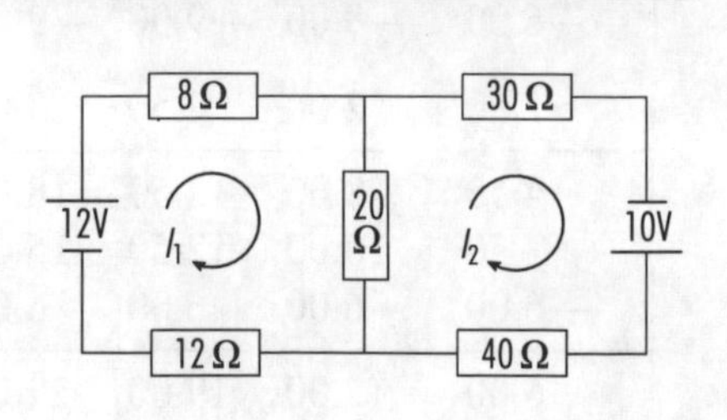

$$8I_1+20(I_1-I_2)+12I_1=12$$
and
$$40I_2+20(I_2-I_1)+30I_2=10.$$

These equations will simplify. Show that they may be reduced to
$$10I_1-5I_2=3$$
and
$$-2I_1+9I_2=1.$$

Now solve this pair of equations for I_1 and I_2.

2 A uniform heavy beam of length 20 m has a bending moment, M. x is the distance measured from one end and M is related to x by $M=2x(20-x)$. Construct a table of values of x and M using intervals of 2 m all along the beam. Plot a graph of M against x. From your graph, what is the maximum bending moment and where does it occur?

Another beam is shorter at 15 m. Its bending moment is given by $M=4x(15-x)$. Again, along the length of the beam construct a table of values for M and x. On the same set of axes hence plot a graph of M against x. What is the maximum bending moment for this second beam and where does it occur?

Where are the 2 bending moments equal? What is that common value of M?

3 The resistance, R kΩ, in an electrical circuit is related to its temperature, $T°$ Celsius.

At first this is thought to be $R=200T-450$. Using 3 specimen values of T from 20 to 150 construct a simple table. Hencc plot a graph of R against T.

An improved relation is found to be $R=2T^2+3T-500$. Using values of T from 20 to 150 at intervals of 10 construct a table for this new relationship. On the same pair of axes plot the new graph of R against T. Use your graphs to find at what temperature(s) the predicted resistances might be the same. Also find the values of

i) T when $R=24$ kΩ,

ii) R when $T=135°$ Celsius.

4 The diagram shows the frustum of a cone. Its volume, V, is given by

$$V=\frac{1}{3}\pi h(R^2+Rr+r^2).$$

R is the base radius and r is the top radius. They are separated by a vertical height, h. Suppose $h=1.2$ m and $r=0.5$ m so that

$$V=0.4\pi(R^2+0.5R+0.25).$$

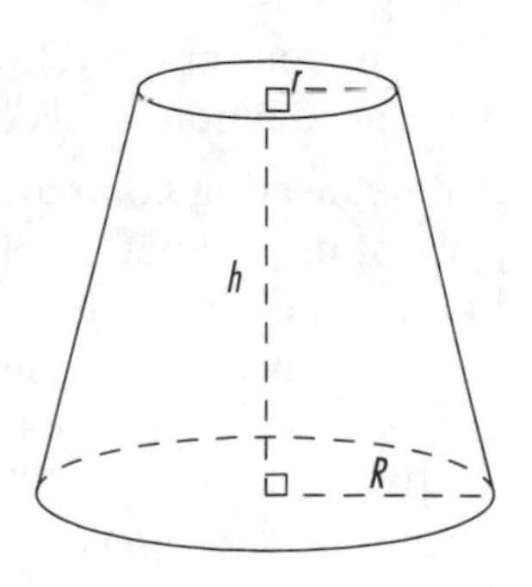

Values of R are to range from 0 m to 1.4 m at intervals of 0.2 m. Construct a table and then plot a graph of V against R.

The volume of a hemisphere of the same base radius, R, is given by $V=\frac{2}{3}\pi R^3$. Using the same values for R construct another table. Plot V against R on the same pair of axes. From your graphs find

i) the value of R (m) where the volumes are equal,

ii) those equal volumes (m^3).

5 There are two vehicles, a van and a car. The van is travelling at a constant speed of $18\,ms^{-1}$. The distance it travels, s metres, is related to time, t seconds, by $s=18t$. Choose 3 specimen values of t up to 20 seconds. Construct a table and then plot a graph of s against t.

The car, from a standing start, attempts to catch the van. The distance it travels, s, is related to time, t, by $s=\frac{4}{3}t^2$. Using values of t from $t=0$ to $t=20$ seconds at intervals of 2 seconds construct a table of values. On the same set of axes plot a graph of s against t. When does the car catch the van? By what distance is it ahead 20 seconds after the start?

Suppose the car had been travelling at $5\ ms^{-1}$ initially so that $s=5t+\frac{4}{3}t^2$.

Construct another table and plot a third graph on the axes. From your graphs find the new time when the overtaking now occurs.

6 A production company has fixed and variable costs which in total are y. (The real £ value of y has been scaled to avoid confusion.) It produces x thousand items each week so that $y=0.5+0.75x$. In the equation which term refers to fixed costs and which term refers to variable costs?

The company's sales income is $y=0.5(x^3-2x^2+x+2)$. Use values of x from $x=0.4$ (i.e. 400 items/week) to $x=2.4$ (i.e. 2400 items/week) at intervals of 0.2. Construct tables for these graphs. Hence plot them on one set of axes.

A break even point occurs where costs and income are equal. At what production levels does the company break even according to your graphs?

From your graph estimate the greatest profit and the worst loss. At what production levels do they occur?

7 An engineering company's position in the market is oriented to the price of its product, £x. It would like to supply more at a higher price, but as the price rises so demand falls away. The market is in equilibrium when supply equals demand. For the following straight line laws one represents supply and one represents demand. They are $y=100+10x$ and $y=350-8x$.

Choose 3 specimen values of x between 5 and 40. Construct tables for these relations and plot the graphs on one set of axes.

Decide which graph represents demand and which represents supply. Decide on the equilibrium price.

Improved research shows that $y=400-1.5x^2$ and $y=50+0.5x^2$ are more likely supply and demand curves. Use values of x from 5 to 25 at intervals of 2.5 to construct the necessary tables. Hence plot these graphs on a fresh pair of axes. Read off from your graph the improved estimate of the equilibrium price.

8 For values of x from $x=-3.5$ to $x=3.5$ radians at intervals of 0.5 construct a table for $y=\sin x$. Plot a graph of this relationship.

It is thought that $y=x-\dfrac{x^3}{6}$ is a close match for $y=\sin x$. Using the same x values construct a table for this relation. Hence plot a graph of y against x on the same pair of axes. Decide whether there actually is a close match.

9 For values of x from $x=-2.0$ to $x=2.0$ radians at intervals of 0.4 construct a table for $y=\cos x$. Plot a graph of this relationship.

It is thought that $y=1-\dfrac{x^2}{2}$ is a close match for $y=\cos x$. Using the same x values construct a table for this relation. Hence plot a graph of y against x on the same pair of axes. Decide whether there actually is a close match.

10 Use your tables from Questions **8** and **9**, extending them to cover from $x=-3$ to $x=3$. On one pair of axes plot the graphs of $y=x-\dfrac{x^3}{6}$ and $y=1-\dfrac{x^2}{2}$.

Where do they intersect? Convert these radian values of x into degrees. One of them is quite accurate whilst the other is not. Decide which is the accurate one.

10 Indices and Logarithms

The objectives of this chapter are to:

1. Recall the laws of indices.
2. Define a logarithm to any base.
3. Convert a simple indicial relationship into a logarithmic relationship and vice versa.
4. Define a common logarithm.
5. Work out common logarithms with a calculator.
6. Sketch the graphs of logarithmic functions.
7. Define a natural (Naperian) logarithm.
8. Evaluate natural logarithms with a calculator.
9. Deduce the laws of logarithms.
10. State that $\log_b 1 = 0$, $\log_b b = 1$ and as $x \to 0$ $\log_b x \to -\infty$.
11. State the relationship between common logarithms and natural logarithms.
12. Solve indicial equations where the indices are linear in one unknown.
13. Solve indicial equations where the indices are quadratic in form.
14. Use logarithms and their laws to evaluate expressions in science and technology.
15. Solve equations involving e^x and $\log_e x$ ($\ln x$).

Introduction

This chapter looks at 3 ideas. They are indices (which you will have seen before), common logarithms and natural logarithms. They are all related and you will see the relationships throughout the chapter.

ASSIGNMENT

This Assignment looks at the power gain of electrical components and systems. There is a formula for power gain. It is a logarithmic ratio of output power, P_0, and input power, P_i.

$$\text{Power gain} = 10\log_{10}\left(\frac{P_0}{P_i}\right).$$

The units for power gain are decibels, dB.

The same formula works for a power loss. We know that $10-35$ is the same as $10+(-35)$, i.e. subtracting 35 is the same as adding -35. Similarly power loss $(-)$ is a negative power gain $(+(-))$.

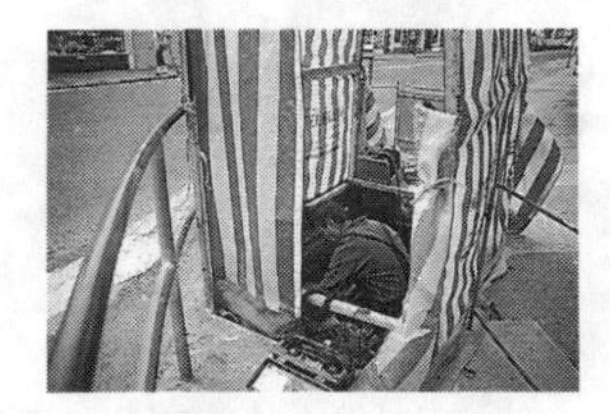

The Assignment is in 2 parts. In the first part there is a calculation based on an amplifier. The second part looks at a simple system involving a series of transmission lines and amplifiers.

The laws of indices

We can write numbers in many forms. One of these is

$$\textbf{NUMBER} = \textbf{BASE}^{\textbf{index}}$$

e.g. $2 = 10^{0.3010}$

The plural of **index** is **indices**. An alternative word for **index** is **power**. It is different from electrical power. We raise the base to a power (or index). When we look at this example the index is rather awkward as a decimal. We are more used to whole numbers or simple fractions like $\frac{1}{2}$ or $\frac{3}{4}$.

As a first step let us recall the basic laws of indices. Whenever we simplify indices they must relate to the **same base**. If the bases are different there is no easy simplification.

We have a look at 6 laws of indices to jog your memory. Each law is followed by a simple example.

LAW 1 $\quad b^m \times b^n = b^{m+n}$

multiplying: add the indices

e.g. $2^4 \times 2^5 = 2^{4+5} = 2^9$ or 512.

LAW 2 $\quad b^m \div b^n = b^{m-n}$

dividing: subtract the indices

e.g. $\dfrac{2^8}{2^5} = 2^{8-5} = 2^3$ or 8.

LAW 3 $\quad (b^m)^n = b^{mn}$

raise to a power: multiply the indices

e.g. $(2^3)^2 = 2^{3\times 2} = 2^6$ or 64.

LAW 4 $\quad b^0 = 1$

raise to a power 0: answer is always 1.

This may seem strange at first glance, but the demonstration is easy. Suppose we have $\frac{2^6}{2^6}$. We can look at this in two different ways.

$$\frac{2^6}{2^6} = 2^{6-6} = 2^0.$$

Also $\dfrac{2^6}{2^6} = \dfrac{2\times2\times2\times2\times2\times2}{2\times2\times2\times2\times2\times2}$

Cancelling 2 six times.

$= \dfrac{1}{1}$

$= 1.$

In each case we started with $\dfrac{2^6}{2^6}$ yet reached answers in different forms. Equating our answers shows that $2^0 = 1$.

LAW 5 $\quad \boldsymbol{b^{-p} = \dfrac{1}{b^p}}$

e.g. $2^{-4} = 2^{0-4} = \dfrac{2^0}{2^4} = \dfrac{1}{2^4}$.

$2^7 = 2^{--7} = 2^{0--7} = \dfrac{2^0}{2^{-7}} = \dfrac{1}{2^{-7}}$.

Using the second law.

Using the fourth law.

LAW 6 $\quad \boldsymbol{b^{m/n} = \sqrt[n]{b^m}}$ ***or*** $\boldsymbol{(\sqrt[n]{b})^m}$

e.g. $2^{5/3}$ may be written as $\sqrt[3]{2^5}$ or $(\sqrt[3]{2})^5$.

We can use the calculator and sequence of buttons learned in Chapter 1.

$\sqrt[3]{2^5}$ has a cube root (i.e. the power $\frac{1}{3}$) and uses

2 x^y 5 =

to display 32 and then

$x^{1/y}$ 3 =

to give an answer of 3.1748 . . .

Alternatively $(\sqrt[3]{2})^5$ uses

2 $x^{1/y}$ 3 =

to display 1.2599 . . . and then

x^y 5 =

to give the same answer of 3.1748 . . .

You will see some similarities between these laws of indices and the laws of logarithms.

Logarithms

We know NUMBER = BASE^{index}. If a number, N, is written as b^x then the index, x, is called the **logarithm**. x is the logarithm of N to the base b.

This is written as $N=b^x$ or $\log_b N=x$.

log is the shortened form of logarithm.

It is important to know that **logarithms are defined only for positive numbers**.

In $\log_b N=x$ the base, b, can take many values. You will find the two important and usual ones on your calculator. These are the bases 10 and e. All the laws of logarithms apply to any base, but we will concentrate on these particular bases.

Common logarithms

Logarithms to the base 10 are called **common logarithms**. Before cheap electronic calculators people used tables of values of common logarithms. Now we can find values by simply pressing buttons on a calculator. We replace b with 10 in our basic definitions so that $N=10^x$ and $\log_{10}N=x$. The notation can be shortened by omitting 10 from $\log_{10}N$. It is understood that $\log N$ means the common logarithm of N.

Examples 10.1

Find the values of

i) $\log 2$, ii) $\log 4$, iii) $\log 10$, iv) $\log 100$, v) $\log 0.1$, vi) $\log 1$.

The calculator button we need is log. Just input the number and press log to display each answer.

i) $\log 2$ uses the buttons 2 log to give 0.3010 (4 dp).

ii) $\log 4=0.6021$ (4 dp).

iii) $\log 10=1$.

iv) $\log 100=2$.

v) $\log 0.1=-1$.

vi) $\log 1=0$.

In place of $\log N=x$ we can think of the alternative version, $N=10^x$. Then we can re-write each question part to show the logarithm in the index position.

i) $2=10^{0.3010}$.

ii) $4=10^{0.6021}$.

iii) $10=10^1$.

iv) $100=10^2$.

v) $0.1=10^{-1}$.

vi) $1=10^0$.

All these examples started with a number, N, and gave a logarithm, x. Suppose we start with x and attempt to find the number, N, according to $N = 10^x$. Above the log button on your calculator you should find the inverse function, 10^x. This used to be called the **antilog**. The next set of examples uses this inverse function.

Examples 10.2

Find the numbers whose common logarithms are

i) 0.4771, ii) -0.3010, iii) 3.7000.

The order of calculator buttons involves inputting the value and then pressing inv log. The effect of inv log is to apply 10^x.

i) $\log N = 0.4771$ is the same as $N = 10^{0.4771}$.
The order of calculator buttons is

0.4771 inv log

to display 2.99985...
Correct to 3 decimal places this is $N = 3$.

ii) $\log N = -0.3010$ is the same as $N = 10^{-0.3010}$ or $\dfrac{1}{10^{0.3010}}$.
Using the same calculator order we get $N = 0.500$ (3 dp).

iii) $\log N = 3.700$ is the same as $N = 10^{3.7000}$ so $N = 5012$ (4sf).

EXERCISE 10.1

Find the values of

1 $\log 5.75$
2 $\log 13.40$
3 $\log 1000$
4 $\log 26.80$
5 $\log 200$
6 $\log 0.50$
7 $\log 10^6$
8 $\log 0.25$
9 $\log -0.25$
10 $\log 1.76$

Find the numbers with the following common logarithms Give your answers to 3 significant figures where appropriate.

11 1.23
12 -0.55
13 0.6021
14 0.1761
15 2.46
16 0.699
17 -1.25
18 1.658
19 3.69
20 1.949

Common logarithmic graph

It is always interesting to see a diagram of a function. We can use the calculator's log button to create a table.

N	0.05	0.1	0.5	1.0	10.0	50.0	100.0
$\log N$	-1.301	-1.000	-0.301	0.000	1.000	1.699	2.000

The specimen values of N have not been chosen for regularity. Instead they have been chosen so you can see the tremendous change in N and how this affects $\log N$. The graph of $y = \log N$ is shown in Fig. 10.1 .

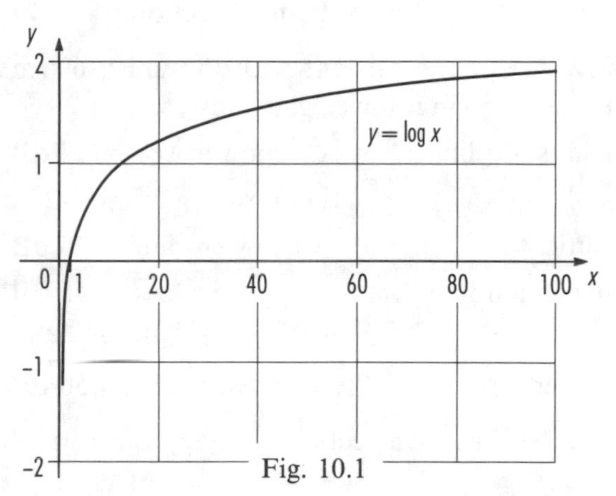

Fig. 10.1

Notice the graph crosses the horizontal axis at 1, i.e. $\log 1 = 0$. Another important feature is that $\log 10 = 1$. Generally we find when the number, N, and the base, b, are the same then the logarithm is 1, i.e. $\log_b b = 1$.

As N increases the gradient becomes less steep, tending to flatten a little. As N decreases, approaching 0, the graph tends to $-\infty$. We write this mathematically; as $x \to 0$ then $\log_b x \to -\infty$. The vertical axis acts as an asymptote. An asympote is a line a graph approaches but never quite touches. In this case the log graph approaches the vertical axis but never quite touches it.

ASSIGNMENT

1. We are going to look at an amplifier's power gain. Suppose its input power is 175 mW and its output power is 10 W. When we introduced our Assignment we had an output power, P_0, and input power, P_i.

Using our values we have $P_0 = 10\,\text{W}$ and $P_i = 175\,\text{mW} = 0.175\,\text{W}$. Notice how we amend the input power so that the units are consistently watts, W. We can re-write our formula too. We have learned that for common logarithms we do not need to include the base, 10.

Now $\quad$ power gain $= 10\log\left(\dfrac{P_0}{P_i}\right)$

becomes $\quad$ power gain $= 10\log\left(\dfrac{10}{0.175}\right)$

$$= 10\log(57.143)$$
$$= 10 \times 1.757$$
$$= 17.6 \text{ dB}.$$

2. In this second part we are going to combine 4 different sections, 2 transmission lines and 2 amplifiers. We will input a signal of 600 mW into the first section of transmission line. The question is what happens to that power input by the end of the system? The output from one section of the system is the input into the next section.

Firstly let us be consistent with the units, 600 mW = 0.6 W. Suppose we have the following power gains/losses:

Transmission line 1, power loss = 35 dB (i.e. power gain = −35 dB)
Amplifier 1, power gain = 25 dB
Transmission line 2, power loss = 45 dB (i.e. power gain = −45 dB)
Amplifier 2, power gain = 50 dB

We can look at each section applying our formula in turn.

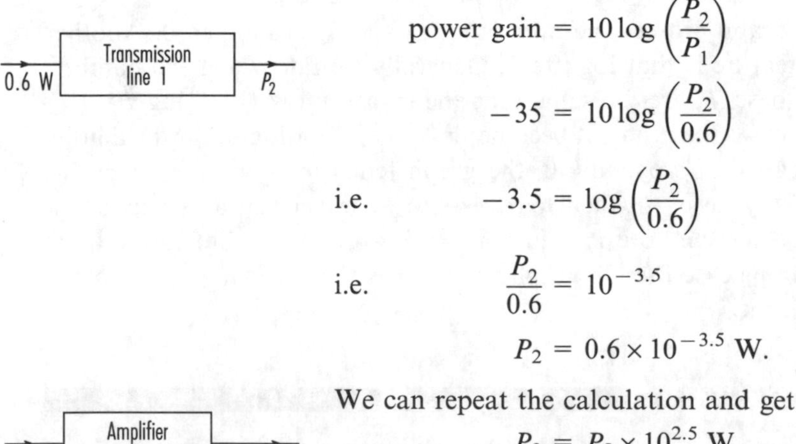

$$\text{power gain} = 10\log\left(\frac{P_2}{P_1}\right)$$
$$-35 = 10\log\left(\frac{P_2}{0.6}\right)$$
i.e.
$$-3.5 = \log\left(\frac{P_2}{0.6}\right)$$
i.e.
$$\frac{P_2}{0.6} = 10^{-3.5}$$
$$P_2 = 0.6 \times 10^{-3.5} \text{ W}.$$

We can repeat the calculation and get

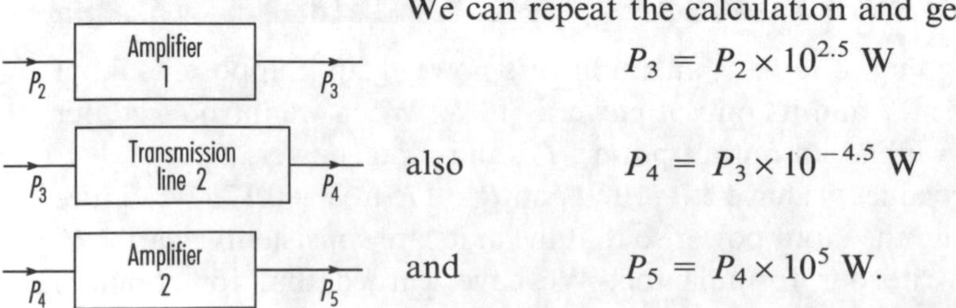

$$P_3 = P_2 \times 10^{2.5} \text{ W}$$
also
$$P_4 = P_3 \times 10^{-4.5} \text{ W}$$
and
$$P_5 = P_4 \times 10^5 \text{ W}.$$

As an exercise you should check these calculations for yourself. We can link together all the calculations:

$$P_5 = P_4 \times 10^5$$
$$= (P_3 \times 10^{-4.5}) \times 10^5.$$

Substituting for P_4.

$$= (P_2 \times 10^{2.5}) \times 10^{-4.5} \times 10^5$$

Substituting for P_3.

$$= (0.6 \times 10^{-3.5}) \times 10^{2.5} \times 10^{-4.5} \times 10^5$$
$$= 0.6 \times 10^{-3.5+2.5-4.5+5}$$
$$= 0.6 \times 10^{-0.5}$$
$$= 0.6 \times 0.316$$
$$= 0.190 \text{ W} \qquad \text{or } 190 \text{ mW.}$$

You can see that the initial input of 600 mW has been reduced to a final output of 190 mW. In the combination of all the calculations you can spot the power gains/losses in each index.

For an extension to this problem you might like to replace an amplifier so you achieve neither a power gain nor loss.

Natural logarithms

Logarithms to the base e are called **natural** or **Naperian** logarithms. Again before cheap electronic calculators people used tables of values of natural logarithms. Now we can find values by simply pressing buttons on a calculator. We replace b with e in our basic definitions so that

$N = e^x$ and $\log_e N = x$.

The notation can be shortened from $\log_e N$. It is understood that $\ln N$ means the natural logarithm of N.

e^x is the **exponential** function. We will look at this in detail in Chapter 11.

Numerical calculations can be performed using any type of logarithms. However, natural logarithms are the more widely used type in mathematics, science and technology. We can check out the value of e (i.e. e^1) by using a calculator.

Input 1 inv ln to display 2.71828 . . . Often this is quoted correct to 3 decimal places as 2.718.

We look at some values in the next set of examples.

Examples 10.3

Find the values of

i) ln 2, ii) ln 4, iii) ln 2.718, iv) ln 7.389, v) ln 0.35, vi) ln 1.

The calculator button we need is ln. Just input the number and press ln to display each answer.

i) ln 2 uses the buttons 2 ln to give 0.693 (3 dp).

ii) $\ln 4 = 1.386$ (4 sf).

iii) $\ln 2.718 = 1$ (3 sf).

iv) $\ln 7.389 = 2$ (3 sf).

v) $\ln 0.35 = -1.0498$ (4 dp).

vi) $\ln 1 = 0$.

In place of $\ln N = x$ we can think of the alternative version, $N = e^x$. Now for each example part we give the alternative version, showing the logarithm in the index position.

i) $2 = e^{0.693}$.

ii) $4 = e^{1.386}$.

iii) $2.718 = e^1$.

iv) $7.389 = e^2$.

v) $0.35 = e^{-1.0498}$.

vi) $1 = e^0$.

All these examples started with a number, N, and gave a logarithm, x. Suppose we start with x and attempt to find the number, N, according to $N = e^x$. Above the ln button on your calculator you should find the inverse function, e^x. This used to be called the **antilog**. The next set of examples uses this inverse function.

Examples 10.4

Find the numbers whose natural logarithms are

i) 0.4771, ii) −0.3010, iii) 3.7000.

The order of calculator buttons involves inputting the value and then pressing inv ln. The effect of inv ln is to apply e^x.

i) $\ln N = 0.4771$ is the same as $N = e^{0.4771}$.
The order of calculator buttons is

0.4771 inv ln

to display 1.61139...
Correct to 3 decimal places this is $N = 1.611$

ii) $\ln N = -0.3010$ is the same as $N = e^{-0.3010}$ or $\dfrac{1}{e^{0.3010}}$.
Using the same calculator order we get $N = 0.740$ (3 dp).

iii) $\ln N = 3.700$ is the same as $N = e^{3.7000}$ so $N = 40.45$ (4sf).

EXERCISE 10.2

Find the values of

1 $\ln 15.75$

2 $\ln 3.40$

3 $\ln 1000$

4 $\ln 6.80$

5 $\ln 250$

6 $\ln 0.15$

7 $\ln e^6$

8 $\ln 0.3679$

9 $\ln -0.75$

10 $\ln 175$

Find the numbers with the following natural logarithms
Give your answers to 3 significant figures where appropriate.

11 1.23

12 -0.55

13 0.6021

14 0.1761

15 2.46

16 0.699

17 -1.25

18 1.658

19 3.69

20 1.949

Natural logarithmic graph

Again, a diagram of this function should be interesting. We can use the calculator's ln button to create a table.

N	0.05	0.1	0.5	1.0	10.0	50.0	100.0
$\ln N$	-2.996	-2.303	-0.693	0.000	2.303	3.912	4.605

From the above table of values you can see that for very large changes in N there are relatively small changes in $\ln N$. The graph of $y = \ln N$ is shown in Fig. 10.2.

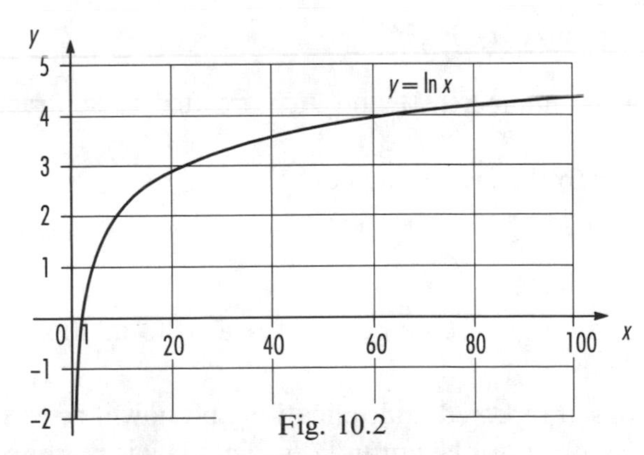

Fig. 10.2

Notice the graph crosses the horizontal axis at 1, i.e. $\ln 1 = 0$. Also $\ln e = 1$. The characteristics are much the same as for the common logarithmic graph. As N increases the gradient becomes less steep, tending to flatten a little. As N decreases, approaching 0, the graph tends to $-\infty$. The way we write this mathematically is $N \to 0$ then $\ln N \to -\infty$. The vertical axis acts as an asymptote.

For comparison the graphs of these different logarithmic functions are plotted together in Fig. 10.3. They do change at slightly different rates, but have similar tendencies.

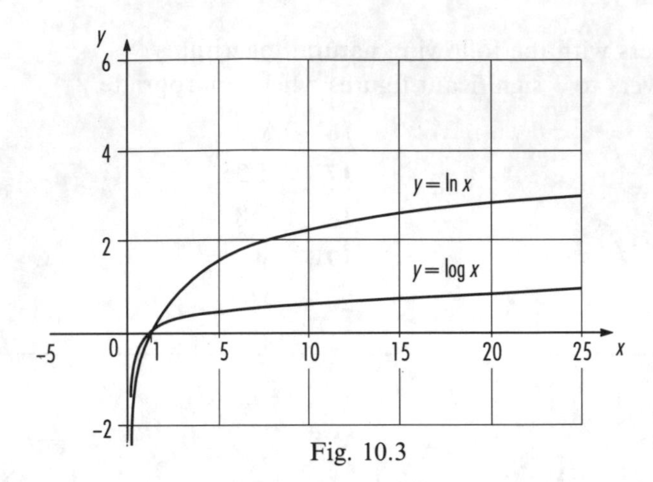

Fig. 10.3

Laws of logarithms

These laws apply to any logarithmic base. We can quote them generally.

1 $\quad \mathbf{\log_b(MN) = \log_b M + \log_b N}$

i.e. multiplying the numbers, M and N, is related to adding their logs. This is similar to the law of indices

$$\mathbf{b^m \times b^n = b^{m+n}}$$

2 $\quad \mathbf{\log_b\left(\frac{M}{N}\right) = \log_b M - \log_b N}$

i.e. dividing the numbers, M and N, is related to subtracting their logs.
This is similar to the law of indices

$$\mathbf{b^m \div b^n = b^{m-n}}$$

3 $\quad \mathbf{\log_b(N^a) = a\log_b N}$

i.e. raising a number, N, to a power is related to multiplying its log, $\log_b N$, by the power.

This law looks at repeated multiplication and how it affects the logs. It is similar to our third law of indices. That law is not quoted again to avoid confusion with the letters we are using here.

The laws may be re-written simply in both our bases.

Common logs, i.e. base 10	*Natural logs, i.e. base e*
$\log MN = \log M + \log N$	$\ln MN = \ln M + \ln N$
$\log\frac{M}{N} = \log M - \log N$	$\ln\frac{M}{N} = \ln M - \ln N$
$\log N^p = p\log N$	$\ln N^p = p\ln N$

Indicial equations

We can use the laws of logarithms to help solve some indicial equations. An indicial equation involves the unknown (e.g. x) in the index position.

Examples 10.5

Solve the indicial equations i) $3^x = 7$
and ii) $3^x = 6$.

The hint before these examples was to use the laws of logarithms. We can choose logarithms to either base 10 or e. For both these examples it is the third law that looks the closest one of them all.

i) In this case we choose the base e.

For $\quad 3^x = 7$

take natural logarithms of both sides,

To maintain the balance.

i.e.

$$\ln 3^x = \ln 7$$

$$x \ln 3 = \ln 7$$

3rd law.

$$x(1.0986) = 1.9459$$

$$x = \frac{1.9459}{1.0986}$$

$$x = 1.77 \quad (3 \text{ sf}).$$

ii) In this case we choose the base 10.

For $\quad 3^x = 6$

take common logarithms of both sides,

To maintain the balance.

i.e.

$$\log 3^x = \log 6$$

$$x \log 3 = \log 6$$

3rd law.

$$x(0.4771) = 0.7782$$

$$x = \frac{0.7782}{0.4771}$$

$$x = 1.63 \quad (3 \text{ sf}).$$

As an exercise you should repeat these 2 examples. For $3^x = 7$ use common logarithms and for $3^x = 6$ use natural logarithms. In each case you will find the different bases do not affect the final answers.

Examples 10.6

Solve the indicial equations i) $3^x = 15.4^{x-2}$,
ii) $3^x = 15 \times 4^{x-2}$.

Like the previous examples we can use either common or natural logarithms. We choose to use common logarithms.

i) Starting with $3^x = 15.4^{x-2}$
take common logarithms of both sides

$$\begin{aligned}
\text{i.e.}\quad \log 3^x &= \log 15.4^{x-2}\\
x\log 3 &= (x-2)\log 15.4 && \text{3rd law.}\\
x(0.4771) &= (x-2)1.1875\\
x(0.4771-1.1875) &= -2\times 1.1875 && \text{2nd law.}\\
\text{i.e.}\quad x(-0.7104) &= -2.3750\\
x &= \frac{-2.3750}{-0.7104} && \text{3rd law.}\\
x &= 3.34 \quad (3\text{ sf}).
\end{aligned}$$

ii) Starting with $3^x = 15\times 4^{x-2}$

take common logarithms of both sides,

$$\begin{aligned}
\text{i.e.}\quad \log 3^x &= \log(15\times 4^{x-2})\\
&= \log 15+\log 4^{x-2}\\
\text{i.e.}\quad x\log 3 &= \log 15+(x-2)\log 4\\
x(0.4771) &= 1.1761+(x-2)0.6021\\
x(0.4771-0.6021) &= 1.1761-(2\times 0.6021)\\
\text{i.e.}\quad x(-0.1250) &= 1.1761-1.2042\\
x &= \frac{-0.0281}{-0.1250}\\
x &= 0.225 \quad (3\text{ sf}).
\end{aligned}$$

Examples 10.7

Solve the indicial equations i) $6^{2x} = 5\times 6^x$,
ii) $8^{2x}-9\times 8^x+20=0$.

One of the laws we remembered at the beginning of the chapter is particularly useful:

$$(b^m)^n = b^{mn} = b^{nm} = (b^n)^m$$

i) In $6^{2x} = 5\times 6^x$

we may think of 6^{2x} as $(6^x)^2$

Now let $q=6^x$ so the original equation becomes

$$\begin{aligned}
q^2 &= 5q\\
\text{i.e.}\quad q^2-5q &= 0\\
q(q-5) &= 0\\
\therefore\quad q &= 0 \text{ or } q-5 = 0\\
q &= 0,\ 5.
\end{aligned}$$

When $q=0$ we have $6^x=0$. This equation has no solution because logarithms are defined only for positive numbers. Of course 0 is **not** a positive number. Also when $q=5$ we have $6^x=5$.

Take common logarithms of both sides

i.e. $\log 6^x = \log 5$

$$x \log 6 = \log 5$$

$$x(0.7782) = 0.6990$$

$$x = \frac{0.6990}{0.7782}$$

$$x = 0.898 \quad \text{(3 sf)}.$$

ii) In $8^{2x} - 9 \times 8^x + 20 = 0$

think of 8^{2x} as $(8^x)^2$

Now let $q = 8^x$ so the original equation becomes

$$q^2 - 9q + 20 = 0$$

This quadratic equation will factorise,

i.e. $(q-4)(q-5) = 0$

$\therefore$ $q - 4 = 0$ or $q - 5 = 0$

$$q = 4,\ 5.$$

When $q = 4$ we have $8^x = 4$.

Take common logarithms of both sides,

i.e. $\log 8^x = \log 4$

$$x \log 8 = \log 4$$

$$x(0.9031) = 0.6021$$

$$x = \frac{0.6021}{0.9031}$$

$$x = 0.667 \quad \text{(3 sf)}.$$

Also when $q = 5$ we have $8^x = 5$.

Using common logarithms in much the same way gives

$$x = \frac{0.6990}{0.9031}$$

$$x = 0.774 \quad \text{(3 sf)}.$$

EXERCISE 10.3

Solve thc indicial equations for x in each case.

1 $2^x = 8$
(You should be able to spot this solution.)

2 $5^x - 14 = 0$

3 $2 \times 5^x - 14 = 0$

4 $4^x = 6^{x-1}$

5 $3^x = 7.5^{x-2}$

6 $2^x \times 3^x = 4^{x+1}$

7 $3 \times 4^x = 6^{x-1}$

8 $2^x \times 5^{x+2} = 6^{x-1}$

9 $4^{2x} = 5 \times 4^x$

10 $3^{2x} - 10 \times 3^x + 21 = 0$

Change of base

There is a link between common and natural logarithms. We need the option of being able to convert between them.

Consider $\log N = x.$

We can re-write this in its alternative form,

i.e. $N = 10^x.$

3rd law.

Now take natural logarithms of both sides,

i.e. $\ln N = \ln 10^x$

$\ln N = x \ln 10$

i.e. $\dfrac{\ln N}{\ln 10} = x$

i.e. $\dfrac{\ln N}{\ln 10} = \log N.$

We may re-write this as

$$\log N = \frac{\ln N}{2.3026}$$

or $\log N = 0.4343 \ln N.$

We can use this formula in the next example.

Example 10.8

If we have $\ln N = 1.6094$ what is the value of $\log N$?

Using $\log N = \dfrac{\ln N}{\ln 10}$

we get $\log N = \dfrac{1.6094}{2.3026}$

i.e. $\log N = 0.699$ (3 sf).

A quick calculator check reveals the value of N to be 5. You can see from this example that not knowing it to be 5 was no handicap.

We can look again to create the opposite conversion.

Consider $\ln N = x.$

We can re-write this in its alternative form,

i.e. $N = e^x.$

Now take common logarithms of both sides,

i.e. $\log N = \log e^x$

$\log N = x \log e$

3rd law.

i.e. $\dfrac{\log N}{\log e} = x$

i.e. $\dfrac{\log N}{\log e} = \ln N.$

We may re-write this as

$$\ln N = \frac{\log N}{0.4343}$$

or $$\ln N = 2.3026 \log N$$

We can use this formula in the next example.

Example 10.9

If we have $\log N = 0.7910$ what is the value of $\ln N$?

Using $$\ln N = \frac{\log N}{\log e}$$

we get $$\ln N = \frac{0.7910}{0.4343}$$

i.e. $$\ln N = 1.821 \qquad \text{(4 sf).}$$

A quick calculator check reveals the value of N to be 6.18. Again you can see that not knowing it to be 6.18 was no handicap.

EXERCISE 10.4

Use the conversion formula to find the value of $\log N$ in each case.

1 $\ln N = 2.5$
2 $\ln N = 5.175$
3 $\ln N = -0.35$
4 $\ln N = 4.60$
5 $\ln N = 1.123$

Use the other conversion formula to find the value of $\ln N$ in each case.

6 $\log N = 2.5$
7 $\log N = -0.65$
8 $\log N = 5.0$
9 $\log N = 7.38$
10 $\log N = 3.0$

Now let us look at some examples applying these techniques to practical problems.

Example 10.10

In a cooling system every second $V\,\text{m}^3$ of cooling fluid flows a distance x m according to $V = 3.95x^{2.55}$. In the design prototype x may be adjusted to test various specifications. Using this formula estimate the volume of cooling fluid when $x = 0.95$ m. Also find the value of x needed for a flow of $7.50\,\text{m}^3$.

There are several ways to solve this problem. This method uses techniques we have learned in this chapter.

Given $x = 0.95$ we substitute into our formula

$$V = 3.95x^{2.55}$$

to get $$V = 3.95(0.95)^{2.55}.$$

We can find the value of this using the calculator.

Input 0.95, press the x^y button and then 2.55 followed by = to display 0.877...

Then
$$V = 3.95 \times 0.877\ldots$$
$$= 3.47\text{ m}^3 \qquad (3\text{ sf})$$

i.e. a volume of 3.47 m^3 of cooling fluid flows each second for this design specification.

The second part of the problem uses the substitution of $V=7.50$ into our formula.

$$V = 3.95x^{2.55}$$

becomes
$$7.50 = 3.95x^{2.55}$$

i.e.
$$\frac{7.50}{3.95} = x^{2.55}$$
$$1.899 = x^{2.55}.$$

Take common logarithms of both sides to give

$$\log 1.899 = \log x^{2.55}$$

i.e.
$$\log 1.899 = 2.55 \log x$$

i.e.
$$\frac{0.278}{2.55} = \log x$$
$$0.109 = \log x$$

i.e.
$$x = 10^{0.109}$$
$$x = 1.29\text{ m}.$$

Example 10.11

In a submarine cable the speed of the signal, v, is related to the radius of the cable's covering, R, by $v = \frac{25}{R^2} \ln\left(\frac{R}{5}\right)$.

Plot a graph of v against R using values of R from R = 2 mm to 10 mm at intervals of 1 mm. With the aid of the graph find the maximum signal speed. For what value of R does this occur?

Our first step is to construct a table of values. This has quite a few extra rows of working. Each row is labelled to aid the explanation.

R	2	3	4	5	6	7	8	9	10	①
$\frac{R}{5}$	0.400	0.600	0.800	1.000	1.200	1.400	1.600	1.800	2.000	②
$\ln\left(\frac{R}{5}\right)$	−0.916	−0.511	−0.223	0.000	0.182	0.336	0.470	0.588	0.693	③
R^2	4	9	16	25	36	49	64	81	100	④
$\frac{25}{R^2}$	6.250	2.778	1.563	1.000	0.694	0.510	0.391	0.309	0.250	⑤
v	−5.73	−1.42	−0.35	0.00	0.13	0.17	0.18	0.18	0.17	⑥

We can emphasise the meaning of each row.
Row ① is the specimen values of R.
Row ② shows each value of R divided by 5.
Row ③ takes the natural logarithms of Row ②.
Row ④ shows each specimen value of R squared.
Row ⑤ uses Row ④, dividing 25 by each value.
Row ⑥ is the product of Rows ⑤ and ③, the values of v.

Fig. 10.4 shows the plot of v against R. You can see that above the horizontal axis the curve bends only slightly. If you plot this for yourself take care.

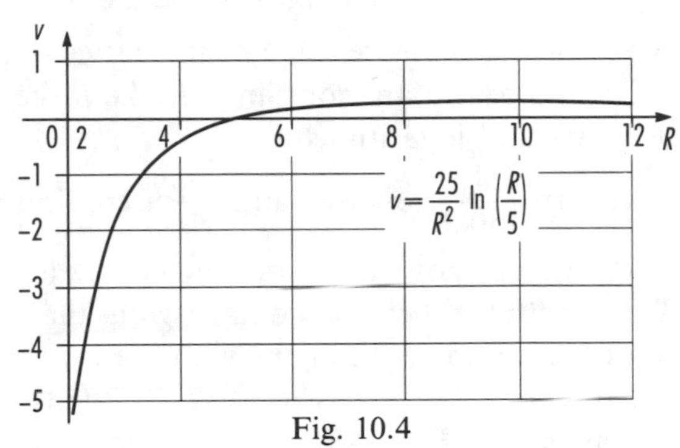

Fig. 10.4

Because of the slight bend Fig. 10.5 looks in more detail at this section of the graph.

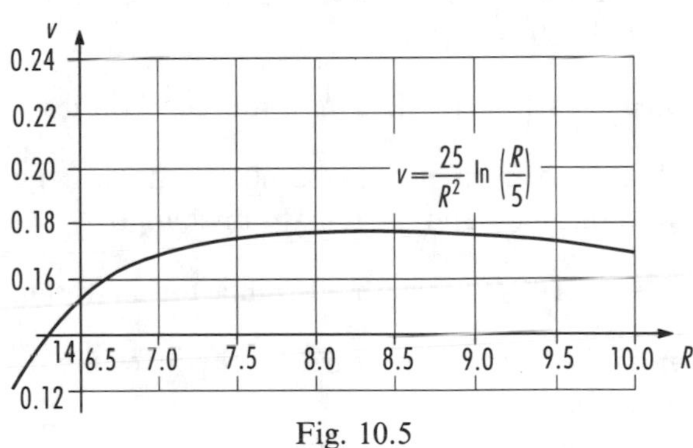

Fig. 10.5

The maximum signal speed occurs at the peak of the curve. Here a tangent to the curve is horizontal. From our graph we see the maximum signal speed is approximately $v = 0.18$ occurring at $R = 8.2$ mm, each correct to 2 significant figures.

EXERCISE 10.5

1 The power gain of an amplifier relates input power, P_1, to output power, P_0, by

$$\text{Power gain} = 10 \log \frac{P_0}{P_1}$$

The units for power gain are decibels, dB.
Calculate the power gain if $P_0 = 18$ W and $P_1 = 300$ mW. For an input power of 300 mW and a power gain of 20 dB what must be the output power?

2 The cutting speed, S m/min, and tool life, T min, of a machine tool for roughing cuts in steel are related by $ST^n = c$. c is a constant. Take common logarithms of both sides to make n the subject of this formula.

3 In a transmitting antenna the attenuation, A, is given by $A = 10\log\left(1+\frac{R_L}{R_A}\right)$. If $R_L = 2.75\,\Omega$ is the loss resistance and $R_A = 90\,\Omega$ is the antenna loss calculate A.

4 Atmospheric pressure, p cm of mercury, is related to height, h m, by $pe^{ah} = c$. a and c are constants. Make h the subject of this formula using natural logarithms.

Given $a = \frac{1}{15000}$, $c = 76.2$ and $p = 50$ cm find h.

5 The diagram shows a belt in contact with a pulley. The length of belt in contact with the pulley subtends an angle θ radians with the centre. The coefficient of friction is $\mu = 0.32$. The tension on the taut side is $T_1 = 34.75$ N and on the slack side is $T_0 = 22.50$ N.

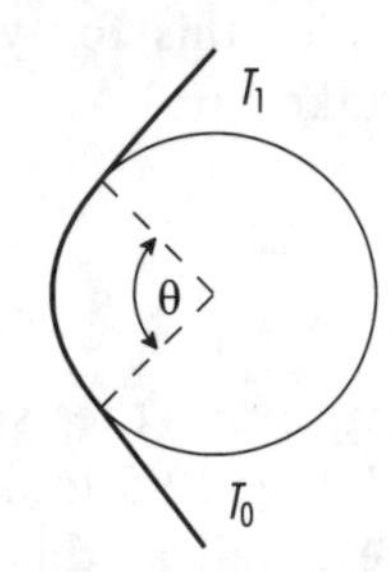

If $\theta = \frac{1}{\mu}\ln\left(\frac{T_1}{T_0}\right)$ calculate θ for these values.

Alternatively if θ is to be 60° what must be the new coefficient of friction if T_1 and T_0 remain unchanged?

6 $A = P\left(1+\frac{R}{100}\right)^n$ is the formula for compound interest.

n is the number of years for the investment,
R is the interest rate,
P is the principal sum invested,
A is the total amount of the principal sum and interest.

The accountant of a local engineering firm earns a profit-related bonus of £10 000. He invests it at an interest rate of 8.5%. How many years will it be before his original bonus is doubled (i.e. $A = £20\,000$)?

If there is an alternative investment with an interest rate of 10.5% re-calculate the new time to double his money.

What would be the interest rate if his original investment doubled in 5 years?

7 The diagram shows a submarine signalling cable. The radius of the core is r and the radius of the covering is R. The ratio of these radii is given by $x = \frac{r}{R}$. The speed of the signal, v, is given by $v = x^2 \ln\frac{1}{x}$. Use a law of logarithms to re-arrange this formula into $v = -x^2 \ln x$.

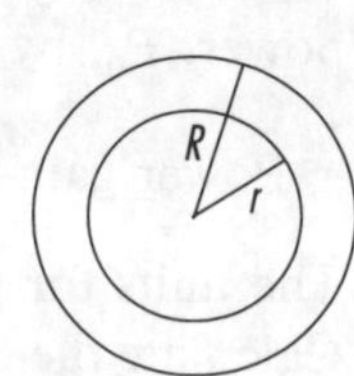

For values of x from 0.1 to 0.8 plot a graph of v against x. When $v=0.125$ what is the value of x? Given the value of r as 5.5 mm what is the associated radius of the covering for this signal speed?

8 A local engineering firm uses the formula $V=P\left(1-\frac{R}{100}\right)^T$ to calculate depreciation costs for its equipment.
T is the working life of the equipment (years),
R is the rate of depreciation,
P is the original equipment cost,
V is the remaining second-hand value of the equipment.
The firm buys some capital equipment for £17 500 with an expected working life of 4 years. The second-hand value when the firm sells it is approximately £5500. What rate of depreciation has the firm used (correct to 2 significant figures)?

9 The electrical current, C, in a wire is related to time, t, by

$$kt = -\ln(1+C).$$

Show that this formula may be re-written as

$$t = \frac{1}{k}\ln\left(\frac{1}{1+C}\right).$$

10 The diagram shows an electrical circuit with an inductor, L, and resistor, R, connected in series to an emf, E. There is a time constant, T, given by $T=\frac{L}{R}$. Calculate the value of T.

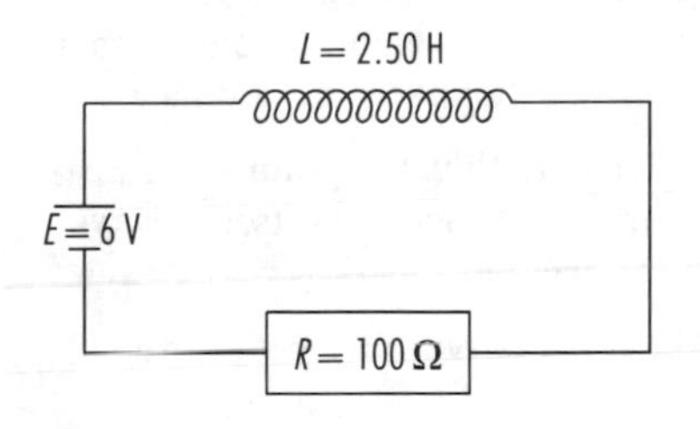

The formula for the current, i, flowing in this circuit during time, t, is given by $i=\frac{E}{R}(1-e^{-t/T})$.
Make t the subject of this formula using natural logarithms. If $i=51$ mA $(=0.051$ A) calculate the value of time t (seconds).

11 Graphs of Logarithms and Exponentials

The objectives of this chapter are to:

1. Evaluate exponentials.
2. Understand the graphical connection between $y = \ln x$ and $y = e^x$.
3. Plot the graphs of $y = e^x$ and $y = e^{-x}$.
4. Determine the gradient of an exponential curve and recognise that it is proportional to the ordinate.
5. Understand that the gradient of an exponential function is itself always an exponential function.
6. Sketch the curves of $y = e^x$, $y = e^{-x}$, $y = 1 - e^x$, $y = 1 - e^{-x}$, $y = ae^{bx}$, $y = ae^{-bx}$, $y = a(1 - e^{-bx})$.
7. Relate such exponential curves to practical applications such as population growth, radioactive decay, growth or decay of an electric current.
8. Draw the graphs of experimental data.
9. Use logarithms to reduce laws of the types $y = ae^{bx}$, $y = ax^b$ and $y = ab^x$ to straight line form.
10. Tabulate values of $\ln x$ (and $\log x$) and $\ln y$ (and $\log y$).
11. Plot a straight line graph to verify a relationship.
12. Determine whether experimental results obey laws and estimate constants from the graph.
13. Use the straight line graph to determine intermediate values.
14. Use log-linear graph paper to plot straight line graphs for $y = e^x$, $y = e^{-x}$, $y = ab^x$ and $y = ae^{bx}$.
15. Use log-log graph paper to plot a straight line graph for $y = ax^b$.

Introduction

In this chapter we are going to link together the natural logarithm, $\ln x$, and the exponential function, e^x. We will apply them in various forms and graphical sketches. By changing the scales on the axes we will present some logarithmic graphs as straight lines instead of curves.

ASSIGNMENT

The Assignment for this chapter involves a belt around part of a pulley. They are in contact along the arc AB. θ is the angle subtended at the centre of the pulley, i.e. $\angle AOB = \theta$. There is friction between the belt and the pulley. The coefficient of friction is μ. This means the tensions on either side of the pulley are different.

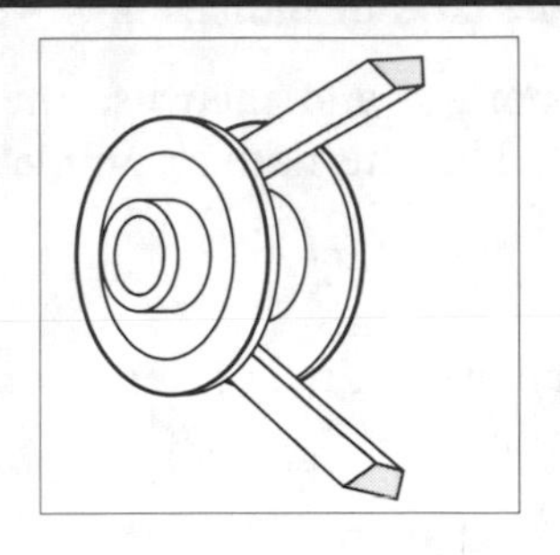

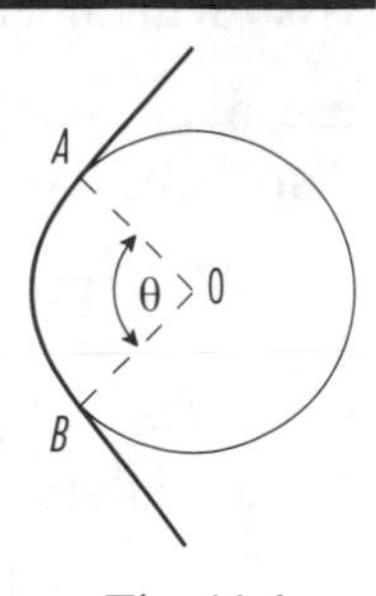

Fig. 11.1

The exponential function

In Chapter 10 we looked at natural (or Naperian) logarithms. They are logarithms to the base e,

i.e. $\log_e N = x$ or $\ln N = x$ and $N = e^x$.

The value of e is 2.71828…, often shortened to 2.718. When no index is written we understand the index is 1. This means we understand e to be e^1. Using a calculator we can find the value of e for ourselves. Often e^x is shown above the ln button. The calculator order is

1 inv ln

to display 2.71828…

The inv button together with the ln button shows that e^x is the inverse function of ln. This is why it is often found above the ln button on a calculator.

Examples 11.1

Find the values of i) $e^{2.5}$, ii) $e^{-2.5}$, iii) $3.74e^{-2.5}$.

i) For $e^{2.5}$ the calculator order is

2.5 inv ln

to display 12.18 (4 sf).

ii) Comparing $e^{-2.5}$ with Example 11.1i) what changes is the sign of the index. Notice how we make that change in the calculator order

2.5 +/− inv ln

to display 0.0821 (4 dp).

Also remember the laws of indices. $e^{-2.5}$ is $\frac{1}{e^{2.5}}$. This means we can use our result from $e^{2.5}$ and then use the 1/x button to reach our display of 0.0821. The alternative complete order for $e^{-2.5}$ is

2.5 inv ln 1/x

iii) For $3.74e^{-2.5}$ again we build on Example 11.1ii), multiplying by 3.74. The calculator order is

2.5 +/– inv ln

to display 0.08... and

× 3.74 =

to display a final answer of 0.307 (3 dp).

Remember that the laws of indices we recalled in Chapter 10 continue to apply for similar bases.

Examples 11.2

Find the values of i) $e^{2.5} \times e^{1.7}$, ii) $\frac{e^{2.7}}{e^{1.7}}$, iii) $(e^{2.5})^4$.

We apply the appropriate law of indices **before** using the calculator. This saves time and effort.

i) For $e^{2.5} \times e^{1.7}$ the bases are the same so we can add the indices, i.e. $e^{2.5} \times e^{1.7} = e^{2.5+1.7} = e^{4.2}$.
Now the calculator order is

4.2 inv ln

to display 66.69 (4 sf).

ii) Again, using another law of indices $\frac{e^{2.5}}{e^{1.7}} = e^{2.5-1.7} = e^{0.8}$.
The calculator order is

0.8 inv ln

to display 2.226 (4 sf).

iii) Yet another law of indices gives $(e^{2.5})^4 = e^{2.5 \times 4} = e^{10}$.
The calculator order is

10 inv ln

to display 22026 (5 sf).
We can look again at this example using the power button x^y. This time we find the value of $e^{2.5}$ and then raise that value to the power 4.
The alternative order is

2.5 inv ln x^y 4

ASSIGNMENT

Let us take a first look at our pulley problem. The friction between the belt and the pulley is important. The tension of the belt before it passes around the pulley differs from the tension afterwards. On either side of the pulley they are related by the formula $T_1 = T_0 e^{\mu\theta}$. T_1 is the tension on the taut side and T_0 is the tension on the slack side. Suppose the slack side tension is 22.50 N, the coefficient of friction is 0.35 and the angle subtended is 80°, i.e. $T_0 = 22.50$, $\mu = 0.35$ and $\theta = 80°$. For the formula to work we need θ to be in radians. We saw how to convert degrees to radians in Chapter 4, i.e.

$$\theta = 80° = \frac{80 \times \pi}{180} = 1.396 \text{ radians.}$$

We can substitue our values into the formula so that

$$\begin{aligned} T_1 &= 22.50\, e^{0.35 \times 1.396} \\ &= 22.50\, e^{0.489} \\ &= 22.50 \times 1.630 \end{aligned}$$

i.e. $T_1 = 36.7$ N.

This means the tension on the taut side of the pulley is 36.7 N.

EXERCISE 11.1

Find the values of

1 $e^{1.7}$

2 $e^{7.4}$

3 e^{-1}

4 e^0

5 $2e^{7.4}$

6 $\dfrac{2e^{7.4}}{e^{1.7}}$

7 $2e^{1.7} \times 3e^{7.4}$

8 $7.4\,e^2$

9 $\dfrac{7.4}{e^2}$

10 $7.4\,e^{-2}$

11 $\dfrac{7.4}{e^{-2}}$

12 $\dfrac{e^{2.35}}{e^{1.2}}$

13 $\dfrac{e^{2.35}}{e^{-1.2}}$

14 $(e^{2.35})^2$

15 $(e^2)^{2.35}$

16 $(5e^{2.35})^2$

17 $5(e^{2.35})^2$

18 $\dfrac{e^{-1.2}}{e^{-2.35}}$

19 $\dfrac{1}{4e^{1.75}}$

20 $\dfrac{1}{4e^{-1.75}}$

Exponential graphs

Exponential graphs involve either e^x or e^{-x}. e^x represents exponential **growth** because the index is **positive**. e^{-x} represents exponential **decay** because the index is **negative**. It is interesting to see a diagram of a function. As we have done so many times we can construct a table and plot a graph of each function. The table is built up using specimen values of x and the inv ln calculator buttons.

$y = e^x$

x	-2.5	-2.0	-1.0	0	1.0	2.0	2.5
e^x	0.082	0.135	0.368	1	2.718	7.389	12.182

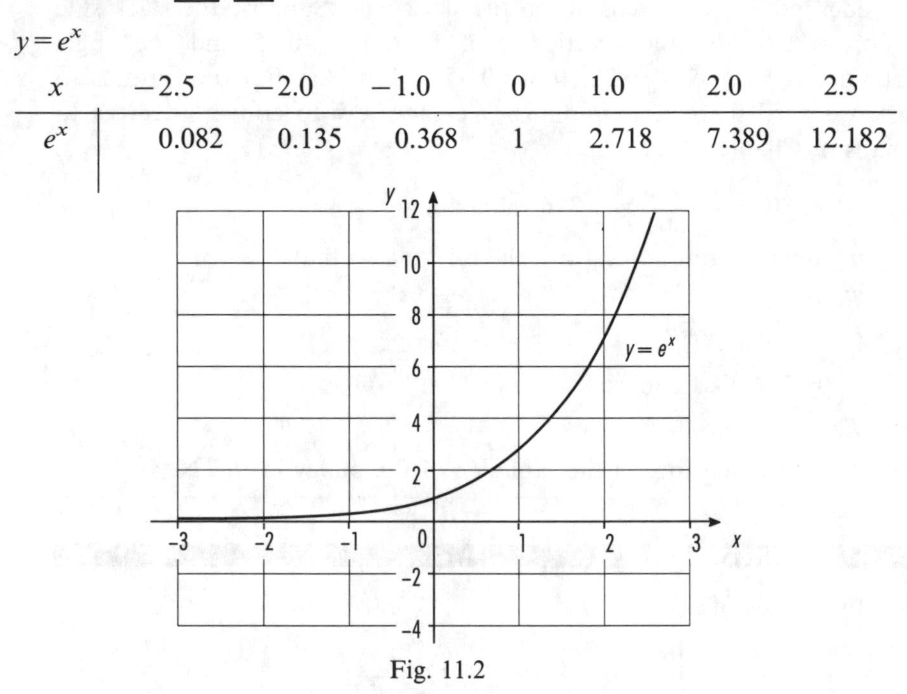

Fig. 11.2

The graph is shown in Fig. 11.2. Notice how it approaches the x-axis but never quite touches it. This means the horizontal axis is an asymptote to the curve. An asymptote is a straight line that a graph gets closer and closer to without actually touching it. Also notice that the curve crosses the vertical axis at $y = 1$. The gradient of the curve is always positive. Remember that gradient is $\text{gradient} = \dfrac{\text{vertical change}}{\text{horizontal change}}$

$y = e^{-x}$

x	-2.5	-2.0	-1.0	0	1.0	2.0	2.5
e^{-x}	12.182	7.389	2.718	1	0.368	0.135	0.082

You can see a pattern when comparing this table of values with the previous table. The graph is shown in Fig. 11.3. Again, notice how the x-axis acts as an asymptote. Again we see it crosses the vertical axis at $y = 1$. This time the gradient of the curve is always negative.

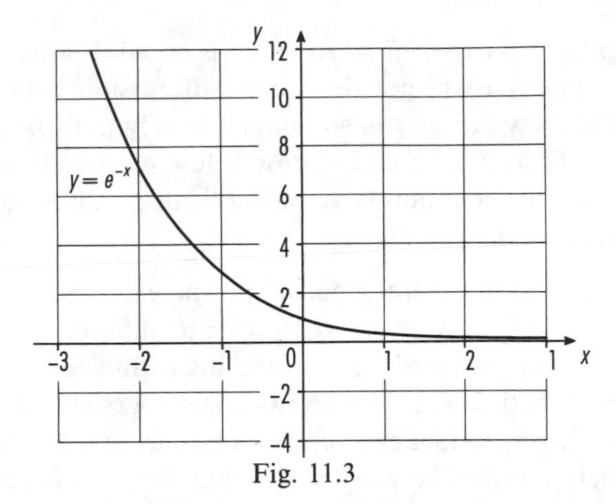

Fig. 11.3

Fig. 11.4 brings these curves together on one pair of axes. It emphasises that each curve is a reflection of the other in the vertical axis. This is hinted at by the pattern of numbers in the 2 tables.

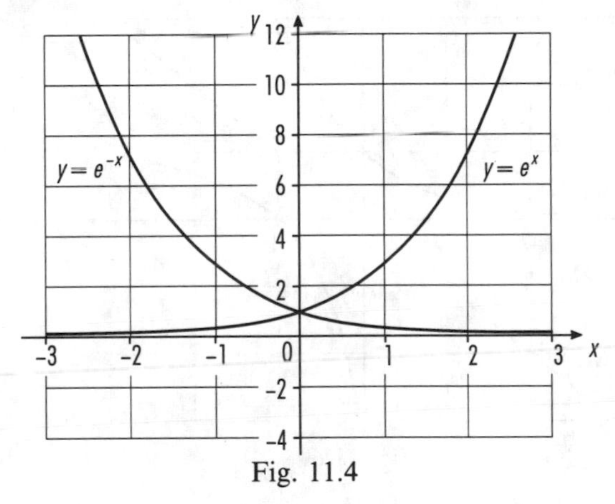

Fig. 11.4

We can look more closely at the gradients of these curves. Fig. 11.5 shows some specimen tangents. With care you find the gradients are related to the curve. (The quality and accuracy of your own graphs are important

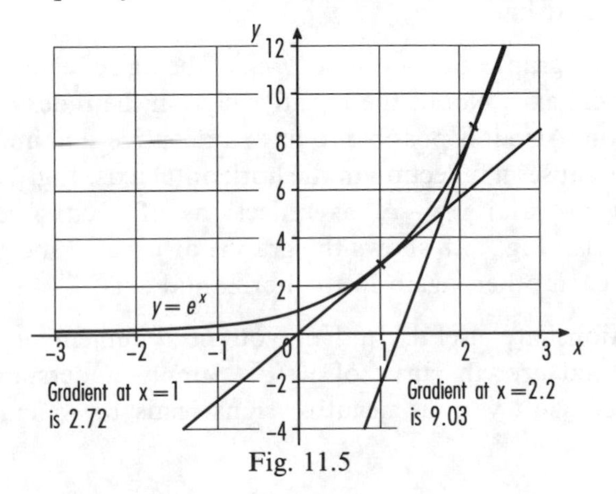

Fig. 11.5

here.) The gradient of an exponential function is equal to an exponential function. At this stage we will not deepen the discussion with a precise rule. As an exercise for yourself you should accurately plot the graphs of $y=e^x$ and $y=e^{-x}$. Then you should choose a few points on each curve and find the gradients at those points. Each gradient should have the same numerical value as y at that point.

Let us look at the graphs of natural logarithmic and exponential functions together. We know that $\ln N=x$ and $N=e^x$ are different forms of a relationship. The first one is based on a natural logarithm. The second one is based on an exponential. Fig. 11.6 shows them together with the line $y=x$ on one pair of axes. In fact each graph is a reflection of the other in the line $y=x$. Such a reflective property occurs for a function and its inverse function. For yourself look at the asymptotes: the x-axis for one function, the y-axis for the other function. One of the graphs crosses an axis at $(0, 1)$, the other at $(1, 0)$.

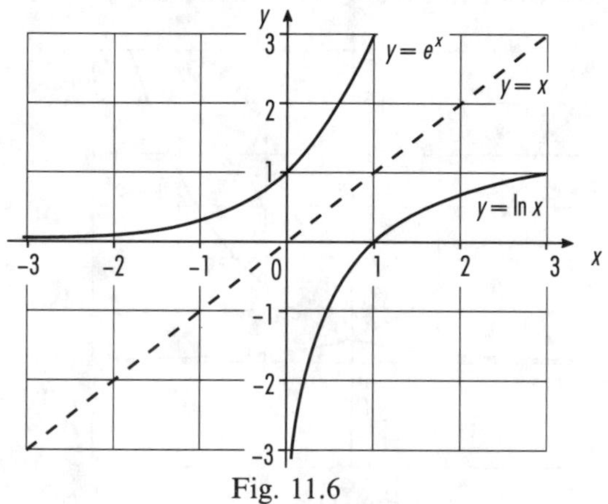

Fig. 11.6

The same situation applies to the common logarithmic function. As an exercise for yourself plot on one pair of axes the graphs of $y=\log x$ and $y=10^x$. If you then plot the graph of $y=x$ on those axes you will notice the reflections in that line.

Fig. 11.4 shows the graphs of $y=e^x$ and $y=e^{-x}$ being reflections of each other in the vertical axis. Notice the negative sign in the index associated with this reflection. Also it is possible to have a negative sign immediately before the e. This causes a reflection in the horizontal axis. Fig. 11.7 shows the graphs of $y=e^x$ and $y=-e^x$ as reflections of each other in the horizontal axis. Also Fig. 11.8 shows the graphs of $y=e^{-x}$ and $y=-e^{-x}$ as reflections of each other, again in the horizontal axis.

All these reflections are useful. In fact you need remember only the original exponential growth curve of $y=e^x$. Simply understanding the reflections created with various negative signs leads to so many other curves.

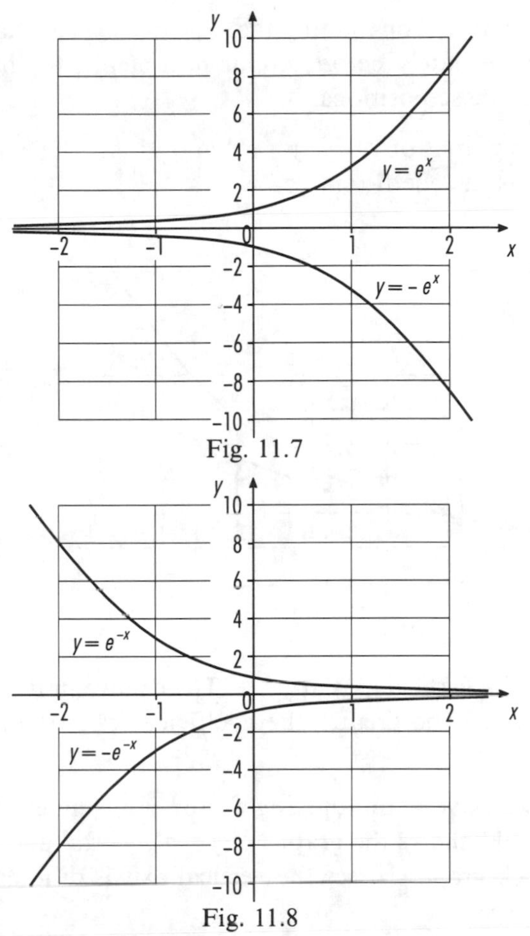

Fig. 11.7

Fig. 11.8

We can apply a similar building technique to sketch the graph of $y=1-e^x$. We do it in stages. Firstly sketch the graph of $y=e^x$. Reflect that curve in the horizontal axis to get $y=-e^x$. If we re-write $y=1-e^x$ as $y=1+(-e^x)$ the final stage is reached by an upward shift of 1. Fig. 11.9 shows these graphs of $y=1-e^x$.

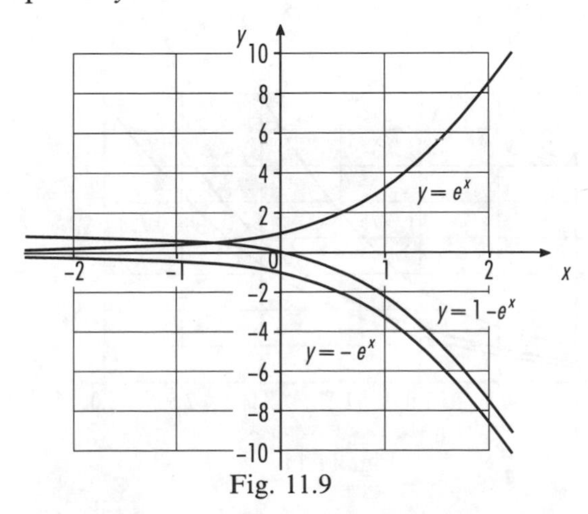

Fig. 11.9

So far we have concentrated on altering the +/– signs. These have been in the index and/or immediately before e. Our next step is to change some numbers in either of these positions.

Fig. 11.10 shows the graphs of $y=e^x$, $y=e^{2x}$ and $y=e^{0.75x}$, i.e. our indices have 1, 2 and 0.75 as the coefficients of x.

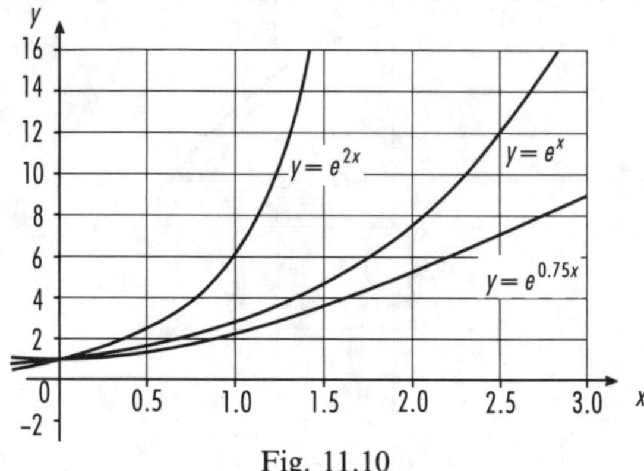

Fig. 11.10

Each graph crosses the vertical axis at $y=1$. The change in the gradient is evident for each curve. The greater the coefficient of x the steeper the curve.

What happens if we have a multiplying factor immediately before the exponential? Fig. 11.11 shows the graphs of $y=e^x$, $y=2e^x$ and $y=0.75e^x$. This time where each graph crosses the vertical axis is different. In each case $x=0$, but in turn

$$y = e^0 \quad = 1$$
$$y = 2e^0 \quad = 2\times 1 \quad = 2$$
$$y = 0.75e^0 \quad = 0.75\times 1 \quad = 0.75.$$

There are changes in the gradient. The changes are not as great as those caused by numerical changes to the index.

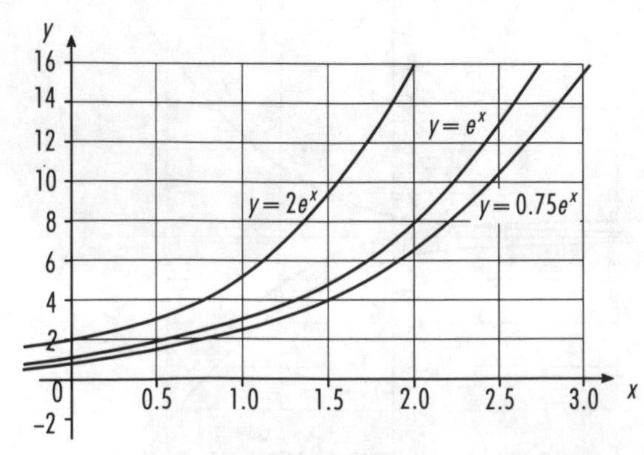

Fig. 11.11

We have looked at changes related to the curve for **exponential growth**, $y=e^x$. The same types of change apply to **exponential decay**, $y=e^{-x}$. Remember that each exponential growth curve has a decay curve as a reflection in the vertical axis.

Example 11.3

Plot the graph of $y=\frac{1}{2}(1+e^{-2x})$. Using the graph with a tangent at the point where $x=-0.3$ estimate the gradient.

We construct our table in the usual way with some specimen values of x. Once we calculate each value of $1+e^{-2x}$ we simply multiply by $\frac{1}{2}$ to get each value of y. In Fig. 11.12 we plot y vertically against x horizontally.

x	-1.0	-0.8	-0.6	-0.4	-0.2	0	0.2	0.4	0.6
$-2x$	2.0	1.6	1.2	0.8	0.4	0	-0.4	-0.8	-1.2
e^{-2x}	7.389	4.953	3.320	2.226	1.492	1.000	0.670	0.449	0.301
$1+e^{-2x}$	8.389	5.953	4.320	3.226	2.492	2.000	1.670	1.449	1.301
y	4.19	2.98	2.16	1.61	1.25	1.00	0.84	0.72	0.65

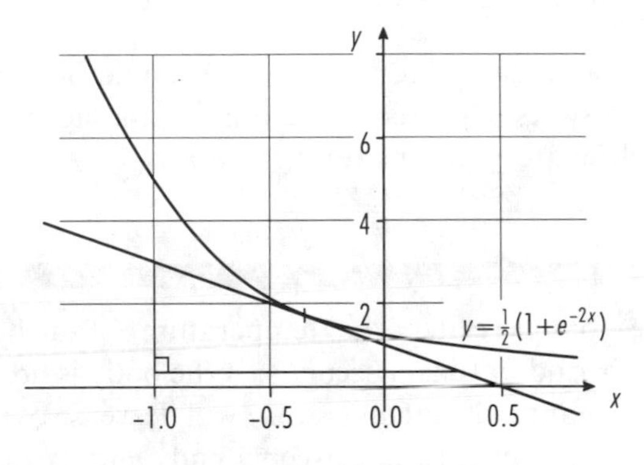

Fig. 11.12

At $x=-0.3$ we draw a tangent to the curve and estimate the gradient. Remember that gradient $=\dfrac{\text{vertical change}}{\text{horizontal change}}$. As usual, we get these change values by drawing a right-angled triangle. If you do this for yourself accurately you should aim for a gradient of approximately -1.8.

EXERCISE 11.2

Use one set of axes for each of Questions **1**–**10**. Sketch the following curves, with intermediate ones if you wish.

1 $y=e^{-2x}$

2 $y=-e^{2x}$

3 $y=-2e^{-2x}$

4 $y=3(1-e^x)$

5 $y=3(1+e^{-x})$

6 $y=1+e^{2x}$

7 $y=1-e^{-x}$

8 $y=3(1+e^{2x})$

9 $y=2(1-3e^x)$

10 $y=2(1+3e^{-x})$

Use one set of axes for each of Questions **11–16**, and values of x from $x=-2.5$ to $x=2.5$ at intervals of 0.5. Plot the following curves.

11 $y=e^{0.5x}$

12 $y=2e^{0.5x}$

13 $y=1+e^{2x}$

14 $y=1+2e^{3x}$

15 $y=2(1+e^{3x})$

16 $y=3(1+2e^{-0.5x})$

This final selection of 4 questions is based on some of the curves you have plotted accurately in Questions **11–16**.

17 By drawing a tangent to the curve $y=e^{0.5x}$ at the point where $x=1.5$ calculate this gradient of the curve.

18 For the curve $y=2e^{0.5x}$ at the point where $x=1.5$ draw a tangent and find its gradient.

19 With the aid of a tangent at $x=1.8$ find the gradient of the curve $y=1+2e^{3x}$ at that point.

20 With the aid of a tangent at $x=1.8$ find the gradient of the curve $y=2(1+e^{3x})$ at that point.

Your answers to Question **20** and to Question **19** should be identical. The closeness of your 2 answers depends on the quality and accuracy of your own graphs. How close are your answers?

Example 11.4

Newton's law of cooling relates the temperature of a body to its surroundings over a period of time, t seconds. If the body is hotter than its surrounding environment then its temperature will decrease over time. Let T°C be the difference in temperature between a body and its environment. Suppose that the relation is $T=400e^{-t/1500}$.

Plot a graph of T°C against t seconds at intervals of 60 seconds (i.e. every minute) for the first 10 minutes.

i) What is the initial temperature difference?

ii) How long will it be before the value of T has fallen by 25%?

iii) What is the temperature gradient at 8 minutes?

As usual our first step is to construct a table of values.

t	0	60	120	180	240	300	360	⟷
$e^{-t/1500}$	1.000	0.961	0.923	0.887	0.852	0.819	0.787	
T	400	384	369	355	341	327	315	↕

	420	480	540	600	⟷
	0.756	0.726	0.698	0.670	
	302	290	279	268	↕

The middle line of working, $e^{-t/1500}$, is quoted to 3 decimal places. In truth all the decimal places have been retained in the table calculations; only for the temperature, T, have the answers been approximated.

The graph is plotted in Fig. 11.13 from where we have deduced our answers.

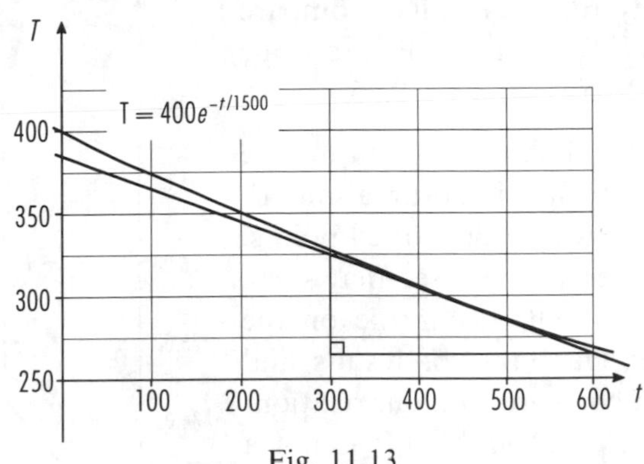

Fig. 11.13

i) The initial temperature is the temperature at $t=0$, i.e. 400°C.

ii) If the temperature has been reduced by 25% this means we are interested in 75% of the original temperature,

i.e. $\quad 400 \times \dfrac{75}{100} = 300°\text{C}.$

From the graph, when $T=300$°C, $t=432$ seconds, i.e. 7 minutes 12 seconds. You can check this result for yourself. Use $T=300$ in the equation

$$T = 400e^{-t/1500}.$$

Take natural logarithms of both sides to find $t=431.5$.

iii) We find the gradient by drawing a tangent to the curve at $t=480$ seconds (i.e. 8 minutes). We complete a right-angled triangle and use

$$\text{gradient} = \frac{\text{vertical change}}{\text{horizontal change}}.$$

This gives a value of about -0.2°C/second. The minus sign indicates a cooling effect. The body's temperature tends towards the temperature of its surrounding environment. If left alone for long enough the body would cool to that temperature.

ASSIGNMENT

Let us take another look at our pulley problem. This time we are going to plot a graph of results. The following table relates the angle, θ (radians), with the belt tension on the taut side, T_1.

θ	1.00	1.10	1.20	1.25	1.30	1.40	1.50
T_1	32.0	33.3	33.9	34.9	35.5	36.9	38.2

We know from our previous look at the Assignment that $T_1 = T_0 e^{\mu\theta}$. T_0 is the tension on the slack side and μ is the coefficient of friction. Using our graph we are going to see how closely the table values compare with $T_1 = T_0 e^{\mu\theta}$. Remember $T_0 = 22.50$ N and $\mu = 0.35$.

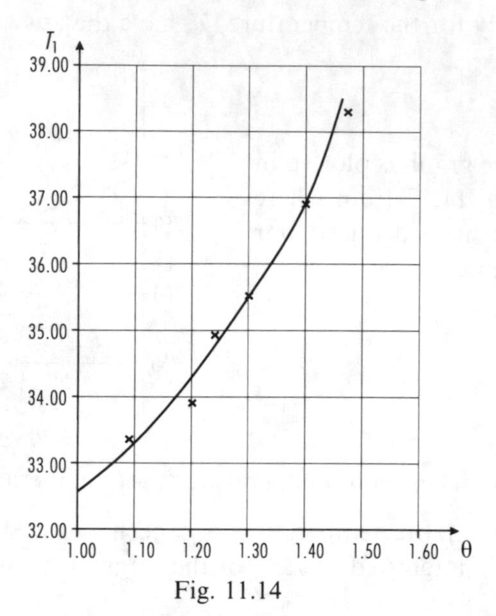

Fig. 11.14

Fig. 11.14 shows the graph. We have attempted to draw a smooth curve through the plotted points. However you can see that some of our results do *not* lie on the curve. Our table of results and graph compare with our relationship, $T_1 = 22.50e^{0.35\theta}$. It is not a totally accurate comparison because not all points lie on the curve. At this stage we have no way of finding the true values for T_0 to replace 22.50 N and μ to replace 0.35.

EXERCISE 11.3

1 $p = 76e^{-h/15000}$ relates the pressure, p (cm of mercury), to the height, h (m), above sea-level. Construct a table of values for h and p. Use values of h from sea-level up to and including 6000 m at intervals of 500 m. Plot a graph of p against h. Drawing a tangent at $h = 2500$ m find the gradient of the curve at this point.

2 The voltage, v, across the inductor, L, in the electrical circuit drops exponentially over time, t. The relation is $v = Ee^{-t/T}$ where $T = \frac{L}{R}$. If L is 2.50 H and R is 125 Ω calculate T. The emf is given by $E = 6.2$ V. For values of t from 0 to 0.05 s at intervals of 0.01 s construct a table of values.

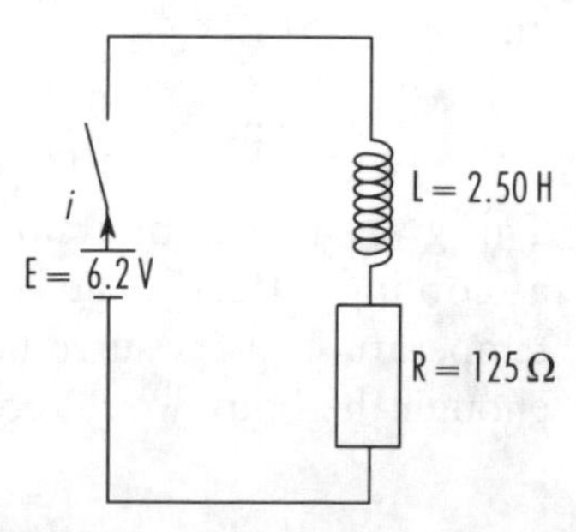

Plot the exponential curve of v against t. Using your graph
 i) draw a tangent at $t = 0.025$ and so find the gradient,
 ii) find the time taken to halve the initial voltage.

3 A bacterial growth, B, is related exponentially to time, t hours, by $B = B_0 e^{ct}$. $B_0 = 1.75 \times 10^3$ and $c = 0.8$. Using this relation and these values construct a table and plot an exponential curve. Use values of t from $\frac{1}{2}$ hour to 5 hours at intervals of $\frac{1}{2}$ hour.

From your graph read off the number of bacteria at $t=\frac{1}{2}$ hour. By what time has this number trebled?

4 The relationship of $N=N_0\mathrm{e}^{-ct}$ relates radioactivity, N, to time, t, years. N_0 is the initial level of radioactivity. The half-life of thorium-228 is 1.9 years. This means that after 1.9 years the value of N is $\frac{1}{2}N_0$. Calculate the value of c. Use this value of c to help construct a table of values of N and t. Use values of t from 0 to 5 years at intervals of 0.5 years. All your values of N will be multiples of N_0. Extract this as a common factor when you plot N against t.

5 In the accompanying electrical circuit the current, i A, is related to time, t s, by $i=\frac{E}{R}(1-\mathrm{e}^{-40t})$.
If $R=100\,\Omega$ and $E=6$ V construct a table of values for i and t. Use values of t from 0 to 0.06 s at intervals of 0.01 s. Hence plot a graph of i against t. What is the current after 0.045 s?

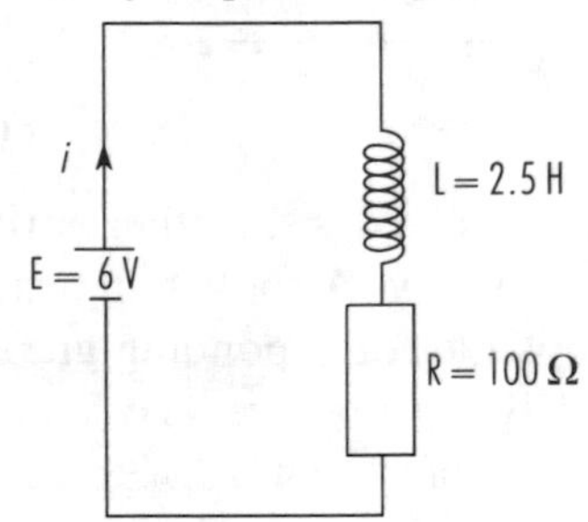

Find the rate of change of current with respect to time at $t=0.025$ s by drawing a tangent to the curve at this point.

Logarithmic graphs

We know the basic shapes of curves involving logarithms and exponentials. A curve is not as simple to draw as a straight line. The gradient of a curve changes continually whilst the gradient of a straight line is constant. We would save ourselves time and effort if we concentrated on straight lines. In fact we can do this for various curves. For exponential curves we can do it using logarithms. We can use logarithms to any base.

The 3 types we will look at involve natural logarithms.

1. Suppose we have $\boldsymbol{y=ae^{bx}}$ where a and b are constant, numerical, values.

Take natural logarithms of both sides,

i.e. $\ln y = \ln(ae^{bx})$

$= \ln a + \ln e^{bx}$ — First log law.

$= \ln a + bx \ln e$ — Third log law.

$= \ln a + bx$ — $\ln e = 1$.

i.e. $\ln y = bx + \ln a.$

Let us compare this relation with the usual one for a straight line,

i.e. $Y = MX + C$.

The comparison is based on pattern and position. On the left sides we have $\ln y$ and Y. Hence $\ln y$ is plotted vertically like Y in Figs. 11.15.

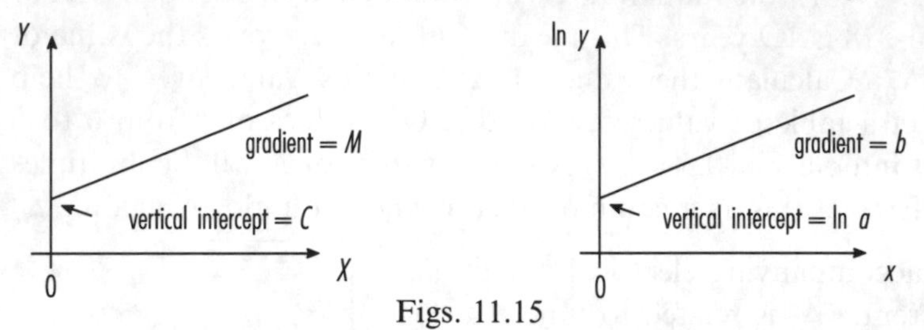

Figs. 11.15

On the right sides the other variables are x and X. Hence as usual we plot x horizontally. Also MX indicates that M is the gradient, so bx indicates that b is the corresponding gradient. Remember this may be negative just as easily as positive. To complete the comparison we have vertical intercepts of C and $\ln a$. These may be negative or zero or positive. This comparison shows we have reduced $y = ae^{bx}$ to a straight line in the form $\ln y = bx + \ln a$.

Example 11.5

Suppose we have the following table of values

t	0	60	120	180	240	300	360	420	480	540	600
T	400	384	369	355	341	327	315	302	290	279	268

It is thought that t and T are related by $T = ae^{bt}$. Check that this is true and find the values of a and b.

You will remember part of this table from a few pages ago. We know that T and t are related according to an exponential graph. We know as well that accurate drawings are difficult. Thus we will look to attempt a straight line based on our new theory.

Using $T = ae^{bt}$

take natural logarithms of both sides

i.e.
$$\begin{aligned} \ln T &= \ln(ae^{bt}) \\ &= \ln a + \ln e^{bt} && \text{First log law.} \\ &= \ln a + bt \ln e && \text{Third log law.} \\ &= \ln a + bt && \ln e = 1. \end{aligned}$$

i.e. $\ln T = bt + \ln a$.

We plot $\ln T$ vertically and t horizontally. Thus we need another row in our table of values.

t	0	60	120	180	240	300	360	420	480	540	600
T	400	384	369	355	341	327	315	302	290	279	268
$\ln T$	5.99	5.95	5.91	5.87	5.83	5.79	5.75	5.71	5.67	5.63	5.59

The gradient is b and the vertical intercept is $\ln a$.

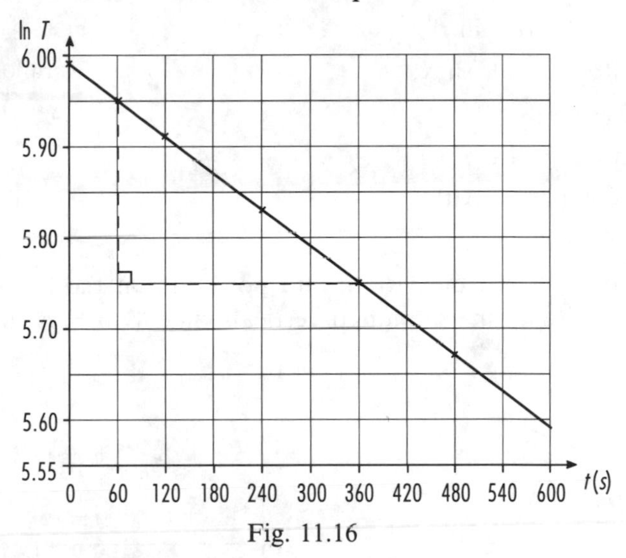

Fig. 11.16

Fig. 11.16 shows the graph is indeed a straight line. You might have expected this from the pattern of values of $\ln T$. From our graph we can read off the vertical intercept as 5.99,

i.e. $\ln a = 5.99$

i.e. $a = e^{5.99} = 399.$

Also $\quad \text{gradient} = \dfrac{\text{vertical change}}{\text{horizontal change}}$

i.e.
$$b = \frac{5.95 - 5.75}{60 - 360}$$
$$= \frac{0.20}{-300}$$

i.e.
$$b = -\frac{1}{1500}.$$

We may link these values together to give

$$\ln T = -\frac{1}{1500}t + 5.99.$$

A preferred version uses our original relation of
$T = ae^{bt}$

i.e. $$T = 399e^{-t/1500}$$

You can see this is only slightly different from our original relation with 399 in place of 400. Such a minor change is acceptable.

2. Suppose we have $\boldsymbol{y = ax^b}$ where a and b are constant, numerical, values.

Take natural logarithms of both sides,

i.e. $$\begin{aligned} \ln y &= \ln(ax^b) \\ &= \ln a + \ln x^b \\ &= \ln a + b\ln x. \end{aligned}$$

First log law.
Third log law.

Writing this as

$$\ln y = b\ln x + \ln a$$

again we can compare it with

$$Y = MX + C.$$

The comparison is based on pattern and position. On the left sides we have $\ln y$ and Y. Hence $\ln y$ is plotted vertically like Y in Figs. 11.17.

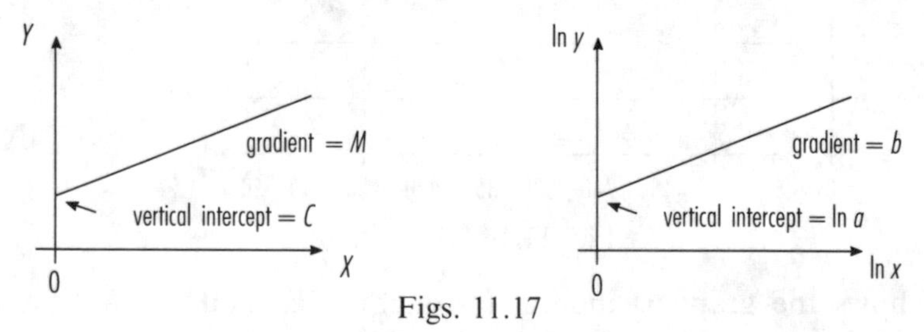

Figs. 11.17

On the right sides the other variables are $\ln x$ and X. Hence we plot $\ln x$ horizontally. Also MX indicates that M is the gradient, so $b\ln x$ indicates that b is the corresponding gradient. Remember this may be negative just as easily as positive. To complete the comparison we have vertical intercepts of C and $\ln a$. These may be negative or zero or positive. This comparison shows we have reduced $y = ax^b$ to a straight line in the form $\ln y = b\ln x + \ln a$.

Example 11.6

Suppose we have the following table of values

x	1	2	3	4	5
y	1.50	8.50	23.00	48.00	84.10

It is thought that x and y are related by $y = ax^b$. Check that this is true and find the values of a and b.

We look to plot a straight line based on this second new piece of theory.

Using $\quad y = ax^b$

we take natural logarithms of both sides and eventually get

$$\ln y = b \ln x + \ln a.$$

Hence we must use our table of values to calculate $\ln x$ and $\ln y$.

$\ln x$	0	0.69	1.10	1.39	1.61
$\ln y$	0.41	2.14	3.14	3.87	4.43

We plot $\ln y$ vertically and $\ln x$ horizontally in Fig. 11.18. The points do not quite all lie on one straight line. We need to draw a line of best fit through them. If they do not all lie on the line ensure there are some points above and some below the line.

Suppose all the spare points are above your line. This means the line is too low and needs a slight shift vertically upwards. Suppose all the spare points are below your line. This means the line is too high and needs a slight shift vertically downwards.

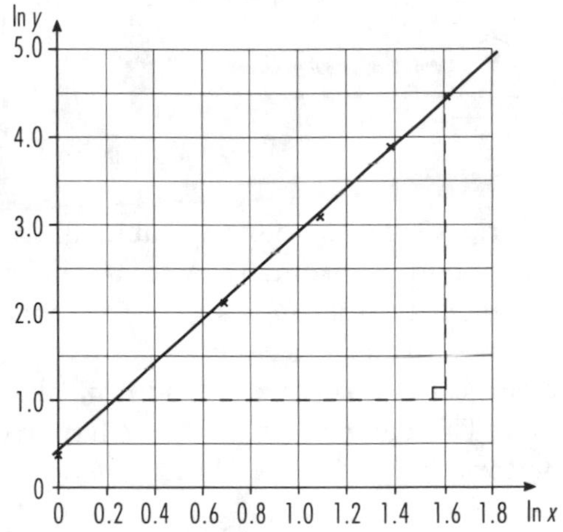

Fig. 11.18

Let us return to our example. From our graph the vertical intercept is 0.45, i.e.

$$\ln a = 0.45$$
$$a = e^{0.45} = 1.57.$$

Also $\quad$ gradient $= \dfrac{\text{vertical change}}{\text{horizontal change}}$

i.e. $\quad b = \dfrac{4.40 - 1.00}{1.60 - 0.22}$

$$= \frac{3.40}{1.38}$$
$$= 2.46.$$

These values give the relationship $y = 1.57x^{2.46}$.

3. Suppose we have $\boldsymbol{y = ab^x}$ where a and b are constant, numerical, values.

Take natural logarithms of both sides

i.e. $\quad \ln y = \ln(ab^x)$

$$= \ln a + \ln b^x \qquad \text{First log law.}$$
$$= \ln a + x \ln b. \qquad \text{Third log law.}$$

Writing this as

$$\ln y = (\ln b)\,x + \ln a$$

again we can compare it with

$$Y = MX + C.$$

The comparison is based on pattern and position. On the left sides we have $\ln y$ and Y. Hence $\ln y$ is plotted vertically like Y in Figs. 11.19.

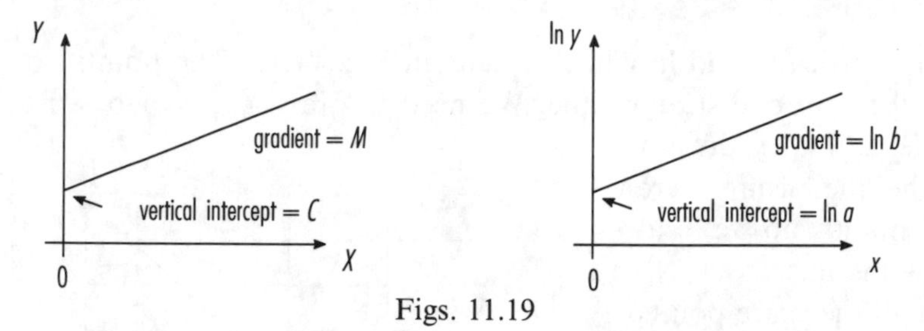

Figs. 11.19

On the right sides the other variables are x and X. Hence we continue to plot them both horizontally. Also MX indicates that M is the gradient, so $(\ln b)\,x$ indicates that $\ln b$ is the corresponding gradient. Remember this may be negative just as easily as positive. To complete the comparison we have vertical intercepts of C and $\ln a$. These may be negative or zero or positive. This comparison shows we have reduced $y = ab^x$ to a straight line in the form $\ln y = (\ln b)\,x + \ln a$.

Example 11.7

Suppose we have the following table of values

x	0	2	4	5	6	8	10
y	3.00	2.40	1.97	1.77	1.59	1.10	1.05

It is thought that x and y are related by $y = ab^x$. Check that this is true and find the values of a and b.
One of the table values is probably inaccurate. Decide which one it is from your graph.

We will look to plot a straight line based on this third new piece of theory.
Using $\qquad y = ab^x$
we take natural logarithms of both sides and eventually get

$$\ln y = (\ln b)\,x + \ln a.$$

Hence we must use our table of values to calculate $\ln y$.

x	0	2	4	5	6	8	10	⟷
y	3.00	2.40	1.97	1.77	1.59	1.10	1.05	
$\ln y$	1.099	0.875	0.678	0.571	0.464	0.095	0.049	↕

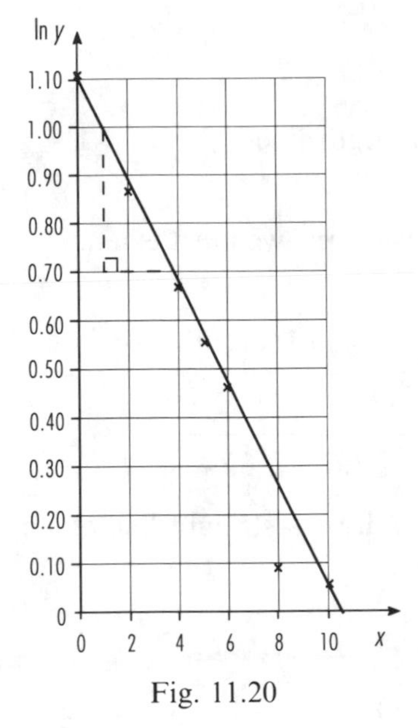

Fig. 11.20

We plot $\ln y$ vertically and x horizontally in Fig. 11.20. Not all the points lie on a straight line. We need to draw a line of best fit through them.

From our graph we can read off the vertical intercept as 1.10,

i.e. $\ln a = 1.10$

i.e. $a = e^{1.10} = 3.00.$

Also $\text{gradient} = \dfrac{\text{vertical change}}{\text{horizontal change}}$

i.e.
$$\ln b = \frac{1.00 - 0.70}{1.0 - 3.8} = \frac{0.30}{-2.8} = -0.11$$

i.e. $b = e^{-0.11} = 0.9.$

We can link these values together with our relationship to get

$$y = 3.00 \times 0.9^x.$$

In addition, from our graph we see that the point where $x = 8$ is probably inaccurate. This is because it lies so far from the straight line.

ASSIGNMENT

Let us take a third look at our pulley problem. Again we are going to plot a graph of results, but this time as a straight line. The table relates the angle, θ (radians), with the belt tension on the taut side, T_1.

θ	1.00	1.10	1.20	1.25	1.30	1.40	1.50
T_1	32.0	33.3	33.9	34.9	35.5	36.9	38.2

We know that $T_1 = T_0 e^{\mu\theta}$. T_0 is the tension on the slack side and μ is the coefficient of friction. Previously we thought $T_0 = 22.50$ N and $\mu = 0.35$. Using our graph we are going to see how closely the table values compare with $T_1 = 22.50e^{0.35\theta}$. Now we are going to use a straight line to check these values.

Using $T_1 = T_0 e^{\mu\theta}$

we take natural logarithms of both sides

i.e.
$$\begin{aligned} \ln T_1 &= \ln (T_0 e^{\mu\theta}) \\ &= \ln T_0 + \ln (e^{\mu\theta}) \\ &= \ln T_0 + \mu\theta \ln e \\ &= \ln T_0 + \mu\theta. \end{aligned}$$

First log law.
Third log law.
$\ln e = 1$.

We can re-write this as

$$\ln T_1 = \mu\theta + \ln T_0.$$

Comparing it with the standard equation for a straight line

$$Y = MX + C$$

we plot $\ln T_1$ vertically against θ horizontally. This means we must extend our table of values to

θ	1.00	1.10	1.20	1.25	1.30	1.40	1.50	$\longleftrightarrow$
T_1	32.0	33.3	33.9	34.9	35.5	36.9	38.2	
$\ln T_1$	3.47	3.51	3.52	3.55	3.57	3.61	3.64	$\updownarrow$

Fig. 11.21 shows the graph. We have attempted to draw a straight line of best fit. Not all our results lie on the straight line.

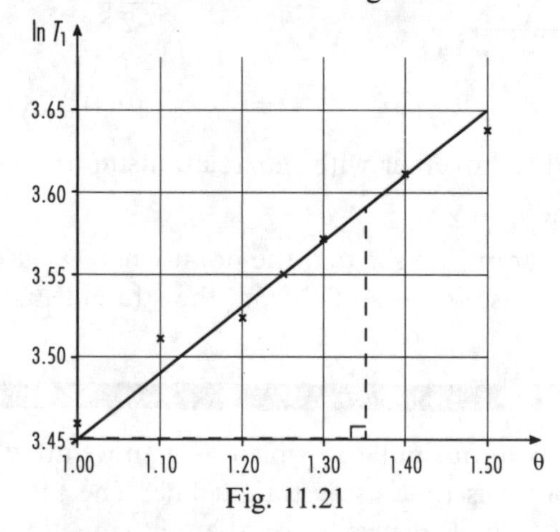

Fig. 11.21

Immediately the straight line with points lying on either side casts doubt on the accuracy of our original $T_1 = 22.50e^{0.35\theta}$. It is not a totally accurate relation. However it is of more use than our previous exponential curve. Let us see if we can improve on our original values for T_0 and μ.

We cannot read off the vertical intercept because we have not included where $\theta = 0$. Instead let us start with the gradient.

$$\text{Gradient} = \frac{\text{vertical change}}{\text{horizontal change}}$$

i.e.

$$\mu = \frac{3.59 - 3.45}{1.35 - 1.00}$$

$$= \frac{0.14}{0.35}$$

$$= 0.40.$$

We can substitute this value of $\mu = 0.40$ into

$$\ln T_1 = \mu\theta + \ln T_0$$

to get

$$\ln T_1 = 0.40\theta + \ln T_0.$$

To find the value of $\ln T_0$ we choose any point on the line and substitute for those values,

$$\begin{aligned} \text{i.e.} \qquad 3.45 &= 0.40 \times 1.00 + \ln T_0 \\ 3.45 &= 0.40 + \ln T_0 \\ 3.45 - 0.40 &= \ln T_0 \\ 3.05 &= \ln T_0 \\ \text{i.e.} \qquad T_0 &= e^{3.05} = 21.12\,\text{N}. \end{aligned}$$

Hence our new complete relationship is $T_1 = 21.12e^{0.40\theta}$.

EXERCISE 11.4

1 $p = p_0 e^{-kH}$ relates the pressure, p (cm of mercury), to the height, H (m), above sea-level. p_0 and k are constants. Use the table of values for H and p and logarithms to plot an appropriate straight line graph.

H	0	500	1000	2000	4000	8000	10000
p	75.8	73.3	70.9	66.3	58.1	44.5	38.9

From your graph find the values of p_0 and k.

2 The voltage, v, across the inductor, L, in the electrical circuit drops exponentially over time, t s. The relation is $v = Ee^{-t/T}$ where the emf, E, and T are constants. Use the table of values for t and v and logarithms to plot an appropriate straight line graph.

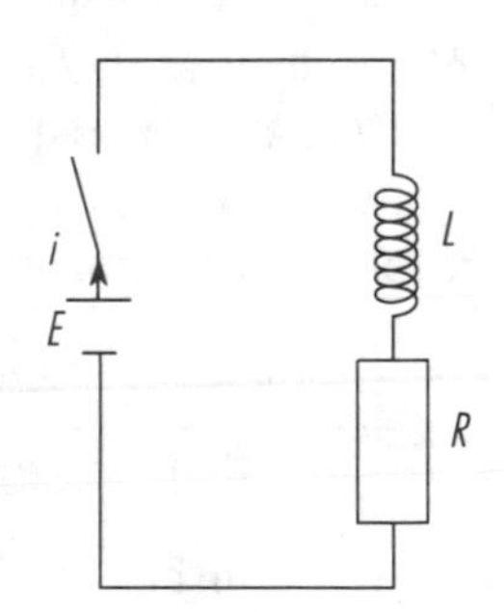

t	0	0.010	0.020	0.025	0.030	0.040	0.050
v	5.50	2.95	1.60	1.15	0.85	0.45	0.25

Use your graph to find the values of E and T. How long did it take to halve the initial voltage?

3 Gas pressure, p bars, and temperature, $T\,°\text{K}$, are thought to be related by $T = \alpha p^{\gamma}$. α and γ are constants to be found from your graph. Attempt to plot a straight line graph using logarithms and this table.

p	1	1.5	2.0	2.5	5.0	7.5
T	510	570	610	650	780	860

4 b, the intensity of light varies with V, the voltage according to $b = V^c$. c is a constant. Take natural logarithms of both sides of this equation. Hence extend the table so that you can plot a straight line graph.

V	14	16	18	20	22	24
$b\ (\times 10^{-3})$	5.82	4.49	3.57	2.90	2.41	2.04

From your graph find the value of c.

5 $V=a^R$ relates a variable resistance, R, with an output voltage, V, for an amplifier. a is a constant. Using logarithms, the table of values and a straight line graph show that this is true.

R	35	40	45	50	55	60	65
V	7.70	10.30	13.80	18.40	24.70	33.00	44.10

From your graph find the value of a.

6 It is thought that a bacterial growth, B, is related exponentially to time, t hours, by $B=B_0e^{ct}$. B_0 and c are constants. Use the table of values and an appropriate straight line graph to find B_0 and c.

t	0.25	0.50	1.00	1.50	2.00	2.50	4.00	5.00
$B\ (\times 10^3)$	1.93	2.33	3.39	4.93	7.17	10.43	32.14	68.00

7 A microwave oven includes an element with a non-linear resistance, $R\,\Omega$. The current, I A, and the voltage, V, are thought to be related by $I=VR^k$. k is a constant. Take natural logarithms of both sides of this equation. Hence extend your table so that you can plot a straight line graph. From your graph find the value of k.

R	2	4	6	8	10	12
I	5.70	16.00	29.40	45.25	63.00	83.00

8 The relationship of $N=N_0e^{-ct}$ relates radioactivity, N, to time t years. N_0 is the initial level of radioactivity. The half-life of thorium-228 is 1.9 years. This means that after 1.9 years the value of N is $\frac{1}{2}N_0$. Calculate the value of c. Use this value of c to help construct a table of values of N and t. Use values of t from 0 to 5 years at intervals of 0.5 years. All your values of N will be multiples of N_0. Extract this as a common factor when you plot your graph. This question appeared in Exercise 11.3. That plot of N against t produced an exponential curve. This time attempt to plot $\ln N$ against t and so create a straight line graph.

9 In the accompanying electrical circuit the current, i A, is related to time, t s, by $i=\dfrac{E}{R}(1-e^{-40t})$.

If $R=100\,\Omega$ and $E=5$ V construct a table of values for i and t. Use values of t from 0 to 0.06 s at intervals of 0.01 s.

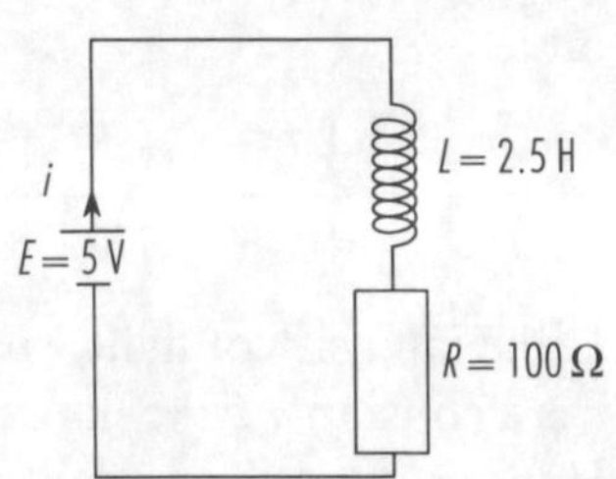

Now re-arrange your original equation into the form $1-20i=e^{-40t}$ Hence include another line in your table for $1-20i$.

Take natural logarithms of both sides of this re-arranged equation. Hence include a further line in your table for $\ln(1-20i)$. Now plot a graph of $\ln(1-20i)$ against t expecting to get a straight line. From your graph what is the current after 0.045 s?

10 A regional engineering company increases its profits each year. During a recent period each year's profits have been 20% higher than those for the previous year. At the end of Year 1 the profits were £150 000. Complete the following table using 3 significant figures.

Year (y)	1	2	3	4	5	6
Profit (£ $\times 10^3$) (P)	150					373

It is thought that P and y are related by $P=P_1r^y$ where P_1 and r are constants. Does a plot of $\ln P$ against y give a straight line? What are the values of P_1 and r?

Logarithmic graph paper

We looked at 3 types of logarithmic graphs in the last section. In each case we reduced them to straight lines using natural logarithms. We plotted the graphs on ordinary graph paper. Instead of using the usual x (horizontally) and y (vertically) scales we used slightly different ones. In each case the correctly chosen different scales on the axes gave us the straight lines.

There is an alternative! Instead of altering the scales on the axes we can alter the graph paper. We can use either natural or common logarithms. Let us use logarithms to the base 10 because powers of 10 are recognised easily (e.g. $10^1=10$, $10^2=100$, $10^3=1000$, etc).

We can work out the following calculator values

x	1	2	3	4	5	6	7	8	9	10
$\log x$	0.000	0.301	0.477	0.602	0.699	0.778	0.845	0.903	0.954	1.000

x	10	20	30	40	50	60	70	80	90	100
$\log x$	1.000	1.301	1.477	1.602	1.699	1.778	1.845	1.903	1.954	2.000

x	100	200	300	400	500	600	700	800	900	1000
$\log x$	2.000	2.301	2.477	2.602	2.699	2.778	2.845	2.903	2.954	3.000

Each pair of table lines shows one logarithmic cycle. In each case:

$$\text{Biggest log} - \text{Smallest log} = 1.000$$

e.g. $1.000 - 0.000 = 1.000$ in the first cycle,

e.g. $2.000 - 1.000 = 1.000$ in the second cycle,

e.g. $3.000 - 2.000 = 1.000$ in the third cycle.

Now look down each column in turn. You can see the number to the left of the decimal point changes from one cycle to another. The numbers to the right of the decimal point remain the same,

e.g. $\log 3 = 0.477$,

$\log 30 = 1.477$

and $\log 300 = 2.477$

The numbers, 3, 30 and 300 have the same position as each other, but in their own cycle.

All these principles continue. From 1,000 to 10,000 is one cycle, 10,000 to 100,000 is another cycle and so on getting larger. Before our first cycle there are other cycles based on decimals. 0.1 to 1.0 is a cycle, 0.01 to 0.1 is another cycle and so on getting smaller.

Now let us look at these alternative types of graph paper. We can use graph paper with only the vertical scale changed. The horizontal scale as before uses equal divisions. This is called **log-linear** (or **semi-log**) graph paper. If we change both scales then we have **log-log** graph paper. Examples of these are shown in Figs. 11.22.

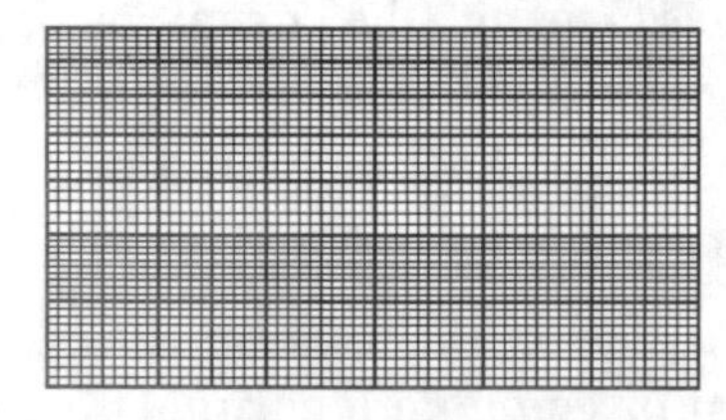

log-linear scales

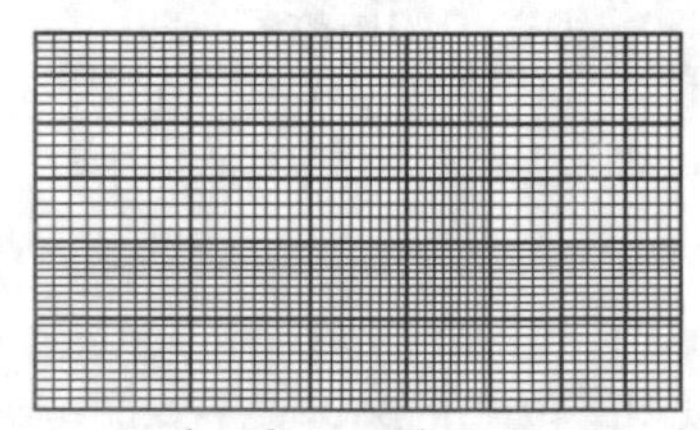

log-log scales

Figs. 11.22

These graph papers are easy to use. You need just a little more care counting the divisions because they can be quite small.

Let us look again at our 3 types of logarithmic graphs, the axes and examples we used.

1. $\boldsymbol{y = aeb^x}$ where a and b are constant, numerical, values. Remember the equivalent logarithmic equation,

$$\ln y = bx + \ln a.$$

Only the vertical axis has changed. Hence we use log-linear graph paper.

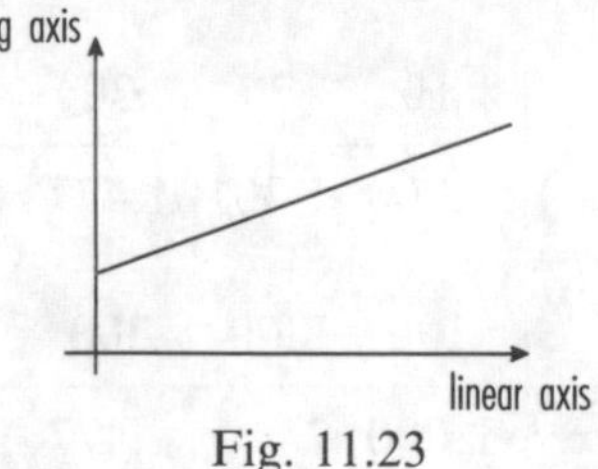

Fig. 11.23

Example 11.8

Suppose we have the following table of values

t	0	60	120	180	240	300	360	420	480	540	600
T	400	384	369	355	341	327	315	302	290	279	268

It is thought that t and T are related by $T=ae^{bt}$. Check that this is true and find the values of a and b.

Now we know this type of exponential relation is a straight line on log-linear graph paper. We plot t horizontally on the linear axis and T vertically on the logarithmic axis. From our earlier work on logarithmic scales we can see that all the values of T are within one cycle. This shown in Fig. 11.24.

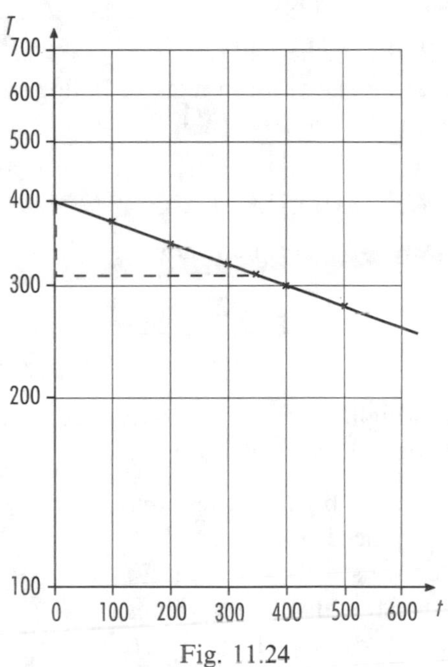

Fig. 11.24

From our graph we can immediately read off the vertical intercept as 400, i.e. $a=400$.

The gradient involves 2 different scales so we must be careful when we use our coordinates.

Using $\quad T = ae^{bt}$

we can take **common** logarithms of both sides to get

$$\log T = \log(ae^{bt})$$

$$= \log a + \log(e^{bt}) \qquad \text{First log law.}$$

$$= \log a + bt \log e \qquad \text{Third log law.}$$

i.e. $$\log T = (0.4343b)t + \log a. \qquad \log e = 0.4343.$$

So $$\text{gradient} = \frac{\text{vertical change}}{\text{horizontal change}}$$

becomes $$0.4343b = \frac{\log 400 - \log 315}{0 - 360}$$

$$= \frac{2.6021 - 2.4983}{-360}$$

$$0.4343\text{b} = \frac{0.1038}{-360}$$

i.e. $$b = -\frac{1}{1507}.$$

We may link these values together to give

$$T = 400e^{-t/1507}.$$

You can see this is only slightly different from our original relation with 1507 in place of 1500.

2. $\boldsymbol{y = ax^b}$ where a and b are constant, numerical, values. Remember the equivalent logarithmic equation,

$$\ln y = b \ln x + \ln a.$$

Fig. 11.25 shows both axes to be logarithmic, compared with $Y = MX + C$.

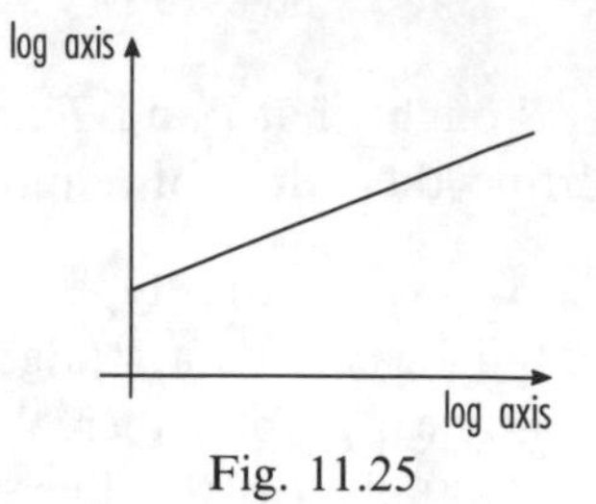

Fig. 11.25

Example 11.9

Suppose we have the following table of values

x	1	2	3	4	5
y	1.50	8.50	23.00	48.00	84.10

It is thought that x and y are related by $y = ax^b$. Check that this is true and find the values of a and b.

We know both scales are logarithmic. Looking at the x values we see they are within one cycle, whilst the y values cover 2 cycles. Fig. 11.26 shows the plot on log-log graph paper.

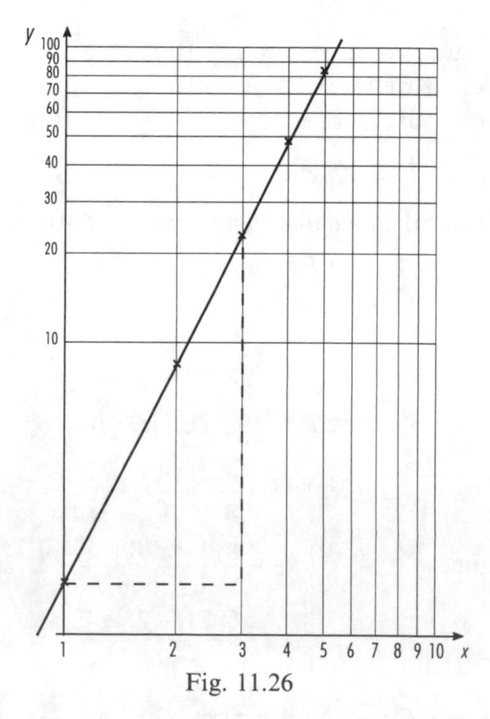

Fig. 11.26

On the horizontal axis we do not have $x = 0$ so we cannot read off the vertical intercept. Therefore we need look at our original relationship again.

Using $\quad y = ax^b$

we can take **common** logarithms of both sides to get eventually

$$\log y = b\log x + \log a.$$

Choosing coordinates on our straight line

$$\text{gradient} = \frac{\text{vertical change}}{\text{horizontal change}}$$

becomes

$$\begin{aligned} b &= \frac{\log 23.00 - \log 1.50}{\log 3 - \log 1} \\ &= \frac{1.3617 - 0.1761}{0.4771 - 0} \\ &= \frac{1.1856}{0.4771} \end{aligned}$$

i.e. $\quad b = 2.48.$

This gives $\log y = 2.48 \log x + \log a$.

Now we can find the value of $\log a$. Substitute a pair of coordinates from our straight line into this equation to get

$$\begin{aligned} \log 40.00 &= 2.48 \times \log 3.75 + \log a \\ 1.6021 &= 2.48 \times 0.5740 + \log a \\ 1.6021 &= 1.4236 + \log a \end{aligned}$$

i.e. $\quad \log a = 0.1785$

$\therefore \quad a = 10^{0.1785} = 1.51.$

These values give us the relationship $y = 1.51x^{2.48}$.

3. $\boldsymbol{y = ab^x}$ where a and b are constant, numerical, values.

Only the vertical axis has changed in Fig. 11.27, compared with $Y = MX + C$. Hence we use log-linear graph paper.

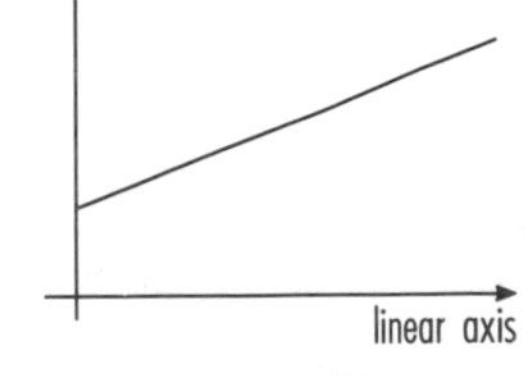

Fig. 11.27

Example 11.10

Suppose we have the following table of values

x	0	2	4	5	6	8	10
y	3.00	2.40	1.97	1.77	1.59	1.10	1.05

It is thought that x and y are related by $y = ab^x$. Check that this is true and find the values of a and b.

One of the table values is probably inaccurate. Decide which one it is from your graph.

From our third new piece of theory we know to plot x horizontally on the linear axis. We plot y vertically on the logarithmic axis, all y values being in one cycle. This is shown in Fig. 11.28. The points do not quite all lie on a straight line. We need to draw a line of best fit.

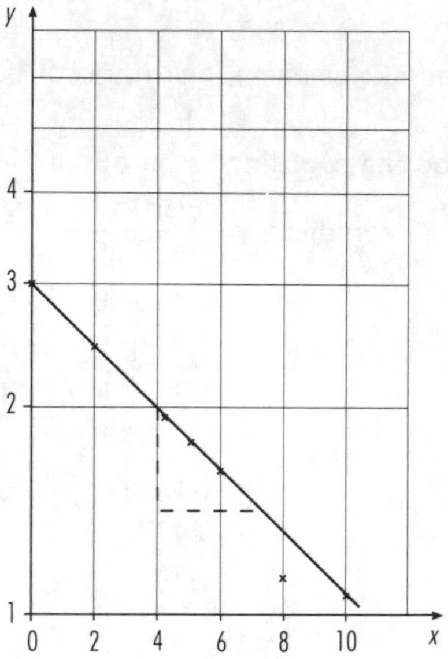

From our graph we can immediately read off the vertical intercept as 3.00, i.e. $a = 3.00$.

Just like the first type of graphs the gradient involves 2 different scales. We must be careful when we use our coordinates.

Fig. 11.28

Using $$y = ab^x$$

we can take **common** logarithms of both sides to get eventually

$$\log y = (\log b)\,x + \log a.$$

So $$\text{gradient} = \frac{\text{vertical change}}{\text{horizontal change}}$$

becomes $$\log b = \frac{\log 2.00 - \log 1.40}{3.80 - 7.20}$$

$$= \frac{0.3010 - 0.1461}{-3.40}$$

$$= \frac{0.1549}{-3.40}$$

i.e. $$\log b = -0.0456$$

$$\therefore \quad b = 10^{-0.0456} = 0.9.$$

We can link these values together in our relationship as

$$y = 3.00 \times 0.9^x.$$

In addition, from our graph we see that the point where $x = 8$ is probably inaccurate. This is because it lies so far from the straight line.

ASSIGNMENT

Let us take a final look at our pulley problem. Again we are going to plot a straight line graph. The table relates the angle, θ (radians), with the belt tension on the taut side, T_1.

θ	1.00	1.10	1.20	1.25	1.30	1.40	1.50
T_1	32.0	33.3	33.9	34.9	35.5	36.9	38.2

We know that $T_1 = T_0 e^{\mu\theta}$. T_0 is the tension on the slack side and μ is the coefficient of friction. Using our straight line graph we are going to see how closely the table values compare with $T_1 = 22.50e^{0.35\theta}$.

We plot θ horizontally on the linear axis and T_1 vertically on the logarithmic axis in Fig. 11.29.

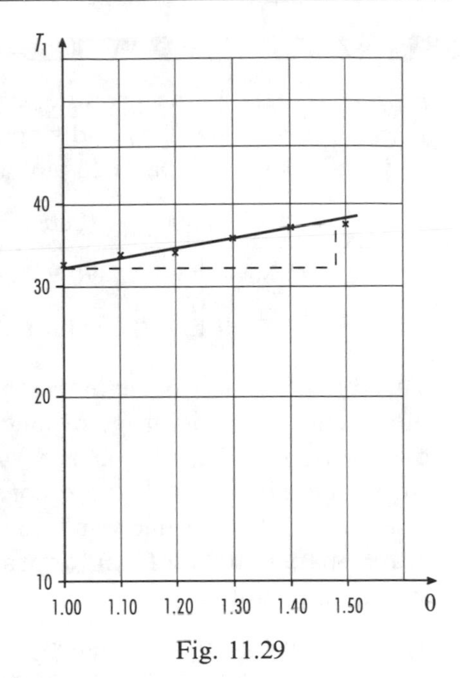

Fig. 11.29

$$\text{Using} \qquad T_1 = T_0 e^{\mu\theta}$$

we can take common logarithms of both sides to get eventually

$$\log T_1 = (0.4343\mu)\theta + \ln T_0.$$

We choose coordinates from our straight line graph so that

$$\text{gradient} = \frac{\text{vertical change}}{\text{horizontal change}}$$

$$\text{becomes} \qquad 0.4343\mu = \frac{\log 38.00 - \log 32.00}{1.48 - 1.02}$$

$$= \frac{1.5798 - 1.5052}{0.46}$$

$$\text{i.e.} \qquad 0.4343\mu = \frac{0.0746}{0.46}$$

$$\therefore \qquad \mu = 0.37.$$

We can substitute this value of $\mu = 0.37$ into

$$\log T_1 = (0.4343\mu)\theta + \log T_0$$

$$\text{to get} \qquad \log T_1 = 0.4343 \times 0.37\theta + \log T_0.$$

To find the value of $\log T_0$ we choose any point on the line and substitute for those values,

$$\text{i.e.} \qquad \log 35.5 = 0.4343 \times 0.37 \times 1.30 + \log T_0$$

$$1.55 = 0.21 + \log T_0$$

$$1.34 = \log T_0$$

$$\text{i.e.} \qquad T_0 = 10^{1.34} = 21.88\,\text{N}.$$

Hence our new complete relationship is $T_1 = 21.88e^{0.37\theta}$.

EXERCISE 11.5

1 $p = p_0 e^{-kH}$ relates the pressure, p (cm of mercury), to the height, H (m), above sea-level. p_0 and k are constants. Use the table of values and log-linear graph paper to plot an appropriate straight line graph.

H	0	500	1000	2000	4000	8000	10000
p	75.8	73.3	70.9	66.3	58.1	44.5	38.9

From your graph find the values of p_0 and k.

2 The voltage, v, across the inductor, L, in the electrical circuit drops exponentially over time, t s. The relation is $v = Ee^{-t/T}$ where the emf, E, and T are constants. Use the table of values and log-linear graph paper to plot an appropriate straight line graph.

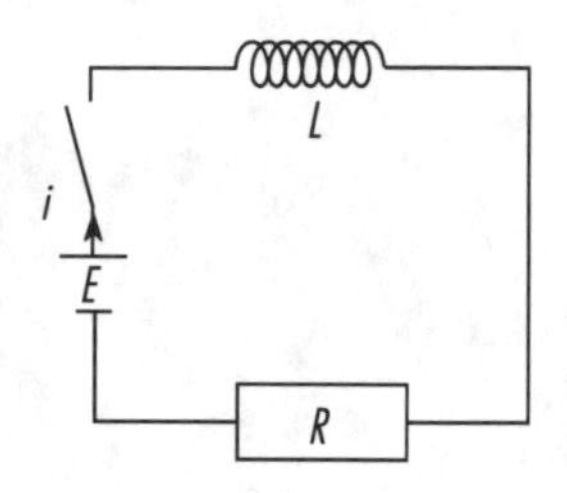

t	0	0.010	0.020	0.025	0.030	0.040	0.050
v	5.50	2.95	1.60	1.15	0.85	0.45	0.25

Use your graph to find the values of E and T. How long did it take to halve the initial voltage?

3 Gas pressure, p bars, and temperature, T°K, are thought to be related by $T = \alpha p^{\gamma}$. α and γ are constants to be found from your graph. Plot a straight line graph using this table.

p	1	1.5	2.0	2.5	5.0	7.5
T	510	570	610	650	780	860

4 b, the intensity of light varies with V, the voltage according to $b = V^c$. c is a constant. Use the table to plot a straight line graph.

V	14	16	18	20	22	24
b ($\times 10^{-3}$)	5.82	4.49	3.57	2.90	2.41	2.04

From your graph find the value of c.

5 $V = a^R$ relates a variable resistance, R, with an output voltage, V, for an amplifier. a is a constant. Using the necessary logarithmic graph paper show this is true.

R	35	40	45	50	55	60	65
V	7.70	10.30	13.80	18.40	24.70	33.00	44.10

From your graph find the value of a.

6 It is thought that a bacterial growth, B, is related exponentially to time, t hours, by $B = B_0e^{ct}$. B_0 and c are constants. Use the table of values and an appropriate straight line graph to find B_0 and c.

t	0.25	0.50	1.00	1.50	2.00	2.50	4.00	5.00
$B\ (\times 10^3)$	1.93	2.33	3.39	4.93	7.17	10.43	32.14	68.00

7 A microwave oven includes an element with a non-linear resistance, $R\,\Omega$. The current, I A, and the resistance, R, are thought to be related by $I = VR^k$. k is a constant. Plot a straight line graph. From your graph find the value of k.

R	2	4	6	8	10	12
I	5.70	16.00	29.40	45.25	63.00	83.00

8 The relationship of $N = N_0e^{-ct}$ relates radioactivity, N, to time t years. N_0 is the initial level of radioactivity. The half-life of thorium-228 is 1.9 years. This means that after 1.9 years the value of N is $\frac{1}{2}N_0$. Calculate the value of c. Use this value of c to help construct a table of values of N and t. Use values of t from 0 to 5 years at intervals of 0.5 years. All your values of N will be multiples of N_0. Extract this as a common factor when you plot your graph on logarithmic type graph paper.

9 In the accompanying electrical circuit the current, i A, is related to time, t s, by $i = \frac{E}{R}(1 - e^{-40t})$. If $R = 100\,\Omega$ and $E = 5$ V construct a table of values for i and t. Use values of t from 0 to 0.06 s at intervals of 0.01 s.

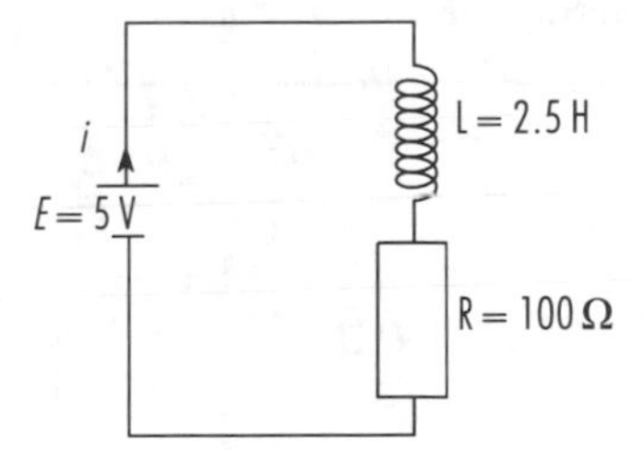

Now re-arrange your original equation into the form $1 - 20i = e^{-40t}$. Hence include another line in your table for $1 - 20i$. Using log-linear graph paper plot $1 - 20i$ vertically against t horizontally. From your graph what is the current after 0.045 s?

10 A regional engineering company increases its profits each year. During a recent period each year's profits have been 20% higher than those for the previous year. At the end of Year 1 the profits were £150 000. Complete the following table using 3 significant figures.

Year (y)	1	2	3	4	5	6
Profit (£ $\times 10^3$) (P)	150					373

It is thought that P and y are related by $P = P_1r^y$ where P_1 and r are constants. Does a plot of P against y on the appropriate logarithmic graph paper give a straight line? What are the values of P_1 and r?

12 Further Graphs

The objectives of this chapter are to:

1. State the relationship between two variables which are proportional.
2. Calculate the coefficient of proportionality from given data.
3. State that for inverse proportionality the product of variables is constant.
4. Solve problems involving proportionality.
5. Draw up suitable tables of values and plot curves of the type $y^2=4ax$, $y=ax^{\frac{1}{2}}$, $y=\frac{a}{x}$
6. Recognise the changes caused by altering the values of a.
7. Recognise the shapes of the curves with equations $\frac{x^2}{a^2}+\frac{y^2}{b^2}=1$, $\frac{x^2}{a^2}-\frac{y^2}{b^2}=1$, $xy=c^2$.
8. Reduce relationships such as $y=ax^2+b$, $y=a+\frac{b}{x}$ to appropriate straight line form.
9. Use the straight line graph form to determine values of constants a and b.
10. Use the graph to determine intermediate values.

Introduction

This chapter starts with simple direct proportionality. The idea is extended, showing examples involving other types of proportion. The early examples are linked to straight line sketches.

The next section looks at the graphs of some standard curves. Some of these curves are re-examined in the final section. They are reduced to straight lines by altering the horizontal axes. Altering the scales on the axes is something we have seen before. We did this with logarithms in Chapter 11.

ASSIGNMENT

The Assignment for this chapter involves a collection of numbers. You can see they are displayed as pairs of coordinates. The aim is to find how these pairs of coordinates are **related**. During the chapter we will attempt to do this using the graphs we introduce.

The pairs of coordinates are:

(1.35, −1.11), (0.40, −0.03), (3.06, −1.54),
(1.74, −1.26), (0.10, 2.34), (2.48, −1.45),
(3.88, −1.64), (0.90, −0.82), (0.15, 1.47).

The first task is to put them in some order. Perhaps a display in a table is the best method of presentation.

x	0.10	0.15	0.40	0.90	1.35	1.74	2.48	3.06	3.88
y	2.34	1.47	−0.03	−0.82	−1.11	−1.26	−1.45	−1.54	−1.64

Proportionality

We can link together any 2 variables by some relation. One of them is an independent variable (e.g. x) and one of them is a dependent variable (e.g. y). In Chapter 2 we saw the independent variable plotted horizontally and the dependent variable plotted vertically. This gave us a straight line. In this section we will see that only some straight lines show 2 variables in proportion. It is possible for curves to show proportion as well.

We can write "y is proportional to x". This means that whatever happens to x in turn affects y.

$\propto$ is the symbol for **is proportional to,**

e.g. $\boldsymbol{y \propto x}$ means **$\boldsymbol{y}$ is proportional to $\boldsymbol{x}$.**

The question now arises: how are they in proportion? Without some numbers we do not know anything about the proportionality. The mention of numbers encourages us to reach for a calculator. Our new $\propto$ symbol is not there. In fact we replace $\propto$ with **=k**. k is the **constant of proportionality** (or **coefficient of proportionality**). $y \propto x$ and $y = kx$ are the same relationship, but $y = kx$ is more useful because of the "=" symbol.

Example 12.1

Generally when running a car the volume of petrol is related to the miles driven. Suppose a car uses 7.5 gallons of petrol whilst travelling 240 miles. How far will it travel on 5.0 gallons?

In this example the distance we can drive the car depends on the volume of petrol in the tank. (We know it depends on many other things too.) The volume of petrol is the independent variable (x) so the distance travelled is the dependent variable (y);

i.e. distance travelled $\propto$ volume of petrol

i.e. $y \propto x$

i.e. $y = kx$

We are given $x = 7.5$ together with $y = 240$ so that

$$240 = k(7.5)$$

$$\frac{240}{7.5} = k$$

i.e. $\qquad k = 32.$

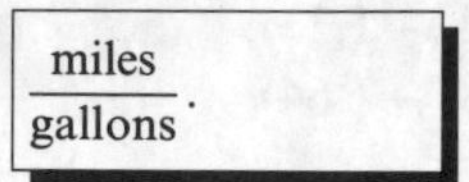

The units for k are miles per gallon (mpg), though knowing this is not vital to the example.

For this particular car we know now that

$$y = kx$$

is $\qquad y = 32x.$

How far it travels on 5.0 gallons of petrol means we can write

$$y = 32 \times 5.0$$
$$= 160$$

i.e. on 5 gallons of petrol the car travels 160 miles.

We also know that if there is no petrol in the fuel tank we cannot drive the car anywhere, i.e. $x = 0$, $y = 0$. Now we have 3 pairs of values for petrol and distance. We can write these as coordinates (x, y); $(0, 0)$, $(5.0, 160)$ and $(7.5, 240)$. In Fig. 12.1 we see them plotted on a pair of axes. They all lie on a straight line passing through the origin. We see that the more petrol we have the greater the distance generally we are able to travel.

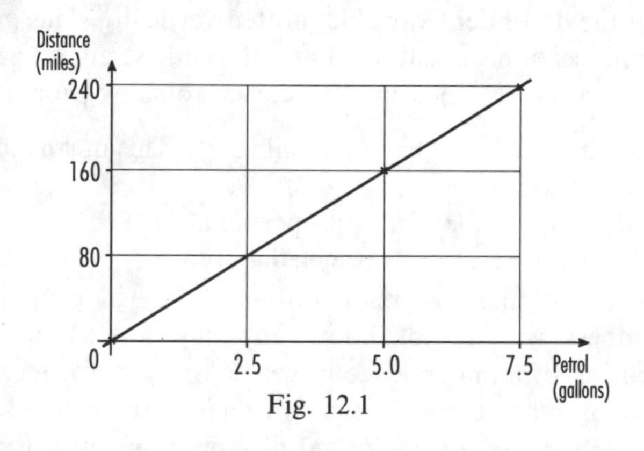

Fig. 12.1

Many short distances around towns are likely to affect our calculation. How somebody drives and in which particular car are some other effects. We can cope with these changes. All we would need to do is re-work our calculation with revised numbers.

Example 12.2

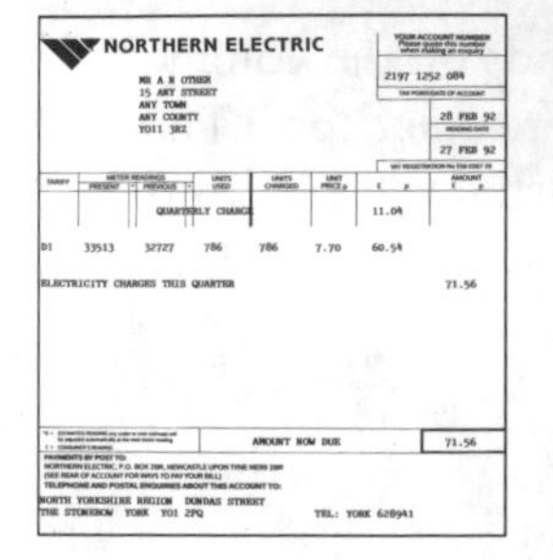
NORTHERN ELECTRIC

MR A N OTHER
15 ANY STREET
ANY TOWN
ANY COUNTY
YO11 3RZ

YOUR ACCOUNT NUMBER
Please quote this number when making an enquiry
2197 1252 084
TAX POINT/DATE OF ACCOUNT
28 FEB 92
READING DATE
27 FEB 92

TARIFF	METER READINGS PRESENT	PREVIOUS	UNITS USED	UNITS CHARGED	UNIT PRICE p	£ p	AMOUNT £ p
	QUARTERLY CHARGE					11.04	
D1	33513	32727	786	786	7.70	60.54	
ELECTRICITY CHARGES THIS QUARTER							71.56

AMOUNT NOW DUE 71.56

PAYMENTS BY POST TO:
NORTHERN ELECTRIC, P.O. BOX 28R, NEWCASTLE UPON TYNE NE99 2BR
(SEE REAR OF ACCOUNT FOR WAYS TO PAY YOUR BILL)
TELEPHONE AND POSTAL ENQUIRIES ABOUT THIS ACCOUNT TO:
NORTH YORKSHIRE REGION DUNDAS STREET
THE STONEBOW YORK YO1 2PQ TEL: YORK 628941

At home the more units of electricity we use the greater the electricity bill. Is the cost of electricity proportional to the number of units used?

In the previous example distance and volume of petrol were in proportion. Also "We see that the more petrol we have the greater the distance generally we are able to travel". At first glance we might expect the cost of electricity to be proportional to the number of units used. However, if you look at a domestic electricity bill there is a quarterly standing charge. If no electricity is used the charge will still appear on the bill. We can look at this as a graph of costs against units of electricity used in Fig. 12.2. This time you can see the straight line graph does *not* pass through the origin. This means the cost of electricity is *not* directly proportional to the number of units used. In fact the intercept on the vertical axis is the cost of the quarterly charge.

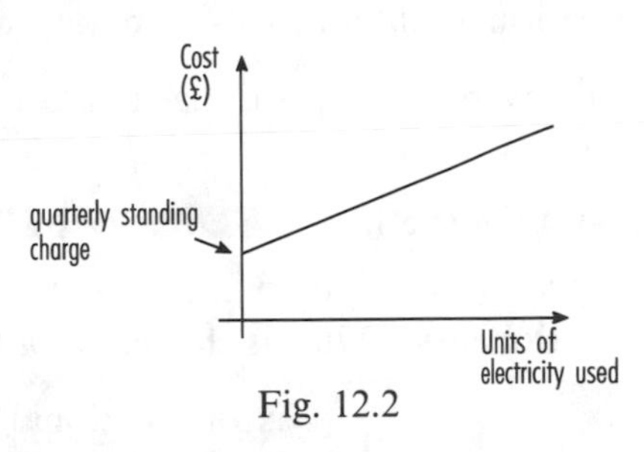

Fig. 12.2

We could extend Example 12.1 to include all a person's motoring costs. Included in the total costs are insurance and the Road Fund Licence, 2 types of standing charge. These must be paid no matter how great or small the distance driven.

$y \propto x$, or $y = kx$, says that y is directly proportional to x. The powers of x and y are 1.

There are many other types of proportionality. The next set of examples looks at a few types.

Example 12.3

One version of Ohm's law says that the current, I, is inversely proportional to the resistance, R. When $R = 5\,\Omega$, $I = 3\,\text{A}$. Find the constant of proportionality. What is the value of I when $R = 9\,\Omega$?

We write this as I is inversely proportional to R

i.e. $$I \propto \frac{1}{R}$$

i.e. $$I = k\left(\frac{1}{R}\right) = \frac{k}{R}$$

> k is the constant of proportionality.

Substituting for $R = 5$ and $I = 3$ gives

$$3 = \frac{k}{5}$$

i.e. $$3 \times 5 = k$$

i.e. $$k = 15.$$

This means $I = \dfrac{15}{R}$ so that when $R = 9$ we have

$$I = \frac{15}{9}$$

i.e. $$I = \frac{5}{3} \text{ or } 1.\bar{6},$$

i.e. the current for this resistance is $1.\bar{6}\,\text{A}$.

Later in this chapter we will look at the type of graph of $I=\frac{k}{R}$. $I=\frac{k}{R}$ may be re-written as $IR=k$, i.e. $IR=\text{constant}$. For any general inversely proportional relation, $y=\frac{k}{x}$, we may write $xy=\text{constant}$. This means that for an inversely proportional relationship the product of the variables is constant.

Example 12.4

y is proportional to x^2. If $y=135$ when $x=4$ find the constant of proportionality. What is the value of x when $y=76.5$?

y is proportional to x^2

is written as $\quad y \propto x^2$

i.e. $\quad y = kx^2$.

> k is the constant of proportionality.

Substituting for $x=4$ and $y=135$ gives

$$135 = k \times 4^2$$

i.e. $$\frac{135}{4^2} = k$$

i.e. $$k = 8.4375.$$

This means $\quad y = 8.4375x^2 \quad$ so that when $y=76.5$ we have

$$76.5 = 8.4375x^2$$

i.e. $$\frac{76.5}{8.4375} = x^2$$

$$9.0\bar{6} = x^2$$

i.e. $$x = \pm\sqrt{9.0\bar{6}}$$

$$x = \pm 3.01.$$

Later in this chapter we will look at a graph of the type $y=kx^2$.

Example 12.5

Without skidding a vehicle negotiates a curved horizontal track. Its velocity, v, is proportional to the (positive) square root of the radius of the curve, $\sqrt{r}$. Write this statement in mathematical terms.
If a vehicle's velocity is $14\,\text{ms}^{-1}$ around a curve of radius 45 m find the coefficient of proportionality.
On the same type of road surface if the radius is 60 m what is the maximum permitted velocity? What must be the curve's radius for no skidding at $10\,\text{ms}^{-1}$?

We have $\quad v \propto \sqrt{r}$

i.e. $\quad v = k\sqrt{r}$.

> k is the constant of proportionality.

Substituting for $v=14$ and $r=45$ gives

$$14 = k\sqrt{45}$$

$$\text{i.e.} \qquad \frac{14}{6.708} = k$$

i.e. $k = 2.087$ is the coefficient of proportionality.

Now to find v given $r=60$ we use the calculator value of k so

$$v = 2.086\ldots \times \sqrt{60}$$

i.e. $v = 16.17 \text{ ms}^{-1}$ is the velocity.

Then to find r given $v=10$ we have

$$10 = 2.086\ldots \times \sqrt{r}$$

$$\text{i.e.} \qquad \frac{10}{2.087} = \sqrt{r}$$

$$\text{i.e.} \qquad (4.792)^2 = r$$

to give $r = 22.96$ m as the curve's radius.

Later in this chapter we will look at a graph of the type $v=k\sqrt{r}$ (or $y=k\sqrt{x}$).

ASSIGNMENT

We can use the table in its raw form and attempt to plot a graph. Remember the table is

x	0.10	0.15	0.40	0.90	1.35	1.74	2.48	3.06	3.88
y	2.34	1.47	−0.03	−0.82	−1.11	−1.26	−1.45	−1.54	−1.64

The plot is shown in Fig. 12.3. It does look as though there is some relationship. At present it is not clear what it might be. However, our

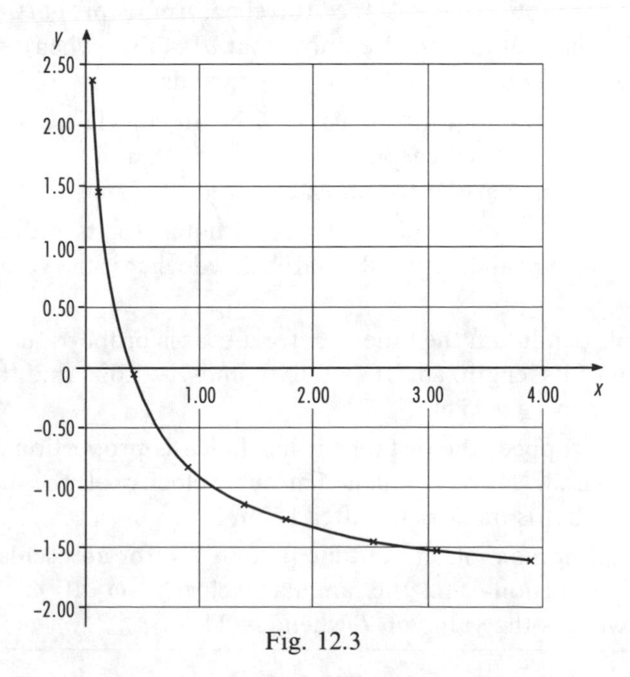

Fig. 12.3

work so far does tell us something. Obviously these points do not lie on a straight line and the graph does not pass through the origin. Hence our knowledge of proportion tells us that x and y are not directly proportional to each other. We may find that there is some other proportional relationship, but at the moment nothing is obvious.

EXERCISE 12.1

1 We know that $y \propto x$ and $y=4$ when $x=10$. Find the coefficient of proportionality. What is the value of y when $x=1.5$?

2 $y=3$ when $x=4$ according to $y \propto x^2$. When $x=32$ what is the value of y?

3 Given that $y \propto x^3$ and $y=80$ when $x=2$ find the constant of proportionality. What is the value of x when $y=16$?

4 $y \propto 1/x$. If $y=8$ when $x=5$ what is the value of y when $x=5.5$?

5 If $y \propto \sqrt{x}$ and $x=6$ when $y=8$ find the coefficient of proportionality. Find a value for y when $x=1$. What would change in this question if $y \propto -\sqrt{x}$?

6 Given that $y \propto 1/x^2$ and $y=8$ when $x=9$ find the values of x when $y=16$.

7 We know that $y \propto \pm\sqrt{x}$. If $y=45$ when $x=9$ calculate the coefficient of proportionality. What can you deduce when $x=-3$?

8 If $y \propto 1/x^2$ find the value of y when $x=2$ given that $y=8$ when $x=5$.

9 If $y \propto \pm 1/\sqrt{x}$ and $x=25$ when $y=30$ find the constant of proportionality. What is the value of x when $y=36$?

10 In a particular motion the distance travelled, d m, is proportional to the time for the motion, t s. We know that $d=480$ m when $t=150$ s. What distance has been travelled in 210 seconds?

11 Hooke's law states that the tension, T N, in an elastic spring is proportional to the extension, x m. Given that $T=20$ N when $x=0.15$ m find the tension for an extension of 0.25 m.

12 The volume, $V\,\text{m}^3$, of a sphere is proportional to its radius, r m, cubed. If $r=0.35$ m and $V=0.18\,\text{m}^3$ are related what is the volume for $r=0.48$ m?

13 For a simple pendulum the time, T s, for a beat is proportional to the square root of its length, l m. If $l=0.65$ m and $T=1.62$ s find the time for a beat when $l=0.95$ m.

14 If a body is dropped, the distance it has fallen is proportional to its velocity squared. Having fallen 5 m the velocity of the body is $9.90\,\text{ms}^{-1}$. What is its velocity after 15 m?

15 In simple harmonic motion the time period, T s, for an oscillation is inversely proportional to the angular velocity, ω. If $\omega=8$ and $T=0.785$ what is the value of T when $\omega=11$?

Some standard curves

There are many interesting curves in Mathematics. Here is a small selection of some standard ones.

1. The parabola, $y^2=4ax$

Usually when we want to plot a graph we have only y on the left-hand side of the relationship. To change our relation into this form we take the square root of both sides, i.e. $y=\pm\sqrt{4ax}$.

a can have a range of values. The form $4a$ is included to make the algebra simpler in Pure Mathematics.

Various parabolas are described in a variety of ways. We saw an alternative version of a parabola in Chapter 7 on Quadratic Equations.

Example 12.6

On one set of axes plot the graphs of

i) $y=\pm\sqrt{x}$ i.e. $y^2=x$,

ii) $y=\pm\sqrt{3x}$ i.e. $y^2=3x$,

iii) $y=\pm\sqrt{\frac{1}{2}x}$ i.e. $y^2=\frac{1}{2}x$.

We choose some specimen values of x and construct our tables as usual. Because we are using $\sqrt{}$ we know that none of the x values can be negative.

x	0	1	2	3	4	5	6	$\leftrightarrow$
$\pm\sqrt{x}$	0	±1.000	±1.414	±1.732	±2.000	±2.236	±2.449	$\updownarrow$
$3x$	0	3	6	9	12	15	18	
$\pm\sqrt{3x}$	0	±1.732	±2.449	±3.000	±3.464	±3.873	±4.243	$\updownarrow$
$\frac{1}{2}x$	0	0.5	1.0	1.5	2.0	2.5	3.0	
$\pm\sqrt{\frac{1}{2}x}$	0	±0.707	±1.000	±1.225	±1.414	±1.581	±1.732	$\updownarrow$

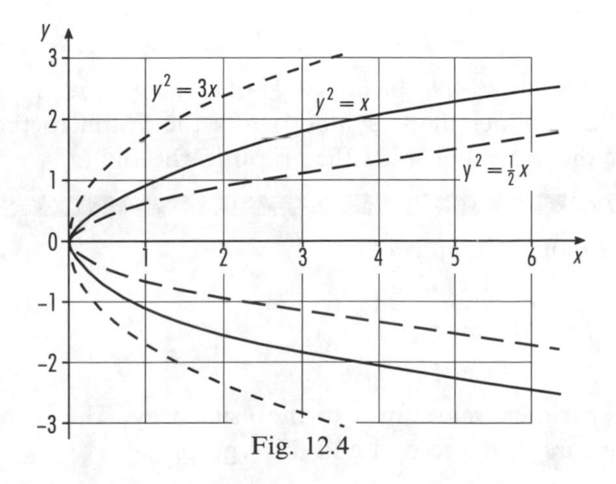

Fig. 12.4

Fig. 12.4 shows the different curvatures of the parabolas. The smaller the value in place of $4a$ the closer the parabola stays to the horizontal axis. The greater the value in place of $4a$ the more the parabola curves away from the horizontal axis.

Example 12.7

On one set of axes plot the graphs of

i) $y = \pm\sqrt{x}$ i.e. $y^2 = x$,
ii) $y = \pm\sqrt{x+1}$ i.e. $y^2 = x+1$,
iii) $y = \pm\sqrt{x-2}$ i.e. $y^2 = x-2$.

We choose some specimen values of x and construct our tables as usual. Because we are using $\sqrt{}$ we know that none of the x values can be.negative.

x	0	1	2	3	4	5	6	$\leftrightarrow$
$\pm\sqrt{x}$	0	± 1.000	± 1.414	± 1.732	± 2.000	± 2.236	± 2.449	$\updownarrow$
$x+1$	1	2	3	4	5	6	7	
$\pm\sqrt{x+1}$	± 1.000	± 1.414	± 1.732	± 2.000	± 2.236	± 2.449	± 2.646	$\updownarrow$
$x-2$	-2	-1	0	1	2	3	4	
$\pm\sqrt{x-2}$	not defined		0	± 1.000	± 1.414	± 1.732	± 2.000	$\updownarrow$

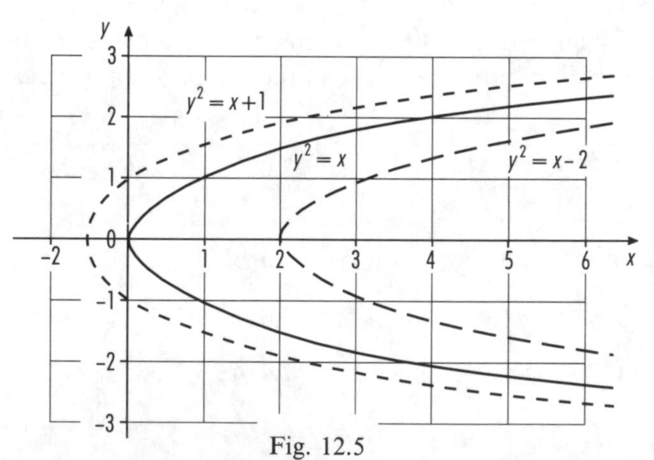

Fig. 12.5

In Fig. 12.5 we see how the basic parabola, $y = \pm\sqrt{x}$ is shifted horizontally. Addition under the $\sqrt{}$ symbol shifts the graph to the left. Subtraction under the $\sqrt{}$ symbol shifts the graph to the right.

Example 12.8

On one set of axes plot the graphs of

i) $y = \pm\sqrt{x}$ i.e. $y^2 = x$,
ii) $y+1 = \pm\sqrt{x}$ i.e. $(y+1)^2 = x$,
iii) $y-2 = \pm\sqrt{x}$ i.e. $(y-2)^2 = x$.

We choose some specimen values of x in the usual way. The $\pm$ option needs a little more care in the second and third graphs.

i) The table for $y = \pm\sqrt{x}$ is the usual one seen in previous examples.

x	0	1	2	3	4	5	6
$\pm\sqrt{x}$	0	±1.000	±1.414	±1.732	±2.000	±2.236	±2.449

ii) For
$$y+1 = \pm\sqrt{x}$$
$$y = -1\pm\sqrt{x}$$
i.e. $\quad y = -1+\sqrt{x}, \qquad y = -1-\sqrt{x}.$

This means there are 2 options for y which we use. The table shows these options.

x	0	1	2	3	4	5	6
$\sqrt{x}$	0	1.000	1.414	1.732	2.000	2.236	2.449
$-1+\sqrt{x}$	-1.000	0	0.414	0.732	1.000	1.236	1.449
$-1-\sqrt{x}$	-1.000	-2.000	-2.414	-2.732	-3.000	-3.236	-3.449

iii) For
$$y-2 = \pm\sqrt{x}$$
$$y = 2\pm\sqrt{x}$$
i.e. $\quad y = 2+\sqrt{x}, \qquad y = 2-\sqrt{x}.$

Again there are 2 options for y. The table shows these options.

x	0	1	2	3	4	5	6
$\sqrt{x}$	0	1.000	1.414	1.732	2.000	2.236	2.449
$2+\sqrt{x}$	2.000	3.000	3.414	3.732	4.000	4.236	4.449
$2-\sqrt{x}$	2.000	1.000	0.586	0.268	0	-0.236	-0.449

In Fig. 12.6 we see how the basic parabola, $y^2 = x$, is shifted. $y = -1\pm\sqrt{x}$ is a vertical shift of -1. $y = 2\pm\sqrt{x}$ is a vertical shift of $+2$.

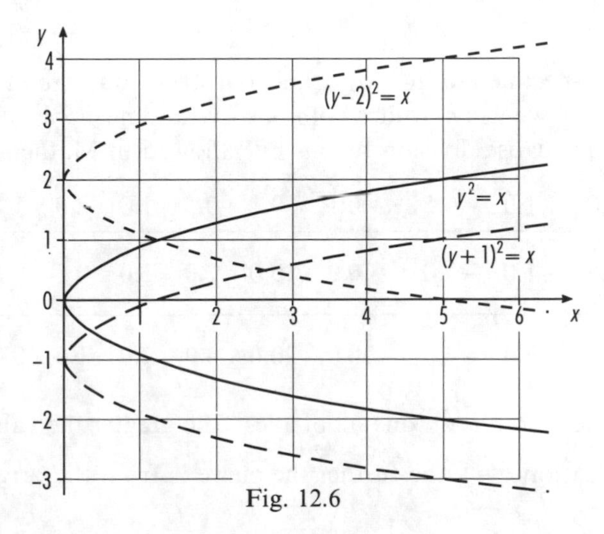

Fig. 12.6

We can continue to investigate and experiment with other parabolas. These examples are just a selection of possibilities.

EXERCISE 12.2

For each question first attempt to sketch the graph. Follow up the sketch with an accurate plot of the graph. In each case the sketch should influence your choice of specimen table values for x.

1 $y=\pm\sqrt{2x}$ or $y^2=2x$.

2 $y=\pm\sqrt{x+3}$ or $y^2=x+3$.

3 $y^2=x$ and $y^2=-x$.

4 $x^2=y$ and $x^2=-y$.

5 $(y-2)^2=x+3$.

2. The rectangular hyperbola, $y=\frac{a}{x}$ or $y=\frac{c^2}{x}$ or $xy=c^2$

This relationship will remind you of inverse proportionality. An increase in x causes a decrease in y. Similarly, a decrease in x causes an increase in y. Again a can have any value. The next example shows the graph in 2 parts. Both parts are necessary, you **cannot** omit either part. Because the 2 parts are separate, without any connecting line or curve, the rectangular hyperbola is **discontinuous**.

Example 12.9

On one set of axes plot the graphs of

i) $y=\frac{1}{x}$ i.e. $xy=1$,

ii) $y=\frac{2}{x}$ i.e. $xy=2$.

We choose some specimen values of x and construct our tables in the usual way. This time we will consider both positive and negative values of x. We avoid $x=0$ because division by 0 is not allowed in Mathematics.

x	-5.0	-2.0	-1.0	-0.5	-0.1	0.1	0.5	1.0	2.0	5.0	$\longleftrightarrow$
$\frac{1}{x}$	-0.2	-0.5	-1.0	-2.0	-10.0	10.0	2.0	1.0	0.5	0.2	$\updownarrow$
$\frac{2}{x}$	-0.4	-1.0	-2.0	-4.0	-20.0	20.0	4.0	2.0	1.0	0.4	$\updownarrow$

Fig. 12.7 shows the 2 graphs on one set of axes. The greater the value of a in the general equation $y=\frac{a}{x}$ the further the curve is from the origin.

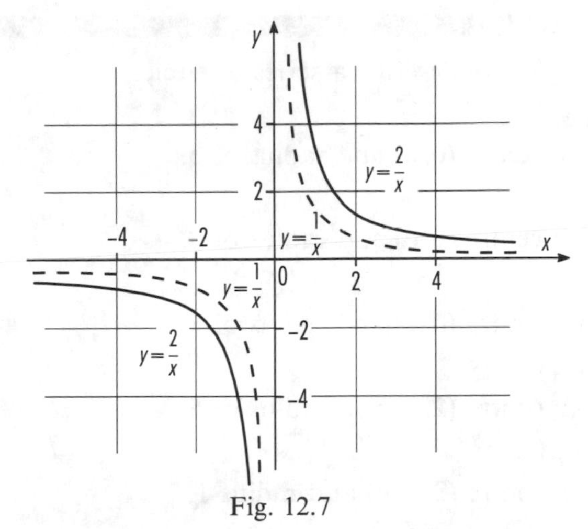

Fig. 12.7

We need not go through as many variations as before. Those we have looked at for the parabola apply in a similar way here. The following exercise looks at a selection.

EXERCISE 12.3

In each question construct a table of values. For convenience you might like to choose specimen values of x from -5 to 5, avoiding $x=0$. Use the table of values to plot the graphs. For reference you will see each question includes $y=\frac{1}{x}$.

1 $y=\frac{1}{x}$ and $y=2+\frac{1}{x}$.

2 $y=\frac{1}{x}$ and $y=\frac{1}{x}-1.5$.

3 $y=\frac{-1}{x}$ and $y=\frac{1}{x}$.

4 $xy=1$ and $xy=0.25$.

5 $y=\frac{1}{x}$ and $y=2+\frac{2}{x}$.

3. The circle, $x^2+y^2=r^2$

r is the radius of the circle.

$x^2+y^2=r^2$ describes a circle with centre (0, 0) and radius r.

More generally $(x-\alpha)^2+(y-\beta)^2=r^2$ describes a circle with centre (α, β) and radius r.

When using $x^2+y^2 = r^2$ we can re-arrange it to get

$$y^2 = r^2-x^2$$

i.e.

$$y = \pm\sqrt{r^2-x^2}.$$

You will find this form more useful when constructing a table of values.

Examples 12.10

In this set of examples we describe a series of circles.

i) $x^2+y^2=9$, i.e. $x^2+y^2=3^2$
is a circle with centre $(0, 0)$ and radius of 3.

ii) $(x-2)^2+y^2=3^2$
is a circle with centre $(2, 0)$ and radius of 3.

iii) $x^2+(y-1)^2=3^2$
is a circle with centre $(0, 1)$ and radius of 3.

iv) $(x-2)^2+(y-1)^2=25$
is a circle with centre $(2, 1)$ and radius of 5.

v) $(x-2)^2+(y+4)^2=1^2$
is a circle with centre $(2, -4)$ and radius 1.

vi) $(x+0.5)^2+(y+3)^2=42.25$
is a circle with centre $(-0.5, -3)$ and radius $\sqrt{42.25}=6.5$.

It is left as an exercise for you to draw these circles. Rather than plot them you may prefer to draw them with a pair of compasses.

Example 12.11

Plot the circle $(x-2)^2+y^2=3^2$.

We know from the previous set of examples this circle has centre $(2, 0)$ and a radius of 3. This means the horizontal diameter stretches 3 units either side of 2, i.e. from $x=-1$ to $x=5$. The vertical diameter stretches 3 units either side of 0, i.e. from $y=-3$ to $y=3$. The values of x we have just deduced will influence our specimen table values.
Also we can re-arrange our equation,

$$\text{i.e.} \qquad (x-2)^2+y^2 = 3^2$$

$$\text{becomes} \qquad y^2 = 3^2-(x-2)^2$$

$$\therefore \qquad y = \pm\sqrt{9-(x-2)^2}.$$

The table is constructed below.

x	-1.00	-0.50	0	0.50	1.00	1.50	2.00	$\longleftrightarrow$
$x-2$	-3.00	-2.50	-2.00	-1.50	-1.00	-0.50	0	
$(x-2)^2$	9.00	6.25	4.00	2.25	1.00	0.25	0	
$9-(x-2)^2$	0	2.75	5.00	6.75	8.00	8.75	9.00	
y	0	± 1.66	± 2.24	± 2.60	± 2.83	± 2.96	± 3.00	$\updownarrow$

	2.50	3.00	3.50	4.00	4.50	5.00	$\longleftrightarrow$
	0.50	1.00	1.50	2.00	2.50	3.00	
	0.25	1.00	2.25	4.00	6.25	9.00	
	8.75	8.00	6.75	5.00	2.75	0	
	± 2.96	± 2.83	± 2.60	± 2.24	± 1.66	0	$\updownarrow$

Fig. 12.8 shows the circle. It confirms the centre as $(2, 0)$ and the radius as 3.

Also you can use the diagram to check the points where it passes through the coordinate axes. These are consistent with the initial circle equation for this example.

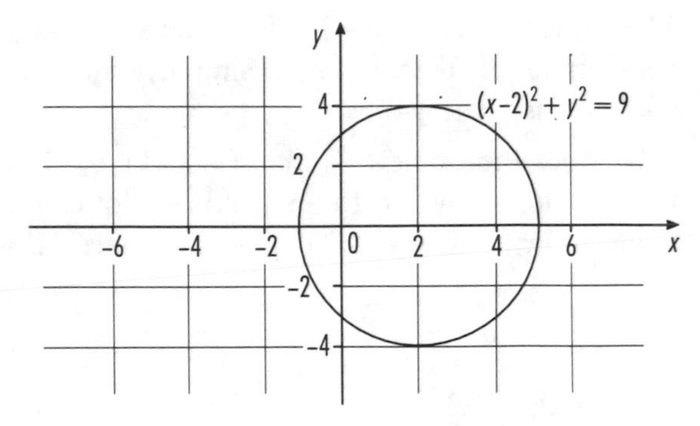

Fig. 12.8

It crosses the x-axis when $y=0$,

$$\begin{aligned}
\text{i.e.} \quad (x-2)^2+y^2 &= 9 \\
\text{becomes} \quad (x-2)^2 &= 9 \\
\therefore \quad x-2 &= \pm 3 \\
\text{i.e.} \quad x &= 2-3, \quad 2+3 \\
\text{i.e.} \quad x &= -1, \quad 5.
\end{aligned}$$

This is confirmed by Fig. 12.8.

Also it crosses the y-axis when $x=0$,

$$\begin{aligned}
\text{i.e.} \quad (x-2)^2+y^2 &= 9 \\
\text{becomes} \quad (0-2)^2+y^2 &= 9 \\
\text{i.e.} \quad 4+y^2 &= 9 \\
y^2 &= 9-4 = 5 \\
\therefore \quad y &= \pm\sqrt{5} = \pm 2.24.
\end{aligned}$$

Again this is confirmed by Fig. 12.8.

EXERCISE 12.4

For Questions **1–10** describe the circles by giving their centres and radii.

1 $x^2+y^2=36$

2 $x^2+y^2=18$

3 $\left(x+\frac{1}{2}\right)^2+y^2=36$

4 $\left(x-\frac{1}{2}\right)^2+y^2=25$

5 $x^2+(y-3)^2=49$

6 $x^2+\left(y+\frac{1}{4}\right)^2=6.25$

7 $\left(x+\frac{1}{4}\right)^2+\left(y-\frac{1}{3}\right)^2=9$

8 $(x-2)^2+(y-2.5)^2=1$

9 $(x+2)^2+(y+1.7)^2=4$

10 $\left(x-\frac{2}{5}\right)^2+\left(y+\frac{3}{4}\right)^2=1$

11 The circle $(x-2)^2+(y-2.5)^2=25$ cuts the x-axis when $y=0$. Find the coordinates of these 2 points. Similarly find the coordinates where the circle crosses the y-axis.

12 Why does the circle $(x-2)^2+(y-2.5)^2=1$ not intersect either axis? Check your answer with an accurate plot of this circle. Use specimen values of x from $x=-3$ to $x=3$ at intervals of 0.5.

4. The ellipse, $\frac{x^2}{a^2}+\frac{y^2}{b^2}=1$

a and b are constant values. A simple description of an ellipse is either an oval or a cross-section through a rugby ball.

$\frac{x^2}{a^2}+\frac{y^2}{b^2}=1$ lies on the coordinate axes. Its centre is at the origin. The extreme values of x are $2a$ apart, i.e. at $(-a, 0)$ and $(a, 0)$, either side of the centre. The extreme values of y are $2b$ apart, i.e. at $(0, -b)$ and $(0, b)$, either side of the centre.

If $2a>2b$ then $2a$ is called the **major axis** of the ellipse. This means $2b$ is called the **minor axis**.

If $2a<2b$ then $2b$ is the **major axis** of the ellipse. Hence $2a$ is the **minor axis**.

If $2a=2b$ then the **major and minor axes are equal**. The ellipse has degenerated into a circle of radius a (or b).

Examples 12.12

i) $\frac{x^2}{25}+\frac{y^2}{1}=1,\qquad$ i.e. $\frac{x^2}{5^2}+\frac{y^2}{1^2}=1$

is an ellipse with centre $(0, 0)$.

The major axis is $2a=2\times 5=10$.

The minor axis is $2b=2\times 1=2$.

$a=5$, $b=1$.

ii) $\frac{x^2}{4}+\frac{y^2}{9}=1,\qquad$ i.e. $\frac{x^2}{2^2}+\frac{y^2}{3^2}=1$

is an ellipse with centre $(0, 0)$

The major axis is $2b=2\times 3=6$.

The minor axis is $2a=2\times 2=4$.

$a=2$, $b=3$.

Figs. 12.9 shows the ellipses.

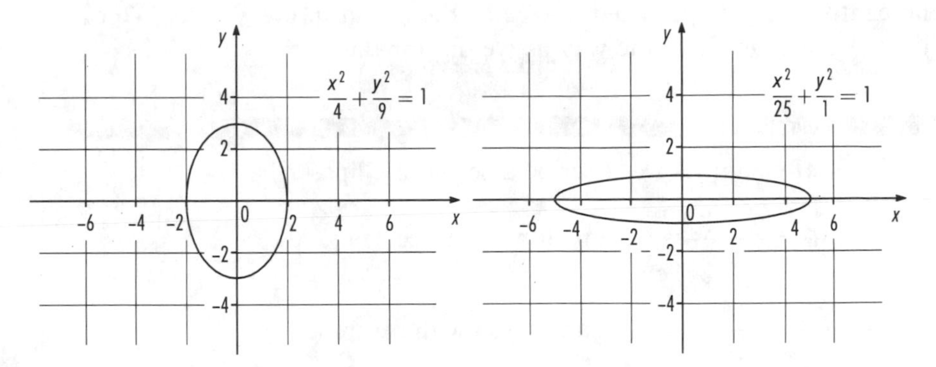

Figs. 12.9

Example 12.13

Plot thc graph of $\frac{x^2}{4}+\frac{y^2}{9}=1$.

We can re-arrange this equation,

i.e. $$\frac{y^2}{9} = 1-\frac{x^2}{4}$$

$$y^2 = 9\left(1-\frac{x^2}{4}\right)$$

i.e. $$y = \pm\sqrt{9\left(1-\frac{x^2}{4}\right)}$$

$$y = \pm 3\sqrt{1-\frac{x^2}{4}}.$$

We know the range of values of x is from $-a$ to a. In Example 12.12ii) we deduced a to be 2. This means that x ranges from -2 to 2. We construct the table of values in the usual way.

x	-2.00	-1.50	-1.00	-0.50	0	0.50	1.00	1.50	2.00
x^2	4.00	2.25	1.00	0.25	0	0.25	1.00	2.25	4.00
$\frac{x^2}{4}$	1.00	0.56	0.25	0.06	0	0.06	0.25	0.56	1.00
$1-\frac{x^2}{4}$	0	0.44	0.75	0.94	1.00	0.94	0.75	0.44	0
$\pm\sqrt{1-\frac{x^2}{4}}$	0	± 0.66	± 0.87	± 0.97	± 1.00	± 0.97	± 0.87	± 0.66	0
y	0	± 1.98	± 2.60	± 2.90	± 3.00	± 2.90	± 2.60	± 1.98	0

This plot is left as an exercise for you. Your own plot should confirm what you saw in Figs. 12.9.

The centre of an ellipse need not be at the origin in every case. We can shift the centre in the same way as we did for the circle.

Examples 12.14

In this set of examples we describe a series of ellipses.

i) $\dfrac{(x-\alpha)}{a^2}+\dfrac{(y-\beta)}{b^2}=1$ is an ellipse with centre (α, β).

ii) $\dfrac{(x-2)^2}{16}+\dfrac{(y-1)^2}{25}=1$ is an ellipse with centre (2, 1).

$a^2=16$ and $b^2=25$.

Thus the major axis is of length $2b=2\times 5=10$ and the minor axis is of length $2a=2\times 4=8$.

iii) $\dfrac{(x+\frac{1}{2})^2}{4}+\dfrac{y^2}{9}=1$ is an ellipse with centre $(-\frac{1}{2}, 0)$.

$a^2=4$ and $b^2=9$.
Thus the major axis is of length $2b=2\times 3=6$ and the minor axis is of length $2a=2\times 2=4$.

iv) $\dfrac{x^2}{49}+\dfrac{(y-5)^2}{36}=1$ is an ellipse with centre (0, 5).

$a^2=49$ and $b^2=36$.

Thus the major axis is of length $2a=2\times 7=14$ and the minor axis is of length $2b=2\times 6=12$.

EXERCISE 12.5

For the following ellipses find the lengths of the major and minor axes. Distinguish between them.

1 $\dfrac{x^2}{16}+\dfrac{y^2}{25}=1$

2 $\dfrac{x^2}{4}+\dfrac{y^2}{5}=1$

3 $\dfrac{x^2}{1}+\dfrac{y^2}{5}=1$

4 $\dfrac{x^2}{2}+y^2=1$

5 $\dfrac{x^2}{81}+\dfrac{y^2}{36}=1$

6 For each of Questions **1–5** you will have an idea of the range of values for x. Use these values to construct a table for each ellipse. Hence plot a graph in each case.

5. The hyperbola, $\frac{x^2}{a^2} - \frac{y^2}{b^2} = 1$

a and b are constant values. Notice the similarity between the equations for the ellipse and the hyperbola. They differ by only a "$-$" sign.

Do not confuse the rectangular hyperbola with the hyperbola, though you may spot some similarities in the shapes. For the rectangular hyperbola the horizontal and vertical axes act as asymptotes. For the hyperbola the straight lines $y = \pm\frac{bx}{a}$ act as asymptotes. Again, just like the rectangular hyperbola, the hyperbola is discontinuous.

The next example looks at the basic shape of a hyperbola.

Example 12.15

Plot the graph of $\frac{x^2}{16} - \frac{y^2}{25} = 1$.

We can re-arrange this equation to get

$$\frac{x^2}{16} - \frac{y^2}{25} - 1 = 0$$

i.e.
$$\frac{x^2}{16} - 1 = \frac{y^2}{25}$$

$$25\left(\frac{x^2}{16} - 1\right) = y^2$$

i.e.
$$y = \pm\sqrt{25\left(\frac{x^2}{16} - 1\right)}$$

i.e.
$$y = \pm 5\sqrt{\frac{x^2}{16} - 1}.$$

We know we cannot take the square root of a negative number.

This means that $\frac{x^2}{16} - 1 \geqslant 0$

i.e.
$$\frac{x^2}{16} \geqslant 1$$

i.e.
$$x^2 \geqslant 16$$

i.e.
$$x \geqslant 4, \quad x \leqslant -4.$$

We have constructed so many tables in this chapter that this one is omitted. As an exercise for yourself you should construct the table for this hyperbola. You can choose the specimen values for x having seen Fig. 12.10. Compare your own plot of the graph with Fig. 12.10. It shows the hyperbola does not exist for values of x between -4 and 4. As well as the hyperbola the asymptotes, $y = \frac{5x}{4}$ and $y = \frac{-5x}{4}$, are drawn.

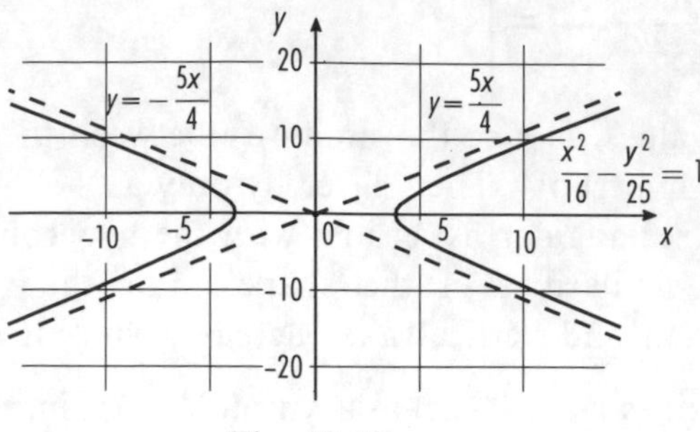

Fig. 12.10

Remember the gradients of these asymptotes are found from the values of a and b for the hyperbola. They are important. The smaller the size of the gradients the tighter the curvature of the hyperbola. For asymptotes with gradients of larger size the curvature of the hyperbola is more gentle.

ASSIGNMENT

We have looked at the Assignment twice already. The second time we plotted the values and drew a smooth curve through the points. Let us have another look at Fig. 12.3. We see that it does not look anything like any of our standard curves. This **suggests** that it is not one of them. We would have to investigate each of the previous 5 types in greater detail to confirm our suspicion.

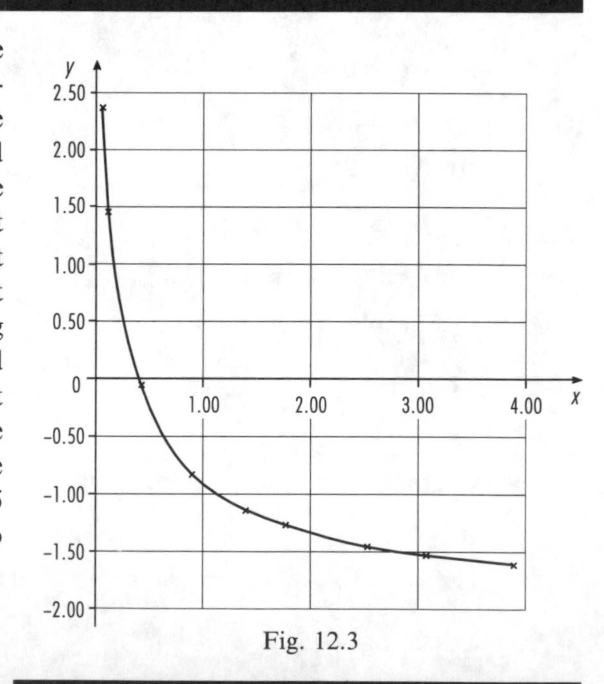

Fig. 12.3

EXERCISE 12.6

1 Construct the table and plot the hyperbola $\frac{x^2}{4}-y^2=1$.

Use values of x from -5 to -2 and from 2 to 5 at intervals of 0.5.

2 On the same pair of axes plot the hyperbolas $x^2-y^2=1$ and $\frac{x^2}{4}-y^2=1$. In the first case use values of x from -5 to -1 and from 1 to 5 at intervals of 0.5. In the second case use values of x from -5 to -2 and from 2 to 5 at the same intervals.

3 What are the equations of the asymptotes to the hyperbolas

i) $\frac{x^2}{4}-\frac{y^2}{9}=1$,

ii) $\frac{x^2}{9}-\frac{y^2}{4}=1$?

4 Plot the hyperbola $\frac{x^2}{4}-\frac{y^2}{9}=1$ and its asymptotes.

5 Plot the hyperbola $\frac{x^2}{9}-\frac{y^2}{4}=1$ and the ellipse $\frac{x^2}{9}+\frac{y^2}{4}=1$ on one pair of axes. What are the coordinates at the points of intersection?

Non-linear laws reduced to a linear form

In Chapter 11 we looked at a selection of curves plotted on the usual axes. Then we saw how to re-draw them as straight lines on amended axes. Those examples were based on logarithms. Not all curves can be simply reduced to straight lines. Not all reductions need logarithms. For each trial we are going to test our assignment table of values against a possible straight line type relationship. The values may all lie on a straight line. Alternatively we may be able to draw a line of best fit. In either of these cases we suggest the values are related by the particular law. If there is no close resemblance to a straight line then that particular law fails. Let us test a few of the more obvious types.

1. $y=\frac{a}{x}+b$, i.e. $y=a\left(\frac{1}{x}\right)+b$

We compare this form with the standard straight line $Y=MX+C$. We change the horizontal axis, plotting $\frac{1}{x}$ in place of X. The gradient is a and the vertical intercept is b.

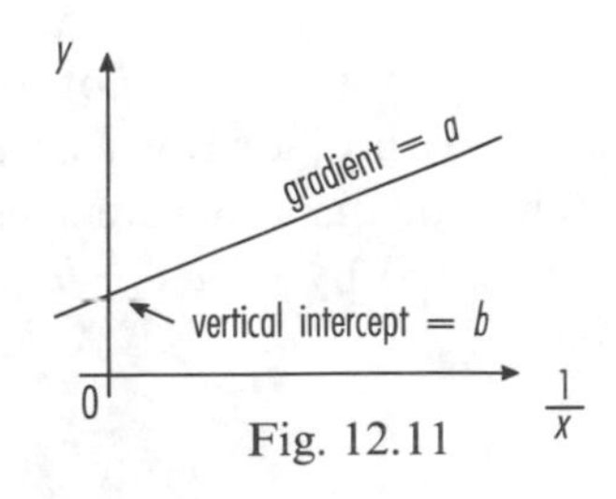

Fig. 12.11

We have met a similar relationship before, but with $b=0$, earlier in this chapter. $y=\frac{a}{x}$ was used for inverse proportionality.

Now to our original assignment values we add another row in the table for $\frac{1}{x}$.

x	0.10	0.15	0.40	0.90	1.35	1.74	2.48	3.06	3.88	
$\frac{1}{x}$	10.00	6.67	2.50	1.11	0.74	0.57	0.40	0.33	0.26	↔
y	2.34	1.47	−0.03	−0.82	−1.11	−1.26	−1.45	−1.54	−1.64	↕

Fig. 12.12 shows our plot of y against $\frac{1}{x}$ with no obvious straight line. This means that our values are *not* related by $y=\frac{a}{x}+b$.

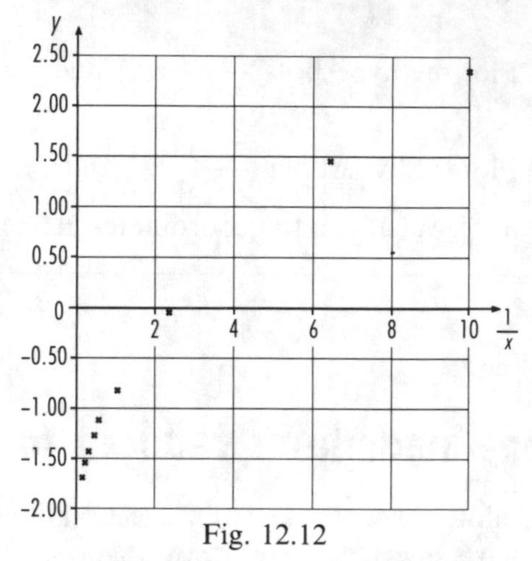

Fig. 12.12

2. $y=ax^2+b$

We compare this form with the standard straight line $Y=MX+C$. We change the horizontal axis, plotting x^2 in place of X. The gradient is a and the vertical intercept is b.

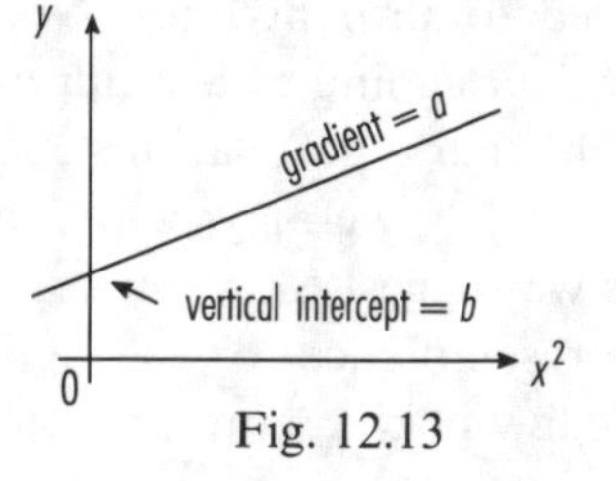

Fig. 12.13

To our original assignment values we add another row in the table for x^2.

x	0.10	0.15	0.40	0.90	1.35	1.74	2.48	3.06	3.88	
x^2	0.01	0.02	0.16	0.81	1.82	3.03	6.15	9.36	15.05	↔
y	2.34	1.47	−0.03	−0.82	−1.11	−1.26	−1.45	−1.54	−1.64	↕

Fig. 12.14 shows our plot of y against x^2 with no obvious straight line. This means that our values are *not* related by $y=ax^2+b$.

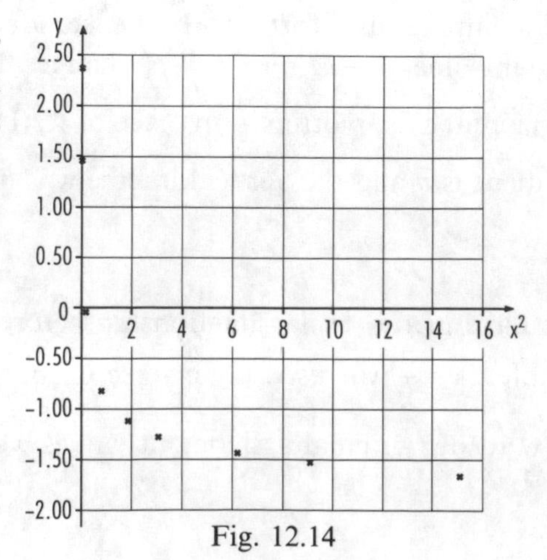

Fig. 12.14

3. $y=\frac{a}{x^2}+b$, i.e. $y=a\left(\frac{1}{x^2}\right)+b$

We compare this form with the standard straight line $Y=MX+C$. We change the horizontal axis, plotting $\frac{1}{x^2}$ in place of X. The gradient is a and the vertical intercept is b.

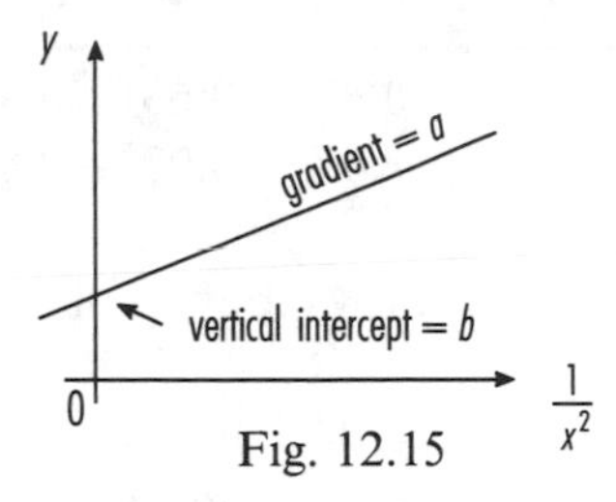

Fig. 12.15

To our original assignment values we add another row in the table for $\frac{1}{x^2}$.

x	0.10	0.15	0.40	0.90	1.35	1.74	2.48	3.06	3.88	
$\frac{1}{x^2}$	100.00	44.44	6.25	1.23	0.55	0.33	0.16	0.11	0.07	↔
y	2.34	1.47	−0.03	−0.82	−1.11	−1.26	−1.45	−1.54	−1.64	↕

Fig. 12.16 shows our plot of y against $\frac{1}{x^2}$ with no obvious straight line. This means that our values are *not* related by $y=\frac{a}{x^2}+b$.

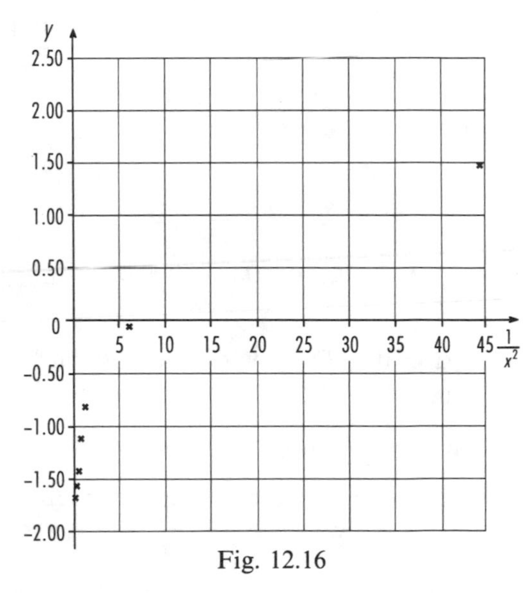

Fig. 12.16

4. $y=a\sqrt{x}+b$

We compare this form with the standard straight line $Y=MX+C$. We change the horizontal axis, plotting $\sqrt{x}$ in place of X. The gradient is a and the vertical intercept is b.

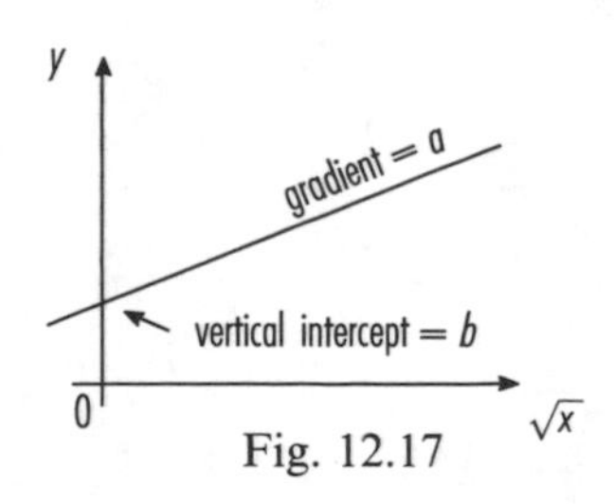

Fig. 12.17

To our original assignment values we add another row in the table for $\sqrt{x}$.

x	0.10	0.15	0.40	0.90	1.35	1.74	2.48	3.06	3.88	
$\sqrt{x}$	0.32	0.39	0.63	0.95	1.16	1.32	1.57	1.75	1.97	$\leftrightarrow$
y	2.34	1.47	−0.03	−0.82	−1.11	−1.26	−1.45	−1.54	−1.64	$\updownarrow$

Fig. 12.18 shows our plot of y against $\sqrt{x}$ with no obvious straight line. This means that our values are *not* related by $y=a\sqrt{x}+b$.

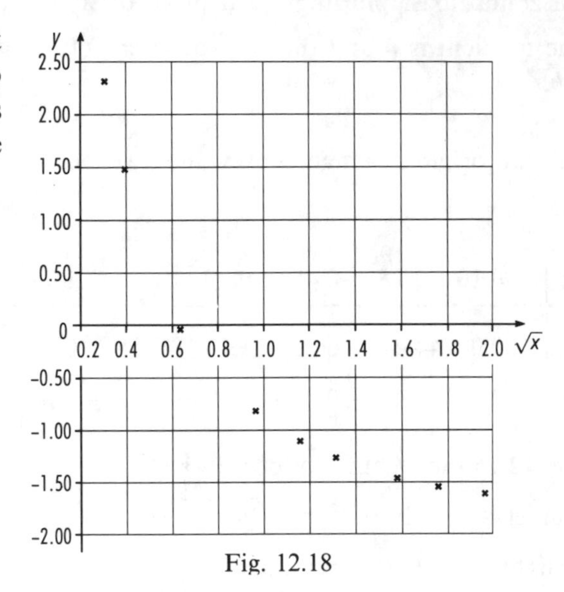

Fig. 12.18

5. $y=\dfrac{a}{\sqrt{x}}+b$, i.e. $y=a\left(\dfrac{1}{\sqrt{x}}\right)+b$

We compare this form with the standard straight line $Y=MX+C$. We change the horizontal axis, plotting $\dfrac{1}{\sqrt{x}}$ in place of X. The gradient is a and the vertical intercept is b.

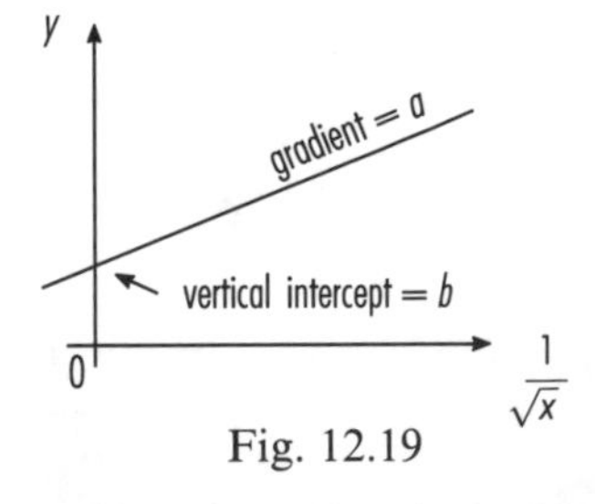

Fig. 12.19

To our original assignment values we add another row in the table for $\dfrac{1}{\sqrt{x}}$.

x	0.10	0.15	0.40	0.90	1.35	1.74	2.48	3.06	3.88	
$\dfrac{1}{\sqrt{x}}$	3.16	2.58	1.58	1.05	0.86	0.76	0.64	0.57	0.51	$\leftrightarrow$
y	2.34	1.47	−0.03	−0.82	−1.11	−1.26	−1.45	−1.54	−1.64	$\updownarrow$

Fig. 12.20 shows our plot of y against $\frac{1}{\sqrt{x}}$. Immediately we see we can draw a straight line of best fit. All the values are very close to our line. This indicates that our values *are* related by $y = \frac{a}{\sqrt{x}} + b$. We can read off the vertical intercept, b, as -2.40. In the usual way we can draw a right-angled triangle and calculate the gradient, a, to be 1.5. These values mean we can write our complete relation as $y = \frac{1.5}{\sqrt{x}} - 2.4$.

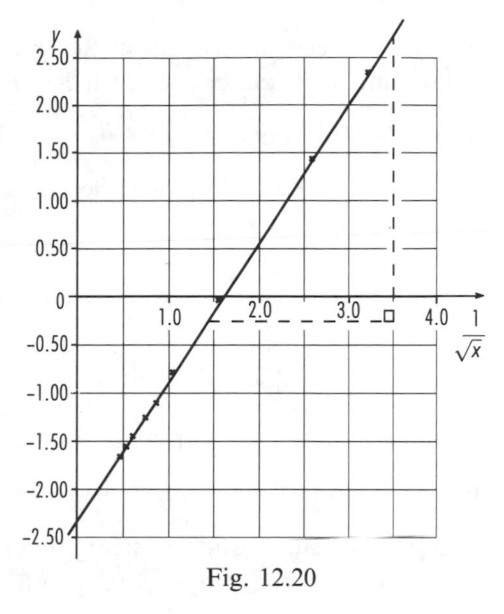

Fig. 12.20

Now the final exercise will give you some practice at finding various relationships. In each case the relation is suggested. Firstly all you need to do is calculate the extra table row of values. Then you need to plot the graph to confirm the relation. Finally you should be able to determine the values for gradient and vertical intercept from your graph.

EXERCISE 12.7

1 Performance trials for a medium size 5 door saloon car recorded the following results. They were from a standing start, relating distance travelled, s m, to time, t s.

t (s)	1	2	4	6	8	10
s (m)	1.45	5.95	24.00	54.05	96.25	150.00

It is thought that these values are related by $s = mt^2$. Using a straight line graph check that this is true. From your graph find the value of m.

2 The time period, T s, for a simple pendulum is related to the length of the pendulum, l m. It is thought that $T = k\sqrt{l} + c$. Using the test results draw a straight line graph to show that this is true for $c = 0$.

l (m)	0.950	0.960	0.980	0.990	0.995	1.000
T (s)	1.949	1.962	1.977	1.989	1.995	2.002

From your graph find the value of k correct to 2 decimal places.

3 Tests have been carried out in the treatment of a bacterial infection. The number of bacteria recorded is B $(\times 10^3)$ during time t hours. They are probably related by $B = \frac{a}{\sqrt{t}} + b$ for the new ointment. Using the table of data test whether this relationship is true within experimental limits.

t	12	18	24	60	120	180
B $(\times 10^3)$	500	380	310	140	50	10

What are the values of a and b?

4 The rate of heat energy transfer for a piece of brass is Q. It is thought to be related to the thickness of the brass, x, by $Q = \frac{m}{x} + c$. Using the following results and a straight line graph test whether this relationship is true. From your graph find values for m and c.

x (m)	0.050	0.075	0.100	0.150	0.200	0.250
Q (W)	25 000	17 000	12 500	8 500	6 500	5 000

5 One particular machine in an engineering workshop costs £C to lease each week. The costs are related to the number of hours/week, t, that the machine is worked. It is thought that $C = at^3 + b$. Using the data from the first six weeks plot a straight line graph. From your graph find the values of a and b.

t (hour)	35.0	37.5	40.0	42.0	46.0	50.0
C (£)	2600	3100	3700	4200	5400	6800

6 We know the power, P, across a resistor, R, is related to the current, I, by $P = RI^2$. Use the following test results to draw a straight line graph.

I (A)	1.20	1.25	1.40	1.50	1.60	1.75
P (W)	4.30	4.70	5.90	6.75	7.70	9.20

From your graph check that this resistor is approximately 3 Ω.

7 For a gas, pressure, p, is thought to be inversely proportional to volume, v. We may write this as $p = \frac{k}{v}$. Using the table of results draw a straight line graph.

v	10.0	12.5	15.0	18.0	20.0	25.0
p	200	160	130	110	100	80

From your graph find the value of k.

8 μ is the coefficient of friction between a pulley and a belt. $v\,\mathrm{ms}^{-1}$ is the velocity of the pulley. It is thought that they are related according to $\mu = a\sqrt{v} + b$. Use the table of results and an appropriate straight line graph to check this.

μ	0.223	0.249	0.258	0.273	0.286	0.301
v (ms^{-1})	2.5	5.0	6.0	8.0	10.0	12.5

From your graph what are the values of a and b?

9 For a van the distance, s m, it travels is related to its velocity, $v\,\mathrm{ms}^{-1}$, by $s = mv^2 + c$. Using the available data and a straight line graph check that this is true.

v (ms^{-1})	4	6	8	10	15	20
s (m)	5.45	15.00	28.50	45.75	106.00	189.75

Find values for m and c from your graph.

10 The second moment of area, I, about xy is related to the outer diameter, D, of the ring by $I = aD^4 + b$. Using a straight line graph based on the table of results find values for a and b.

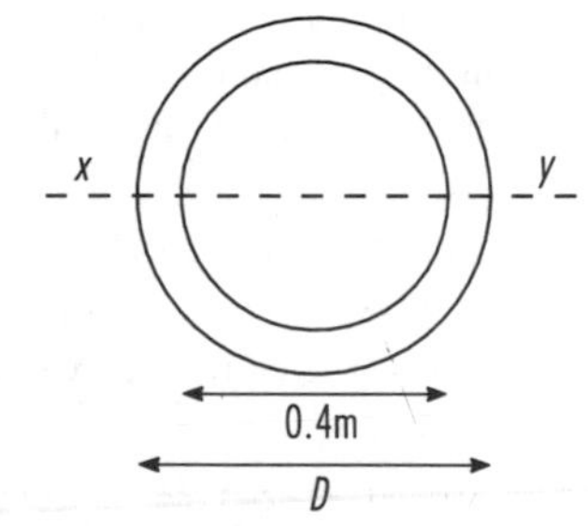

D	0.50	0.55	0.60	0.65	0.70	0.75	0.80
I ($\times 10^{-3}$)	1.81	3.24	5.11	7.51	10.53	14.27	18.85

13 Vectors

The objectives of this chapter are to:

1 Define a scalar.
2 Define a vector.
3 Define a negative vector, $-\boldsymbol{a}$.
4 Define the addition of two or more vectors.
5 Resolve a vector into two component parts at right angles to each other.
6 Calculate the magnitude and direction of the resultant of two component parts at right angles to each other.
7 Resolve a system of vectors into component parts and calculate the totals of the horizontal and vertical components.
8 Calculate the magnitude and direction of the resultant of the two total component parts calculated.
9 Define the result of multiplying a vector by a scalar.
10 State the parallelogram law for the addition of any two vectors.
11 Deduce the resultant of $\boldsymbol{a}-\boldsymbol{b}$ and relate the result to the second diagonal of the parallelogram.
12 Define the unit triad $\boldsymbol{i}, \boldsymbol{j}, \boldsymbol{k}$ in three dimensions.
13 Express any vector $\boldsymbol{r}$ in the form $x\boldsymbol{i}+y\boldsymbol{j}+z\boldsymbol{k}$.
14 Solve simple problems involving addition or subtraction of vectors.

Introduction

In this chapter we will concentrate on vectors. However, to begin with we do distinguish between scalars and vectors. There are many examples of vectors in engineering and science, and many different ways of representing them. As a start we will use a 3 dimensional system. All the axes will be mutually perpendicular.

ASSIGNMENT

Two problems form the Assignments for this chapter.

1.

A construction firm is currently engaged in building a bungalow on a remote site. It is accessible only across a river, 120 m wide, with a downstream current of $4\ \mathrm{ms}^{-1}$. A boat with a top speed of $5\ \mathrm{ms}^{-1}$ is available for transporting materials. For operational efficiency should the boat cross the river by the shortest route or in the fastest possible time?

2. An aircraft is capable of flying at 600 kmh^{-1} in still air. It is attempting to fly due East but is being blown off course by a wind from the North East at 75 kmh^{-1}. What is the resultant velocity?

Vectors and scalars

We may divide physical quantities into two groups: **scalar quantities** and **vector quantities**. Temperature is one type of **scalar quantity**, e.g. 32°F, 100°C.

A scalar is defined as a quantity that has magnitude only. It is the numerical value (or **size** or **magnitude** or **modulus**) of 32 or 100 that is important on the particular temperature scale. 32 and 100 are said to be **scalars**.

A vector is defined in terms of its magnitude and direction. Hence a vector includes a scalar in part of its definition. A person may have a mass of 80 kg and hence a weight of 80g N (approximately 800 N taking the acceleration due to gravity, g, as 10 ms^{-2}). The mass is a scalar because it acts in no particular direction. This contrasts with the weight. It is a vector because it has direction, acting vertically downwards.

There are some subtle differences between scalars and vectors. We look at these in the following examples.

Examples 13.1

i) A speed of 15 ms^{-1} is a scalar, whereas a velocity of 15 ms^{-1} due East is a vector. In fact speed is the magnitude of the velocity vector.

ii) A distance of 7.2 m is a scalar, whereas a displacement of 7.2 m horizontally to the left is a vector. Similarly distance is the magnitude of the displacement vector.

We may represent a vector in a diagram by a directed line in various ways. The length of the line is the magnitude of the vector and its direction is the direction of the vector (see Fig. 13.1).

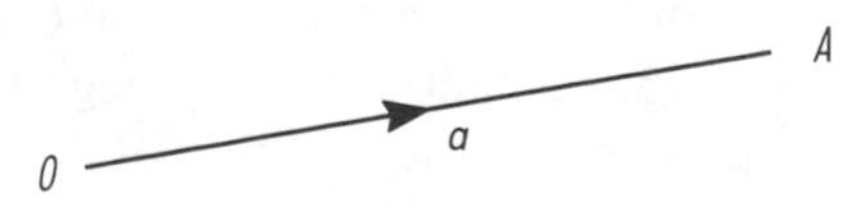

Fig. 13.1

We start at O and finish at A, shown by the directional arrow. Textbooks denote this vector by $\boldsymbol{OA}$ or $\boldsymbol{a}$ in bold type. In hand-written form they would be $\overrightarrow{OA}$ or $\underset{\sim}{a}$. $\boldsymbol{OA}$ (or $\overrightarrow{OA}$) means that the vector starts at O and ends at A. Really $\boldsymbol{a}$ is a ***free vector*** because its start and finish points are not given.

Equal vectors

Two vectors are equal if they have the same magnitude and the same direction, as in Figs. 13.2.

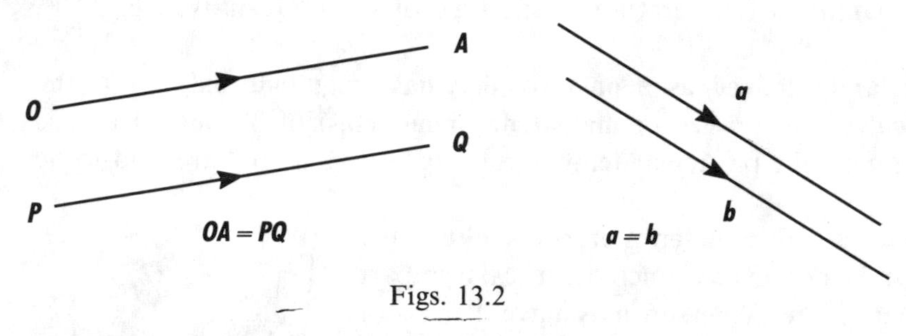

Figs. 13.2

We know already that both $\boldsymbol{a}$ and $\boldsymbol{b}$ are **free vectors** because the starting and finishing points are not specified. A free vector may be shifted provided its magnitude and direction remain unaltered. In some cases we re-label fixed vectors as free vectors and shift them too.

The direction is important. $\boldsymbol{OA}$ is different from $\boldsymbol{QP}$ because their directions are *not* the same.

Negative vectors

Let us look at two vectors $\boldsymbol{x}$ and $\boldsymbol{y}$ that are parallel and have equal magnitude. Figs. 13.3 show the two possibilities.

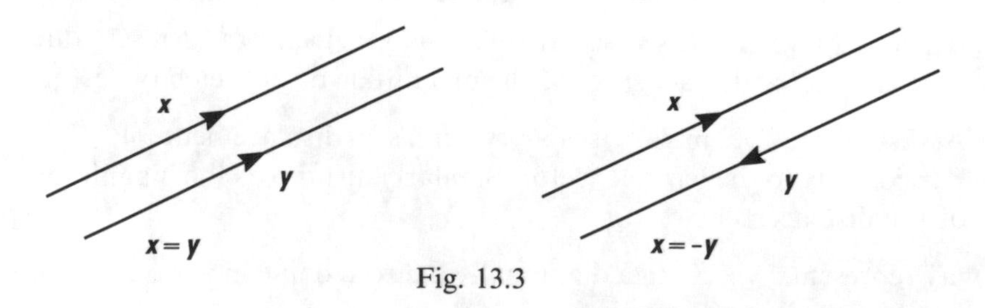

Fig. 13.3

The second option, of opposite direction, shows the negation of a vector, i.e. a minus sign means 'in the opposite direction' to the original vector.

Subtraction is a simplified form of "addition of a negative quantity", i.e. $\boldsymbol{a}-\boldsymbol{b}=\boldsymbol{a}+(-\boldsymbol{b})$. We interpret this as the addition of $\boldsymbol{a}$ and a vector in the opposite direction to $\boldsymbol{b}$.

Scale drawing

We can use an accurate scale drawing to represent a vector, e.g. A velocity of $15\,ms^{-1}$ due East can be represented by

Using the same scale the next line represents $15\,ms^{-1}$ due West.

15 ms⁻¹

Figs. 13.4

If $\quad \boldsymbol{v}_1 = 15\,ms^{-1}$ due East

and $\quad \boldsymbol{v}_2 = 15\,ms^{-1}$ due West

then $\quad \boldsymbol{v}_1 \neq \boldsymbol{v}_2$ because the directions are different.

In fact these opposite directions are interpreted with a minus sign as $\boldsymbol{v}_1 = -\boldsymbol{v}_2$. Both examples have the same length of line, i.e. $|\boldsymbol{v}_1| = |\boldsymbol{v}_2|$. The horizontal lines represent the size of each vector. A simpler way is to ignore the vector symbols and write this as $v_1 = v_2$. (Notice the lack of bold type.)

Similarly $\boldsymbol{OA} = -\boldsymbol{AO}$ and so $OA = AO$.

ASSIGNMENT

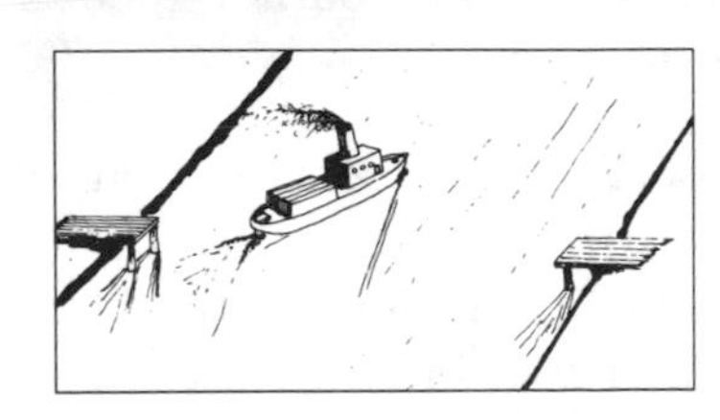

We can return to our first problem. For the bungalow problem we have 2 different choices. Should we steer the boat directly across the river and allow the current to take it downstream as in Fig. 13.5a? Alternatively, should we steer it partially upstream? Then the downstream current can bring it to the other bank. This ought to be directly opposite its starting point, Fig. 13.5b. The calculations support the reasoning behind the scale drawing.

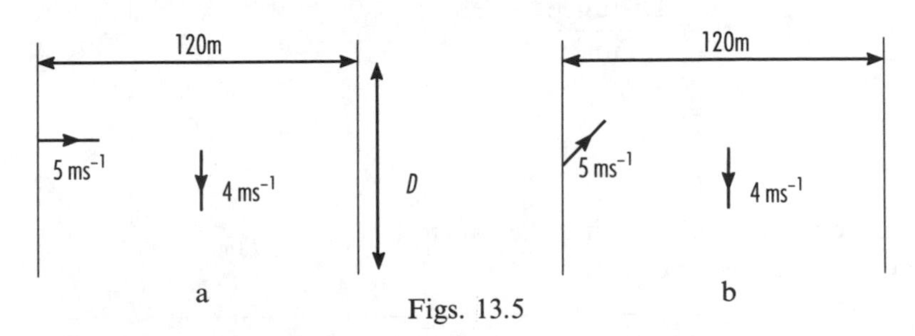

Figs. 13.5

Let us look at the first option. We attempt to steer the boat directly across the river and let the current take it downstream. This means the vertical and horizontal velocities and displacements (from Fig. 13.5a) will be in the same ratio. In scale terms, for every $5\,\text{ms}^{-1}$ across the river the current will take the boat at $4\,\text{ms}^{-1}$ downstream, i.e. for each second the distances are 5 m across the river and 4 m downstream. We compare these as a ratio of

$$\frac{\text{downstream}}{\text{across the river}}$$

so that $$\frac{D}{120} = \frac{4}{5}$$

i.e. $$D = 120 \times \frac{4}{5}$$

$$D = 96\,\text{m}.$$

This means the boat reaches the other side of the river at a distance of 96 m downstream.

Also, using $\text{Speed} = \dfrac{\text{Distance}}{\text{Time}}$

horizontally $5 = \dfrac{120}{t}$ or vertically $4 = \dfrac{96}{t}$

i.e. $5t = 120$ $\quad 4t = 96$

$t = \dfrac{120}{5}$ $\quad t = \dfrac{96}{4}$

i.e. $t = 24\,\text{s}$ in both cases.

This means the journey time is consistent at 24 s across the river and downstream.

Now for the second option. We attempt to steer the boat partially upstream. Part of its velocity cancels out the effect of the current. The remaining part moves the boat directly across the river as though the motion was in still water. This produces a right-angled triangle with $5\,\text{ms}^{-1}$ as the hypotenuse (see Fig. 13.6). According to Pythagoras' theorem, the unknown side is $3\,\text{ms}^{-1}$, i.e. the boat moves directly across the river with a reduced velocity of $3\,\text{ms}^{-1}$.

Again using $\text{Speed} = \dfrac{\text{Distance}}{\text{Time}}$

i.e. $$3 = \frac{120}{t}$$

$$t = 40\ \text{s}$$

i.e. the journey time is 40 s taking the shortest distance directly across the river.

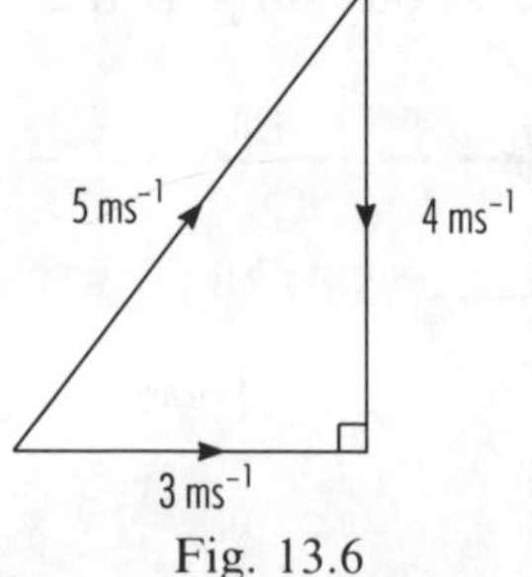

Fig. 13.6

Equivalent vectors

Let us look at a person walking 2 km due South and then 4 km due East. This may be drawn as in Fig. 13.7.

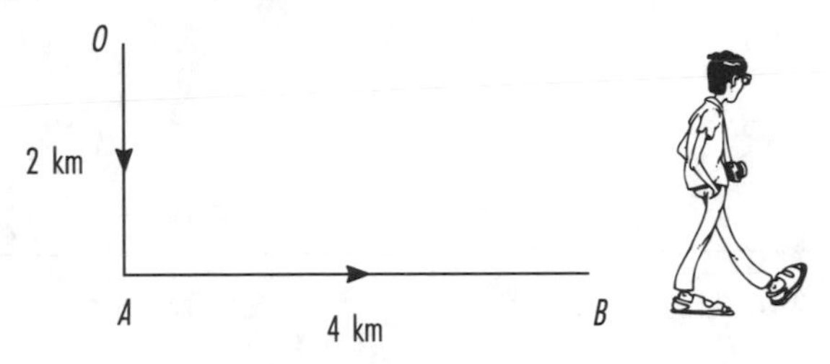

Fig. 13.7

Alternatively the person might have walked directly along *OB* to reach the same point, *B*. This is the same resultant vector or net displacement (see Fig. 13.8).

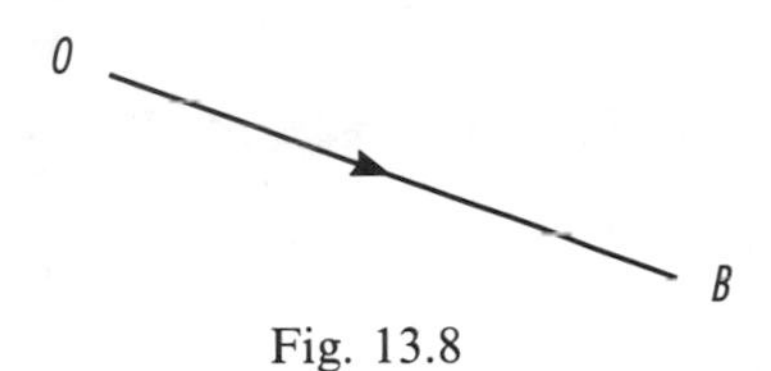

Fig. 13.8

i.e. ***OA*** + ***AB*** = ***OB***
i.e. ***OA*** 'together with' ***AB*** 'is the same as' ***OB***.

The common letter either side of + allows this addition.

When two sides of a triangle are taken in order their resultant is represented by the third side. This is called the **triangle law of vectors**.

Suppose the person continues walking from *B* for another 2 km in a North Westerly (i.e. N 45° W) direction. The extended diagram becomes that in Fig. 13.9.

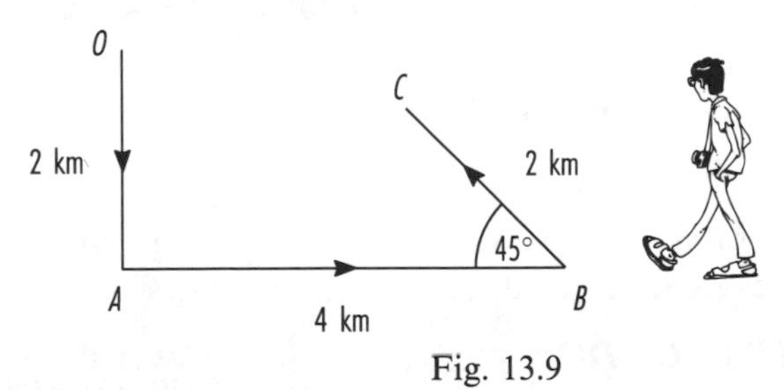

Fig. 13.9

The person started at *O* and finished at *C*. This gives the resultant vector as ***OC***. If we take all stages of the walk in turn, *O* to *A*, *A* to *B* and *B* to *C* we get

$$\boldsymbol{OA}+\boldsymbol{AB}+\boldsymbol{BC}=\boldsymbol{OC}.$$

For vectors in pairs

$\boldsymbol{OA}+\boldsymbol{AB}=\boldsymbol{OB}$ or $\boldsymbol{AB}+\boldsymbol{BC}=\boldsymbol{AC}$
and $\boldsymbol{OB}+\boldsymbol{BC}=\boldsymbol{OC}$ and $\boldsymbol{OA}+\boldsymbol{AC}=\boldsymbol{OC}$.

Using the common letter either side of +.

Thus we may add vectors in pairs, either the first pair or the last pair. In the case of some long expression we may add any intermediate pair. Figs. 13.11 look at the combinations of pairs of vectors.

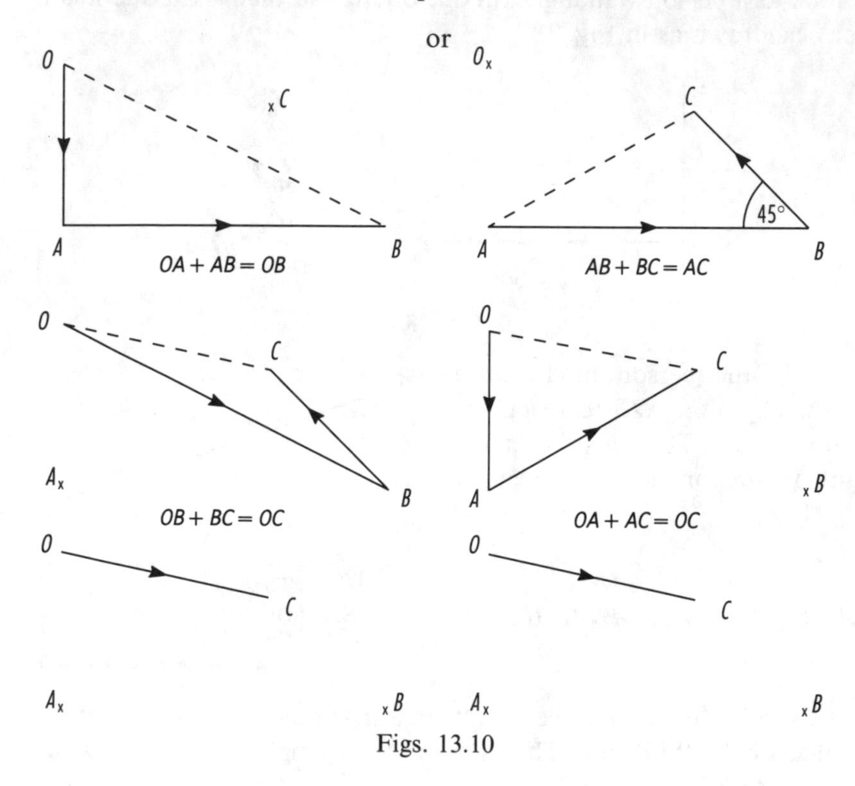

Figs. 13.10

OC is the third side of both triangles OBC and OAC. We see how the triangle law of vectors is obeyed.

Examples 13.2

i) Simplify $\mathbf{AB}+\mathbf{BC}-\mathbf{DC}$,
i.e. find the resultant of $\mathbf{AB}+\mathbf{BC}-\mathbf{DC}$.

ii) Simplify $\mathbf{AB}+\mathbf{BC}+\mathbf{DC}$.

i) The first pair of vectors are separated by a + sign with a common letter on either side. We deal with this pair first.

Now $\quad \mathbf{AB}+\mathbf{BC}-\mathbf{DC}$

$= \mathbf{AC}-\mathbf{DC}$ — Using the common B either side of +.

$= \mathbf{AC}+\mathbf{CD}$ — $\mathbf{CD} = -\mathbf{DC}$.

$= \mathbf{AD}$.

ii) For this second example once more we can look at the first pair of vectors.

This time $\quad \mathbf{AB}+\mathbf{BC}+\mathbf{DC}$

$= \mathbf{AC}+\mathbf{DC}$.

We cannot simplify this result any further because there is no common letter either side of +. Immediately to the left of the + is C and to the right is D.

Example 13.3

Given $\boldsymbol{OA}=\boldsymbol{a}$, $\boldsymbol{OB}=\boldsymbol{b}$ and $\boldsymbol{OC}=\boldsymbol{c}$ find expressions for $\boldsymbol{AB}$ and $\boldsymbol{CA}$ using $\boldsymbol{a}$, $\boldsymbol{b}$ and $\boldsymbol{c}$.

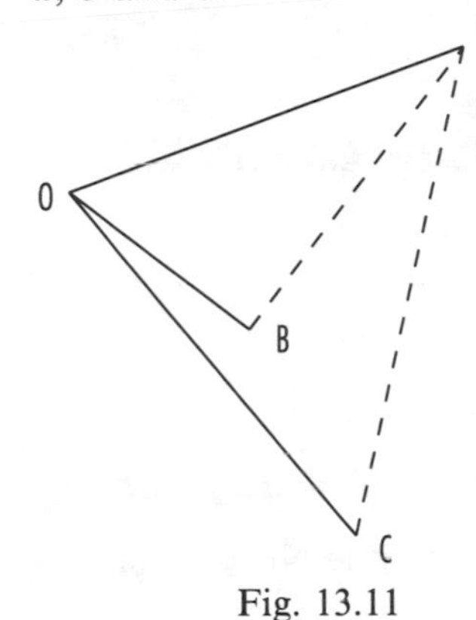

Fig. 13.11

We can write the direct vector $\boldsymbol{AB}$ in terms of other vectors involving O,

i.e.
$$\begin{aligned}\boldsymbol{AB} &= \boldsymbol{AO}+\boldsymbol{OB}\\ &= -\boldsymbol{OA}+\boldsymbol{OB}\\ &= -\boldsymbol{a}+\boldsymbol{b}\end{aligned}$$
or $\boldsymbol{b}-\boldsymbol{a}$.

Similarly
$$\begin{aligned}\boldsymbol{CA} &= \boldsymbol{CO}+\boldsymbol{OA}\\ &= -\boldsymbol{OC}+\boldsymbol{OA}\\ &= -\boldsymbol{c}+\boldsymbol{a}\end{aligned}$$
or $\boldsymbol{a}-\boldsymbol{c}$.

Example 13.4

Simplify $\boldsymbol{KL}+\boldsymbol{MN}-\boldsymbol{ON}+\boldsymbol{OK}-\boldsymbol{ML}$.

There are many ways to simplify this expression of vectors. To avoid confusion we find it easier to make sure they are all positive.

$$\begin{aligned}&\boldsymbol{KL}+\boldsymbol{MN}-\boldsymbol{ON}+\boldsymbol{OK}-\boldsymbol{ML}\\ &= \boldsymbol{KL}+\boldsymbol{MN}+(\boldsymbol{NO}+\boldsymbol{OK})+\boldsymbol{LM}\\ &= \boldsymbol{KL}+(\boldsymbol{MN}+\boldsymbol{NK})+\boldsymbol{LM}\\ &= \boldsymbol{KL}+\boldsymbol{MK}+\boldsymbol{LM}\\ &= \boldsymbol{KL}+(\boldsymbol{LM}+\boldsymbol{MK})\\ &= \boldsymbol{KM}+\boldsymbol{MK}\\ &= \boldsymbol{KM}-\boldsymbol{KM}\\ &= \boldsymbol{0}.\end{aligned}$$

> $\boldsymbol{NO}=-\boldsymbol{ON}$ and $\boldsymbol{LM}=-\boldsymbol{ML}$.

> Common letters either side of +.

> $\boldsymbol{MK}=-\boldsymbol{KM}$.

0 represents the **zero vector** (or **null vector**) which has neither magnitude nor direction.

The brackets highlight the order of addition, but are not essential to the mathematics.

ASSIGNMENT

We can apply the principles of scale drawing and equivalent vectors to our aircraft problem. The resultant velocity is the added effect of the wind velocity onto the aircraft's velocity. In Fig. 13.12 we use accurate drawing and measurement. The third side of the triangle is this resultant velocity. It is approximately 550 ms^{-1} inclined at an angle of 5.5° to the original easterly direction.

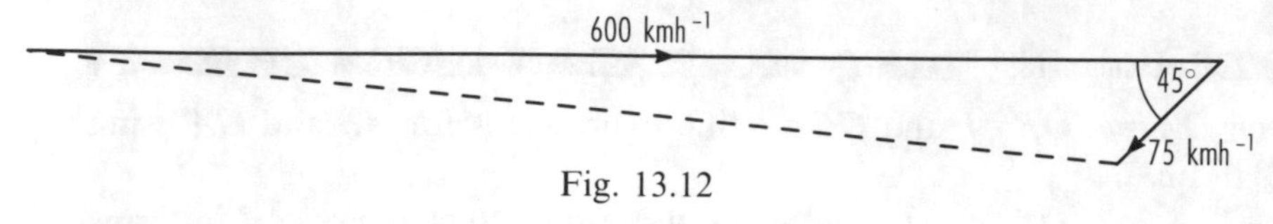

Fig. 13.12

EXERCISE 13.1

1 Decide whether the following are scalars or vectors:
- i) 3.7 m
- ii) 24 ms^{-1} vertically up
- iii) a weight of 19 N
- iv) 0.36 m due North
- v) a crowd of 17 362 people.

2 By drawing find the resultant of a person walking 5 m due North, 2 m due West, 3 m due South and 2 m due East.

3 Simplify the following vectors:
- i) $\mathbf{CA}+\mathbf{AD}+\mathbf{DB}$
- ii) $\mathbf{DB}+\mathbf{CA}+\mathbf{AD}$
- iii) $\mathbf{AB}-\mathbf{CB}$
- iv) $\mathbf{AD}-\mathbf{DC}+\mathbf{CE}-\mathbf{FE}$
- v) $\mathbf{BA}-\mathbf{EA}+\mathbf{FE}$

4 Draw diagrams to represent the following relationships:
- i) $\mathbf{AB}+\mathbf{BC}=\mathbf{AC}$
- ii) $\mathbf{CA}+\mathbf{AB}+\mathbf{BC}=\mathbf{0}$

5 PQRS is a square. Decide, stating your reason, whether the following statements are true or false:
- i) $\mathbf{PQ}=\mathbf{SR}$
- ii) $PQ=SR$
- iii) $\mathbf{PQ}=\mathbf{RS}$
- iv) $\mathbf{SR}=\mathbf{RQ}$
- v) $\mathbf{PQ}+\mathbf{QR}+\mathbf{RS}=\mathbf{PS}$

Components and resultants

A vector may exist in any direction. Sometimes it is easier to consider it in parts, i.e. in **components**. Together these components have exactly the same effect as the original vector. The most useful way is to **resolve** a vector into 2 components that are at 90° (at right-angles) to each other.

These components are perpendicular components. Then we can use some basic trigonometry to evaluate them. Frequently we use horizontal and vertical directions for the components. However, there are other helpful alternatives. They can involve directions that are parallel and perpendicular to an inclined plane.

Figs. 13.13 show a vector $\boldsymbol{d}$ in two different, but equivalent, ways. In one of them it is inclined at an angle θ above the horizontal. In the other it is shown in its equivalent perpendicular components.

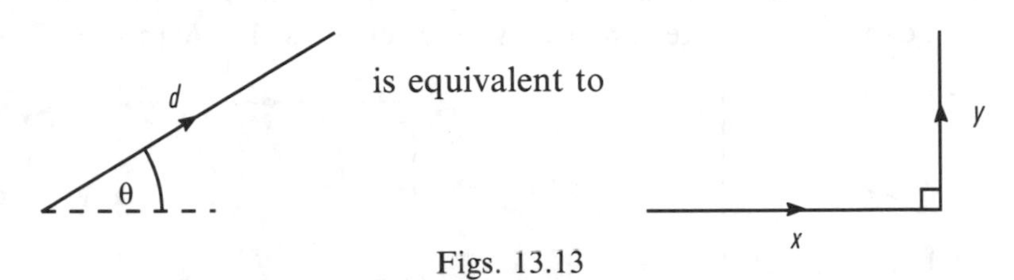

Figs. 13.13

We may superimpose these two figures to form a right-angled triangle. Using some simple trigonometry from Chapter 5 we have

$$\cos\theta = \frac{x}{d} \qquad \text{and} \qquad \sin\theta = \frac{y}{d}$$

$$\text{i.e.} \qquad x = d\cos\theta \qquad\qquad \text{i.e.} \qquad y = d\sin\theta$$

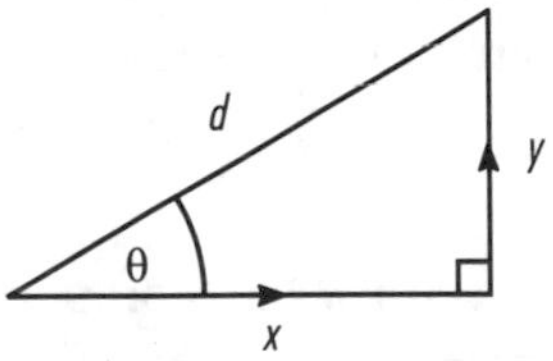

Fig. 13.14

The diagram shows $d\cos\theta$ acting horizontally together with $d\sin\theta$ acting vertically. They have the same effect as the original $\boldsymbol{d}$.

In Fig. 13.13 we started with a vector and split it into 2 perpendicular components. Instead Fig. 13.15 shows us starting with the same perpendicular x and y. We may apply Pythagoras' theorem to find the third side, say d,

$$\text{i.e.} \qquad d^2 = x^2 + y^2$$

$$\text{i.e.} \qquad d = \sqrt{x^2 + y^2}.$$

$$\text{Also} \qquad \tan\theta = \frac{y}{x}.$$

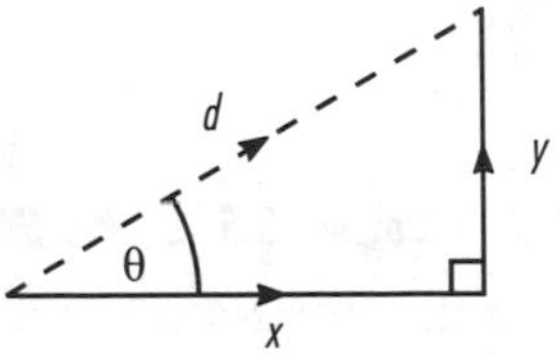

Fig. 13.15

We can look at pairs of perpendicular vectors in several ways. Because the start and finish points for $\boldsymbol{y}$ are not specified it is a free vector. Similarly $\boldsymbol{x}$ is a free vector. We know already that free vectors may be shifted. In Figs. 13.16 we have 2 possibilities. In each case we find the resultant of $\boldsymbol{x}$ and $\boldsymbol{y}$ by completing the triangle.

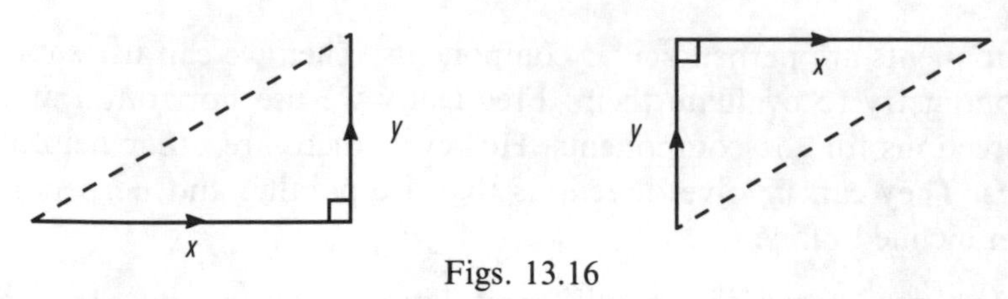

Figs. 13.16

Figs. 13.17 show the vectors $\boldsymbol{x}$ and $\boldsymbol{y}$ acting at a point. In one case they are coming from a point. In the other case they are going to a point. We find their resultant by completing the rectangle. The diagonal is the resultant vector. You can see there are similarities between Figs. 13.16 and 13.17.

Figs.13.17

Equally valid, though generally not as useful, are non-perpendicular components. Right-angled trigonometry and Pythagoras' theorem cannot be applied if the components are not perpendicular. By way of example are Figs. 13.18 and 13.19:

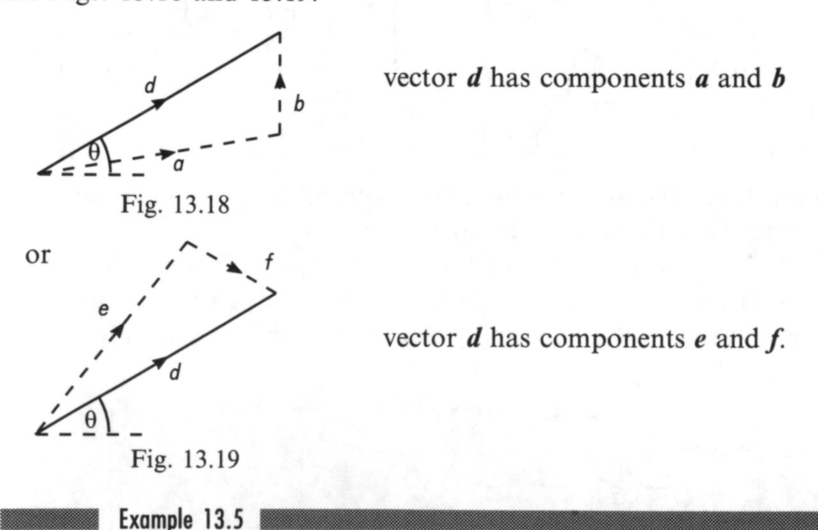

vector $\boldsymbol{d}$ has components $\boldsymbol{a}$ and $\boldsymbol{b}$

Fig. 13.18

or

vector $\boldsymbol{d}$ has components $\boldsymbol{e}$ and $\boldsymbol{f}$.

Fig. 13.19

Example 13.5

Evaluate the horizontal and vertical components of a velocity vector of 5ms^{-1} inclined at 30° above the horizontal.

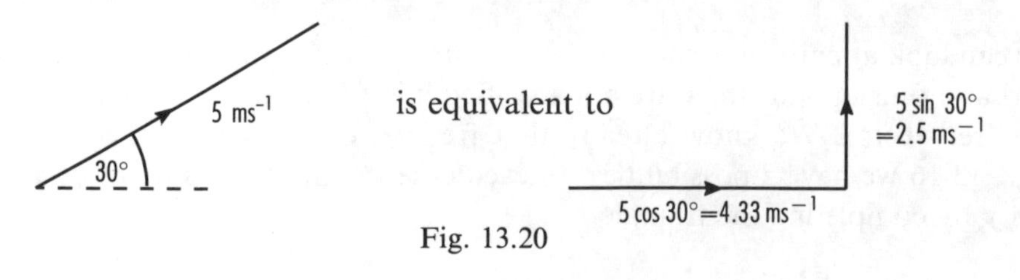

Fig. 13.20

Example 13.6

A force of 14 N acts horizontally as shown below. Find its components parallel and perpendicular to the plane inclined at 60° to the horizontal.

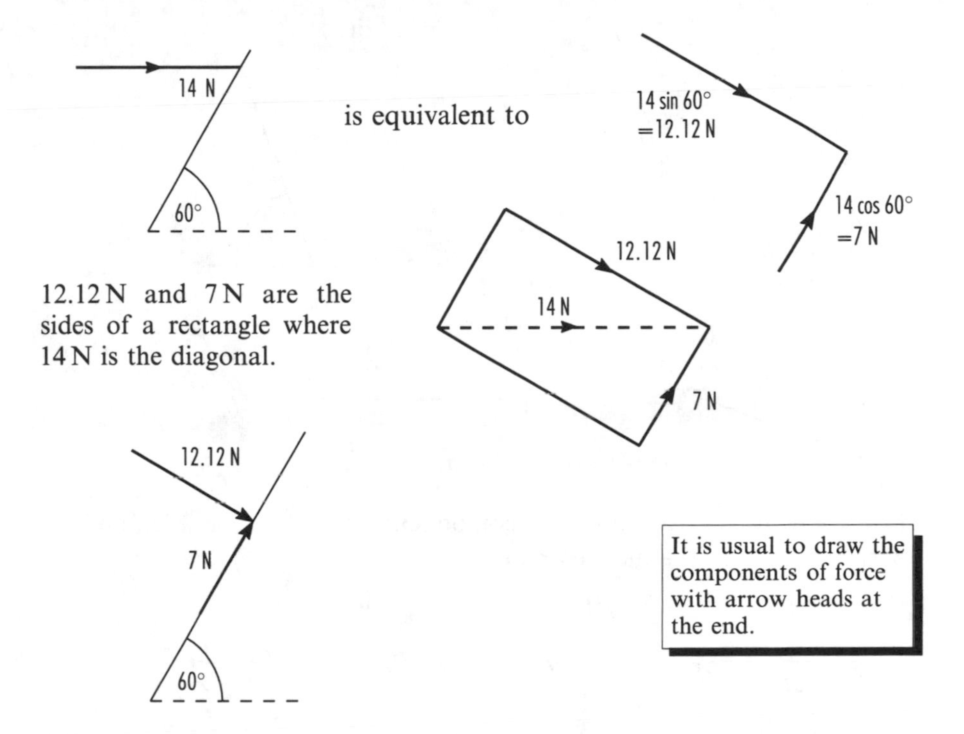

12.12 N and 7 N are the sides of a rectangle where 14 N is the diagonal.

It is usual to draw the components of force with arrow heads at the end.

Figs. 13.21

Now we look at an example with the two perpendicular components given. We find the magnitude and direction of the original vector.

Example 13.7

Find the magnitude and direction of the resultant velocity given its horizontal and vertical components of $12\,\text{ms}^{-1}$ and $8\,\text{ms}^{-1}$ shown in Fig. 13.22.

$8\,\text{ms}^{-1}$

θ

$12\,\text{ms}^{-1}$

Fig. 13.22

$$\text{Magnitude} = \sqrt{12^2 + 8^2}$$
$$= 14.42\,\text{ms}^{-1}$$
$$\tan\theta = \frac{8}{12}$$
$$\theta = 33.69^\circ$$

i.e. the resultant velocity vector is $14.42\,\text{ms}^{-1}$ at 33.69° above the horizontal.

EXERCISE 13.2

1 Calculate the horizontal and vertical components of the following vectors:

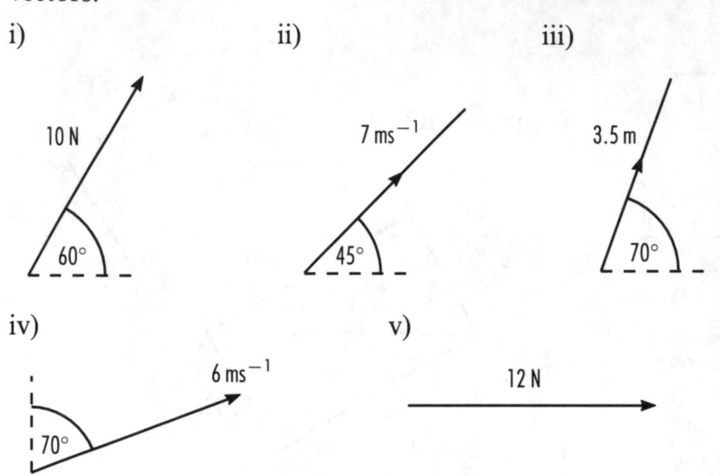

2 For each force calculate the components that are parallel and perpendicular to the inclined plane:

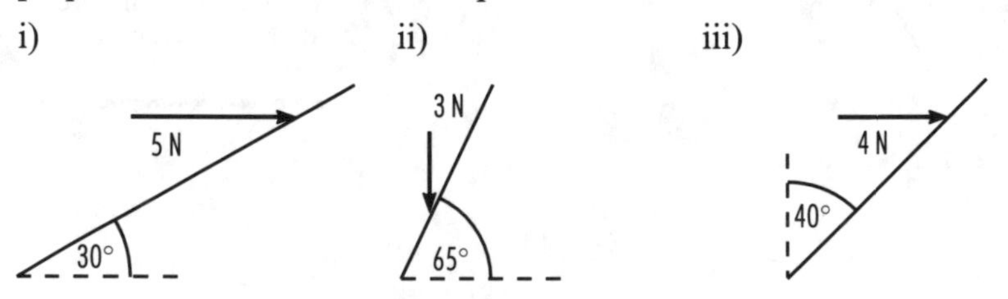

3 Determine the magnitude and direction of the resultant vector given the perpendicular components:

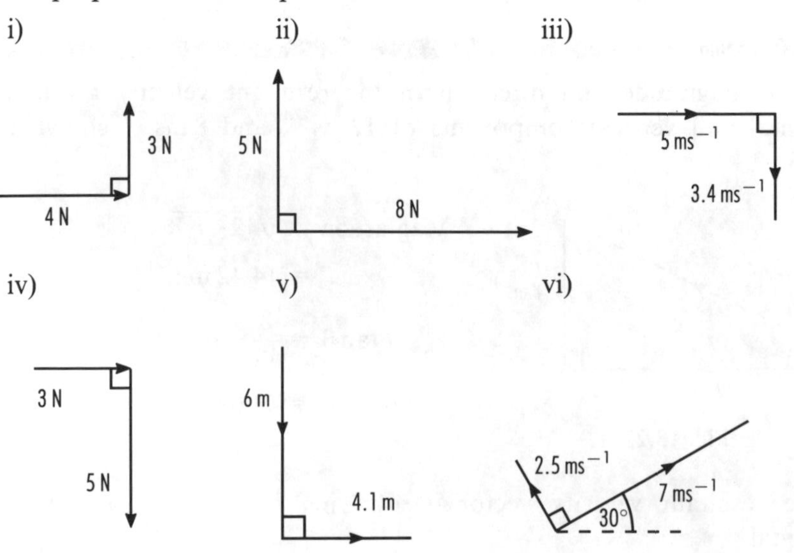

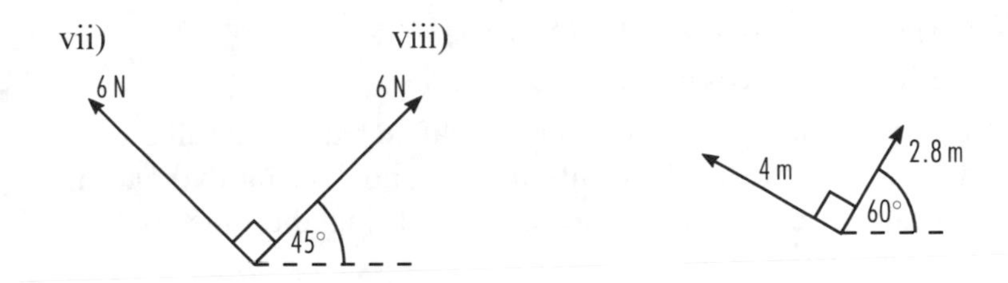

These two techniques of finding components and finding a resultant may be extended to find the resultant of many vectors.

In Fig. 13.23 we look at a system of vectors all applied at some point.

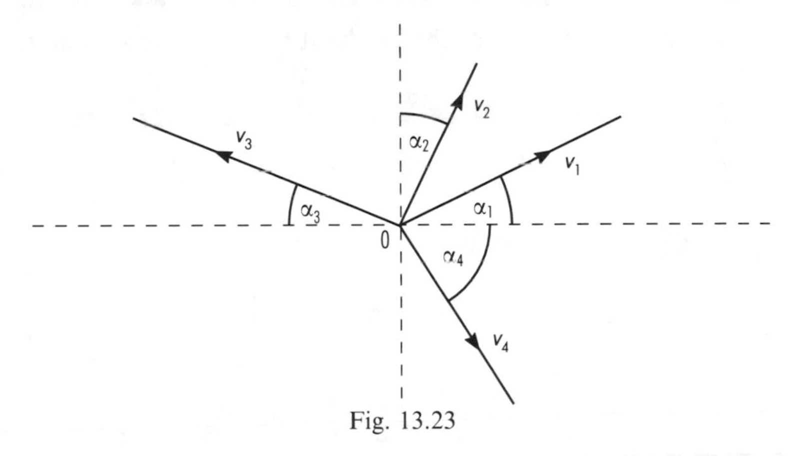

Fig. 13.23

Each of them, $\boldsymbol{v}_1$, $\boldsymbol{v}_2$, $\boldsymbol{v}_3$, $\boldsymbol{v}_4$, may be resolved into horizontal and vertical components, as in Fig. 13.24.

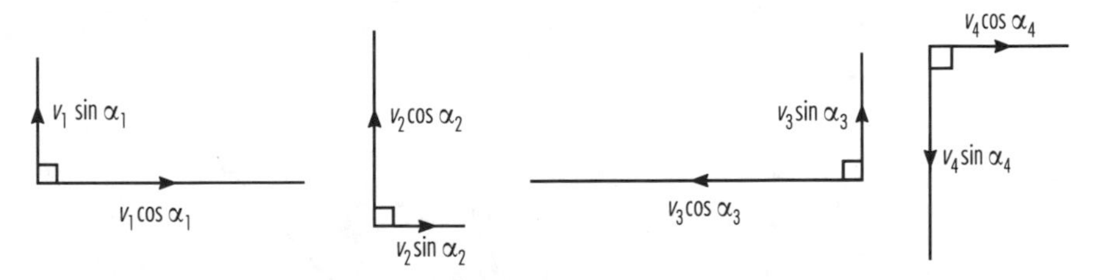

Figs. 13.24

We see that not all the horizontal directions are to the right. Similarly not all the vertical directions are upwards. We need to decide which are to be the positive directions. We are going to be consistent with axes and graphical work. Horizontally right and vertically up will be taken as positives. Let the resultant of all these vectors be represented by a single vector with components (Fig. 13.25) X and Y.

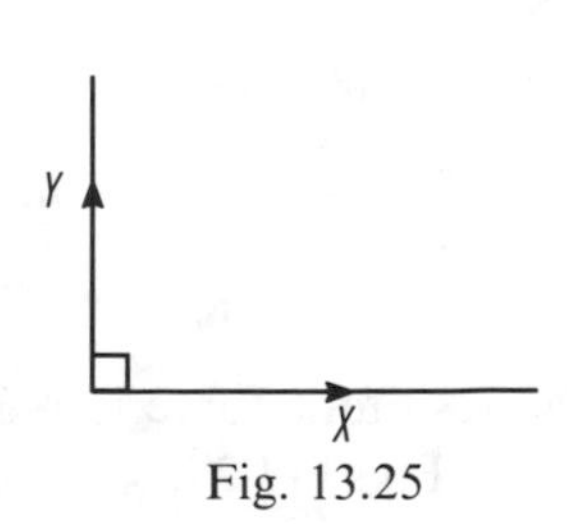

Fig. 13.25

Then $\quad X = v_1 \cos\alpha_1 + v_2 \sin\alpha_2 - v_3 \cos\alpha_3 + v_4 \cos\alpha_4$

and $\quad Y = v_1 \sin\alpha_1 + v_2 \cos\alpha_2 + v_3 \sin\alpha_3 - v_4 \sin\alpha_4$

In a numerical example X and Y can be simplified into one value each. Then we find the magnitude of the resultant of X and Y using Pythagoras' theorem. We find the direction of the resultant using the tangent from trigonometry.

Sometimes we find the value of X or Y may turn out to be negative. The magnitude remains the same. The negative sign means the true direction is opposite to the one we have chosen, i.e. horizontally to the left instead of the right, or vertically down instead of up.

Example 13.8

Find the magnitude and direction of the resultant force for the system in Fig. 13.26.

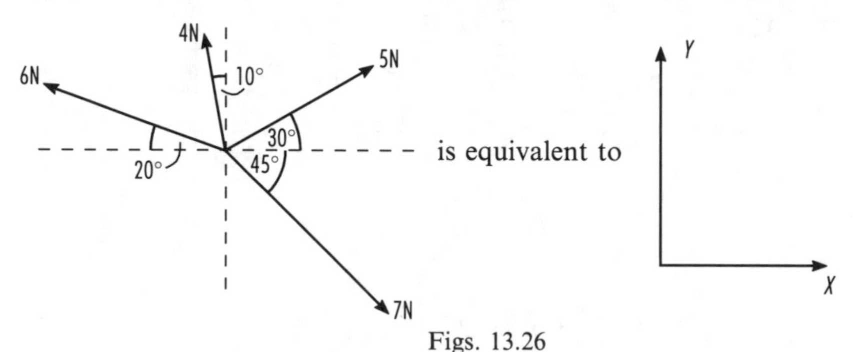

Figs. 13.26

Now $\quad X = 5\cos 30° - 4\sin 10° - 6\cos 20° + 7\cos 45°$

$\quad\quad = 2.947\,\text{N}$

and $\quad Y = 5\sin 30° + 4\cos 10° + 6\sin 20° - 7\sin 45°$

$\quad\quad = 3.542\,\text{N}.$

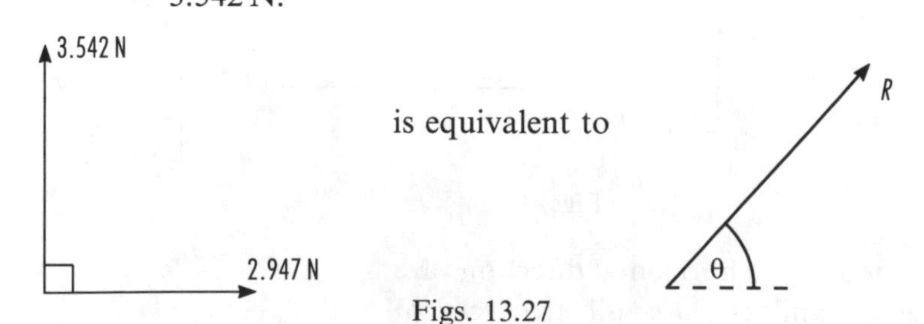

Figs. 13.27

By Pythagoras' theorem

$$R = \sqrt{2.947^2 + 3.542^2} \qquad \tan\theta = \frac{3.542}{2.947}$$

$$= 4.61\,\text{N} \qquad \theta = 50.24°$$

i.e. the resultant force is 4.61 N inclined at 50.24° above the horizontal, shown in Figs. 13.27.

Example 13.9

For the velocity vectors in Figs. 13.28 find the magnitude and direction of their resultant.

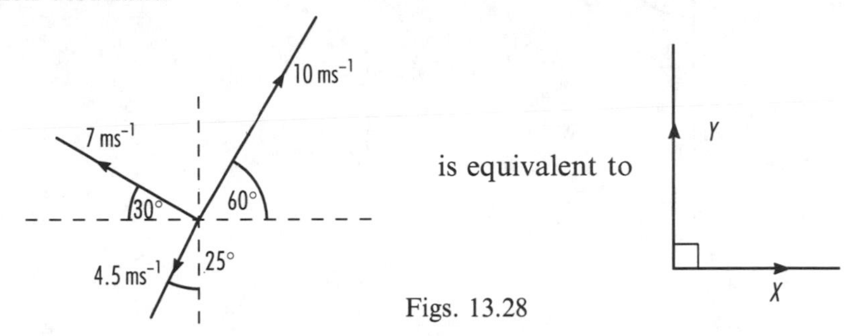

Figs. 13.28

Now $\quad X = 10\cos 60° - 7\cos 30° - 4.5\sin 25°$

$\qquad = -2.964\ \text{ms}^{-1}$

and $\quad Y = 10\sin 60° + 7\sin 30° - 4.5\cos 25°$

$\qquad = 8.082\ \text{ms}^{-1}.$

The negative value for X means that it should be directed towards the left whilst retaining the same magnitude.

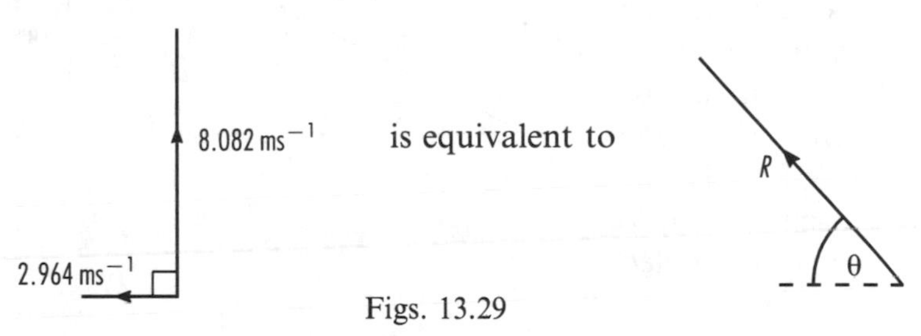

Figs. 13.29

By Pythagoras' theorem

$$R = \sqrt{2.964^2 + 8.082^2}$$
$$= 8.61\ \text{ms}^{-1}$$

$$\tan\theta = \frac{8.082}{2.964}$$
$$\theta = 69.86°$$

i.e. the resultant velocity is $8.61\ \text{ms}^{-1}$ inclined at 69.86° above the horizontal as shown in Figs. 13.29.

ASSIGNMENT

We have developed the theory a little further. Now it is possible to re-consider the Assignment problems from the beginning of the chapter.

In the construction problem we have two options. The first one is to cross the river as quickly as possible. This means letting the current take the boat downstream. Fig. 13.30 shows the resultant velocity, $\boldsymbol{v}_1$, along the hypotenuse of the right-angled triangle.

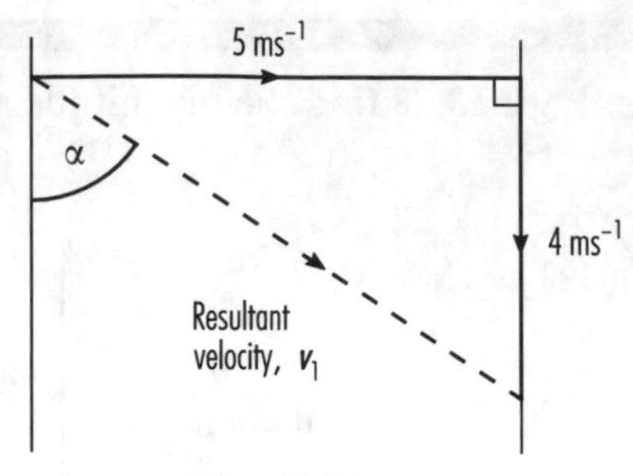

Fig. 13.30

By Pythagoras' theorem

$$v_1 = \sqrt{5^2 + 4^2} \qquad \tan\alpha = \frac{5}{4} = 1.25$$
$$= 6.40\,\mathrm{ms}^{-1} \qquad \alpha = 51.34^\circ$$

i.e. the boat's resultant velocity is $6.40\,\mathrm{ms}^{-1}$ at 51.34° to the river bank.

For the second option we attempt to steer the boat partially upstream. The downstream effect of the current means the resultant velocity, $\boldsymbol{v}_2$, is directly across the river. In this case Fig. 13.31 shows it is the $5\,\mathrm{ms}^{-1}$ that forms the hypotenuse of the right-angled triangle.

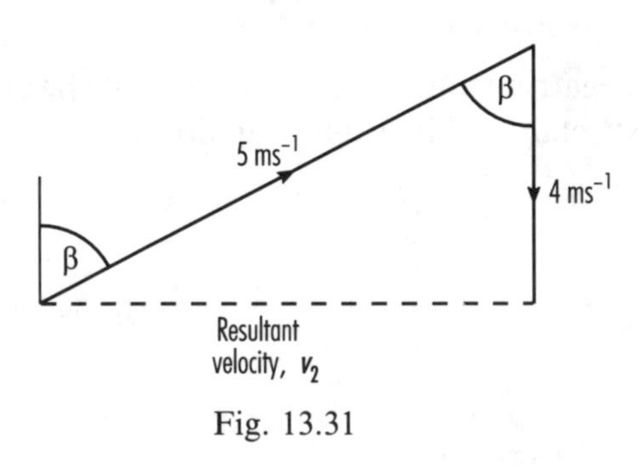

Fig. 13.31

The velocities of $5\,\mathrm{ms}^{-1}$ and $4\,\mathrm{ms}^{-1}$ in Figs. 13.32 are not perpendicular. This means a little more care is necessary with the components and the final resultant, i.e.

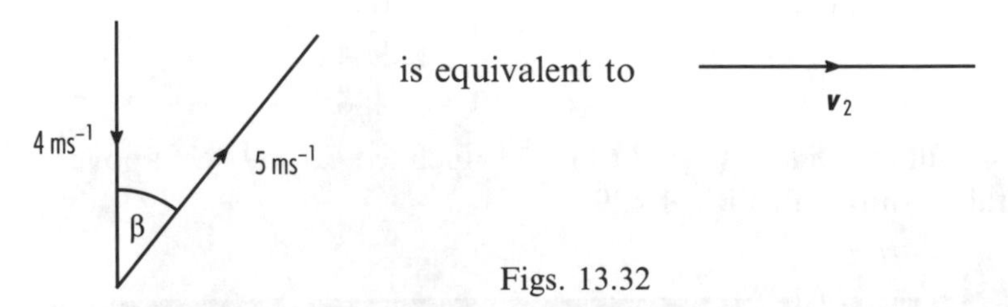

Figs. 13.32

Now across the river $\quad v_2 = 5\sin\beta$

and downstream $\quad 0 = 5\cos\beta - 4$

The second equation gives a solution of $\beta = 36.87°$. By substitution the first equation gives $v_2 = 3\,\mathrm{ms}^{-1}$, i.e. the helmsman must attempt to steer the boat at 36.87° to the bank. This is because the effect of the current produces a resultant velocity of $3\,\mathrm{ms}^{-1}$ directly across the river.

We can apply this last technique to the aircraft problem. The velocities of $600\,\text{kmh}^{-1}$ due East and $75\,\text{kmh}^{-1}$ from the North East are shown in Fig. 13.33.

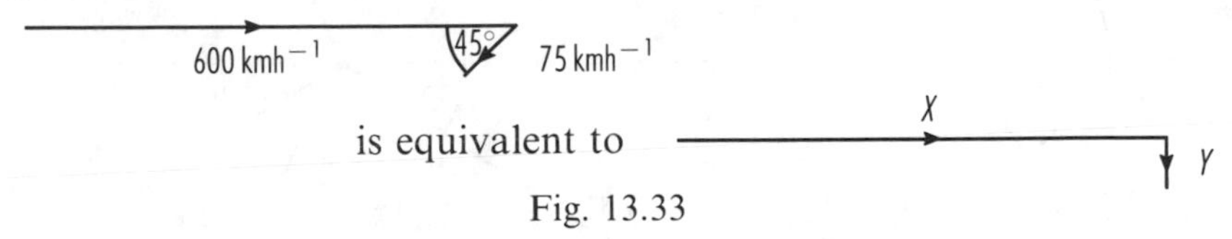

Fig. 13.33

Our choice of Y vertically downwards is consistent with our problem. Alternatively we could have chosen vertically upwards as positive. We would interpret the resulting negative sign as the 'opposite direction'.

Now horizontally $\quad X = 600 - 75\cos 45^\circ$
$\qquad\qquad\qquad\quad = 546.97\,\text{kmh}^{-1}$

and vertically $\quad Y = 75\sin 45^\circ$
$\qquad\qquad\qquad\quad = 53.03\,\text{kmh}^{-1}$.

Now we can combine our 2 perpendicular components to find their resultant. Thus

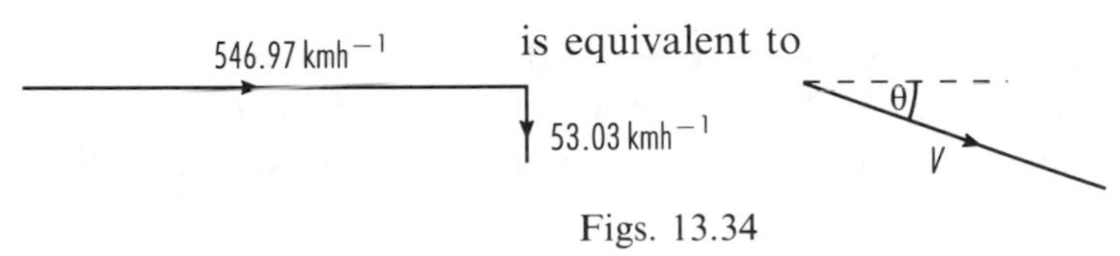

Figs. 13.34

By Pythagoras' theorem

$$V = \sqrt{546.97^2 + 53.03^2} \qquad\qquad \tan\theta = \frac{53.03}{546.97}$$

$$= 549.5\,\text{kmh}^{-1}. \qquad\qquad \theta = 5.54^\circ.$$

The bearing is $\quad 90^\circ + \theta$
$\qquad\qquad\qquad = 095.54^\circ$

i.e. the aircraft's resultant velocity is $549.5\,\text{kmh}^{-1}$ on a bearing of 095.54°.

EXERCISE 13.3

1 In each case determine the magnitude and direction of the resultant vector:

i) ii)

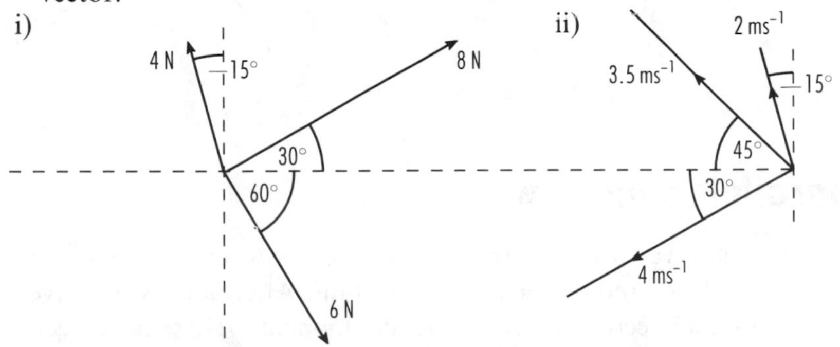

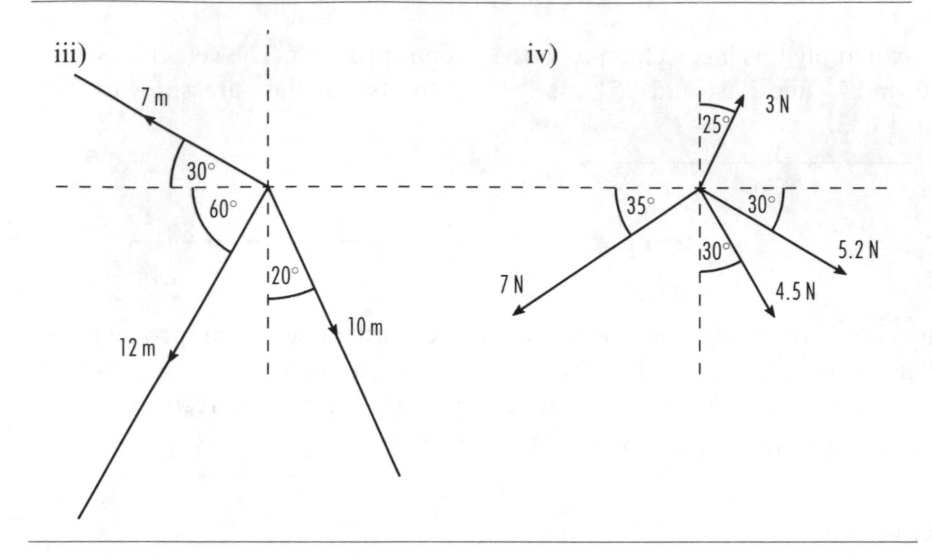

Scalar multiplication

Scalar multiplication of a vector is simple arithmetical multiplication. By way of example in Figs. 13.35 $3\boldsymbol{a}$ is a vector in the same direction as $\boldsymbol{a}$ but three times the length. Also $-3\boldsymbol{a}$ is the same length as $3\boldsymbol{a}$ but in the opposite direction.

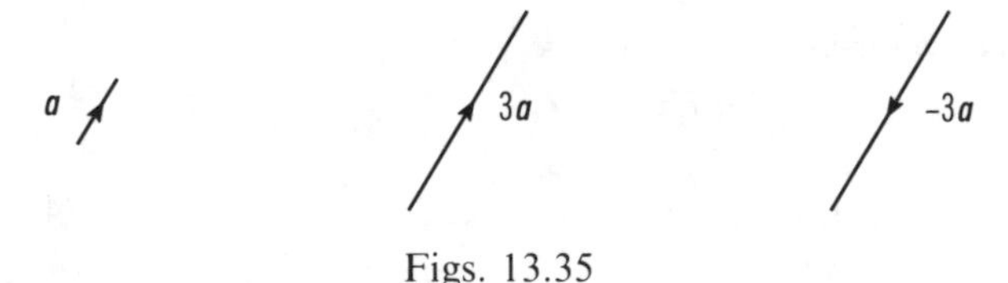

Figs. 13.35

These are free vectors so we have not given any start or finish points.

In contrast in Figs. 13.36 we have $\boldsymbol{OA}$, $3\boldsymbol{OA}$ and $-3\boldsymbol{OA}$ where those points are important.

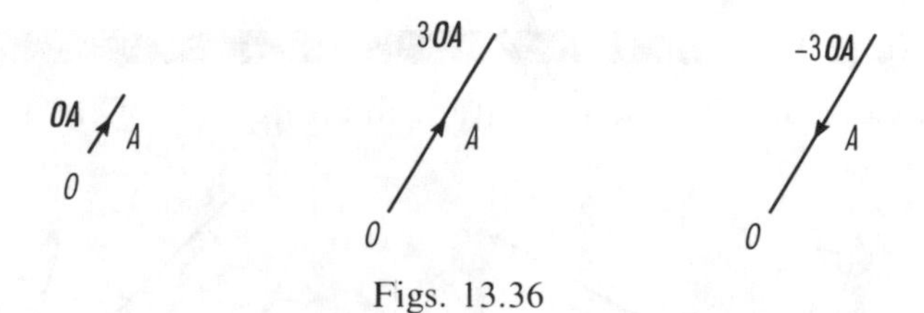

Figs. 13.36

The parallelogram law

In earlier examples we formed a triangle by vectors when one followed another. The third side represented their resultant. Alternatively we have looked at an original vector in terms of its components. These have been equivalent systems. More recently our examples have concentrated on horizontal and vertical components for easier calculation.

Now let us look at vectors at a point with any angle between them. **The parallelogram law, for two vectors acting at a point**, uses vector simplification similar to the triangle law where one vector follows another. It is based upon the opposite sides of a parallelogram being equal and parallel, and that free vectors may be shifted.

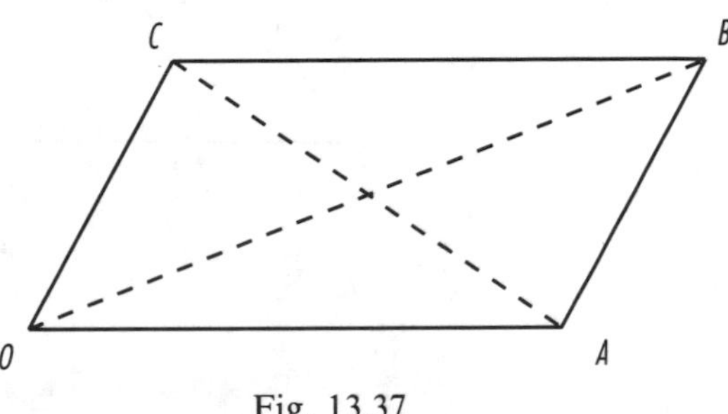

Fig. 13.37

We may think of the parallelogram $OABC$ in Fig. 13.37 as pairs of triangles:
either triangles OAB and OCB with a common side OB
or triangles OAC and BAC with a common side AC.

Let us consider two vectors, $\boldsymbol{a}$ and $\boldsymbol{b}$, acting at one point. Because they are free vectors acting at a point either of them may be shifted. Figs. 13.38 shows the 2 possible variations.

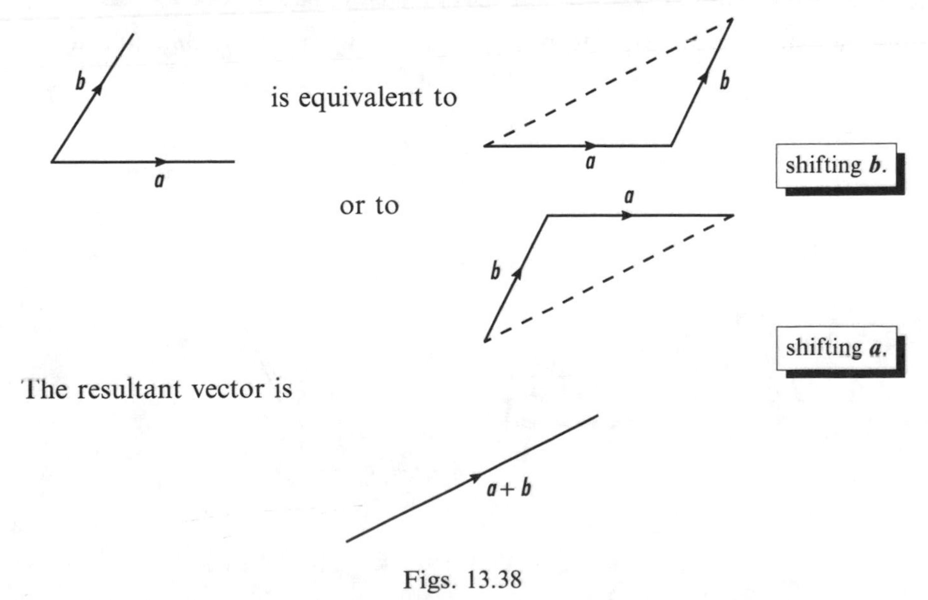

Figs. 13.38

We may apply the parallelogram law to vector subtraction. We represent $\boldsymbol{a}-\boldsymbol{b}$ as $\boldsymbol{a}+(-\boldsymbol{b})$ for vectors acting at one point.

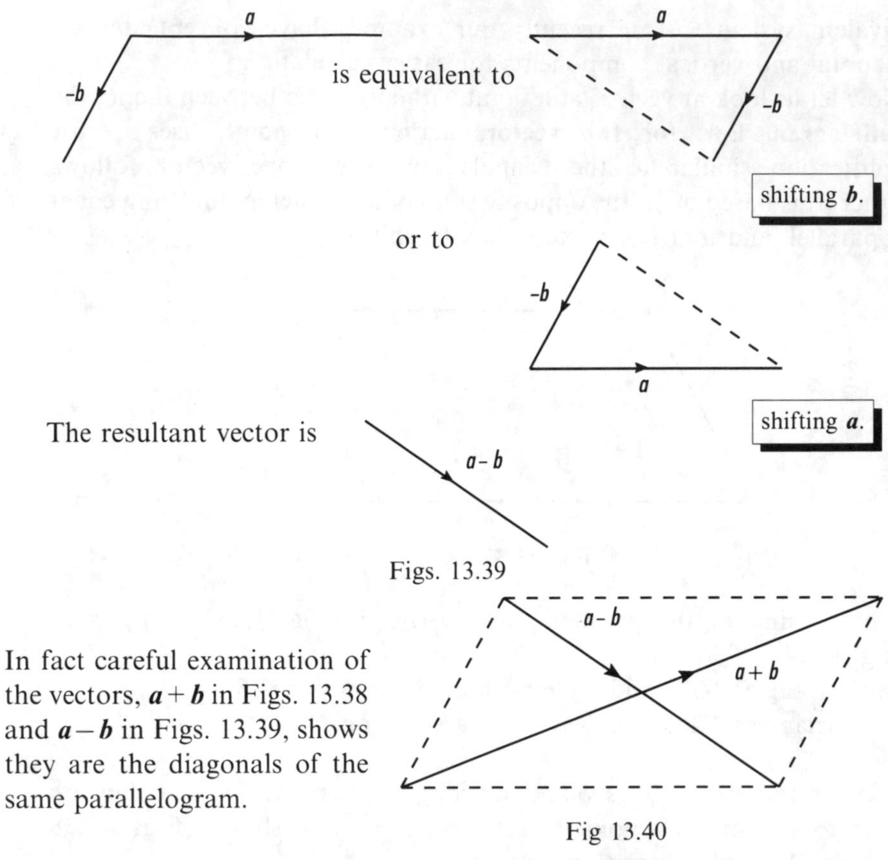

Figs. 13.39

In fact careful examination of the vectors, $\boldsymbol{a}+\boldsymbol{b}$ in Figs. 13.38 and $\boldsymbol{a}-\boldsymbol{b}$ in Figs. 13.39, shows they are the diagonals of the same parallelogram.

Fig 13.40

Example 13.10

Apply the parallelogram law to represent $\boldsymbol{a}+3\boldsymbol{b}$ where $\boldsymbol{a}$ and $3\boldsymbol{b}$ act at a point O.

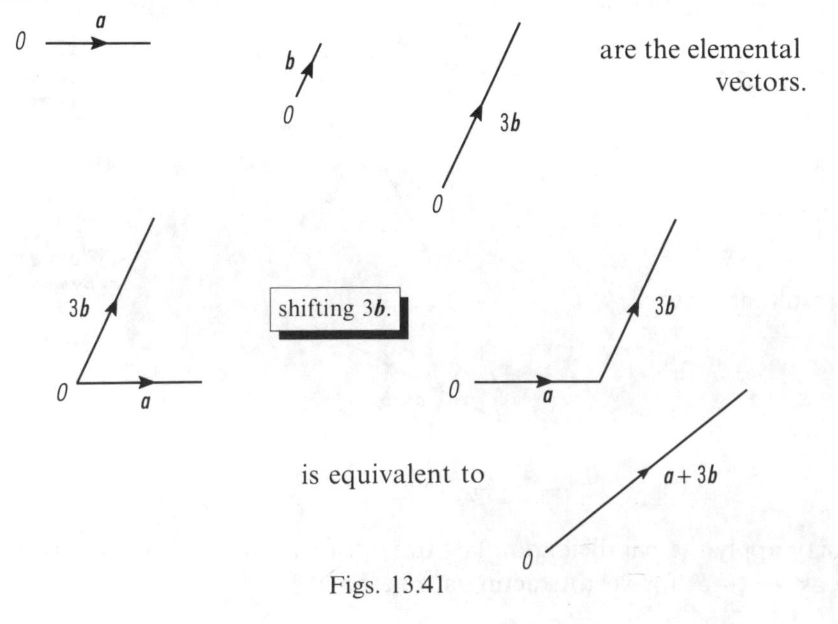

Figs. 13.41

Example 13.11

Represent $2\boldsymbol{a}-3\boldsymbol{b}$, where the elemental vectors act at O, using a parallelogram.

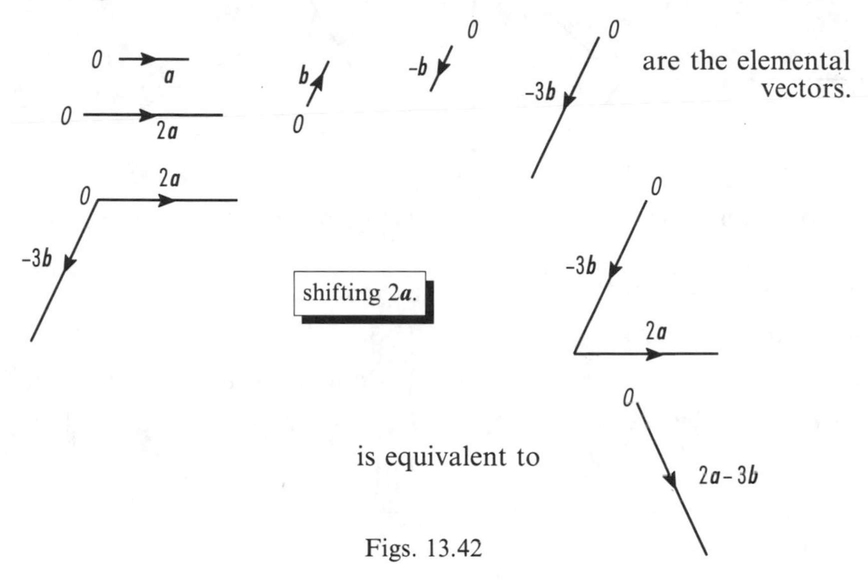

Figs. 13.42

When we deal with more than two vectors the parallelogram law may be used repeatedly on pairs of vectors. We apply this technique in the next example.

Example 13.12

Use the parallelogram law to represent $\boldsymbol{a}+\boldsymbol{b}+2\boldsymbol{d}$ where

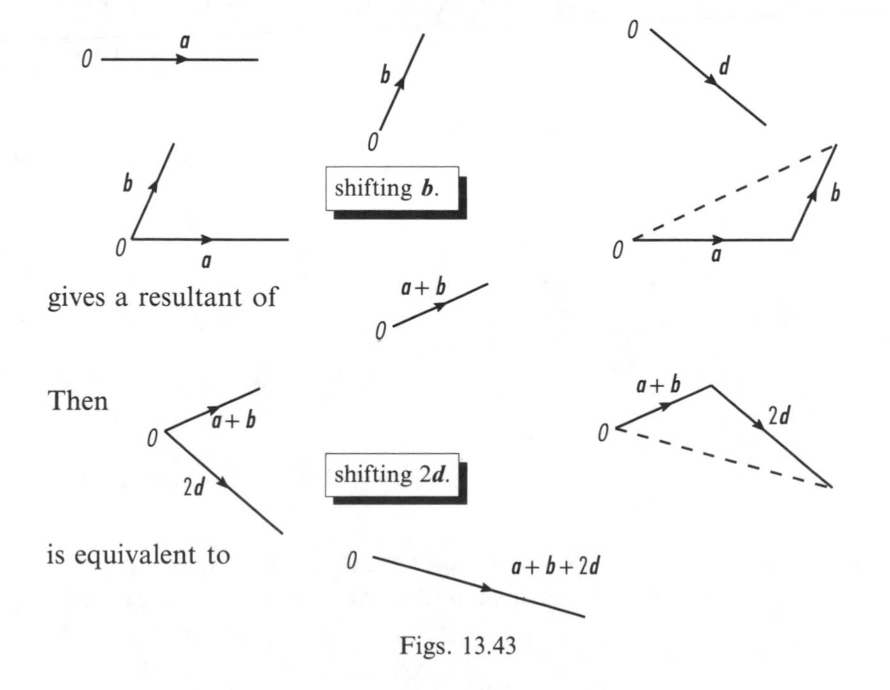

Figs. 13.43

EXERCISE 13.4

Let $\boldsymbol{a}$, $\boldsymbol{b}$, $\boldsymbol{c}$, $\boldsymbol{d}$ and $\boldsymbol{e}$ be represented by lines of 4 cm in the directions shown:

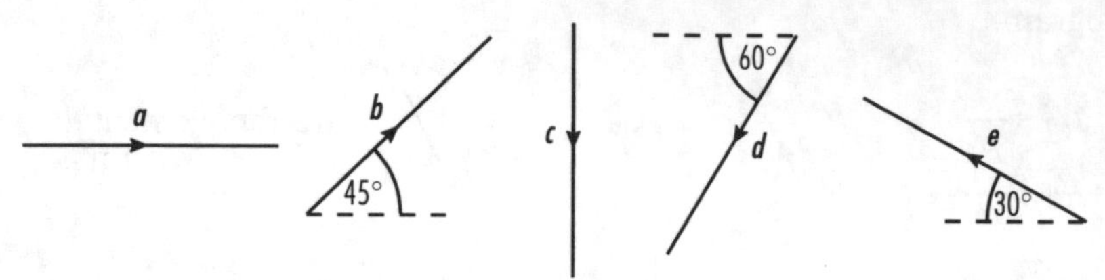

In the following questions use the parallelogram law to represent the vectors.

1 $\boldsymbol{b}+\boldsymbol{c}$

2 $\boldsymbol{b}+\boldsymbol{c}+\boldsymbol{d}$

3 $2\boldsymbol{a}+\boldsymbol{d}$

4 $\boldsymbol{b}-\boldsymbol{e}$

5 $\boldsymbol{a}+\boldsymbol{b}+\boldsymbol{c}$

6 $0.5\boldsymbol{b}-\boldsymbol{a}$

7 $3\boldsymbol{e}+2\boldsymbol{d}$

8 $2\boldsymbol{c}-3\boldsymbol{d}+0.5\boldsymbol{a}$

9 $\boldsymbol{a}+\boldsymbol{b}+\boldsymbol{c}+\boldsymbol{d}+\boldsymbol{e}$

10 $\boldsymbol{a}-\boldsymbol{b}+\boldsymbol{c}-\boldsymbol{d}+\boldsymbol{e}$

Rectangular form

Rectangular (or **Cartesian**) form expresses a vector in terms of components that are mutually perpendicular. A sheet of paper has only two dimensions. The components usually act along axes that are horizontal and vertical. It is difficult to draw 3-dimensional axes on paper. In the reality of the 3-dimensional world the axes are defined by the 'right-handed screw rule' shown in Fig. 13.44.

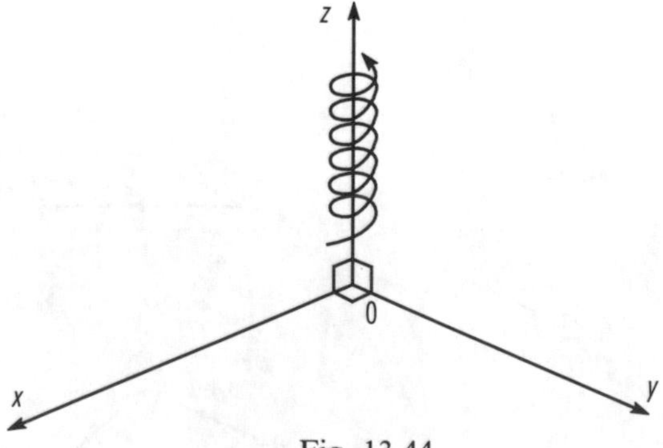

Fig. 13.44

If you turned a screwdriver from the Ox axis towards Oy you could drive a screw upwards along Oz.

A vector is such a versatile tool that we can extend it to many more dimensions. For more than 3 dimensions it is not possible to create a diagram.

To identify the direction of the components we need some standard notation. We use a unit vector in the direction of each axis. As the name suggests a **unit vector** has a size of 1.

$\boldsymbol{i}$ is the unit vector in the Ox direction
$\boldsymbol{j}$ is the unit vector in the Oy direction
$\boldsymbol{k}$ is the unit vector in the Oz direction

$\boldsymbol{a}=2\boldsymbol{i}+4\boldsymbol{j}+5\boldsymbol{k}$ is an example of a vector. It has component lengths 2 along Ox, 4 along Oy and 5 along Oz. It is the vector line joining the origin, $O(0,0,0)$ to $A(2,4,5)$, shown in Fig. 13.45.

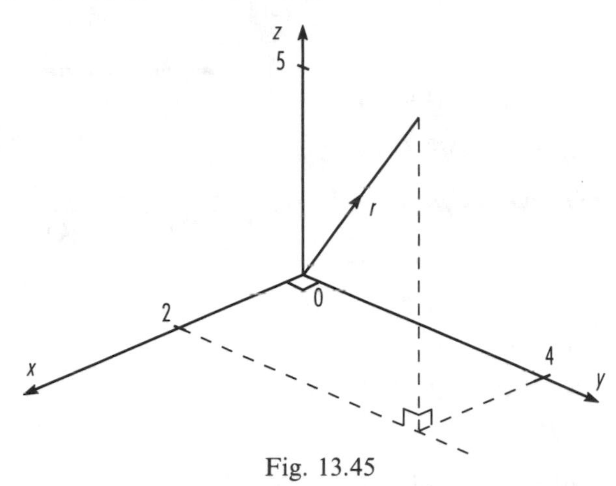

Fig. 13.45

Our example has positive values. As might be expected some components may be zero and/or some may be negative.

Rectangular form is just one form of vector representation. Other useful ones are cylindrical polars and spherical polars, though the mathematics becomes too involved for this level. All the general theories developed throughout the chapter apply to any of these vector forms.

We will concentrate on the rectangular system. The next 2 sets of examples look at some simple calculations. In Examples 13.13 we look at scalar multipliers, both positive and negative ones. In Examples 13.14 we look at addition and subtraction. Just using the usual rules of algebra we add/subtract **like terms**, i.e. we add/subtract the **same type of components**.

Examples 13.13

Given $\boldsymbol{a}=2\boldsymbol{i}+4\boldsymbol{j}+5\boldsymbol{k}$ represent in rectangular form

i) $-\boldsymbol{a}$, ii) $3\boldsymbol{a}$, iii) $-3\boldsymbol{a}$, iv) $0.2\boldsymbol{a}$.

i) $-\boldsymbol{a} = -(2\boldsymbol{i}+4\boldsymbol{j}+5\boldsymbol{k})=-2\boldsymbol{i}-4\boldsymbol{j}-5\boldsymbol{k}$.

ii) $3\boldsymbol{a} = 3(2\boldsymbol{i}+4\boldsymbol{j}+5\boldsymbol{k})=6\boldsymbol{i}+12\boldsymbol{j}+15\boldsymbol{k}$.

iii) $-3\boldsymbol{a} = -3(2\boldsymbol{i}+4\boldsymbol{j}+5\boldsymbol{k})=-6\boldsymbol{i}-12\boldsymbol{j}-15\boldsymbol{k}$.

iv) $0.2\boldsymbol{a} = 0.2(2\boldsymbol{i}+4\boldsymbol{j}+5\boldsymbol{k})=0.4\boldsymbol{i}+0.8\boldsymbol{j}+\boldsymbol{k}$.

Examples 13.14

Given $\boldsymbol{a}=2\boldsymbol{i}+4\boldsymbol{j}+5\boldsymbol{k}$ and $\boldsymbol{b}=6\boldsymbol{i}-3\boldsymbol{k}$ in rectangular form represent as simply as possible

i) $\boldsymbol{a}+\boldsymbol{b}$, ii) $\boldsymbol{b}-\boldsymbol{a}$, iii) $2\boldsymbol{a}+3\boldsymbol{b}$, iv) $\boldsymbol{a}-5\boldsymbol{b}$.

i) $$\begin{aligned}\boldsymbol{a}+\boldsymbol{b} &= (2\boldsymbol{i}+4\boldsymbol{j}+5\boldsymbol{k})+(6\boldsymbol{i}-3\boldsymbol{k})\\ &= (2+6)\boldsymbol{i}+(4+0)\boldsymbol{j}+(5-3)\boldsymbol{k}\\ &= 8\boldsymbol{i}+4\boldsymbol{j}+2\boldsymbol{k}\end{aligned}$$

An equally valid solution is $2(4\boldsymbol{i}+2\boldsymbol{j}+\boldsymbol{k})$ where a common factor of 2 is extracted from all the components.

ii) $$\begin{aligned}\boldsymbol{b}-\boldsymbol{a} &= (6\boldsymbol{i}-3\boldsymbol{k})-(2\boldsymbol{i}+4\boldsymbol{j}+5\boldsymbol{k})\\ &= (6-2)\boldsymbol{i}+(0-4)\boldsymbol{j}+(-3-5)\boldsymbol{k}\\ &= 4\boldsymbol{i}-4\boldsymbol{j}-8\boldsymbol{k}\end{aligned}$$

or $-4(\boldsymbol{i}-\boldsymbol{j}-2\boldsymbol{k})$.

iii) $$\begin{aligned}2\boldsymbol{a}+3\boldsymbol{b} &= 2(2\boldsymbol{i}+4\boldsymbol{j}+5\boldsymbol{k})+3(6\boldsymbol{i}-3\boldsymbol{k})\\ &= 4\boldsymbol{i}+8\boldsymbol{j}+10\boldsymbol{k}+18\boldsymbol{i}-9\boldsymbol{k}\\ &= (4+18)\boldsymbol{i}+(8+0)\boldsymbol{j}+(10-9)\boldsymbol{k}\\ &= 22\boldsymbol{i}+8\boldsymbol{j}+\boldsymbol{k}\end{aligned}$$

iv) $$\begin{aligned}\boldsymbol{a}-5\boldsymbol{b} &= (2\boldsymbol{i}+4\boldsymbol{j}+5\boldsymbol{k})-5(6\boldsymbol{i}-3\boldsymbol{k})\\ &= 2\boldsymbol{i}+4\boldsymbol{j}+5\boldsymbol{k}-30\boldsymbol{i}+15\boldsymbol{k}\\ &= (2-30)\boldsymbol{i}+(4+0)\boldsymbol{j}+(5+15)\boldsymbol{k}\\ &= -28\boldsymbol{i}+4\boldsymbol{j}+20\boldsymbol{k}\end{aligned}$$

or $4(-7\boldsymbol{i}+\boldsymbol{j}+5\boldsymbol{k})$.

EXERCISE 13.5

Given $\boldsymbol{a}=2\boldsymbol{i}+0.5\boldsymbol{j}-3\boldsymbol{k}$, $\mathrm{b}=-\boldsymbol{i}+\boldsymbol{j}+4\boldsymbol{k}$ and $\boldsymbol{c}=6\boldsymbol{j}-5\boldsymbol{k}$ evaluate and simplify the following vectors in $\boldsymbol{i}, \boldsymbol{j}, \boldsymbol{k}$ form.

1 $-\boldsymbol{b}$

2 $2\boldsymbol{a}$

3 $\boldsymbol{b}+\boldsymbol{c}$

4 $-\boldsymbol{b}+2\boldsymbol{a}$

5 $3\boldsymbol{b}+4\boldsymbol{c}$

6 $\boldsymbol{b}-2\boldsymbol{a}$

7 $\boldsymbol{c}+\boldsymbol{b}+\boldsymbol{a}$

8 $\boldsymbol{a}+2\boldsymbol{b}+3\boldsymbol{c}$

9 $2\boldsymbol{a}+0.4\boldsymbol{b}+0.5\boldsymbol{c}$

10 $3\boldsymbol{c}-2\boldsymbol{b}+4\boldsymbol{c}$

ASSIGNMENT

And so once again the theory allows us to re-consider the two problems posed at the beginning of the chapter.

Firstly we will consider our construction problem. In option one for the fastest crossing we ended 96 m downstream.

By considering a unit vector $\boldsymbol{i}$ across the river and a unit vector $\boldsymbol{j}$ upstream we may interpret the velocities as $5\boldsymbol{i}$ and $-4\boldsymbol{j}$ (see Fig. 13.46). Using simple vector addition their resultant is

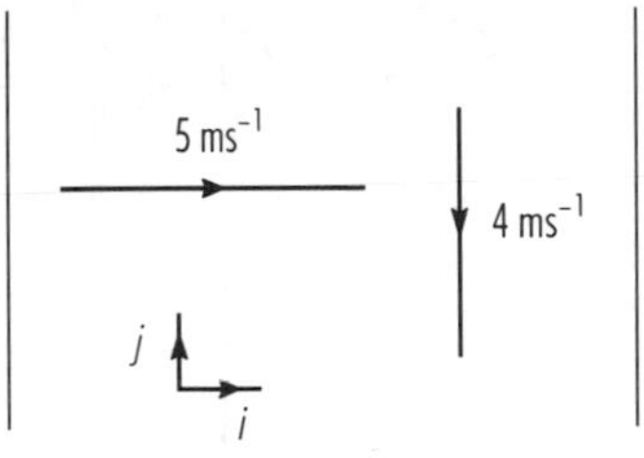

Fig. 13.46

$$\boldsymbol{v}_1 = 5\boldsymbol{i} - 4\boldsymbol{j} \text{ ms}^{-1}.$$

We can find the magnitude and direction of $\boldsymbol{v}_1$ using Pythagoras' theorem and trigonometry as before.

For the direct crossing

$$\boldsymbol{v}_2 = (5 \sin \beta)\boldsymbol{i} + (5 \cos \beta - 4)\boldsymbol{j} \text{ ms}^{-1}$$

We have $5 \cos \beta - 4 = 0$ because $\boldsymbol{v}_2$ must act directly across the river. Again the values of $\boldsymbol{v}_2$ and β are as before.

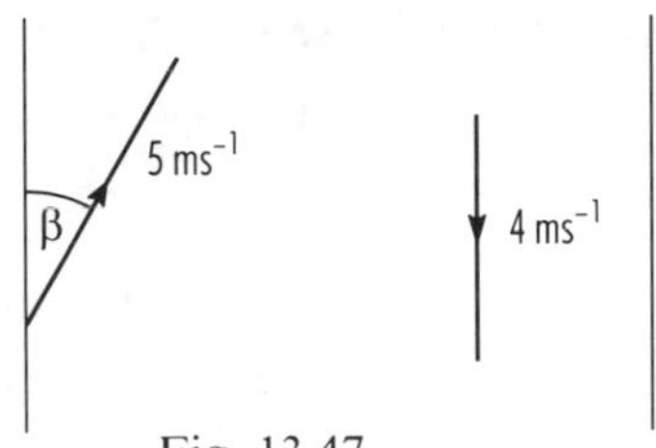

Fig. 13.47

Secondly it is just as easy to consider the aircraft with a unit vector $\boldsymbol{i}$ due East and a unit vector $\boldsymbol{j}$ due North.

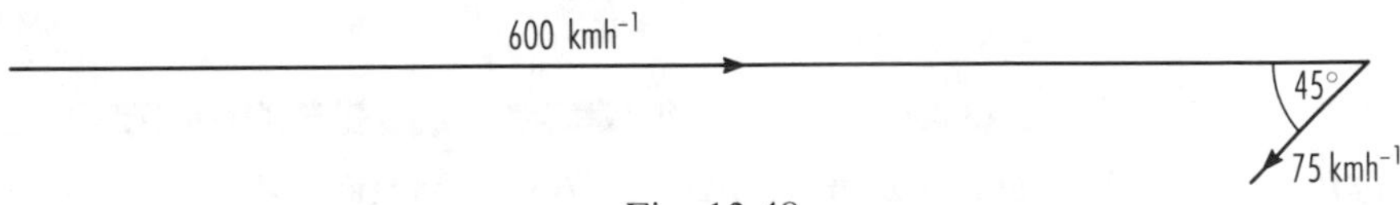

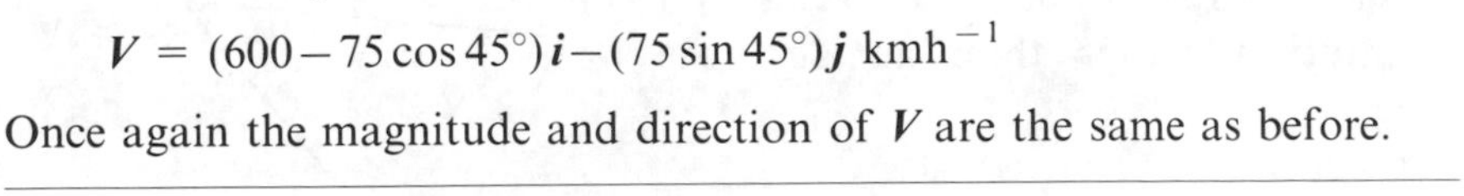

Fig. 13.48

Then we can represent the overall velocity by

$$\boldsymbol{V} = (600 - 75 \cos 45°)\boldsymbol{i} - (75 \sin 45°)\boldsymbol{j} \text{ kmh}^{-1}$$

Once again the magnitude and direction of $\boldsymbol{V}$ are the same as before.

The following examples are to consolidate the theory we have developed during this chapter. They concentrate on rectangular form before calculating any magnitudes or directions.

Example 13.15

Triangle OAB has vertices (0, 0), (2, −3) and (4, 5) respectively. Calculate the lengths of sides OA and AB.

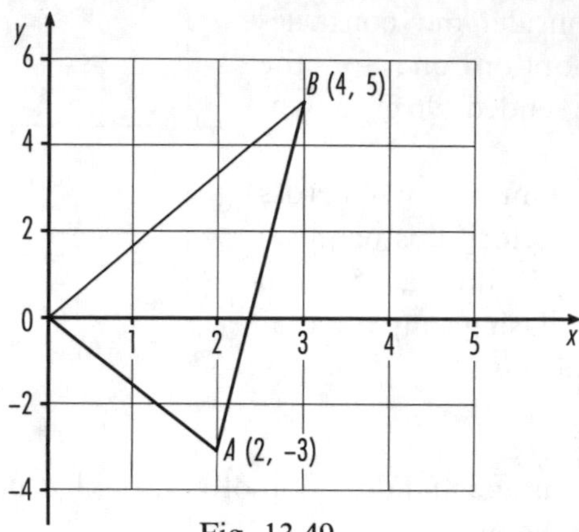

Fig. 13.49

In vector form $\quad \boldsymbol{OA} = 2\boldsymbol{i} - 3\boldsymbol{j}$

The magnitude of $\boldsymbol{OA}$, or $|\boldsymbol{OA}|$, or OA

$$= \sqrt{2^2 + (-3)^2}$$

$$= 3.61 \text{ units.}$$

Pythagoras' theorem.

Also

$$\boldsymbol{AB} = \boldsymbol{AO} + \boldsymbol{OB}$$
$$= -\boldsymbol{OA} + \boldsymbol{OB}$$
$$= -(2\boldsymbol{i} - 3\boldsymbol{j}) + (4\boldsymbol{i} + 5\boldsymbol{j})$$
$$= 2\boldsymbol{i} + 8\boldsymbol{j}.$$

Then

$$|\boldsymbol{AB}| = \sqrt{2^2 + 8^2}$$
$$= 8.25 \text{ units.}$$

Example 13.16

Calculate the length of side AB in triangle OAB with vertices $A\,(2, -3, 1)$ and $B\,(4, 5, 7)$.

To draw an accurate 3-dimensional diagram in 2 dimensions is difficult. Fig. 13.50 is a representation rather than an accurate drawing.

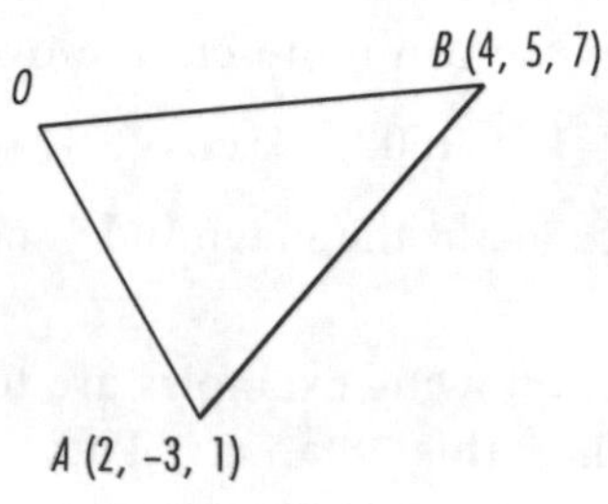

Fig. 13.50

In vector form

$$\boldsymbol{AB} = \boldsymbol{AO} + \boldsymbol{OB}$$
$$= -\boldsymbol{OA} + \boldsymbol{OB}$$
$$= -(2\boldsymbol{i} - 3\boldsymbol{j} + \boldsymbol{k}) + (4\boldsymbol{i} + 5\boldsymbol{j} + 7\boldsymbol{k})$$
$$= 2\boldsymbol{i} + 8\boldsymbol{j} + 6\boldsymbol{k}.$$

Then $|\boldsymbol{AB}| = \sqrt{2^2 + 8^2 + 6^2}$

$$= 10.20.$$

Extending Pythagoras' theorem.

EXERCISE 13.6

1 Triangle OXY has vertices given by the origin, $X(-1,4)$ and $Y(3,10)$. Calculate the lengths of the triangle's sides.

2 In triangle MNO calculate the length of MN given that $\boldsymbol{OM}=2\boldsymbol{i}-3\boldsymbol{j}+4\boldsymbol{k}$ and $\boldsymbol{ON}=\boldsymbol{i}-5\boldsymbol{k}$.

3 In rectangular form find the resultant force of $-\boldsymbol{i}+2\boldsymbol{j}$, $3\boldsymbol{i}-4\boldsymbol{k}$, $6\boldsymbol{j}+2\boldsymbol{k}$ and $\boldsymbol{i}-5\boldsymbol{j}+7\boldsymbol{k}$ N. What is its magnitude?

4 Four forces are in equilibrium, i.e. the sum of the forces is zero, $\Sigma\boldsymbol{F}=0$. They are $2\boldsymbol{i}+\boldsymbol{j}$, $6\boldsymbol{i}-3\boldsymbol{j}$, $4\boldsymbol{j}$ and $a\boldsymbol{i}+b\boldsymbol{j}$ N. Find the values of a and b.

5 $ABCD$ is a quadrilateral with $A(1,0,3)$, $B(-2,4,9)$, $C(2,5,-1)$ and $D(-3,8,6)$. Calculate the lengths of $\boldsymbol{AB}$ and $\boldsymbol{BC}$. Also calculate the perimeter of $ABCD$.

6 A ring of weight 4 N is pulled horizontally by a force of 7 N. What is the magnitude and direction of the resultant force?

7 It is mid-February and snow is falling vertically at a gentle $2\,\text{ms}^{-1}$. The weather changes with a horizontal breeze now blowing at $1.8\,\text{ms}^{-1}$. What is the magnitude and direction of a snowflake's velocity now?

8 The figure shows a rowing boat travelling directly across a river at $10\,\text{ms}^{-1}$. There is a northerly current of $8\,\text{ms}^{-1}$. Calculate the magnitude and direction of the resultant velocity. The river width is 200 m. How far downstream will the boat be when it reaches the other bank?

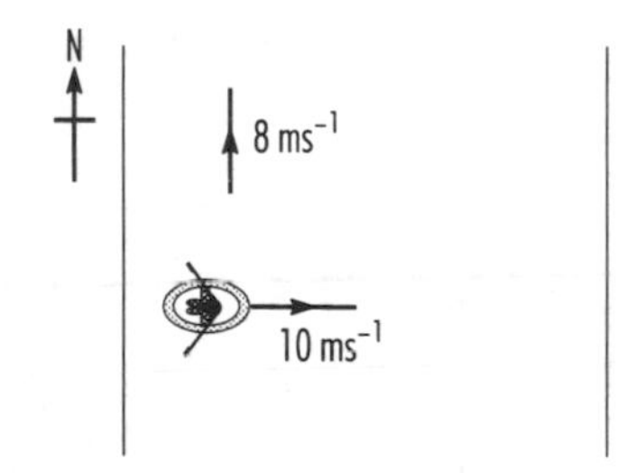

9 A failed engine has left a boat at sea at the mercy of the wind, blowing South East at 12ms^{-1}, and the current running South West at $18\,\text{ms}^{-1}$. In which direction is it drifting? Dangerous rocks are close and the engineer manages a temporary repair just in time. Headed due South at $10\,\text{ms}^{-1}$ calculate the actual resultant velocity in these weather conditions.

10 A surveyor starts from his mobile office, O, and walks 100 m due West. He turns and walks 50 m South West followed by 120 m due South, reaching a tower crane, T. Take due North as $\boldsymbol{j}$ and due East as $\boldsymbol{i}$. Represent each of these displacements in rectangular vector form. Write down their resultant $\boldsymbol{OT}$ in the same form. Calculate the direct distance of the crane's base from the office. If the cab on the crane is 35 m above ground level find its elevation from the surveyor's office.

14 Determinants and Matrices

The objectives of this chapter are to:

1. Recognise the notation for a determinant.
2. Evaluate a second order (i.e. 2×2) determinant.
3. Describe the meaning of a determinant whose value is zero.
4. Solve simultaneous linear equations with two unknowns using determinants.
5. Recognise the notation for a matrix.
6. Calculate the sum and difference of two matrices.
7. Define a unit matrix.
8. Calculate the product of two 2×2 matrices.
9. Demonstrate that the product of two matrices is, in general, non-commutative.
10. Define a singular matrix.
11. Obtain the inverse of a 2×2 matrix.
12. Solve simultaneous linear equations with two unknowns using matrices.
13. Relate the use of matrices to simple technical problems.

Introduction

We have 2 main separate sections in this chapter. The first one looks at **determinants** and the second one at **matrices**. **Matrices** is the plural of **matrix**. As you work through the second section you will notice some similarities with the first section. The different notations are important, though they do both appear as types of arrays. Determinants and matrices are not the same. They have different mathematical meanings. However there is a connection. You can find the determinant of a matrix, but not the other way about.

A major use for both determinants and matrices is the solution of simultaneous linear equations.

ASSIGNMENT

This Assignment looks at a manufacturing company producing window frames. It produces both hardwood and uPVC frames at its two sites. To cover the country they operate from factory units in northern and southern England. Later in this chapter we will look at the company's attempts to meet a production target.

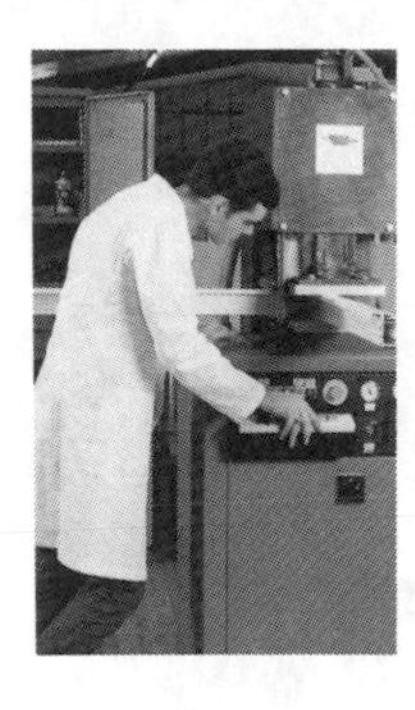

The company has secured an order from a builder operating in the Midlands. It has to be integrated within the existing long term manufacturing contracts. The order is for 540 uPVC and 295 hardwood window frames. Because the 2 factories are differently equipped their daily production capacities are different. The following table shows the capacity figures for each type of window frame.

Daily Production	uPVC	Hardwood
Northern Factory	80	50
Southern Factory	65	30

We want to know how to schedule this order using both sites. Suppose the order will use x days of production at the northern factory and y days at the southern factory. Combining the daily production capacities with these days gives

Order Production	uPVC	Hardwood
Northern Factory	$80x$	$50x$
Southern Factory	$65y$	$30y$
Total	540	295

Having set out the problem we will return to it once we have looked at some relevant Mathematics

Determinants

A **determinant** is an array of numbers (or letters) in rows and columns. Rows are horizontal, $\longleftrightarrow$, and columns are vertical, $\updownarrow$. Determinant is often shortened to either **det** or **Δ**. All determinants are **square**, i.e. they have the same number of rows as columns.

In this chapter we will look at second order determinants. Second order means there are 2 rows, and hence 2 columns.

A 2×2 determinant (i.e. a 2 by 2 determinant) is another way of writing a second order determinant.

$\begin{vmatrix} a & b \\ d & e \end{vmatrix}$ is a general example of a second order determinant.

Its value is $ae - bd$, $\begin{vmatrix} a & b \\ d & e \end{vmatrix} = ae - bd.$

We multiply the values in the leading diagonal $\begin{vmatrix} a \searrow & \\ & e \end{vmatrix}$.

Then we multiply the values in the trailing diagonal $\begin{vmatrix} & \nearrow b \\ d \swarrow & \end{vmatrix}$

and subtract the results,

i.e. $\nwarrow\!\!\!\searrow \times \; - \; \times \swarrow\!\!\!\nearrow$

Examples 14.1

Find the values of the second order determinants

i) $\begin{vmatrix} 12 & 6 \\ 5 & 4 \end{vmatrix}$, ii) $\begin{vmatrix} 1 & 7 \\ -6 & 21 \end{vmatrix}$, iii) $\begin{vmatrix} 1 & 0 \\ -6 & 21 \end{vmatrix}$, iv) $\begin{vmatrix} -12 & -6 \\ -5 & -4 \end{vmatrix}$,

v) $\begin{vmatrix} 0 & 7 \\ -6 & -21 \end{vmatrix}$, vi) $\begin{vmatrix} 0.6 & -0.1 \\ 0.2 & -0.9 \end{vmatrix}$.

We simply follow the order of operations we have shown before these examples. We have included brackets to highlight the order, though they are not really necessary.

i)
$$\begin{vmatrix} 12 & 6 \\ 5 & 4 \end{vmatrix} = (12 \times 4) - (6 \times 5)$$
$$= 48 - 30$$
$$= 18.$$

ii)
$$\begin{vmatrix} 1 & 7 \\ -6 & 21 \end{vmatrix} = (1 \times 21) - (7 \times -6)$$
$$= 21 - (-42)$$
$$= 21 + 42$$
$$= 63.$$

iii)
$$\begin{vmatrix} 1 & 0 \\ -6 & 21 \end{vmatrix} = (1 \times 21) - (0 \times -6)$$
$$= 21 - 0$$
$$= 21.$$

iv)
$$\begin{vmatrix} -12 & -6 \\ -5 & -4 \end{vmatrix} = (-12 \times -4) - (-6 \times -5)$$
$$= 48 - 30$$
$$= 18.$$

v)
$$\begin{vmatrix} 0 & 7 \\ -6 & 21 \end{vmatrix} = (0 \times 21) - (7 \times -6)$$
$$= 0 - (-42)$$
$$= 0 + 42$$
$$= 42.$$

vi) $\begin{vmatrix} 0.6 & -0.1 \\ 0.2 & -0.9 \end{vmatrix} = (0.6 \times -0.9) - (-0.1 \times 0.2)$

$= -0.54 - (-0.02)$
$= -0.54 + 0.02$
$= -0.52.$

The order of writing the rows and columns is important. In the next set of examples we look at some interchanges.

Examples 14.2

Find the values of the second order determinants

i) $\begin{vmatrix} 1 & -2 \\ 4 & -1 \end{vmatrix}$, ii) $\begin{vmatrix} -2 & 1 \\ -1 & 4 \end{vmatrix}$, iii) $\begin{vmatrix} 4 & -1 \\ 1 & -2 \end{vmatrix}$.

i) $\begin{vmatrix} 1 & -2 \\ 4 & -1 \end{vmatrix} = (1 \times -1) - (-2 \times 4)$

$= -1 + 8$
$= 7.$

ii) In this second example we have interchanged the columns.

$\begin{vmatrix} -2 & 1 \\ -1 & 4 \end{vmatrix} = (-2 \times 4) - (1 \times -1)$

$= -8 + 1$
$= -7.$

It looks like interchanging the columns of the original determinant changes the +/− sign of that determinant.

iii) For the third example we have interchanged the rows.

$\begin{vmatrix} 4 & -1 \\ 1 & -2 \end{vmatrix} = (4 \times -2) - (-1 \times 1)$

$= -8 + 1$
$= -7.$

It looks like interchanging the rows of the original determinant changes the +/− sign of that determinant.

Examples 14.2 looked at determinants of specific values. We can follow up with a look at our general second order determinant. (The same principles apply to determinants of any order.)

We know $\begin{vmatrix} a & b \\ d & e \end{vmatrix} = ae - bd.$

Interchanging the columns gives $\begin{vmatrix} b & a \\ e & d \end{vmatrix} = bd - ae$

$= -(ae - bd).$

This general case confirms our ideas in the numerical example.

Interchanging the original rows gives
$$\begin{vmatrix} d & e \\ a & b \end{vmatrix} = db - ea$$
$$= bd - ae$$
$$= -(ae - bd).$$

We now have a general rule.

If any two rows (or two columns) are interchanged then the sign of the determinant changes +/−.

Now let us see what happens when we make 2 changes. We will interchange the columns and then interchange the rows,

i.e. $\begin{vmatrix} a & b \\ d & e \end{vmatrix}$ changes to $\begin{vmatrix} b & a \\ e & d \end{vmatrix}$ and then $\begin{vmatrix} e & d \\ b & a \end{vmatrix}$.

The algebraic value is
$$\begin{vmatrix} e & d \\ b & a \end{vmatrix} = ea - db$$
$$= ae - bd.$$

You can see we have returned to our original value. These changes are similar to the rules for signs generally in algebra. Remember $(-)(-) = +$.

Examples 14.3

Find the values of the second order determinants

i) $\begin{vmatrix} 2 & 2 \\ 7 & 7 \end{vmatrix}$, ii) $\begin{vmatrix} 2 & -3 \\ 2 & -3 \end{vmatrix}$.

i)
$$\begin{vmatrix} 2 & 2 \\ 7 & 7 \end{vmatrix} = (2 \times 7) - (2 \times 7)$$
$$= 14 - 14$$
$$= 0.$$

ii)
$$\begin{vmatrix} 2 & -3 \\ 2 & -3 \end{vmatrix} = (2 \times -3) - (-3 \times 2)$$
$$= -6 - (-6)$$
$$= -6 + 6$$
$$= 0.$$

We can look at the general cases of Examples 14.3, writing both rows the same or both columns the same.

For both columns the same
$$\begin{vmatrix} a & a \\ d & d \end{vmatrix} = ad - ad = 0.$$

For both rows the same $\begin{vmatrix} a & b \\ a & b \end{vmatrix} = ab - ab = 0.$

If any rows (or columns) are the same then the determinant has a value of 0.

Examples 14.4

Find the values of the second order determinants

i) $\begin{vmatrix} 5 & 9 \\ -2 & 6 \end{vmatrix}$, ii) $\begin{vmatrix} 5 & -2 \\ 9 & 6 \end{vmatrix}$.

i) $$\begin{vmatrix} 5 & 9 \\ -2 & 6 \end{vmatrix} = (5 \times 6) - (9 \times -2)$$
$$= 30 - (-18)$$
$$= 30 + 18$$
$$= 48.$$

ii) $$\begin{vmatrix} 5 & -2 \\ 9 & 6 \end{vmatrix} = (5 \times 6) - (-2 \times 9)$$
$$= 48, \text{ as before.}$$

We can look at the general cases of Examples 14.4, writing the rows as columns, and hence the columns as rows.

We know that $\begin{vmatrix} a & b \\ d & e \end{vmatrix} = ae - bd$

Re-writing the rows as columns and the columns as rows we have

$$\begin{vmatrix} a & d \\ b & e \end{vmatrix} = ae - db = ae - bd, \text{ as before.}$$

Writing the rows as columns and the columns as rows leaves the value of the determinant unchanged.

Example 14.5

Find the value of the second order determinant $\begin{vmatrix} 2 & 3 \\ 12 & 18 \end{vmatrix}$.

We can find the value in the usual way,

$$\begin{vmatrix} 2 & 3 \\ 12 & 18 \end{vmatrix} = (2 \times 18) - (3 \times 12)$$
$$= 36 - 36$$
$$= 0.$$

A good look at the numbers arrayed in our determinant will show this answer might have been expected.

You can see the second row of this determinant is 6 times the first row,

i.e. $\quad \text{Row } ② = 6 \times \text{Row } ①$

or $\quad \frac{1}{6} \times \text{Row } ② = \text{Row } ①.$

Similarly the second column is a multiple of the first column,

i.e. $\quad \text{Column } ② = 1.5 \times \text{Column } ①$

or $\quad \frac{1}{1.5} \times \text{Column } ② = \text{Column } ①.$

We will see this pattern of multiples later in some pairs of simultaneous linear equations

EXERCISE 14.1

Find the values of the second order determinants.

1 $\begin{vmatrix} 2 & 1 \\ 3 & 4 \end{vmatrix}$

2 $\begin{vmatrix} 1 & 5 \\ 9 & 2 \end{vmatrix}$

3 $\begin{vmatrix} -6 & 3 \\ 2 & 5 \end{vmatrix}$

4 $\begin{vmatrix} 1 & 1 \\ -2 & 3 \end{vmatrix}$

5 $\begin{vmatrix} 0 & 4 \\ -7 & 0 \end{vmatrix}$

6 $\begin{vmatrix} \frac{1}{2} & \frac{1}{3} \\ -\frac{1}{2} & 0 \end{vmatrix}$

7 $\begin{vmatrix} 4 & 0 \\ 0 & -7 \end{vmatrix}$

8 $\begin{vmatrix} 56 & -9 \\ -12 & 3 \end{vmatrix}$

9 $\begin{vmatrix} 8 & 25 \\ 2 & 32 \end{vmatrix}$

10 $\begin{vmatrix} 0.7 & 0.5 \\ -0.25 & 0.4 \end{vmatrix}$

Simultaneous equations

We learned how to solve simultaneous linear equations in Chapter 9. The next example ought to refresh your memory.

Example 14.6

Solve the pair of simultaneous equations $5x - 3y = 21$
and $7x + 8y = 5.$

We can choose to eliminate either x or y in each case. Our choice is to eliminate y.

In $\quad 5x - 3y = 21$

and $\quad 7x + 8y = 5 \quad$ the y terms are not the same.

To make them the same we multiply the first equation by 8 and the second equation by 3. Then they will be $24y$, 24 being the lowest common multiple (LCM) of 3 and 8.

$$5x-3y = 21 \qquad \times 8$$

and

$$7x+8y = 5 \qquad \times 3$$

become

$$40x-24y = 168$$

and

$$21x+24y = 15$$

Add

$$40x+21x = 168+15$$

i.e.

$$61x = 183$$

$$\frac{61x}{61} = \frac{183}{61}$$

$$x = 3.$$

To find y we substitute this value into one or other of our original equations. Let us substitute into the second equation for a change to get

$$7(3)+8y = 5$$

i.e.

$$21+8y = 5$$

$$8y = 5-21$$

$$8y = -16$$

$$\frac{8y}{8} = \frac{-16}{8}$$

$$y = -2.$$

Our complete solution is $x=3$, $y=-2$.

We check our solution with the other equation. Substitute both values into the left-hand side so that

$$5x-3y = 5(3)-3(-2) = 15+6 = 21.$$

This is consistent with our second equation, so confirming our solution.

We can follow this numerical example with a general example. For later consistency we write all the terms on one side of the "=" sign. Because this general case involves so many letters we have highlighted $\boldsymbol{x}$ and $\boldsymbol{y}$ in bold type.

Generally

$$a\boldsymbol{x}+b\boldsymbol{y}+c = 0$$

and

$$d\boldsymbol{x}+e\boldsymbol{y}+f = 0.$$

Suppose we wish to eliminate $\boldsymbol{x}$. We multiply the first equation by d and the second equation by a. This makes both coefficients of $\boldsymbol{x}$ the same,

i.e.

$$ad\boldsymbol{x}+bd\boldsymbol{y}+cd = 0$$

and

$$ad\boldsymbol{x}+ae\boldsymbol{y}+af = 0.$$

We subtract the equations to eliminate $\boldsymbol{x}$,

$$bd\boldsymbol{y} - ae\boldsymbol{y} + cd - af = 0$$

i.e. $$(bd - ae)\boldsymbol{y} = af - cd.$$

From this line we can find a solution for $\boldsymbol{y}$.

If we wish to eliminate $\boldsymbol{y}$ our multiplication is different. We multiply the first equation by e and the second equation by b. This makes both coefficients of $\boldsymbol{y}$ the same,

i.e. $$ae\boldsymbol{x} + be\boldsymbol{y} + ce = 0$$

and $$bd\boldsymbol{x} + be\boldsymbol{y} + bf = 0.$$

We subtract the equations to eliminate $\boldsymbol{y}$,

$$ae\boldsymbol{x} - bd\boldsymbol{x} + ce - bf = 0$$

i.e. $$(ae - bd)\boldsymbol{x} = bf - ce.$$

Again, from this line we can find a solution for $\boldsymbol{x}$. We are going to re-arrange the solution lines for $\boldsymbol{x}$ and $\boldsymbol{y}$. From the new patterns we will be able to spot some determinants.

Now $$(ae - bd)\boldsymbol{x} = bf - ce$$

becomes $$\frac{(ae - bd)\,\boldsymbol{x}}{bf - ce} = 1$$

Dividing both sides by $bf - ce$.

i.e. $$\frac{\boldsymbol{x}}{bf - ce} = \frac{1}{ae - bd}.$$

Dividing both sides by $ae - bd$.

In determinant form this is

$$\frac{\boldsymbol{x}}{\begin{vmatrix} b & c \\ e & f \end{vmatrix}} = \frac{1}{\begin{vmatrix} a & b \\ d & e \end{vmatrix}}.$$

Also $$(bd - ae)\boldsymbol{y} = af - cd$$

i.e. $$-(ae - bd)\boldsymbol{y} = af - cd$$

Needing consistent brackets, $ae - bd$.

becomes $$\frac{-(ae - bd)\boldsymbol{y}}{af - cd} = 1$$

Dividing both sides by $af - cd$.

i.e. $$\frac{-\boldsymbol{y}}{af - cd} = \frac{1}{ae - bd}$$

Dividing both sides by $ae - bd$.

In determinant form this is

$$\frac{-\boldsymbol{y}}{\begin{vmatrix} a & c \\ d & f \end{vmatrix}} = \frac{1}{\begin{vmatrix} a & b \\ d & e \end{vmatrix}}.$$

We can link our answers together to get

$$\frac{\boldsymbol{x}}{\begin{vmatrix} b & c \\ e & f \end{vmatrix}} = \frac{-\boldsymbol{y}}{\begin{vmatrix} a & c \\ d & f \end{vmatrix}} = \frac{1}{\begin{vmatrix} a & b \\ d & e \end{vmatrix}}.$$

We can see these determinants in the original pair of equations,

$$ax+by+c = 0$$

and $$dx+ey+f = 0.$$

For $\begin{vmatrix} b & c \\ e & f \end{vmatrix}$ in $\dfrac{x}{\begin{vmatrix} b & c \\ e & f \end{vmatrix}}$ visually we strike out the column involving x and highlight the remainder in

$$ax+by+c = 0$$

and $$dx+ey+f = 0.$$

For $\begin{vmatrix} a & c \\ d & f \end{vmatrix}$ in $\dfrac{-y}{\begin{vmatrix} a & c \\ d & f \end{vmatrix}}$ we include a minus sign for consistency in this method. Visually we strike out the column involving y and highlight the remainder in

$$ax+by+c = 0$$

and $$dx+ey+f = 0.$$

For $\begin{vmatrix} a & b \\ d & e \end{vmatrix}$ in $\dfrac{1}{\begin{vmatrix} a & b \\ d & e \end{vmatrix}}$ visually we strike out the third column and highlight the remainder in

$$ax+by+c = 0$$

and $$dx+ey+f = 0.$$

Remember a, b, c, d, e and f represent pure numbers. They may be positive, zero or negative.

Now we can apply this new theory to a pair of simultaneous linear equations.

Example 14.7

Solve the pair of simultaneous equations $5x-3y = 21$
and $7x+8y = 5$.

Our first step is to gather all terms in each equation to one side of the "=" sign,

i.e. $$5x-3y-21 = 0$$

and $$7x+8y-5 = 0.$$

For the determinant under x we ignore those numbers associated with x and highlight $\begin{vmatrix} -3 & -21 \\ 8 & -5 \end{vmatrix}$.

For the determinant under y we ignore those numbers associated with y and highlight $\begin{vmatrix} 5 & -21 \\ 7 & -5 \end{vmatrix}$.

For the determinant under the 1 we ignore the pure numbers and highlight $\begin{vmatrix} 5 & -3 \\ 7 & 8 \end{vmatrix}$.

Linking these all together we have

$$\frac{x}{\begin{vmatrix} -3 & -21 \\ 8 & -5 \end{vmatrix}} = \frac{-y}{\begin{vmatrix} 5 & -21 \\ 7 & -5 \end{vmatrix}} = \frac{1}{\begin{vmatrix} 5 & -3 \\ 7 & 8 \end{vmatrix}}$$

i.e. $$\frac{x}{15--168} = \frac{-y}{-25--147} = \frac{1}{40--21}$$

$$\frac{x}{183} = \frac{-y}{122} = \frac{1}{61}.$$

We have 3 fractions all equal to each other. Using the first and last ones we have

$$\frac{x}{183} = \frac{1}{61}$$

i.e. $$x = 183 \times \frac{1}{61} = 3.$$

Using the middle and last fractions we have

$$\frac{-y}{122} = \frac{1}{61}$$

i.e. $$y = -122 \times \frac{1}{61} = -2.$$

Our complete solution is $x=3$, $y=-2$.

We check our solution with one of the original equations. Substituting both values into the left-hand side of the first equation gives

$$5x-3y-21 = 5(3)-3(-2)-21 = 15+6-21 = 0.$$

These values are consistent, confirming our solution.

Examples 14.8

Solve the pairs of simultaneous equations

i) $4x+y-14 = 0$ and $4x+y-9 = 0$; ii) $4x+y-14 = 0$ and $8x+2y-28 = 0$.

i) This first pair of equations has the same terms for x and the same terms for y. However the third terms are different. Remember that the solutions for x and y must apply to both equations simultaneously. How can $4x+y$ give the value of 0 when combined with the different -14 and -9? This should not happen. These equations are **inconsistent**. They have no simultaneous solution. As we saw in Chapter 9 they are a pair of parallel lines.
In determinant form we have

$$\frac{x}{\begin{vmatrix} 1 & -14 \\ 1 & -9 \end{vmatrix}} = \frac{-y}{\begin{vmatrix} 4 & -14 \\ 4 & -9 \end{vmatrix}} = \frac{x}{\begin{vmatrix} 4 & 1 \\ 4 & 1 \end{vmatrix}}.$$

It is the last determinant that causes the problem.

$\begin{vmatrix} 4 & 1 \\ 4 & 1 \end{vmatrix} = (4 \times 1) - (1 \times 4) = 0$ and then $\frac{1}{0}$ is *not* allowed in Mathematics.

ii) Look closely at the second pair of equations. Really we have only one equation because $8x + 2y - 28 = 0$ is just twice $4x + y - 14 = 0$. As we have only one equation repeated and not a pair of them we are unable to solve them simultaneously. Each equation is **dependent** on the other. Remember because we have 2 unknowns, x and y, we need at least 2 different equations.

In determinant form we have

$$\frac{x}{\begin{vmatrix} 1 & -14 \\ 2 & -28 \end{vmatrix}} = \frac{-y}{\begin{vmatrix} 4 & -14 \\ 8 & -28 \end{vmatrix}} = \frac{1}{\begin{vmatrix} 4 & 1 \\ 8 & 2 \end{vmatrix}}.$$

This time all 3 determinants have values of 0. (You can check this for yourself as an exercise.) Again $\frac{1}{0}$ is *not* allowed in Mathematics.

From Example 14.8 we can define 2 rules where there is no solution. **If only the determinant under the 1 is zero then the equations are inconsistent. If all 3 determinants are zero then the equations are dependent**.

ASSIGNMENT

Earlier in this chapter we set out the order production figures in total and for both factories. We may combine them to create 2 linear equations

$$80x + 65y = 540$$

and

$$50x + 30y = 295.$$

We can cope with these figures on a calculator. However there are common factors of 5 in each equation that may be cancelled. Also we may bring all the terms to the left-hand side. These changes give

$$16x + 13y - 108 = 0$$

and

$$10x + 6y - 59 = 0.$$

In determinant form we can write

$$\frac{x}{\begin{vmatrix} 13 & -108 \\ 6 & -59 \end{vmatrix}} = \frac{-y}{\begin{vmatrix} 16 & -108 \\ 10 & -59 \end{vmatrix}} = \frac{1}{\begin{vmatrix} 16 & 13 \\ 10 & 6 \end{vmatrix}}$$

i.e.

$$\frac{x}{-767 - -648} = \frac{-y}{-944 - -1080} = \frac{1}{96 - 130}$$

$$\frac{x}{-767 + 648} = \frac{-y}{-944 + 1080} = \frac{1}{96 - 130}$$

$$\frac{x}{-119} = \frac{-y}{136} = \frac{1}{-34}.$$

Using the first and last fractions we have

$$\frac{x}{-119} = \frac{1}{-34}$$

i.e. $$x = -119 \times \frac{1}{-34} = 3.5.$$

Using the middle and last fractions we have

$$\frac{-y}{136} = \frac{1}{-34}$$

i.e. $$y = -136 \times \frac{1}{-34} = 4.$$

Now we have a complete solution. Together the factories can meet the order for the builder. The northern factory must use 3.5 days and the southern factory 4 days of production time.

EXERCISE 14.2

Using determinants solve the following pairs of simultaneous linear equations.

1 $x - y = 19$ and $x + y = 1$

2 $3x + y = 7$ and $x + y = 11$

3 $2x + y = 9$ and $x - y = 0$

4 $x - 2y = -4$ and $x + 2y = 36$

5 $3x - y = 2$ and $5x + 2y = 40$

6 $-3x + 8y = 5$ and $x - 5y = 3$

7 $3x + 2y = 2$ and $x + 3y = 6.5$

8 $x + \frac{1}{2}y = 26$ and $\frac{1}{3}x - y = 4$

9 $2x - 5y = 9$ and $7x - y = 15$

10 $\frac{1}{2}x - 2y = 2.5$ and $\frac{1}{3}x + y = 0.5$

Matrices

A **matrix** is an array of numbers or letters. **Matrix** is the singular and **matrices** are plural. A matrix is *not* the same as a determinant. You cannot find the value of a matrix. Each number (or letter) is an **element** in the matrix rather than some part of an overall value. Again we have rows and columns. For matrices we are allowed different numbers of rows and columns. An $m \times n$ (i.e. m by n) matrix has m rows and n columns. We write that the order of the matrix is $m \times n$. The order of writing rows and then columns is important.

Where the numbers of rows and columns are the same we have a **square** matrix. You will remember that a determinant must be square. We will see later that we can find the determinant of a square matrix.

The next set of examples looks at matrices of different orders.

Examples 14.9

In this set of examples we state the orders of the matrices.

i) $\begin{pmatrix} 2 & 3 & 1 \\ 0 & -5 & 6 \end{pmatrix}$ has 2 rows and 3 columns.

It is a 2×3 matrix.

ii) $\begin{pmatrix} 4 & 0 & 1 \\ 0 & 0 & -4 \\ 2 & 0 & 7.5 \\ 1 & -7 & 6 \end{pmatrix}$ has 4 rows and 3 columns.

It is a 4×3 matrix.

iii) $\begin{pmatrix} 1 & 2 \\ 3 & 7 \end{pmatrix}$ has 2 rows and 2 columns.

It is a 2×2 matrix.

Because the numbers of rows and columns are the same it is also a square matrix.

iv) (4) has 1 row and 1 column.

It is a 1×1 matrix.

Again this is a square matrix.

v) $(1 \quad 2 \quad 3)$ has 1 row and 3 columns.

It is a 1×3 matrix.

Where there is only 1 row, a matrix can be called a **row vector**.

vi) $\begin{pmatrix} 5 \\ -2 \\ 0 \end{pmatrix}$ has 3 rows and 1 column.

It is a 3×1 matrix.

Where there is only 1 column, a matrix can be called a **column vector**.

The **null** or **zero** matrix has only elements of 0. For example

$\begin{pmatrix} 0 & 0 & 0 \\ 0 & 0 & 0 \\ 0 & 0 & 0 \end{pmatrix}$ is a 3×3 null matrix.

The **unit** matrix has elements of 1 along the leading diagonal and elements of 0 elsewhere. For example $\begin{pmatrix} 1 & 0 & 0 & 0 \\ 0 & 1 & 0 & 0 \\ 0 & 0 & 1 & 0 \\ 0 & 0 & 0 & 1 \end{pmatrix}$ is a 4×4 unit matrix.

It is possible for matrices to be equal. Firstly they must have the same order. Then each corresponding element must be the same. When checking equality it is simply a matter of comparing the matrices.

Examples 14.10

In these examples we look at matrices that are *not* equal.

i) $\begin{pmatrix} 2 & 3 & 1 \\ 0 & -5 & 6 \end{pmatrix}$ and $\begin{pmatrix} 2 & 3 & 1 \\ 0 & 5 & 6 \end{pmatrix}$ are *not* equal.

In the second row, second column one element is -5 and the other is 5.

ii) $\begin{pmatrix} 2 & 3 & 1 \\ 0 & -5 & 6 \end{pmatrix}$ and $\begin{pmatrix} 2 & 0 \\ 3 & -5 \\ 1 & 6 \end{pmatrix}$ are *not* equal.

The first matrix has order 2×3 and the second matrix has order 3×2.

When we looked at determinants we calculated their values. We can do more with matrices. First of all let us look at addition and subtraction. We can only add/subtract matrices of exactly the same order, i.e. the same number of rows and the same number of columns. In Examples 14.11 we look at some possible additions/subtractions.

Examples 14.11

i) We can add/subtract a 2×3 matrix and a 2×3 matrix.

ii) We *cannot* add/subtract a 2×3 matrix (2 rows) and a 3×3 matrix (3 rows). This is because of the different numbers of rows.

iii) We *cannot* add/subtract a 2×3 matrix and a 3×2 matrix. This is because neither the numbers of rows nor columns correspond.

EXERCISE 14.3

What are the orders of the following matrices?

1 $\begin{pmatrix} 2 & 1 \\ 3 & 6 \end{pmatrix}$

2 $\begin{pmatrix} 1 & 2 & 3 \\ 4 & -5 & 6 \end{pmatrix}$

3 $\begin{pmatrix} 1 & 9 \\ 2 & -17 \\ 4 & 6 \end{pmatrix}$

4 $\begin{pmatrix} 5 \\ 9 \end{pmatrix}$

5 $\begin{pmatrix} -1 & 4 & -7 \\ 2 & -5 & 8 \\ 3 & 6 & 9 \end{pmatrix}$

6 $\begin{pmatrix} 2 & 1 & 4 \\ 1 & -3 & 9 \end{pmatrix}$

7 $\begin{pmatrix} 6 & 19 & 7 \\ 1 & -2 & 0 \end{pmatrix}$

8 (8)

9 $(6 \quad 6 \quad 6)$

10 $\begin{pmatrix} p & q \\ r & s \end{pmatrix}$

11 Using the matrices in Questions **1–10** decide which matrices can be added/subtracted.

When we add/subtract matrices we do this with corresponding elements in the matrices. In this chapter we are going to concentrate on 2×2 matrices, but the principles apply generally. Adding/subtracting 2×2 matrices means the answer will be another 2×2 matrix.

Examples 14.12

If $\boldsymbol{A} = \begin{pmatrix} 1 & 12 \\ 9 & 7 \end{pmatrix}$, $\boldsymbol{B} = \begin{pmatrix} -3 & 6 \\ 2 & 0 \end{pmatrix}$ and $\boldsymbol{C} = \begin{pmatrix} -4 & -5 \\ 10 & 11 \end{pmatrix}$ write the following as one simplified 2×2 matrix

i) $\boldsymbol{A}+\boldsymbol{B}$, ii) $\boldsymbol{A}-\boldsymbol{C}$, iii) $\boldsymbol{A}+\boldsymbol{B}+\boldsymbol{C}$, iv) $\boldsymbol{A}-\boldsymbol{B}-\boldsymbol{C}$.

Notice how the matrices are represented by capital letters in **bold** type. This is due to their connection with vectors

i) For $\quad \boldsymbol{A}+\boldsymbol{B} = \begin{pmatrix} 1 & 12 \\ 9 & 7 \end{pmatrix} + \begin{pmatrix} -3 & 6 \\ 2 & 0 \end{pmatrix}$

we add $\quad \begin{pmatrix} \text{row 1, col 1} & \text{row 1, col 2} \\ \text{row 2, col 1} & \text{row 2, col 2} \end{pmatrix}$

to get $\quad \begin{pmatrix} 1+(-3) & 12+6 \\ 9+2 & 7+0 \end{pmatrix}$

$$= \begin{pmatrix} -2 & 18 \\ 11 & 7 \end{pmatrix}.$$

ii) For $\quad \boldsymbol{A}-\boldsymbol{C} = \begin{pmatrix} 1 & 12 \\ 9 & 7 \end{pmatrix} - \begin{pmatrix} -4 & -5 \\ 10 & 11 \end{pmatrix}$

we subtract $\quad \begin{pmatrix} \text{row 1, col 1} & \text{row 1, col 2} \\ \text{row 2, col 1} & \text{row 2, col 2} \end{pmatrix}$

to get $\quad \begin{pmatrix} 1-(-4) & 12-(-5) \\ 9-10 & 7-11 \end{pmatrix}$

> $1-(-4) = 1+4$ and $12-(-5) = 12+5$.

$$= \begin{pmatrix} 5 & 17 \\ -1 & -4 \end{pmatrix}.$$

iii) For $\boldsymbol{A}+\boldsymbol{B}+\boldsymbol{C}$, already we have part of this result for $\boldsymbol{A}+\boldsymbol{B}$.

$$\boldsymbol{A}+\boldsymbol{B}+\boldsymbol{C} = \begin{pmatrix} -2 & 18 \\ 11 & 7 \end{pmatrix} + \begin{pmatrix} -4 & -5 \\ 10 & 11 \end{pmatrix}$$

$$= \begin{pmatrix} -2+(-4) & 18+(-5) \\ 11+10 & 7+11 \end{pmatrix}$$

$$= \begin{pmatrix} -6 & 13 \\ 21 & 18 \end{pmatrix}.$$

Alternatively we could have completed the calculation in one move as

$$\boldsymbol{A}+\boldsymbol{B}+\boldsymbol{C} = \begin{pmatrix} 1 & 12 \\ 9 & 7 \end{pmatrix} + \begin{pmatrix} -3 & 6 \\ 2 & 0 \end{pmatrix} + \begin{pmatrix} -4 & -5 \\ 10 & 11 \end{pmatrix}$$

$$= \begin{pmatrix} 1+(-3)+(-4) & 12+6+(-5) \\ 9+2+10 & 7+0+11 \end{pmatrix}$$

$$= \begin{pmatrix} -6 & 13 \\ 21 & 18 \end{pmatrix}.$$

iv) We may write $\boldsymbol{A}-\boldsymbol{B}-\boldsymbol{C}$ as $\boldsymbol{A}-\boldsymbol{C}-\boldsymbol{B}$ or $(\boldsymbol{A}-\boldsymbol{C})-\boldsymbol{B}$ and use our result for $\boldsymbol{A}-\boldsymbol{C}$.

Then
$$(\boldsymbol{A}-\boldsymbol{C})-\boldsymbol{B} = \begin{pmatrix} 5 & 17 \\ -1 & -4 \end{pmatrix} - \begin{pmatrix} -3 & 6 \\ 2 & 0 \end{pmatrix}$$

$$= \begin{pmatrix} 5-(-3) & 17-6 \\ -1-2 & -4-0 \end{pmatrix}$$

$5-(-3)=5+3.$

$$= \begin{pmatrix} 8 & 11 \\ -3 & -4 \end{pmatrix}.$$

Alternatively we could have completed the calculation in one move as

$$\boldsymbol{A}-\boldsymbol{B}-\boldsymbol{C} = \begin{pmatrix} 1 & 12 \\ 9 & 7 \end{pmatrix} - \begin{pmatrix} -3 & 6 \\ 2 & 0 \end{pmatrix} - \begin{pmatrix} -4 & -5 \\ 10 & 11 \end{pmatrix}$$

$$= \begin{pmatrix} 1-(-3)-(-4) & 12-6-(-5) \\ 9-2-10 & 7-0-11 \end{pmatrix}$$

$$= \begin{pmatrix} 1+3+4 & 6+5 \\ 7-10 & 7-11 \end{pmatrix}$$

$$= \begin{pmatrix} 8 & 11 \\ -3 & -4 \end{pmatrix}.$$

There are two other simple arithmetic operations we can perform with matrices. Each one is the reverse process of the other. For the first one we can multiply a matrix by a scalar, i.e. **scalar multiplication**. For this process each element in the matrix is multiplied by the scalar.

Example 14.13

If $\boldsymbol{B} = \begin{pmatrix} -3 & 6 \\ 0 & 2 \end{pmatrix}$ then we can write down the matrix for $4\boldsymbol{B}$.

4 is the scalar multiplier.

$$\text{Then} \quad 4\boldsymbol{B} = 4\begin{pmatrix} -3 & 6 \\ 0 & 2 \end{pmatrix}$$

$$= \begin{pmatrix} 4\times -3 & 4\times 6 \\ 4\times 0 & 4\times 2 \end{pmatrix}$$

$$= \begin{pmatrix} -12 & 24 \\ 0 & 8 \end{pmatrix}.$$

The reverse process to scalar multiplication is **removing a common factor**. We must remove that factor from *all* the elements.

Examples 14.14

If possible remove a factor from the matrices

i) $\begin{pmatrix} -3 & 6 \\ 9 & 27 \end{pmatrix}$, ii) $\begin{pmatrix} -3 & 6 \\ 0 & 27 \end{pmatrix}$, iii) $\begin{pmatrix} -3 & 6 \\ 0 & 2 \end{pmatrix}$.

i) In $\begin{pmatrix} -3 & 6 \\ 9 & 27 \end{pmatrix}$ each term has a common factor of 3.

We may write the matrix as the alternative $3\begin{pmatrix} -1 & 2 \\ 3 & 9 \end{pmatrix}$.

ii) Again, in $\begin{pmatrix} -3 & 6 \\ 0 & 27 \end{pmatrix}$ we may remove a common factor of 3.

We remember that $0\times 3=0$.

The alternative matrix is $3\begin{pmatrix} -1 & 2 \\ 0 & 9 \end{pmatrix}$.

iii) In $\begin{pmatrix} -3 & 6 \\ 0 & 2 \end{pmatrix}$ we could remove a common factor of 3 from only some of the elements. The element 2 is not an obvious multiple of 3. We would not benefit from removing that factor of 3. If it really was necessary we might do so, though this would introduce a fraction as

$3\begin{pmatrix} -1 & 2 \\ 0 & \frac{2}{3} \end{pmatrix}$.

EXERCISE 14.4

Let $A=\begin{pmatrix}2 & 6\\5 & 9\end{pmatrix}$, $B=\begin{pmatrix}7 & -3\\0 & -5\end{pmatrix}$, $C=\begin{pmatrix}4 & -6\\2 & \frac{1}{2}\end{pmatrix}$, $D=\begin{pmatrix}-10 & 13\\8 & -1\end{pmatrix}$

and $Z=\begin{pmatrix}0 & 0\\0 & 0\end{pmatrix}$.

For the following matrix additions and subtractions give each answer as a single 2×2 matrix.

1 $A+B$

2 $A-B$

3 $A+B+C$

4 $A-B+D$

5 $A-B-D$

6 $B-A$

7 $B+A$

8 $2D-3C$

9 $2A+B+Z$

10 $\frac{1}{2}C+D-Z$

11 $3Z+B-D$

12 $A+2B-3C$

13 $C-\frac{1}{2}A$

14 $5D-3C+B$

15 $B-A+2C-D$

In Exercise 14.4 we have demonstrated 2 simple relationships. As might be expected the answers to Questions **1** and **7** are the same. The order of addition does not affect the answer. Generally we may write

$$A+B=B+A.$$

This says that matrix addition is commutative or matrix addition obeys the commutative law.

Also answers to Questions **2** and **6** differ, but only by a minus sign. The order of subtraction is important.

Generally we may write

$$A-B\neq B-A$$

$$\text{or } A-B=-(B-A).$$

Matrix subtraction is not commutative, i.e. matrix subtraction does not obey the commutative law.

Matrix multiplication

The order in which we multiply matrices is important. Generally matrix multiplication is *not* commutative, though there are a few exceptions, i.e.

$$AB\neq BA.$$

The size (i.e. order) of the matrices is also vital. Suppose we have 2 matrices, A and B, and wish to multiply them together as AB. This is possible only if the number of columns of $\boldsymbol{A}$ is the same as the number of rows of $\boldsymbol{B}$.

If $\boldsymbol{A}$ is an $m \times n$ matrix then we need $\boldsymbol{B}$ to be an $n \times p$ matrix to calculate $\boldsymbol{AB}$.

Notice the diagonal pattern.

The values for m and p do not matter. They may be either the same or different. If they are different we would be *unable* to calculate $\boldsymbol{BA}$. If $m \neq p$ then $\boldsymbol{B}$ being an $n \times p$ matrix and
$\boldsymbol{A}$ being an $m \times n$ matrix show the wrong diagonal pattern.

In this next set of examples we look at which matrices we may multiply together.

Examples 14.15

$\boldsymbol{A} = \begin{pmatrix} 2 & 3 & 1 \\ 0 & -5 & 6 \end{pmatrix}$ is a 2×3 matrix,

$\boldsymbol{B} = \begin{pmatrix} 4 & 0 & 1 \\ 0 & 0 & -4 \\ 2 & 0 & 7.5 \\ 1 & -7 & 6 \end{pmatrix}$ is a 4×3 matrix,

$\boldsymbol{C} = \begin{pmatrix} 1 & 2 \\ 3 & 7 \end{pmatrix}$ is a 2×2 matrix and

$\boldsymbol{D} = \begin{pmatrix} 5 \\ -2 \\ 0 \end{pmatrix}$ is a 3×1 matrix.

Remember the column and row rule. The number of columns of the first matrix must be the same as the number of rows of the second matrix. Now we can decide which matrices we might be able to multiply together.

Matrices $\boldsymbol{A}$ and $\boldsymbol{B}$ each have 3 columns whilst matrix $\boldsymbol{D}$ has 3 rows. We may multiply $\boldsymbol{AD}$ and $\boldsymbol{BD}$.

Matrix $\boldsymbol{C}$ has 2 columns whilst matrix $\boldsymbol{A}$ has 2 rows. We may multiply $\boldsymbol{CA}$.

You might like to check for yourself whether there are any other possible multiplications.

EXERCISE 14.5

Decide which, if any, of the following matrices may be multiplied together. In each case be careful to write down your answer in the correct order.

$$\boldsymbol{A} = \begin{pmatrix} 2 & 1 \\ 3 & 6 \end{pmatrix}, \boldsymbol{B} = \begin{pmatrix} 1 & 2 & 3 \\ 4 & -5 & 6 \end{pmatrix}, \boldsymbol{C} = \begin{pmatrix} 5 \\ 9 \end{pmatrix} \text{ and } \boldsymbol{D} = \begin{pmatrix} 1 & 9 \\ 2 & -17 \\ 4 & 6 \end{pmatrix}.$$

For the rest of this chapter we are going to use 2×2 and 2×1 matrices. Then we will use them to solve pairs of simultaneous linear equations.

The general idea when multiplying matrices is to multiply particular elements and add those results. We multiply row elements from the first matrix with column elements from the second matrix. Let us have a look at some general cases before the numerical examples.

Suppose we have $\boldsymbol{A} = \begin{pmatrix} a_1 & b_1 \\ d_1 & e_1 \end{pmatrix}$ and $\boldsymbol{B} = \begin{pmatrix} a_2 \\ d_2 \end{pmatrix}$.

$\boldsymbol{A}$ is a 2×2 matrix and

number of columns is 2.

$\boldsymbol{B}$ is a 2×1 matrix.

number of rows is 2.

The column and row rule allows us to multiply $\boldsymbol{AB}$. The matrix $\boldsymbol{AB}$ will be of order 2×1. These are the number of rows from $\boldsymbol{A}$ and the number of columns from $\boldsymbol{B}$. We will look at $\boldsymbol{AB}$ in stages:

First row $$\begin{pmatrix} a_1 & b_1 \\ & \end{pmatrix}\begin{pmatrix} a_2 \\ d_2 \end{pmatrix} = \begin{pmatrix} a_1a_2 + b_1d_2 \\ \end{pmatrix}$$

Second row $$\begin{pmatrix} & \\ d_1 & e_1 \end{pmatrix}\begin{pmatrix} a_2 \\ d_2 \end{pmatrix} = \begin{pmatrix} \\ d_1a_2 + e_1d_2 \end{pmatrix}.$$

We link together these stages to give a complete matrix solution of

$$\boldsymbol{AB} = \begin{pmatrix} a_1 & b_1 \\ d_1 & e_1 \end{pmatrix}\begin{pmatrix} a_2 \\ d_2 \end{pmatrix} = \begin{pmatrix} a_1a_2 + b_1d_2 \\ d_1a_2 + e_1d_2 \end{pmatrix}.$$

Look closely at the final matrix. You should notice the pattern of letters and subscripts. It is consistent with the original matrices.

Examples 14.16

Multiply together the matrices

i) $\begin{pmatrix} 1 & 2 \\ 3 & 4 \end{pmatrix}\begin{pmatrix} 5 \\ 6 \end{pmatrix}$ and ii) $\begin{pmatrix} -3 & 14 \\ -7 & -2 \end{pmatrix}\begin{pmatrix} 0 \\ 1 \end{pmatrix}$.

We use the rules involving multiplication and addition that we have just looked at generally.

i) $$\begin{pmatrix} 1 & 2 \\ 3 & 4 \end{pmatrix}\begin{pmatrix} 5 \\ 6 \end{pmatrix} = \begin{pmatrix} 1 \times 5 + 2 \times 6 \\ 3 \times 5 + 4 \times 6 \end{pmatrix}$$
$$= \begin{pmatrix} 5 + 12 \\ 15 + 24 \end{pmatrix}$$
$$= \begin{pmatrix} 17 \\ 39 \end{pmatrix}.$$

ii) $$\begin{pmatrix} -3 & 14 \\ -7 & -2 \end{pmatrix}\begin{pmatrix} 0 \\ 1 \end{pmatrix} = \begin{pmatrix} -3 \times 0 + 14 \times 1 \\ -7 \times 0 + -2 \times 1 \end{pmatrix}$$

$$= \begin{pmatrix} 0+14 \\ 0-2 \end{pmatrix}$$

$$= \begin{pmatrix} 14 \\ -2 \end{pmatrix}.$$

You need to look back at these examples. Notice how the original column and row rule allows us to multiply these matrices. Look at the pattern of numbers in the working of the calculation. Finally check that the first matrix had 2 rows and the second matrix had 1 column. This means our answer must have, and does have, 2 rows and 1 column.

Let us look at 2 more general matrices,

$$\boldsymbol{A} = \begin{pmatrix} a_1 & b_1 \\ d_1 & e_1 \end{pmatrix} \text{ and } \boldsymbol{C} = \begin{pmatrix} a_2 & b_2 \\ d_2 & e_2 \end{pmatrix}.$$

We can check out the column and row rule to allow us to attempt the multiplication. Also we know that $\boldsymbol{AC}$ will be a 2×2 matrix. In stages we have:

First row, first column

$$\begin{pmatrix} a_1 & b_1 \\ & \end{pmatrix} \begin{pmatrix} a_2 & \\ d_2 & \end{pmatrix} = \begin{pmatrix} a_1a_2 + b_1d_2 & \\ & \end{pmatrix}$$

First row, second column

$$\begin{pmatrix} a_1 & b_1 \\ & \end{pmatrix} \begin{pmatrix} & b_2 \\ & e_2 \end{pmatrix} = \begin{pmatrix} & a_1b_2 + b_1e_2 \\ & \end{pmatrix}$$

Second row, first column

$$\begin{pmatrix} & \\ d_1 & e_1 \end{pmatrix} \begin{pmatrix} a_2 & \\ d_2 & \end{pmatrix} = \begin{pmatrix} & \\ d_1a_2 + e_1d_2 & \end{pmatrix}$$

Second row, second column

$$\begin{pmatrix} & \\ d_1 & e_1 \end{pmatrix} \begin{pmatrix} & b_2 \\ & e_2 \end{pmatrix} = \begin{pmatrix} & \\ & d_1b_2 + e_1e_2 \end{pmatrix}.$$

Again we can bring together these stages for one complete solution. In the next set of examples we work in stages and then gather together our complete solutions.

Examples 14.17

Multiply together the matrices

i) $\begin{pmatrix} 1 & 2 \\ 3 & 4 \end{pmatrix} \begin{pmatrix} -3 & 14 \\ -7 & -2 \end{pmatrix}$, ii) $\begin{pmatrix} 0 & 9 \\ 13 & -6 \end{pmatrix} \begin{pmatrix} 5 & 7 \\ 2 & 4 \end{pmatrix}$ and

iii) $\begin{pmatrix} -3 & 14 \\ -7 & -2 \end{pmatrix} \begin{pmatrix} 1 & 2 \\ 3 & 4 \end{pmatrix}$.

We look at the first and second examples in stages, then bring together the complete solutions.

i) In stages we have:

First row, first column

$$\begin{pmatrix} 1 & 2 \\ & \end{pmatrix}\begin{pmatrix} -3 & \\ -7 & \end{pmatrix} = \begin{pmatrix} 1\times -3+2\times -7 & \\ & \end{pmatrix}$$

First row, second column

$$\begin{pmatrix} 1 & 2 \\ & \end{pmatrix}\begin{pmatrix} & 14 \\ & -2 \end{pmatrix} = \begin{pmatrix} & 1\times 14+2\times -2 \\ & \end{pmatrix}$$

Second row, first column

$$\begin{pmatrix} & \\ 3 & 4 \end{pmatrix}\begin{pmatrix} -3 & \\ -7 & \end{pmatrix} = \begin{pmatrix} & \\ 3\times -3+4\times -7 & \end{pmatrix}$$

Second row, second column

$$\begin{pmatrix} & \\ 3 & 4 \end{pmatrix}\begin{pmatrix} & 14 \\ & -2 \end{pmatrix} = \begin{pmatrix} & \\ & 3\times 14+4\times -2 \end{pmatrix}$$

Bringing together all these stages gives us

$$\begin{pmatrix} 1 & 2 \\ 3 & 4 \end{pmatrix}\begin{pmatrix} -3 & 14 \\ -7 & -2 \end{pmatrix} = \begin{pmatrix} -3-14 & 14-4 \\ -9-28 & 42-8 \end{pmatrix}$$

$$= \begin{pmatrix} -17 & 10 \\ -37 & 34 \end{pmatrix}$$

ii) In stages we have:

First row, first column

$$\begin{pmatrix} 0 & 9 \\ & \end{pmatrix}\begin{pmatrix} 5 & \\ 2 & \end{pmatrix} = \begin{pmatrix} 0\times 5+9\times 2 & \\ & \end{pmatrix}$$

First row, second column

$$\begin{pmatrix} 0 & 9 \\ & \end{pmatrix}\begin{pmatrix} & 7 \\ & 4 \end{pmatrix} = \begin{pmatrix} & 0\times 7+9\times 4 \\ & \end{pmatrix}$$

Second row, first column

$$\begin{pmatrix} & \\ 13 & -6 \end{pmatrix}\begin{pmatrix} 5 & \\ 2 & \end{pmatrix} = \begin{pmatrix} & \\ 13\times 5-6\times 2 & \end{pmatrix}$$

Second row, second column

$$\begin{pmatrix} & \\ 13 & -6 \end{pmatrix}\begin{pmatrix} & 7 \\ & 4 \end{pmatrix} = \begin{pmatrix} & \\ & 13\times 7-6\times 4 \end{pmatrix}$$

Bringing together all these stages gives us

$$\begin{pmatrix} 0 & 9 \\ 13 & -6 \end{pmatrix}\begin{pmatrix} 5 & 7 \\ 2 & 4 \end{pmatrix} = \begin{pmatrix} 0+18 & 0+36 \\ 65-12 & 91-24 \end{pmatrix}$$

$$= \begin{pmatrix} 18 & 36 \\ 53 & 67 \end{pmatrix}.$$

iii) For this third example we will bring together all the stages from the beginning.

$$\begin{pmatrix} -3 & 14 \\ -7 & -2 \end{pmatrix}\begin{pmatrix} 1 & 2 \\ 3 & 4 \end{pmatrix} = \begin{pmatrix} -3\times 1+14\times 3 & -3\times 2+14\times 4 \\ -7\times 1-2\times 3 & -7\times 2-2\times 4 \end{pmatrix}$$

$$= \begin{pmatrix} -3+42 & -6+56 \\ -7-6 & -14-8 \end{pmatrix}$$

$$= \begin{pmatrix} 39 & 50 \\ -13 & -22 \end{pmatrix}.$$

Now you should look back at the first and third examples in this set. Notice how the original matrices have been written. This example demonstrates the commutative law failing, $\boldsymbol{AB} \neq \boldsymbol{BA}$, we mentioned earlier.

Most examples of matrix multiplication do *not* obey the commutative law. However there is one exception. It involves the **unit** matrix.

Examples 14.18

Multiply together the matrices

i) $\begin{pmatrix} 5 & 7 \\ 2 & 4 \end{pmatrix}\begin{pmatrix} 1 & 0 \\ 0 & 1 \end{pmatrix}$, ii) $\begin{pmatrix} 1 & 0 \\ 0 & 1 \end{pmatrix}\begin{pmatrix} 5 & 7 \\ 2 & 4 \end{pmatrix}$.

i) $$\begin{pmatrix} 5 & 7 \\ 2 & 4 \end{pmatrix}\begin{pmatrix} 1 & 0 \\ 0 & 1 \end{pmatrix} = \begin{pmatrix} 5\times 1+7\times 0 & 5\times 0+7\times 1 \\ 2\times 1+4\times 0 & 2\times 0+4\times 1 \end{pmatrix}$$

$$= \begin{pmatrix} 5+0 & 0+7 \\ 2+0 & 0+4 \end{pmatrix}$$

$$= \begin{pmatrix} 5 & 7 \\ 2 & 4 \end{pmatrix}.$$

ii) $$\begin{pmatrix} 1 & 0 \\ 0 & 1 \end{pmatrix}\begin{pmatrix} 5 & 7 \\ 2 & 4 \end{pmatrix} = \begin{pmatrix} 1\times 5+0\times 2 & 1\times 7+0\times 4 \\ 0\times 5+1\times 2 & 0\times 7+1\times 4 \end{pmatrix}$$

$$= \begin{pmatrix} 5+0 & 7+0 \\ 0+2 & 0+4 \end{pmatrix}$$

$$= \begin{pmatrix} 5 & 7 \\ 2 & 4 \end{pmatrix}.$$

The pattern of matrix multiplication and the answers applies to any order of matrices. Suppose $\boldsymbol{I}$ is the unit matrix and $\boldsymbol{A}$ is any other matrix of the correct order. Then $\boldsymbol{AI} = \boldsymbol{IA} = \boldsymbol{A}$.

EXERCISE 14.6

Multiply together the following matrices.

1 $\begin{pmatrix} 2 & -1 \\ 4 & 2 \end{pmatrix}\begin{pmatrix} 1 \\ 2 \end{pmatrix}$

2 $\begin{pmatrix} 3 & -2 \\ 5 & 0 \end{pmatrix}\begin{pmatrix} 1 \\ 2 \end{pmatrix}$

3 $\begin{pmatrix} 0 & -5 \\ 6 & -1 \end{pmatrix}\begin{pmatrix} 0 \\ -3 \end{pmatrix}$

4 $\begin{pmatrix} 2 & 5 \\ 6 & 8 \end{pmatrix}\begin{pmatrix} 4 \\ \frac{1}{2} \end{pmatrix}$

5 $\begin{pmatrix} 5 & 2 \\ 1 & 6 \end{pmatrix}\begin{pmatrix} -3 & 2 \\ 4 & 1 \end{pmatrix}$

6 $\begin{pmatrix} 2 & -1 \\ 0 & -2 \end{pmatrix}\begin{pmatrix} 4 & 5 \\ 3 & 1 \end{pmatrix}$

7 $\begin{pmatrix} 1 & 2 \\ -1 & -2 \end{pmatrix}\begin{pmatrix} 11 & 5 \\ 4 & -2 \end{pmatrix}$

8 $\begin{pmatrix} 3 & -2 \\ 5 & 0 \end{pmatrix}\begin{pmatrix} 0 & 2 \\ 3 & -4 \end{pmatrix}$

9 $\begin{pmatrix} 3 & 10 \\ 6 & 6 \end{pmatrix}\begin{pmatrix} 3 & -2 \\ 4 & 1 \end{pmatrix}$

10 $\begin{pmatrix} \frac{1}{2} & 3 \\ 1 & 2 \end{pmatrix}\begin{pmatrix} 7 & 4 \\ 6 & -\frac{1}{2} \end{pmatrix}$

Determinant of a matrix

In the Introduction we mentioned that it is possible to find the determinant of a matrix. Remember a determinant has the same number of rows as columns. This means we can only find the determinant of a square matrix. Simply, if $\boldsymbol{A} = \begin{pmatrix} a_1 & b_1 \\ d_1 & e_1 \end{pmatrix}$ then $\det \boldsymbol{A} = \begin{vmatrix} a_1 & b_1 \\ d_1 & e_1 \end{vmatrix}$.

The usual calculation rules apply to give $\quad \det \boldsymbol{A} = a_1e_1 - b_1d_1$.

Examples 14.19

Calculate the determinants of the matrices

i) $\begin{pmatrix} 5 & 7 \\ 2 & 4 \end{pmatrix}$, ii) $\begin{pmatrix} 0 & 2 \\ 3 & -4 \end{pmatrix}$, iii) (10).

i) $$\Delta = \begin{vmatrix} 5 & 7 \\ 2 & 4 \end{vmatrix} = 5 \times 4 - 7 \times 2$$
$$= 20 - 14$$
$$= 6.$$

ii) $\Delta = \begin{vmatrix} 0 & 2 \\ 3 & -4 \end{vmatrix} = 0 \times -4 - 2 \times 3$

$= 0 - 6$

$= -6.$

iii) $\Delta = |10| = 10$ is a very simple case.

It is possible that the determinant of a matrix may have a value of 0. We need to note this case. It refers to a singular matrix.

A singular matrix has a determinant of zero value.

Examples 14.20

i) If $A = \begin{pmatrix} 4 & 4 \\ 1 & 1 \end{pmatrix}$ then $\det A = \begin{vmatrix} 4 & 4 \\ 1 & 1 \end{vmatrix} = 4 \times 1 - 4 \times 1 = 0.$

ii) If $B = \begin{vmatrix} 4 & 1 \\ 8 & 2 \end{vmatrix}$ then $\det \mathrm{B} = \begin{vmatrix} 4 & 1 \\ 8 & 2 \end{vmatrix} = 4 \times 2 - 1 \times 8 = 0.$

We saw both these determinants in Examples 14.8. They came from "pairs" of simultaneous equations that had no solution. We will meet this situation again when we look at the matrix solution of simultaneous equations.

Transpose of a matrix

When we interchange the rows and columns of a matrix we get the **transpose** of the original matrix.

If $A = \begin{pmatrix} a_1 & b_1 \\ d_1 & e_1 \end{pmatrix}$ then we write the transpose of A as $A^T = \begin{pmatrix} a_1 & d_1 \\ b_1 & e_1 \end{pmatrix}$.

You can transpose any size of matrix, it does *not* have to be square. We concentrate on square matrices because we are working towards finding inverse matrices. In this next set of examples we look at transposing matrices and calculating determinants.

Examples 14.21

Write down the transpose of the following matrices

i) $\begin{pmatrix} 5 & 7 \\ 2 & 4 \end{pmatrix}$ and ii) $\begin{pmatrix} 0 & 2 \\ 3 & -4 \end{pmatrix}$.

Also calculate the determinants of the transposed matrices.

i) If $A = \begin{pmatrix} 5 & 7 \\ 2 & 4 \end{pmatrix}$ then $A^T = \begin{pmatrix} 5 & 2 \\ 7 & 4 \end{pmatrix}$.

Also $\det A^T = \begin{vmatrix} 5 & 2 \\ 7 & 4 \end{vmatrix} = 5 \times 4 - 2 \times 7$

$= 6.$

ii) If $B = \begin{pmatrix} 0 & 2 \\ 3 & -4 \end{pmatrix}$ then $B^T = \begin{pmatrix} 0 & 3 \\ 2 & -4 \end{pmatrix}$.

Also $\det B^T = \begin{vmatrix} 0 & 3 \\ 2 & -4 \end{vmatrix} = 0 \times -4 - 3 \times 2$

$= -6.$

You might like to compare these determinant answers with those of Examples 14.19. You will see they are the same. This is consistent with one of our earlier rules for determinants: **"Writing the rows as columns and the columns as rows leaves the value of the determinant unchanged."**

EXERCISE 14.7

For the following matrices
i) write down their transposes, and
ii) calculate the determinants of the transposed matrices.

1 $\begin{pmatrix} 2 & -1 \\ 4 & 2 \end{pmatrix}$

2 $\begin{pmatrix} 2 & 5 \\ 6 & 8 \end{pmatrix}$

3 $\begin{pmatrix} 3 & 10 \\ 6 & 6 \end{pmatrix}$

4 $\begin{pmatrix} \frac{1}{2} & 3 \\ 1 & 2 \end{pmatrix}$

5 $\begin{pmatrix} 5 & 2 \\ 1 & 6 \end{pmatrix}$

6 $\begin{pmatrix} 11 & 5 \\ 4 & -2 \end{pmatrix}$

7 $\begin{pmatrix} 1 & 0 \\ 0 & 1 \end{pmatrix}$

8 $\begin{pmatrix} 3 & -2 \\ 4 & 1 \end{pmatrix}$

9 $\begin{pmatrix} 1 & 2 \\ -1 & -2 \end{pmatrix}$

10 $\begin{pmatrix} 1 & 4 \\ 2 & -\frac{1}{2} \end{pmatrix}$

Minors and cofactors of a matrix

We are working towards finding the inverse of a matrix. Again we concentrate on the 2×2 matrix.

Firstly let us introduce the double suffix notation for the elements of the matrix, e.g. a_{14}. a_{14} means the element in row 1, column 4.

For our general 2×2 matrix suppose $\boldsymbol{A} = \begin{pmatrix} a_{11} & a_{12} \\ a_{21} & a_{22} \end{pmatrix}$.

Let us find the minor of a_{11}. We strike out the row and column containing a_{11}, i.e. strike out row 1 and column 1. Whatever remains is the minor, i.e. a_{22}.

We can repeat this process for each element in turn. For the minor of a_{21} we strike out row 2 and column 1. What remains is a_{12}. For all our elements:

the minor of a_{11} is a_{22},
the minor of a_{12} is a_{21},
the minor of a_{21} is a_{12}
and the minor of a_{22} is a_{11}.

The cofactor involves the minor together with a sign adjustment. The adjustment uses the addition of the subscripts. -1 is raised to the power of that addition.

The cofactor of a_{11} is $(-1)^{1+1}a_{22} = (-1)^2 a_{22} = a_{22}$,
the cofactor of a_{12} is $(-1)^{1+2}a_{21} = (-1)^3 a_{21} = -a_{21}$,
the cofactor of a_{21} is $(-1)^{2+1}a_{12} = (-1)^3 a_{12} = -a_{12}$
and the cofactor of a_{22} is $(-1)^{2+2}a_{11} = (-1)^4 a_{11} = a_{11}$.

Generally the cofactor of a_{ij} uses $(-1)^{i+j}$.

Cofactor of $a_{ij} = (-1)^{i+j} \times$ minor of a_{ij}.

Inverse of a matrix

Only a square matrix has an inverse. If the original matrix is $\boldsymbol{A}$ then the inverse matrix is $\boldsymbol{A}^{-1}$. $\boldsymbol{A}^{-1}$ is just a symbol; do not write it in any other form. If we multiply together the original and inverse matrices in any order we get the unit matrix, I, i.e.

$$\boldsymbol{A}\boldsymbol{A}^{-1} = \boldsymbol{I} = \boldsymbol{A}^{-1}\boldsymbol{A}.$$

This is a rare exception to the order of matrix multiplication being important.

Only a non-singular matrix has an inverse, i.e. the determinant of the matrix is non-zero. Hence, first find the value of the determinant when attempting to find the inverse of a matrix.

Before we attempt an example we can go through the stages in finding an inverse matrix.

Suppose $\boldsymbol{A} = \begin{pmatrix} a_{11} & a_{12} \\ a_{21} & a_{22} \end{pmatrix}$

Calculate $\det \boldsymbol{A}$. If $\det \boldsymbol{A} \neq 0$ then we can proceed.

For each element find each minor.

Now find each cofactor.

Write down the matrix of cofactors, say $\boldsymbol{B}$.

Transpose $\boldsymbol{B}$ to get $\boldsymbol{B}^T$. $\boldsymbol{B}^T$ is the **adjoint matrix** of $\boldsymbol{A}$.

The shortened form of adjoint matrix is **adj**.

$\boldsymbol{B}^T = \text{adj}\,\boldsymbol{A}$.

The inverse matrix $\boldsymbol{A}^{-1}$ is $\boldsymbol{A}^{-1} = \frac{1}{\det \boldsymbol{A}} \times \text{adj}\,\boldsymbol{A}$.

For **2×2 matrices** only the inverse of $\boldsymbol{A} = \begin{pmatrix} a_{11} & a_{12} \\ a_{21} & a_{22} \end{pmatrix}$ is given by $\boldsymbol{A}^{-1} = \frac{1}{\det \boldsymbol{A}} \begin{pmatrix} a_{22} & -a_{12} \\ -a_{21} & a_{11} \end{pmatrix}$.

Examples 14.22

Find the inverses of the matrices

i) $\begin{pmatrix} 2 & 5 \\ 3 & 8 \end{pmatrix}$ and ii) $\begin{pmatrix} 4 & -7 \\ 0 & 1 \end{pmatrix}$.

i) Let $\boldsymbol{A} = \begin{pmatrix} 2 & 5 \\ 3 & 8 \end{pmatrix}$.

We calculate the determinant of $\boldsymbol{A}$,

i.e. $\det \boldsymbol{A} = \begin{vmatrix} 2 & 5 \\ 3 & 8 \end{vmatrix} = 2 \times 8 - 5 \times 3 = 1.$

Now we find the minors and cofactors.

Element a_{11} is 2, minor is 8, cofactor is $(-1)^{1+1}8 = 8$.

Element a_{12} is 5, minor is 3, cofactor is $(-1)^{1+2}3 = -3$.

Element a_{21} is 3, minor is 5, cofactor is $(-1)^{2+1}5 = -5$.

Element a_{22} is 8, minor is 2, cofactor is $(-1)^{2+2}2 = 2$.

The matrix of cofactors is $\begin{pmatrix} 8 & -3 \\ -5 & 2 \end{pmatrix}$.

Transposing the matrix of cofactors we get

$$\text{adj}\,\boldsymbol{A} \text{ is } \begin{pmatrix} 8 & -5 \\ -3 & 2 \end{pmatrix}.$$

Hence

$$\boldsymbol{A}^{-1} = \frac{1}{1} \times \begin{pmatrix} 8 & -5 \\ -3 & 2 \end{pmatrix}$$

$$= \begin{pmatrix} 8 & -5 \\ -3 & 2 \end{pmatrix}.$$

We ought to test this answer. Remember that $\boldsymbol{A}\boldsymbol{A}^{-1} = \boldsymbol{I}$

Now $\begin{pmatrix} 2 & 5 \\ 3 & 8 \end{pmatrix} \begin{pmatrix} 8 & -5 \\ -3 & 2 \end{pmatrix} = \begin{pmatrix} 16-15 & -10+10 \\ 24-24 & -15+16 \end{pmatrix} = \begin{pmatrix} 1 & 0 \\ 0 & 1 \end{pmatrix}$.

confirms our answer is correct.

ii) Let $A = \begin{pmatrix} 4 & -7 \\ 0 & 1 \end{pmatrix}$.

We calculate the determinant of A,

i.e. $\det A = \begin{vmatrix} 4 & -7 \\ 0 & 1 \end{vmatrix} = 4 \times 1 - -7 \times 0 = 4.$

Now we find the minors and cofactors.

Element a_{11} is 4, minor is 1, cofactor is $(-1)^{1+1}1 = 1$.
Element a_{12} is -7, minor is 0, cofactor is $(-1)^{1+2}0 = 0$.
Element a_{21} is 0, minor is -7, cofactor is $(-1)^{2+1}(-7) = 7$.
Element a_{22} is 1, minor is 4, cofactor is $(-1)^{2+2}4 = 4$.

The matrix of cofactors is $\begin{pmatrix} 1 & 0 \\ 7 & 4 \end{pmatrix}$.

Transposing the matrix of cofactors we get

$$\text{adj}\, A \text{ is } \begin{pmatrix} 1 & 7 \\ 0 & 4 \end{pmatrix}.$$

Hence
$$A^{-1} = \frac{1}{4} \times \begin{pmatrix} 1 & 7 \\ 0 & 4 \end{pmatrix}$$
$$= \begin{pmatrix} 0.25 & 1.75 \\ 0 & 1.00 \end{pmatrix}.$$

Again we can test our answer, this time with $A^{-1}A = I$.

Now
$$\begin{pmatrix} 0.25 & 1.75 \\ 0 & 1.00 \end{pmatrix}\begin{pmatrix} 4 & -7 \\ 0 & 1 \end{pmatrix}$$
$$= \begin{pmatrix} 0.25 \times 4 + 1.75 \times 0 & 0.25 \times -7 + 1.75 \times 1 \\ 0 \times 4 + 1 \times 0 & 0 \times -7 + 1 \times 1 \end{pmatrix}$$
$$= \begin{pmatrix} 1 & 0 \\ 0 & 1 \end{pmatrix} \quad \text{confirms our answer is correct.}$$

EXERCISE 14.8

Find the inverse of each matrix.

1 $\begin{pmatrix} 4 & 3 \\ 5 & 4 \end{pmatrix}$

2 $\begin{pmatrix} 6 & 2 \\ 17 & 6 \end{pmatrix}$

3 $\begin{pmatrix} 4 & 0 \\ -3 & 1 \end{pmatrix}$

4 $\begin{pmatrix} 3 & 2 \\ 4 & 1 \end{pmatrix}$

5 $\begin{pmatrix} 4 & 2 \\ 5 & 3 \end{pmatrix}$

6 $\begin{pmatrix} 10 & 4 \\ 5 & 3 \end{pmatrix}$

7 $\begin{pmatrix} 1 & 0 \\ 0 & 1 \end{pmatrix}$

8 $\begin{pmatrix} 8 & 3 \\ 1 & \frac{1}{2} \end{pmatrix}$

9 $\begin{pmatrix} 5 & 3 \\ 5 & -5 \end{pmatrix}$

10 $\begin{pmatrix} 2 & 1 \\ 4 & -\frac{1}{2} \end{pmatrix}$

Simultaneous equations

Already in this chapter we have looked at several pairs of simultaneous linear equations. An earlier general pair highlighted $\boldsymbol{x}$ and $\boldsymbol{y}$ in bold type. Also in Example 14.6 we had the pair

$$5x - 3y = 21$$

and $$7x + 8y = 5.$$

On the left of these equations we may separate the coefficients from the variables x and y. We write this in matrix form as

$$\begin{pmatrix} 5 & -3 \\ 7 & 8 \end{pmatrix}\begin{pmatrix} x \\ y \end{pmatrix} = \begin{pmatrix} 21 \\ 5 \end{pmatrix}.$$

You can check for yourself that the left-hand side multiplies out to give $5x - 3y$ and $7x + 8y$.

Generally we may write $\boldsymbol{AX} = \boldsymbol{B}$, where $\boldsymbol{A} = \begin{pmatrix} 5 & -3 \\ 7 & 8 \end{pmatrix}$, $\boldsymbol{X} = \begin{pmatrix} x \\ y \end{pmatrix}$ and $\boldsymbol{B} = \begin{pmatrix} 21 \\ 5 \end{pmatrix}$.

Now $\boldsymbol{AX} = \boldsymbol{B}$ is a general equation. Whatever we do to **one** side we must do to the **other** side, i.e. our operation must be consistent throughout the equation. We will multiply both sides by the inverse matrix, $\boldsymbol{A}^{-1}$,

i.e. $\boldsymbol{A}^{-1}\boldsymbol{AX} = \boldsymbol{A}^{-1}\boldsymbol{B}$

i.e. $\boldsymbol{IX} = \boldsymbol{A}^{-1}\boldsymbol{B}$

i.e. $\boldsymbol{X} = \boldsymbol{A}^{-1}\boldsymbol{B}$.

> $A^{-1}A = I$.

> $IX = X$.

Let us look at this in practice.

Example 14.23

Solve the pair of simultaneous equations $5x - 3y = 21$
and $7x + 8y = 5$.

In matrix form these equations are $\begin{pmatrix} 5 & -3 \\ 7 & 8 \end{pmatrix}\begin{pmatrix} x \\ y \end{pmatrix} = \begin{pmatrix} 21 \\ 5 \end{pmatrix}$.

Our general solution is $\boldsymbol{X} = \boldsymbol{A}^{-1}\boldsymbol{B}$. This means we must find the inverse matrix of $\begin{pmatrix} 5 & -3 \\ 7 & 8 \end{pmatrix}$.

Firstly we calculate the determinant of $\boldsymbol{A}$,

i.e. $\det \boldsymbol{A} = \begin{vmatrix} 5 & -3 \\ 7 & 8 \end{vmatrix} = 5 \times 8 - -3 \times 7 = 61.$

Now we find the minors and cofactors.

Element a_{11} is 5, minor is 8, cofactor is $(-1)^{1+1}8 = 8$.
Element a_{12} is -3, minor is 7, cofactor is $(-1)^{1+2}7 = -7$.
Element a_{21} is 7, minor is -3, cofactor is $(-1)^{2+1}(-3) = 3$.
Element a_{22} is 8, minor is 5, cofactor is $(-1)^{2+2}5 = 5$.

The matrix of cofactors is $\begin{pmatrix} 8 & -7 \\ 3 & 5 \end{pmatrix}$.

$$\operatorname{adj} \boldsymbol{A} \text{ is } \begin{pmatrix} 8 & 3 \\ -7 & 5 \end{pmatrix}.$$

Hence $$\boldsymbol{A}^{-1} = \frac{1}{61} \times \begin{pmatrix} 8 & 3 \\ -7 & 5 \end{pmatrix}.$$

Using $$\boldsymbol{X} = \boldsymbol{A}^{-1}\boldsymbol{B}$$

we have
$$\begin{pmatrix} x \\ y \end{pmatrix} = \frac{1}{61} \times \begin{pmatrix} 8 & 3 \\ -7 & 5 \end{pmatrix} \begin{pmatrix} 21 \\ 5 \end{pmatrix}$$
$$= \frac{1}{61} \begin{pmatrix} 168 + 15 \\ -147 + 25 \end{pmatrix}$$
$$= \begin{pmatrix} \dfrac{183}{61} \\ \dfrac{-122}{61} \end{pmatrix}$$
$$\begin{pmatrix} x \\ y \end{pmatrix} = \begin{pmatrix} 3 \\ -2 \end{pmatrix} \qquad \text{or } x = 3,\ y = -2.$$

As usual we can check these solutions by substituting into either of the original equations.

ASSIGNMENT

We return to our pair of simultaneous linear equations from the window frames,

$$80x + 65y = 540$$
and $$50x + 30y = 295.$$

Remember we noticed common factors of 5 in each equation that may be cancelled to give

$$16x + 13y = 108$$
and $$10x + 6y = 59.$$

In matrix form we have

$$\begin{pmatrix} 16 & 13 \\ 10 & 6 \end{pmatrix}\begin{pmatrix} x \\ y \end{pmatrix} = \begin{pmatrix} 108 \\ 59 \end{pmatrix}.$$

We aim to find the inverse matrix of $\begin{pmatrix} 16 & 13 \\ 10 & 6 \end{pmatrix}$.

Let this be $\boldsymbol{A}$.
We calculate the determinant of $\boldsymbol{A}$,

$$\text{i.e.} \quad \det \boldsymbol{A} = \begin{vmatrix} 16 & 13 \\ 10 & 6 \end{vmatrix} = 16 \times 6 - 13 \times 10$$

$$= 96 - 130$$

$$= -34.$$

Now we find the minors and cofactors.
Element a_{11} is 16, minor is 6, cofactor is $(-1)^{1+1}6 = 6$.
Element a_{12} is 13, minor is 10, cofactor is $(-1)^{1+2}10 = -10$.
Element a_{21} is 10, minor is 13, cofactor is $(-1)^{2+1}13 = -13$.
Element a_{22} is 6, minor is 16, cofactor is $(-1)^{2+2}16 = 16$.

The matrix of cofactors is $\begin{pmatrix} 6 & -10 \\ -13 & 16 \end{pmatrix}$.

$$\text{adj}\,\boldsymbol{A} \text{ is } \begin{pmatrix} 6 & -13 \\ -10 & 16 \end{pmatrix}.$$

$$\text{Hence} \quad \boldsymbol{A}^{-1} = -\frac{1}{34} \times \begin{pmatrix} 6 & -13 \\ -10 & 16 \end{pmatrix}.$$

$$\text{Using} \quad \boldsymbol{X} = \boldsymbol{A}^{-1}\boldsymbol{B}$$

$$\text{we have } \begin{pmatrix} x \\ y \end{pmatrix} = \frac{1}{-34} \times \begin{pmatrix} 6 & -13 \\ -10 & 16 \end{pmatrix}\begin{pmatrix} 108 \\ 59 \end{pmatrix}$$

$$= \frac{1}{-34} \times \begin{pmatrix} 648 - 767 \\ -1080 + 944 \end{pmatrix}$$

$$= \begin{pmatrix} \dfrac{-119}{-34} \\ \dfrac{-136}{-34} \end{pmatrix}$$

$$\text{i.e.} \quad \begin{pmatrix} x \\ y \end{pmatrix} = \begin{pmatrix} 3.5 \\ 4 \end{pmatrix} \qquad \text{or } x = 3.5,\ y = 4.$$

You will notice that these values agree with the answers we worked out using the determinant method.

EXERCISE 14.9

In each question write the simultaneous equations in matrix form. Hence solve them for x and y.

1 $x+y=11$ and $x-y=1$

2 $x+3y=8$ and $x-2y=3$

3 $3y=x+1$ and $5y=20-2x$

4 $2x-y=0$ and $4x-5y=-3$

5 $3y-5x=4$ and $y=1$

6 $x+2y=5$ and $2x-y=7$

7 $2x+3y=3$ and $x=0$

8 $7x-6y=18$ and $6y=7x-12$

9 $3x+2y-4=0$ and $x+3y-11=0$

10 $2x-5y-23=0$ and $15y=69-6x$

The final set of exercises involves the solution of pairs of practical simultaneous linear equations. For extra practice you can attempt the solutions using i) determinants
and ii) matrices.

EXERCISE 14.10

1 A lifting machine raises loads, W kg, with effort, E N, according to the equation $xW+y=E$. From a set of test results we know that 230 N will raise 60 kg and 80 N will raise 10 kg. We can use these values to create a pair of simultaneous equations $60x+y=230$
and $10x+y=80$.
Solve these equations for x and y.

2 The velocity of a van, $v\,\text{ms}^{-1}$, is connected to time, t s, by the formula $v=mt+c$. After 20 s the velocity is $7\,\text{ms}^{-1}$ and after 32 s the velocity is $10\,\text{ms}^{-1}$. Substituting these pairs of values into the formula we get

$$7 = 20m+c$$

and $$10 = 32m+c.$$

Solve this pair of simultaneous linear equations.

3 An engineering company's position in the market is oriented to the price of its product, £x. It would like to supply more at a higher price, but as the price rises so demand falls away. The market is in equilibrium when supply equals demand. For the following simultaneous equations representing supply and demand

$$y = 100+10x$$

and $y = 350-8x$ find that equilibrium position.

4 The electrical circuit shows 2 emfs together with various resistors. The sums of the relevant voltages produce the simultaneous equations.

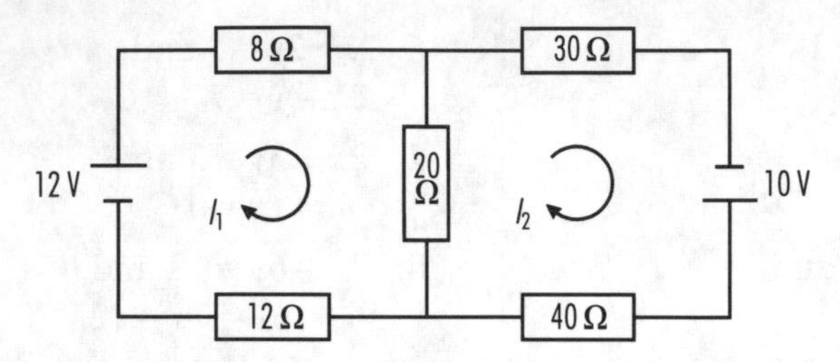

They are in terms of the currents, I_1 and I_2 amps.

$$8I_1 + 20(I_1 - I_2) + 12I_1 = 12$$

and $$40I_2 + 20(I_2 - I_1) + 30I_2 = 10.$$

These equations will simplify. Show that they may be reduced to

$$10I_1 - 5I_2 = 3$$

and $$-2I_1 + 9I_2 = 1.$$

Now solve this pair of equations for I_1 and I_2.

5 A loaded beam has a shear force given by $S = ax + b$ where x is the distance from one end. At a distance of 2 m from that end the shear force is -400 N and at 4.5 m it is 600 N. Substituting these pairs of values into the formula we get

$$-400 = 2a + b$$

and $$600 = 4.5a + b.$$

Solve these simultaneous linear equations for a and b.

15 Series

The objectives of this chapter are to:

1. Define an arithmetic progression.
2. Express an arithmetic progression in terms of the first term, a, and the common difference, d.
3. Determine an expression for the nth term of an arithmetic progression and calculate terms.
4. Calculate the parameters of an arithmetic progression given any two pieces of relevant information.
5. Determine the expression for and calculate the sum of n terms of an arithmetic series.
6. Define a geometric progression.
7. Express a geometric progression in terms of the first term, a, and the common ratio, r.
8. Determine an expression for the nth term of a geometric progression and calculate terms.
9. Determine the expression for and calculate the sum of n terms of a geometric series.
10. State the condition for convergence of a geometric progression.

Introduction

In this chapter we look at progressions and series. Sometimes these labels can be interchanged. We will distinguish between them.

A **progression** is a sequence of numbers (or letters) that follows a particular pattern. The numbers (or letters) are called **terms**. There are many different series. In this chapter we will look at just 2 types. The first of these is the **arithmetic progression** shortened to **AP**. The second is the **geometric progression** shortened to **GP**.

The sum of an arithmetic progression is called an **arithmetic series**. Similarly the sum of a geometric progression is called a **geometric series**.

ASSIGNMENT

The Assignment for this chapter looks at a drilling company sinking bore holes. Before leaving the depot the drilling crew's engineer has a choice of 2 types of drilling gear. Which he chooses is important. Bonuses are only paid to the crew if they reach certain depths after 6, 12 and 24 hours. One set of drilling gear will bore 10 m in the first hour. In each subsequent hour that reduces by 4% of the original 10 m. The other set of drilling gear will also bore 10 m in the first hour. However, from then onwards the hourly reduction is 4% based on the previous hour's drilled depth.

The arithmetic progression

The numbers (or letters) of an AP are in a particular pattern. The pattern involves adjacent terms (i.e. numbers or letters) differing by the same amount. That amount is the common difference. It may be positive or negative.

Adjacent means "next to".

Let us have a look at a few examples.

Examples 15.1

i) 1, 2, 3, 4, ... is an AP.
As we read from left to right each number increases by 1, i.e. the common difference is 1.

ii) 2, 4, 6, 8, ... is an AP.
As we read from left to right each number increases by 2, i.e. the common difference is 2.

iii) 2, -4, 6, -8, ... is *not* an AP.
There is *no* common difference.
2 to -4 is a difference of -6,
-4 to 6 is a difference of $+10$,
6 to -8 is a difference of -14.

iv) 0, -5, -10, ... is an AP.
As we read from left to right each number decreases by 5, i.e. the common difference is -5.

We have some standard notation for APs. The first term is a and the common difference is d.

Let us look at an AP, 2.5, 9.5, 16.5, 23.5, 30.5, ... Obviously the first term is $a=2.5$. Now choose any adjacent terms. In this case you will see the common difference is 7, e.g. for 9.5 and 16.5 we have $16.5-9.5=7$.

Our **parameters** for this AP are the **first term**, $\boldsymbol{a}=2.5$, and the **common difference**, $\boldsymbol{d}=7$.

We may re-write our progression as

$$2.5, \quad 2.5+7, \quad 2.5+7+7, \quad 2.5+7+7+7, \ldots$$

i.e.
$$2.5, \quad 2.5+(1\times 7), \quad 2.5+(2\times 7), \quad 2.5+(3\times 7), \ldots$$

Using our general letters we have

$$a, \quad a+d, \quad a+2d, \quad a+3d, \ldots$$

Notice that $a+2d$ is the 3rd term, i.e. $a+(3-1)d$;
and that $a+3d$ is the 4th term, i.e. $a+(4-1)d$.
If we continue the pattern

$a+14d$ is the 15th term, i.e. $a+(15-1)d$.

If we have n terms then the **nth term** is $\boldsymbol{a+(n-1)d}$.

Examples 15.2

For the AP 1, 9, 17, 25, ...

i) what is the common difference?
ii) what is the 9th term?
iii) which term has a value of 145?

Just looking at the AP we see the first term is 1, i.e. $a=1$.

i) Checking adjacent terms, the common difference is $d=9-1=8$.

ii) The nth term is $a+(n-1)d$
which in this case is $1+(n-1)8$.
For the 9th term we have $1+(9-1)8. = 1+(8)8$
$= 1+64$
$= 65.$

iii) For this final question we use the nth (or general) term, $1+(n-1)8$, to give

$$1+(n-1)8=145$$

i.e. $(n-1)8=144$ — Subtracting 1 from both sides.

i.c. $n-1=18$ — Dividing by 8.

$\therefore$ $n=19$,

i.e. the 19th term is 145.

Examples 15.3

For the AP 3, -2, -7, -12, ...

i) what is the common difference?
ii) what is the 9th term?
iii) which term has a value of 145?

Just looking at the AP we see the first term is 3, i.e. a = 3.

i) Checking adjacent terms, the common difference is
$d = -12--7=-12+7=-5.$

ii) The nth term is $a+(n-1)d$
which in this case is $3-(\text{n}-1)5$.
For the 9th term we have $3-(9-1)5 = 3-(8)5$
$= 3-40$
$= -37.$

iii) For this final question we use the nth (or general) term, $3-(n-1)5$, to give

$$3-(n-1)5=145$$

Subtracting 3 from both sides.

i.e. $\quad -(n-1)5=142.$

If we continue with this calculation two strange things happen. n will turn out to be negative and fractional. Neither of these are possible because we need whole numbers of terms. We count the terms from the first, meaning they are positive whole numbers. The inconsistency in this question is obvious when we look at the original sequence $3, -2, -7, -12, \ldots$ Quickly the terms become negative yet the question asked about the positive value 145. Hence no term in this AP has a value of 145.

The arithmetic series

Now we distinguish between an arithmetic progression and an arithmetic series.

An arithmetic series is the sum of the terms of an arithmetic progression.

We make this distinction in the next examples.

Examples 15.4

i) $1, 2, 3, 4, \ldots$ is an arithmetic progression.

ii) $1+2+3+4 \ldots$ is an arithmetic series.

We can look further at the series using our general AP

$$a,\ a+d,\ a+2d,\ a+3d,\ \ldots,\ a+(n-1)d.$$

The arithmetic series is the addition of all these terms. S_n is the sum of the first n terms. It always starts with the first term in the sequence. We may write this as

$$S_n=a+(a+d)+(a+2d)+(a+3d)+\quad\ldots\quad+(a+(n-1)d).$$

The terms may be added in any order. If we reverse the order

$$S_n=(a+(n-1)d)+(a+(n-2)d)+\quad\ldots\quad+(a+d)+a.$$

We add corresponding terms so combining these 2 expressions for S_n.

On the left we get $2S_n$.

On the right, addition gives

the first terms as $\quad a+(a+(n-1)d) = 2a+(n-1)d.$

the second terms as $\quad (a+d)+(a+(n-2)d) = 2a+(n-1)d.$

If we continue along all n terms on the right each addition of pairs gives $2a+(n-1)d$, i.e. we have n lots of $2a+(n-1)d$. Combining left- and right-hand sides this gives us

$$2S_n = n \times \{2a + (n-1)\,d\}$$

i.e.
$$S_n = \frac{n}{2}\{2a + (n-1)\,d\}.$$

We can easily demonstrate this formula in Examples 15.5.

Examples 15.5

i) For the arithmetic progression 3, −2, −7, −12, ... find the sum of the first 14 terms.
ii) What is the sum of the first 100 positive numbers?

i) For our progression 3, −2, −7, −12, ...
the first term is $a = 3$,
the common difference is $d = -12 - -7 = -12 + 7 = -5$ and
the number of terms is $n = 14$.
We substitute these values into our formula

$$S_n = \frac{n}{2}\{2a + (n-1)\,d\}$$

i.e.
$$\begin{aligned} S_{14} &= \frac{14}{2}\{2(3) + (14-1)(-5)\} \\ &= 7\{6 - 65\} \\ &= -413 \end{aligned}$$

is the sum of the first 14 terms.

ii) The sequence for the first 100 positive numbers is
1, 2, 3, 4, ..., 100.
The first term is $a = 1$,
the common difference is $d = 2 - 1 = 1$ and
the number of terms is $n = 100$.
We substitute these values into our formula

$$S_n = \frac{n}{2}\{2a + (n-1)d\}$$

i.e.
$$\begin{aligned} S_{100} &= \frac{100}{2}\{2(1) + (100-1)1\} \\ &= 50\{2 + 99\} \\ &= 5050 \end{aligned}$$

is the sum of the first 100 positive numbers.

Example 15.6

Fill in the missing terms in the arithmetic progression 52, , , , , 37. What is the sum of the sequence?

We compare our progression

52, , , , , 37

with the general progression

$a,\ a+d,\ a+2d,\ a+3d,\ a+4d,\ a+5d.$

Comparing first and last terms we see that $a=52$ and $a+5d=37$

i.e. $\quad 52+5d = 37$

i.e. $\quad 5d = -15$

$\quad d = -3.$

Now starting at 52 we fill in the missing terms, subtracting 3 each time to get $\quad 52-3=49$

and $49-3=46$

and $46-3=43$

and $43-3=40$

We can check the final term with $\quad 40-3=37.$

Collecting together our answers the four missing terms are 49, 46, 43 and 40.

To find the sum of this arithmetic progression we have two alternatives. We can simply add together all the terms,

$$52+49+46+43+40+37 = 267$$

Alternatively we can use our formula for S_n, i.e.

$$S_n = \frac{n}{2}\{2a+(n-1)d\}$$

becomes

$$S_6 = \frac{6}{2}\{2(52)+(6-1)(-3)\}$$

$$= 3\{104-15\}$$

$$= 267 \quad \text{as before.}$$

We have 6 terms.

Example 15.7

The 4th term of an AP is 16 and the 10th term is 58. What is the

i) first term?

ii) common difference?

iii) sum of the 5th to 10th terms inclusive?

We know the general expression for the nth term is $a+(n-1)d$.

For the 4th term, $n=4$ and $a+(4-1)d=16.$

For the 10th term, $n=10$ and $a+(10-1)d=58.$

Simplifying these equations we get $\quad a+3d=16.$

and $\quad a+9d=58.$

This is a pair of simultaneous linear equations. We have seen various methods for solving these pairs of equations. As an exercise for yourself you should solve them. Your answers should agree with $a=-5$ and $d=7$, i.e. the first term is -5 and the common difference is 7.

Our formula for the sum of a series is $S_n = \frac{n}{2}\{2a+(n-1)d\}$. However it only applies to the first n terms. We are *not* starting with the first term, we are starting with the 5th term. We look at this problem in two different ways.

Firstly let us look at the numbered terms

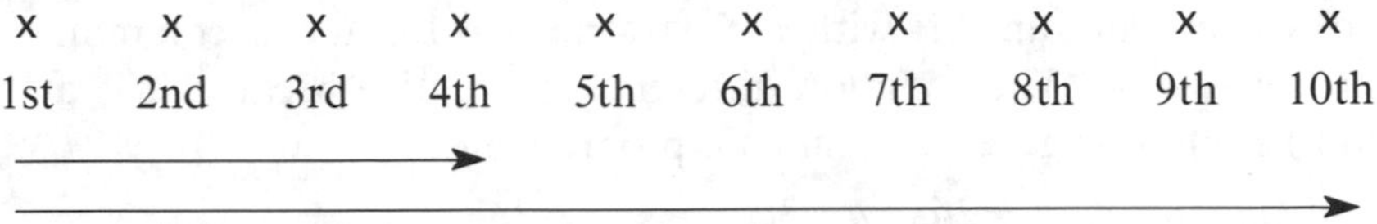

x	x	x	x	x	x	x	x	x	x
1st	2nd	3rd	4th	5th	6th	7th	8th	9th	10th

For the first 4 terms; $n=4$, $a=-5$, $d=7$

and $S_n = \frac{n}{2}\{2a+(n-1)d\}$

becomes $S_4 = \frac{4}{2}\{2(-5)+(4-1)7\}$

$= 2\{-10+21\}$

$= 22.$

For the first 10 terms; $n=10$, $a=-5$, $d=7$

and $S_n = \frac{n}{2}\{2a+(n-1)d\}$

becomes $S_{10} = \frac{10}{2}\{2(-5)+(10-1)7\}$

$= 5\{-10+63\}$

$= 265.$

S_{10} includes all terms from the 1st to the 10th. This means it includes the 1st to the 4th terms of S_4. For the 5th to the 10th terms only we must remove S_4, i.e. required sum $= S_{10}-S_4 = 265-22 = 243$.

Now let us look at an alternative method.

We know that $a=-5$ and $d=7$. If the 4th term is 16 then the 5th term is $16+d=16+7=23$. From here there are six more terms up to and including the 10th term. We can think of the 5th to the 10th terms inclusive as a series of 6 terms with $a=23$ and $d=7$.

Then $S_n = \frac{n}{2}\{2a+(n-1)d\}$

becomes $S_6 = \frac{6}{2}\{2(23)+(6-1)7\}$

$= 3\{46+35\}$

$= 243$ as before.

ASSIGNMENT

Let us take a first look at our drilling company. For the first set of drilling gear there is a constant reduction, 4% of 10 m, i.e. $\frac{4}{100}$ of 10 m

$$\text{i.e.} \quad \frac{4}{100} \times 10 = 0.4 \text{ m}.$$

This looks like being an AP with a first term, $a = 10$, and a common difference, $d = -0.4$. (The common difference is negative because it is a reduction.) From our general arithmetic progression of

$$a, \quad a+d, \quad a+2d, \quad a+3d, \quad \ldots$$

we get $\quad 10, \quad 9.6, \quad 9.2, \quad 8.8, \quad \ldots$

Remember the drilling company pays bonuses to its crews. For this particular job the company will pay extra on the following minimum drilled depths:

i) 50 m in the first 6 hours,
ii) 100 m in the first 12 hours,
iii) 150 in the first 24 hours.

In each of these cases we have to sum the terms based on our formula

$$S_n = \frac{n}{2}\{2a + (n-1)d\}.$$

i) Substituting $n = 6$, $a = 10$, $d = -0.4$ we get

$$\begin{aligned} S_6 &= \frac{6}{2}\{2(10) + (6-1)(-0.4)\} \\ &= 3\{20-2\} \\ &= 54 \text{ m}. \end{aligned}$$

This means the crew have earned their bonus. 54 m exceeds the minimum depth for the first 6 hours.

ii) Substituting $n = 12$, $a = 10$, $d = -0.4$ we get

$$\begin{aligned} S_{12} &= \frac{12}{2}\{2(10) + (12-1)(-0.4)\} \\ &= 6\{20-4.4\} \\ &= 93.6 \text{ m}. \end{aligned}$$

No bonus for the crew this time. They have failed to meet their 12 hour target of 100 m.

iii) Substituting $n = 24$, $a = 10$, $d = -0.4$ we get

$$\begin{aligned} S_{24} &= \frac{24}{2}\{2(10) + (24-1)(-0.4)\} \\ &= 12\{20-9.2\} \\ &= 129.6 \text{ m}. \end{aligned}$$

Again no bonus for the crew. They have missed the 150 m target for the first 24 hours.

We will look at the second set of drilling gear later in the chapter. Perhaps with the second set of drilling gear the crew will meet more of their targets?

EXERCISE 15.1

1 Which of the following are arithmetic progressions or arithmetic series or neither of these?

i) 7, 9, 11, 13, 15, . . . ,

ii) $1+5+9+13+\ \dots,$

iii) $1,\ -\frac{1}{2},\ \frac{1}{4},\ -\frac{1}{8},\ \dots,$

iv) 5, 4, 3, 2, 1,

v) $10+9+8+8+7+\ \dots,$

vi) $1+1+1+1+1+\ \dots,$

vii) $0.1+0.6+1.1+1.6+\ \dots,$

viii) 1, 2, 4, 8, 16, 32, . . .

2 Use your answers to Question **1**. For each arithmetic progression what are the first term, a, and the common difference, d?

3 For the AP 5, 6.5, 8, 9.5, . . . what is the 10th term? Find the sum of the first 10 terms.

4 What is the 15th term of the AP 11, 14, 17, . . . ?

5 What is the 8th term of the AP 1.50, 2.75, 4.00, . . . ?

6 The 6th term of an AP is 8 and the 9th term is 11.75. Find the first term, the common difference and the sum of the first 9 terms.

7 Find the sum of $21+24+27+\ \dots$ to the first 9 terms.

8 Find the sum of the first 20 terms of 25, 20, 15, 10,.. Also find the sum of the 16th to 20th terms inclusive.

9 What is the sum of the first 50 odd numbers? What is the sum of the odd numbers between 0 and 50? Explain any similarity/difference between your answers.

10 The first term of an AP of multiples of 3 is 3 itself. Find the sum of the i) first 10 terms
and ii) next 10 terms.

The geometric progression

The numbers (or letters) of a GP are in a particular pattern Each term is a multiple of the previous term. For successive terms that multiple remains the same. It is the **common ratio**. It may be positive or negative.

Examples 15.8

i) 1, 3, 9, 27, ... is a GP.
As we read from left to right each number is 3 times the previous one, i.e. the common ratio is 3.

ii) 1, -3, 9, -27, ... is a GP.
As we read from left to right each number is -3 times the previous one, i.e. the common ratio is -3.

iii) 1, 0.5, 0.25, ... is a GP.
As we read from left to right each number is 0.5 times the previous one, i.e. the common ratio is 0.5.

iv) -1, 0.6, -0.36, ... is a GP.
As we read from left to right each number is -0.6 times the previous one, i.e. the common ratio is -0.6.

v) 1, -0.6, -0.36, ... is *not* a GP.
If we compare the first and second terms we see that $1 \times \mathbf{-0.6} = -0.6$. If we compare the second and third terms we see that $-0.6 \times \mathbf{0.6} = -0.36$. We have highlighted the different multipliers. For a GP they must be the same. Because they differ by a +/– sign they are *not* a common ratio. Hence this is *not* a GP.

We have some standard notation for GPs. The first term is a (as for APs) and the common ration is r.

Let us look at a GP; $\frac{1}{2}$, 1 2, 4, 8, 16, 32, ...

Obviously the first term is $a = \frac{1}{2}$.

Choose any two successive terms and you will see your second choice is 2 times your first choice in this case;

e.g. 4 and 8 where $8 = 4 \times 2$

e.g. 16 and 32 where $32 = 16 \times 2$

Our **parameters** for this GP are the **first term**, $\boldsymbol{a} = \frac{1}{2}$, and the **common ratio**, $\boldsymbol{r} = 2$.

We may re-write our progression as

$$\frac{1}{2},\ \frac{1}{2} \times 2,\ \frac{1}{2} \times 2 \times 2,\ \frac{1}{2} \times 2 \times 2 \times 2,\ \frac{1}{2} \times 2 \times 2 \times 2 \times 2,\ \ldots$$

i.e. $$\frac{1}{2},\ \frac{1}{2} \times 2,\ \frac{1}{2} \times 2^2,\ \frac{1}{2} \times 2^3,\ \frac{1}{2} \times 2^4,\ \ldots$$

Using our general letters we have

$$a, \quad ar, \quad ar^2, \quad ar^3, \quad ar^4, \quad \ldots$$

Remember only r is raised to the powers.

Notice that ar^2 is the 3rd term, i.e. ar^{3-1}
and that ar^3 is the 4th term, i.e. ar^{4-1}.
If we continue the pattern
ar^8 is the 9th term, i.e. ar^{9-1}.
If we have n terms then the ***n*th term** is ar^{n-1}.

Examples 15.9

For the GP 5, 10, 20, 40, ...
i) what is the common ratio?
ii) what is the 7th term?
iii) which term has a value of 2560?

Just looking at the GP we see the first term is 5, i.e. $a = 5$.

i) After the first term we can choose any other term.
Suppose we choose the second term, $ar = 10$.
We have $5r = 10$
i.e. $r = \frac{10}{5} = 2$
is the common ratio.

ii) The nth term is ar^{n-1}
which in this case is $5 \times 2^{n-1}$.
For the 7th term we have

$$5 \times 2^{7-1} = 5 \times 2^6 = 5 \times 64 = 320.$$

iii) For this final question we use our nth term, $5 \times 2^{n-1}$, so that

$$5 \times 2^{n-1} = 2560$$

i.e. $2^{n-1} = 512$.

Dividing by 5.

Taking common logarithms of both sides we get

$$\log 2^{n-1} = \log 512$$

i.e. $(n-1)\log 2 = \log 512$

Third log law.

i.e. $n - 1 = \frac{2.70927}{0.30103}$

$$n - 1 = 9$$

i.e. $n = 10$.

It is the 10th term that has a value of 2560.

Examples 15.10

For the GP $3, \frac{-3}{4}, \frac{3}{16}, \frac{-3}{64}, \ldots$

i) what is the common ratio?
ii) what is the 6th term?

Just looking at the GP we see the first term is 3, i.e. $a=3$.

i) After the first term we can choose any term, but decide to choose the 2nd term for simplicity so that

$$ar = \frac{-3}{4}$$

$$\text{i.e.} \quad 3r = \frac{-3}{4}$$

$$r = \frac{-1}{4} \quad \text{or } -0.25$$

is the common ratio.

Just as easily we could have used $ar^3 = \frac{-3}{64}$, though this does involve a cube.

ii) The nth term is ar^{n-1} which in this case is $3 \times -\left(\frac{1}{4}\right)^{n-1}$.

For the 6th term we have

$$3 \times \left(\frac{-1}{4}\right)^{6-1} = 3 \times (-0.25)^5$$
$$= -3 \times 9.7656 \times 10^{-4}$$
$$= -0.00293.$$

The geometric series

Now we can distinguish between a geometric progression and a geometric series.

A geometric series is the sum of the terms of a geometric progression.
We make this distinction in the next pair of examples.

Examples 15.11

i) $3, \frac{-3}{4}, \frac{3}{16}, \frac{-3}{64}, \ldots$ is a geometric progression.

ii) $3 - \frac{3}{4} + \frac{3}{16} - \frac{3}{64}, \ldots$ is a geometric series.

We can look further at the series using our general GP

$$a, ar, ar^2, ar^3, \ldots ar^{n-1}.$$

The geometric series is the addition of all these terms. S_n, again, is the sum of the first n terms. It always starts with the first term in the sequence. We may write this as

$$S_n = a + ar + ar^2 + ar^3 + \ldots \quad + ar^{n-1}.$$

This is an equation. Whatever we do to the left we must do to the right-hand side. Suppose we multiply each term by r to get

Remember the laws of indices

$$rS_n = ar + ar^2 + ar^3 + ar^4 + \ldots \quad + ar^n.$$

Now we can subtract our 2 equations involving S_n. Some terms disappear as in simultaneous equations. On the right the ar terms disappear, the ar^2 terms disappear and so on. Only the first, a, and the last, ar^n, remain,

i.e. $S_n - rS_n = a - ar^n$

Common factors on both sides.

i.e. $(1-r)S_n = a(1-r^n)$

i.e. $S_n = \dfrac{a(1-r^n)}{1-r}$

Dividing by $1-r$.

There are other, very similar versions, of this formula.

$S_n = \dfrac{a(-r^n+1)}{-r+1}$

Interchanging first and last terms.

i.e. $S_n = \dfrac{-a(r^n-1)}{-(r-1)}$

Removing "$-$" signs as common factors.

so that $S_n = \dfrac{a(r^n-1)}{r-1}.$

Cancelling the "$-$" signs.

Remember that division by 0 is not allowed in Mathematics. This means that in both cases $r \neq 1$. Provided $r \neq 1$ we can use either formula and so wc demonstrate them both in Examples 15.12.

Examples 15.12

i) For the geometric progression 5, 10, 20, 40, ... find the sum of the first 9 terms.

ii) For the geometric progression 3, $\dfrac{-3}{4}$, $\dfrac{3}{16}$, $\dfrac{-3}{64}$, ... find the sum of the first 6 terms.

i) For our progression 5, 10, 20, 40, ... we know from Example 15.9 that $a=5$ and $r=2$.
Also we are given $n=9$.
We substitute these values into our formula

$$S_n = \frac{a(r^n-1)}{r-1}$$

so that $S_9 = \dfrac{5\,(2^9-1)}{2-1}$

$= \dfrac{5\,(512-1)}{1}$

$= 2555$ is the sum of the first 9 terms.

ii) For our progression $3, \frac{-3}{4}, \frac{3}{16}, \frac{-3}{64}, \ldots$ we know from Example 15.10 that $a=3$ and $r=-0.25$.
Also we are given $n=6$.
We substitute these values into the other version of our formula

$$S_n = \frac{a(1-r^n)}{1-r}$$

so that

$$S_6 = \frac{3(1-(-0.25)^6)}{1-0.25}$$

$$= \frac{3(1-0.000244)}{1+0.25}$$

$$= \frac{3 \times 0.99976}{1.25}$$

i.e. $S_6 = 2.40$ is the sum of the first 6 terms.

Example 15.13

Fill in the missing terms in the geometric progression 7, , , , 567. What is the sum of the sequence?

We compare this GP 7, , , , 567

with the general GP $a, \quad ar, \quad ar^2, \quad ar^3, \quad ar^4$

Comparing the first terms we have $a = 7$.

Comparing the last terms we have $ar^4 = 567$

i.e. $7r^4 = 567$

$$r^4 = 81$$

Hence $r = \pm\sqrt[4]{81}$

i.e. $r = \pm 3$.

It appears we have two options. Starting at 7 we can fill in the missing terms, multiplying by either 3 or -3.

Using $r=3$ we get

$$7 \times 3 = 21$$

$$\text{and } 21 \times 3 = 63$$

$$\text{and } 63 \times 3 = 189.$$

We can check the final term using $189 \times 3 = 567$.

Alternatively, using $r=-3$ we get

$$7 \times -3 = -21$$

$$\text{and } -21 \times -3 = 63$$

$$\text{and } 63 \times -3 = -189.$$

Again we can check the final terms using $-189 \times -3 = 567$.

Collecting together our answers the three missing terms are 21, 63 and 189 or -21, 63 and -189.

To find the sum of the geometric progressions we can consider two alternatives. The first alternative is to add the terms in each case. This is left as an exercise for you to check the answers.

For 7, 21, 63, 189 and 567 we use

$$S_n = \frac{a(1-r^n)}{1-r}$$

with $a=7$, $r=3$ and $n=5$ so that

$$S_5 = \frac{7(1-3^5)}{1-3}$$

$$= \frac{7(1-243)}{-2}$$

$$= 847 \qquad \text{is this sum of 5 terms.}$$

For 7, -21, 63, -189 and 567 we use

$$S_n = \frac{a(1-r^n)}{1-r}$$

with $a=7$, $r=-3$ and $n=5$ so that

$$S_5 = \frac{7(1-(-3)^5)}{1--3}$$

$$= \frac{7(1--243)}{4}$$

$$= 427 \qquad \text{is the other sum of 5 terms.}$$

Examples 15.14

The 3rd term of a GP is 1 and the 8th term is 1024. What is the

i) first term?
ii) common ratio?
iii) sum of the 4th to 8th terms inclusive?

We know the general expression for the nth term is ar^{n-1}.

For the 3rd term, $n=3$ and $ar^2 = 1$.

For the 8th term, $n=8$ and $ar^7 = 1024$.

This is a pair of simultaneous equations. We may eliminate a by dividing them,

i.e. $$\frac{ar^7}{ar^2} = \frac{1024}{1}$$

i.e. $$r^{7-2} = 1024$$

$$r^5 = 1024$$

$$r = \sqrt[5]{1024}$$

$$= 4 \qquad \text{is the common ratio.}$$

We may substitute for $r = 4$ into either of the original equations.

Using $\quad ar^2 = 1$

we get $\quad a4^2 = 1$

i.e. $\quad a = \frac{1}{4^2}$

$\quad = \frac{1}{16}$ is the first term.

To find the sum of the 4th to 8th terms inclusive we use a method from Example 15.7. The sum of the first 8 terms automatically includes the sum of the first 3 terms. We remove those first 3 terms by simple subtraction.

$$\text{Sum required} = S_8 - S_3$$

$$= \frac{\frac{1}{16}(4^8 - 1)}{4 - 1} - \frac{\frac{1}{16}(4^3 - 1)}{4 - 1}$$

$$= \frac{\frac{1}{16}(65536 - 1 - 64 + 1)}{3}$$

Removing common factors.

$$= \frac{1}{48}(65472)$$

$$= 1364.$$

ASSIGNMENT

In this calculation we look at the second set of drilling gear. Remember 10 m is the first hour's drilled depth. This time the 4% reduction is based on the previous hour's drilling. It is associated with GPs. A reduction of 4% means that 96% is retained. $96\% = \frac{96}{100} = 0.96$.

Now we have $a = 10$ and $r = 0.96$.

Just like our previous look at the Assignment we can calculate the drilled depths after 6, 12 and 24 hours.

We use $\quad S_n = \frac{a(1 - r^n)}{1 - r}$

i) Substituting $n = 6$, $a = 10$ and $r = 0.96$ we get

$$S_6 = \frac{10\,(1 - 0.96^6)}{1 - 0.96}$$

$$= \frac{10\,(1 - 0.78276)}{0.04}$$

$$= 54.3 \text{ m}.$$

This means the crew have earned their bonus: 54.3 m exceeds the minimum depth for the first 6 hours.

ii) Substituting $n=12$, $a=10$ and $r=0.96$ we get

$$S_{12} = \frac{10\,(1-0.96^{12})}{1-0.96}$$

$$= \frac{10\,(1-0.61271)}{0.04}$$

$$= 96.8 \text{ m.}$$

No bonus for the crew this time. They have failed to meet their 12 hour target of 100 m.

iii) Substituting $n=24$, $a=10$ and $r=0.96$ we get

$$S_{24} = \frac{10\,(1-0.96^{24})}{1-0.96}$$

$$= \frac{10\,(1-0.37541)}{0.04}$$

$$= 156.1 \text{ m.}$$

Once again the crew get a bonus, exceeding their target of 150 m in the first 24 hours.

Perhaps now we can bring together our results for both sets of drilling gear.

Time	Target	First Set (AP)	Second Set (GP)	Bonus
6	50 m	54 m	54.3 m	YES/YES
12	100 m	93.6 m	96.8 m	NO/NO
24	150 m	129.6 m	156.1 m	NO/YES

EXERCISE 15.2

1 Which of the following are geometric progressions or geometric series or neither of these?

i) $2, 4, 8, 16, \ldots,$

ii) $2, 4, 6, 8, \ldots,$

iii) $2+4+8+16+ \ldots,$

iv) $1, -0.1, 0.01, -0.001, \ldots,$

v) $36, 12, 4, \frac{4}{3}, \ldots,$

vi) $1, -1, 1, -1, \ldots,$

vii) $4-0.4+0.04-0.004 \ldots,$

viii) $-36, -12, 4, \frac{4}{3}, \ldots,$

2 Use your answers to Question **1**. For each progression what is the first term, a, and the common ratio, r?

3 For the GP 18, 6, 2, ... what is the 6th term correct to 3 decimal places? Find the sum of those 6 terms.

4 What is the 8th term of the GP 1.25, 2.5, 5, ...?

5 What is the 7th term of the GP 1, 3, 9, ...?

6 The 2nd term of a GP is 54 and the 4th term is 96. Find the first term, the common ratio and the sum of the first 8 terms.

7 The sum of a geometric series is 1639.5. If the first term of the series is 1.5 and the common ratio is 3 how many terms are there? What is the value of that last term?

8 Find the sum of the first 6 terms of 25, 20, 16, ...

9 i) Why is $1^3, 2^3, 3^3, 4^3, \ldots$ neither a GP nor an AP?
ii) Why is $1^1, 2^2, 3^3, 4^4, \ldots$ neither a GP nor an AP?

10 A sequence of eight numbers starts with 2.
Those eight numbers form an AP.
Also the 1st, 2nd, 4th and 8th numbers form a GP.
Show that the common difference for the AP and the common ratio for the GP both equal 2. What are the sums of the AP and of the GP?

The geometric progression – sum to infinity

We have seen various examples of GPs. It is possible for us to split them into 2 types based on the size of the common ratio. In Examples 15.8i) $(r=3)$ and ii) $(r=-3)$ the common ratios had sizes of 3, i.e. $|r|=3$. If we continue the series in each case the numbers would get bigger and bigger. Hence the sum would get bigger and bigger. In both cases the sequences are **divergent**.

In Examples 15.8iii) $(r=0.5)$ and iv) $(r=-0.\dot{6})$ the common ratios had sizes less than 1, i.e. $|r|<1$ or $-1<r<1$. If we continue the series in each case the size of the terms would get smaller and smaller. As those sizes of terms get smaller they have less and less overall effect on the sum. We only notice small alterations to the sum by considering many decimal places. Eventually the calculator does not display the alterations as they are so small. In these cases the sequences are **convergent**. We are interested in convergent series because we can predict what the sum will approximate to for many terms.

Example 15.15

For the GP 4, 1 $\frac{1}{4}, \frac{1}{16}, \ldots$ let us look at how successive terms affect the sum. The diagram shows the sum, S, levelling off to $5.\bar{3}$ as we use more terms.

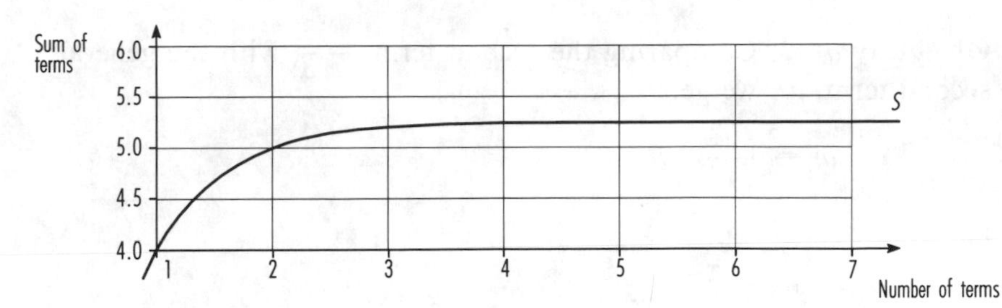

Obviously the first term is 4. From earlier examples we know how to calculate the common ratio. In this case it is $\frac{1}{4}$ i.e. 0.25.

First term only,	Sum	$= 4$
1st and 2nd terms,	Sum $= 4+1$	$= 5$
$n=3$,	$S = 5+\frac{1}{4}$	$= 5.25$
$n=4$,	$S = 5.25+\frac{1}{16}$	$= 5.3125$
$n=5$,	$S = 5.3125+\frac{1}{64}$	$= 5.328125$
$n=6$,	$S = 5.328125+\frac{1}{256}$	$= 5.3320313$
$n=7$,	$S = 5.3320313+\frac{1}{1024}$	$= 5.3330078.$

If we take enough terms we will see this sum approach $5.\bar{3}$,
i.e. for many terms the sum will tend to $5.\bar{3}$,
i.e. as $n \to \infty$, $S \to 5.\bar{3}$.

$\to$ means "tends to" or "approaches".

The formula for the sum to infinity of an infinite convergent GP is

$$S_\infty = \frac{a}{1-r}.$$

Examples 15.16

Find the sum to infinity for the GPs i) $1, 0.5, 0.25, \ldots,$

ii) $3, -\frac{3}{4}, \frac{3}{16}, \ldots$

i) Obviously $a=1$. Comparing the second term, 0.5, with the general second term, ar, we get

$$ar = 0.5$$

i.e. $$1r = 0.5$$

$$r = 0.5.$$

Substituting for a.

Using $S_\infty = \frac{a}{1-r}$

we have $S_\infty = \frac{1}{1-0.5} = 2$ as the sum to infinity.

ii) Obviously $a = 3$. Comparing the second term, $-\frac{3}{4}$, with the general second term, ar, we get

$$ar = -\frac{3}{4}$$

i.e.
$$3r = -\frac{3}{4}$$

$$r = -\frac{1}{4} \text{ or } -0.25.$$

Using
$$S_\infty = \frac{a}{1-r}$$

we have
$$S_\infty = \frac{3}{1 - -0.25} = 2.4 \qquad \text{as the sum to infinity.}$$

EXERCISE 15.3

1 Find the sum to infinity for the GPs

i) 1, 0.6, 0.36, . . . ,

ii) -1, 0.6, -0.36, . . .

2 Why is $\left(-\frac{1}{2}\right)^0, \left(-\frac{1}{2}\right)^1, \left(-\frac{1}{2}\right)^2, \ldots$ a GP? Find the sum to infinity.

3 The first term of a sequence is 50. The common ratio is 0.1. Write down the first four terms. What is the sum to infinity?

4 Find the sum to infinity of $\frac{1}{5}, -\frac{1}{25}, \frac{1}{125}, -\frac{1}{625}, \ldots$

5 A GP has a first term of 4 and a sum to infinity of 20. What is the common ratio?

Let us complete this chapter with a practical example followed by a short exercise.

Example 15.17

A 999 year lease charges £50 in the first year. This rises at 6% pa as a GP. What are the lease charges in the 10th year?

The cost of collection is £5 in the first year. In each of the following years it increases by 50p.

What are the profits in the 1st and 10th years?

Firstly let us look at the lease charges. More usefully we can re-consider the 6% increase. 6% is $\frac{6}{100}$ or 0.06. This means the new charge will be 1.06 (i.e. $1 + 0.06$) of the previous year's charge. Looking at the GP for the early years we have

1st year, charges = £50
2nd year, charges = £50 × 1.06
3rd year, charges = £50 × 1.06^2, and so on.
Notice how the power of 1.06 is 1 less than the number of years so that
10th year, charges = £50 × 1.06^9 = £84.47.

Now we can look at the cost of collection. This time the annual increase is a constant 50p, i.e. £0.50 to be consistent with the initial £5. We have an AP.
1st year, collection costs = £5
2nd year, collection costs = £5 + 0.50, and so on.

The multiple of £0.50 added each year is always 1 less than the number of years so that
10th year, collection costs = £5 + 0.50 × 9 = £9.50.

As we would expect Profit = Lease charges − Collection costs
In the 1st year, Profit = £50 − 5 = £45.
In the 10th year, Profit = £84.47 − 9.50 = £74.97.

EXERCISE 15.4

1 A tennis ball manufacturer is testing some new prototype tennis balls. They are dropped from a height of 2 m. After each bounce they rebound to 45% of the previous height, so forming a GP. Write down the first 4 terms of the sequence correct to 2 decimal places. Calculate the sum of the first 6 terms.

2 A production line shift lasts for $7\frac{1}{2}$ hours. The rate of production is not constant. 15 items are made in the first $\frac{1}{2}$ hour. Over the first 4 hours, at half-hourly intervals, the production forms an AP with a common difference of 2. Over the remaining $3\frac{1}{2}$ hours, at half-hourly intervals, it is an AP with a common difference of −1. How many items are produced during the shift?

3 An engineering firm buys capital equipment for £50 000. It allows a depreciation rate of 15% pa. This means that during the working life of the equipment its value forms a GP. Write down the values of a and r and the first 3 terms of this GP. How long will it be before the equipment has halved in value?

4 A fencing contractor starts with the first fence post at 2 m from the rear wall of a house. Then he measures inaccurately, an extra 0.75% of the original 2 m. What is this extra distance?
At the second post he adds this extra distance to the original 2 m. At the third post he adds this extra distance onto his previous extra. The continual inaccuracy means the distances between the posts form an AP. After erecting 20 posts he should have covered 40 m. What distance has he actually covered?

5 Awkward access means that tarmac should be delivered to site in 10 tonne loads. The job needs 8 loads yet there appears to have been a shortfall. After 8 loads the job is incomplete. How much tarmac has been delivered if the shortfall has been a continual 1.5% so creating

i) an AP,
ii) a GP?

16 Surface Areas and Volumes

The objectives of this chapter are to:

1. Calculate the volumes and surface areas of spheres, cylinders, cones and pyramids.
2. Define frustum.
3. Calculate the surface area and volume of frusta of cones and pyramids.
4. Identify the component basic shapes of composite surface areas and volumes (i.e. squares, rectangles, trapezia, triangles, circles, pyramids, cones, cylinders and spheres.)
5. Use the Prismoidal Rule to calculate volume where appropriate.
6. Calculate the total surface area or volume of composite figures.
7. Draw a diagram from given data to suitable scale.
8. Calculate areas of irregular shapes using the mid-ordinate and trapezoidal rules.
9. Apply Simpson's Rule to the determination of area using $2n$ strips.
10. Use the formulae πab and $\pi(a+b)$ for the area and approximate perimeter of an ellipse.
11. State the theorems of Pappus for the volume of a solid and surface area.
12. Use the theorems of Pappus to calculate volume and surface area where appropriate.

Introduction

In Chapter 6 we developed theory and calculated some areas based on trigonometry and the area of a triangle. We ended that chapter looking at 3 dimensional bodies. Some of this work included the areas of a sector and a circle from Chapter 4.

This chapter has 3 main sections. We start with standard results for a variety of standard 3 dimensional bodies. Using these results we split composite bodies into component sections and apply our standard results. In the second section we look at irregular, non-standard, shapes concentrating on 2 dimensions. Finally we return to volume calculations, completing the chapter with Pappus' theorems.

Before we start any calculations let us remind ourselves of the appropriate units. This is revision.

The standard unit of length is the metre (m). Length is 1 dimensional.

Area is 2 dimensional with standard units of metres squared (m^2). We have seen this in earlier chapters and you will have seen it elsewhere. We treat the surface area of a volume in the same way.
Volume is 3 dimensional with standard units of metres cubed (m^3). Another standard unit is the litre (l). The connection between litres and m^3 is $1000l = 1\,m^3$.

ASSIGNMENTS

There are 2 Assignments for this chapter.

1. A pedestrian area in a new town development has a series of removable bollards and chains. Each bollard is cast as a hollow hexagonal frustum mounted with a solid sphere. Each chain link is elliptical (i.e. shaped like an ellipse) with a solid circular cross-section. For this assignment we are going to find the volumes of a specimen bollard and chain link.

2. In this second assignment we estimate the area under a sine curve. Remember we first saw the sine curve in Chapter 5. You should look back at its shape to refresh your memory.

Volumes and surface areas

In turn we are going to look at some standard 3 dimensional bodies.

1. The sphere

$$\text{Volume} = \frac{4}{3}\pi r^3$$

$$\text{Surface Area} = 4\pi r^2$$

where r is the radius (see Fig. 16.1).

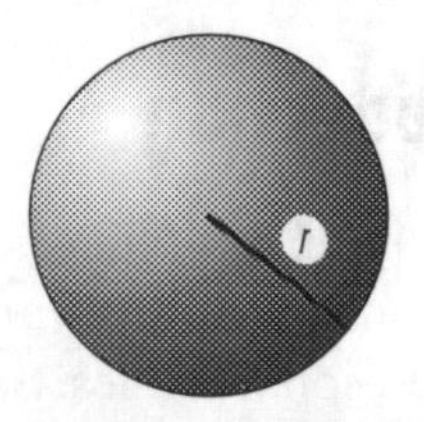

Fig. 16.1

Example 16.1

The volume of a sphere is $2.50\,m^3$. What is the radius of the sphere and its surface area?

Using our formula for volume, V, we have

$$\frac{4}{3}\pi r^3 = 2.50$$

i.e. $$r^3 = 2.50 \times \frac{3}{4\pi}$$

Multiplying through by $\frac{3}{4\pi}$

$$r^3 = 0.5968\ldots$$

$\therefore$ $$r = \sqrt[3]{0.5968\ldots}$$

i.e. $r = 0.84$ m is the radius.

Calculator value is 0.8419...

Also our surface area formula of $4\pi r^2$ becomes

$$\text{Surface area} = 4\pi(0.8419\ldots)^2$$
$$= 8.91\ \text{m}^2.$$

2. The cylinder

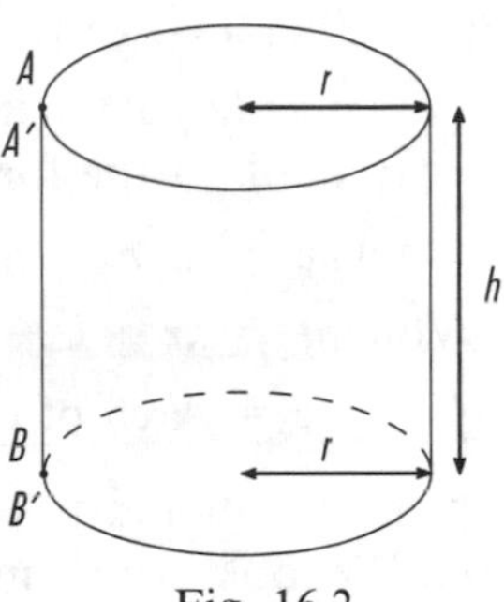

Fig. 16.2

Volume $= \pi r^2 h$

where r is the radius

and h is the height (see Fig. 16.2).

The surface area consists of 2 circular ends and a curved surface. A and A' are the same point on the top circumference. B and B' are, vertically below A and A', together on the base circumference.

If we unwrap the curved area it forms a rectangle $AA'B'B$. One side is h and the other is $2\pi r$ from the circumference of the unwrapped circle.

We can link together these areas.

$$\text{Total surface area} = \text{Area of 2 ends} + \text{Area of curved surface area}$$
$$= 2 \times \pi r^2 + 2\pi r \times h$$
$$= 2\pi r^2 + 2\pi r h$$
$$= 2\pi r(r + h).$$

$2\pi r$ is a common factor.

Example 16.2

Find the volume and surface area of a concrete cylindrical column of height 2.84 m and radius 0.46 m.

How many columns can be cast from 20 m^3 of concrete?

In our formulae we have $h = 2.84$ and $r = 0.46$ so that

$$\text{Volume} = \pi r^2 h$$

becomes $$\text{Volume} = \pi(0.46)^2 2.84$$
$$= 1.89\ \text{m}^3.$$

Calculator value is 1.8879...

Using our surface area formula of $2\pi r(r + h)$ we get

$$\text{Surface area} = 2\pi(0.46)(0.46 + 2.84)$$
$$= 9.54\ \text{m}^2.$$

To find the number of concrete columns we can cast we use

$$\text{No. of columns} = \frac{\text{Volume of concrete}}{\text{Volume of a column}}$$
$$= \frac{20}{1.88\ldots}$$
$$= 10.59\ldots, \qquad \text{i.e. we can cast 10 columns.}$$

The remaining concrete would be waste or spillage.

Now let us return to our cylinder and its volume formula. Notice how the height (or length) of the cylinder is h. The cross-sectional area is a constant circle (area of circle is πr^2).

We can think of the formula

$$\text{Volume} = \pi r^2 h$$

as $\quad \text{Volume} = \pi r^2 \times h$

i.e. $\quad \text{Volume} = \text{Area of constant cross-section} \times \text{Height}.$

This principle applies to all 3 dimensional bodies with a constant cross-section. We look at this in the next example.

Example 16.3

A column of length 3.00 m has a hexagonal cross-section of side 0.80 m. Find the volume of the column. Fig. 16.3 shows the hexagonal cross-section. We can split a hexagon into 6 adjacent equilateral triangles. Using one of the triangle area formulae we get

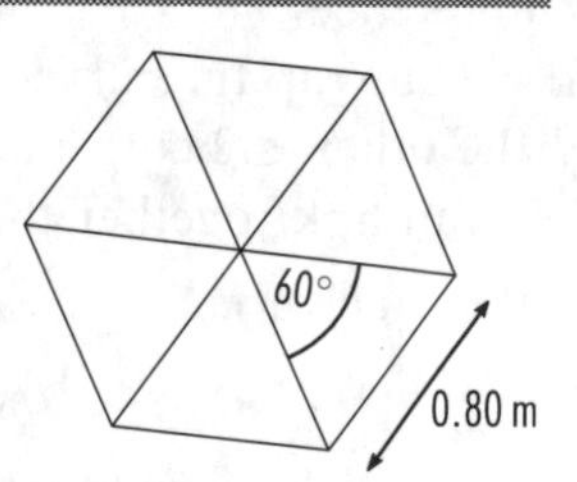

Fig. 16.3

$$\text{Area of triangle} = \frac{1}{2} ab \sin C$$
$$= \frac{1}{2}(0.80)(0.80)\sin 60°.$$
$$\text{Area of hexagon} = 6 \times \text{Area of triangle}$$
$$= 6 \times \frac{1}{2}(0.80)^2 \sin 60°$$
$$= 1.66 \text{ m}^2.$$

Calculator value is 1.662. . .

Now we use our general formula

$$\text{Volume of column} = \text{Area of constant cross-section} \times \text{Height}$$

to get

$$\text{Volume} = 1.66\ldots \times 3.00$$
$$= 4.99 \text{ m}^3.$$

3. The cone

$$\text{Volume} = \frac{1}{3}\pi r^2 h$$

$$\text{Curved surface area} = \pi r l$$

where r is the base radius.

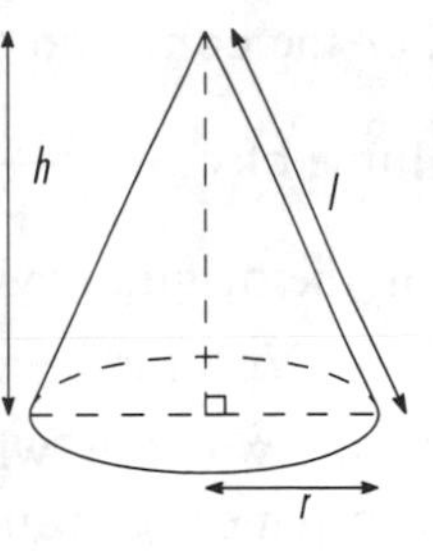

Fig. 16.4

We distinguish between the height, h, (vertical height or altitude) and the slant height, l, of the cone. They are related according to Pythagoras' theorem,

$$l^2 = h^2 + r^2.$$

If we need the total surface area of the cone we must remember to add the area of the circular base, πr^2.

Example 16.4

A cone has an altitude of 2.60 m and a base radius of 0.50 m. Find the volume and the curved surface area of the cone. If the cone came from a cylinder of the same height and the same radius what was the percentage of waste material?

To save us re-drawing a cone we can use Fig. 16.4 with $h = 2.60$ and $r = 0.50$ in our formula

$$\text{Volume} = \frac{1}{3}\pi r^2 h$$

so that $$\text{Volume} = \frac{1}{3}\pi (0.50)^2 2.60$$

$$= 0.68 \text{ m}^3.$$

Before we can find the curved surface area we need to find the slant height, l, of the cone. Using Pythagoras' theorem

$$l^2 = h^2 + r^2$$

we get $$l^2 = 2.60^2 + 0.50^2$$

i.e. $$l^2 = 7.01$$

$\therefore$ $$l = \sqrt{7.01}$$

$$= 2.65 \text{ m}.$$

Calculator value is 2.6476...

Using our surface area formula of $\pi r l$ we get

$$\text{Surface area} = \pi (0.50)(2.6476\ldots)$$

$$= 4.16 \text{ m}^2.$$

For the final section of this example we bring together the cone and the cylinder in Fig. 16.5.

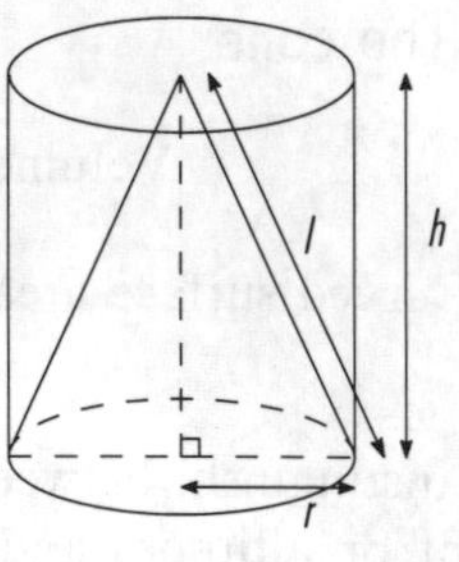

Fig. 16.5

$$\text{Volume of cone} = \frac{1}{3}\pi r^2 h$$

and for the original cylinder

$$\text{Volume} = \pi r^2 h.$$

Simply the waste is what remains once the cone has been removed from the cylinder.

$$\therefore \quad \text{Waste} = \pi r^2 h - \frac{1}{3}\pi r^2 h$$

$$= \left(1 - \frac{1}{3}\right)\pi r^2 h$$

$$= \frac{2}{3}\pi r^2 h.$$

$$\%\text{ of waste} = \frac{\text{Waste}}{\text{Original cylinder}} \times 100$$

$$= \frac{\frac{2}{3}\pi r^2 h}{\pi r^2 h} \times 100$$

Cancelling by $\pi r^2 h$.

$$= \frac{2}{3} \times 100$$

$$= 66.\bar{6}.$$

4. The pyramid

What is a pyramid? A pyramid has a plane base. From the corners of the base come straight lines that meet at some height above it. This means we have a set of sloping sides, often triangular. The cone, with a circular base, has a curved surface in place of the pyramid's sloping sides. A simple example is a witch's hat.

Often the pyramid has a regular base, perhaps a square or rectangle or polygon (e.g. pentagon or hexagon). By way of example Fig. 16.6 shows a square base.

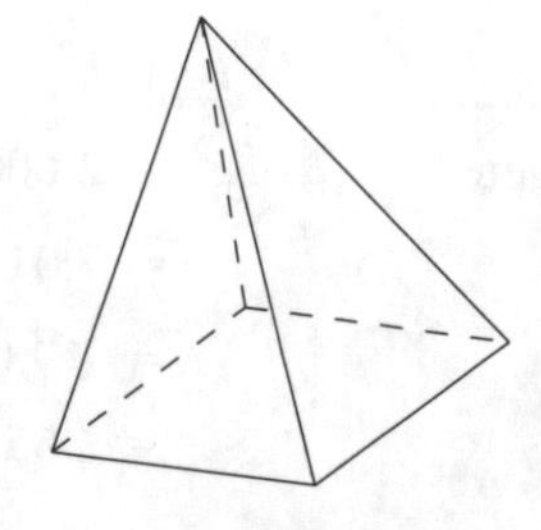
Fig. 16.6

A few pages earlier we extended the idea of a cylinder to other bodies of constant cross-sectional area. In this case the cross-sectional area of the base is *not* constant. It reduces from the area of the base to zero like the cone. However, we can apply a similar principle.

For the cone, $\text{Volume} = \frac{1}{3}\pi r^2 h$

i.e. $\text{Volume} = \frac{1}{3} \times \pi r^2 \times h$

i.e. $\text{Volume} = \frac{1}{3} \times \text{Area of circular base} \times \text{Height}$

We can generalise this to pyramids,

$$\text{Volume of pyramid} = \frac{1}{3} \times \text{Area of base} \times \text{Height}.$$

Example 16.5

A solid pyramid with a slant height of 1.00 m has a square base of side 0.60 m. Find the volume and total surface area of the pyramid.

For reference we draw and label a pyramid in Fig. 16.7. We have some preliminary work to do, rather than straightforward formula substitution.

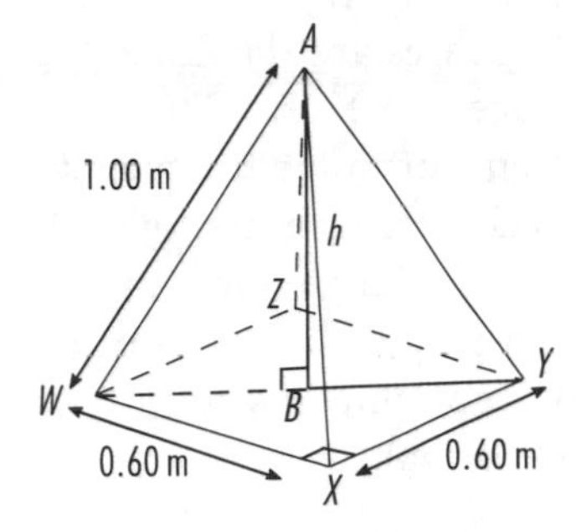

Fig. 16.7

In triangle WXY, by Pythagoras' theorem

$$WY^2 = WX^2 + XY^2$$
$$= 0.60^2 + 0.60^2$$

i.e. $WY^2 = 0.72$

$\therefore$ $WY = \sqrt{0.72}$

$= 0.85$ m.

Calculator value is 0.8485...

The apex, A, of the pyramid is vertically above the centre of the base, B. B is the mid-point of WY,

i.e. $BY = \frac{1}{2} WY$

$= \frac{1}{2} \times 0.8485\ldots$

$= 0.42$ m.

Calculator value is 0.424...

In triangle ABY, by Pythagoras' theorem,

$$AB^2 + BY^2 = AY^2$$

i.e. $h^2 + 0.424\ldots^2 = 1.00^2$

$h^2 = 1.00 - 0.18$

i.e. $h^2 = 0.82$

$\therefore$ $h = \sqrt{0.82}$

$= 0.91$ m is the vertical height of the pyramid.

Calculator value is 0.9055...

We need to calculate the area of the square base. We will use this for both the volume and the surface area.

$$\text{Area of square base} = 0.60^2$$
$$= 0.36 \text{ m}^2.$$

$$\text{Volume of pyramid} = \frac{1}{3} \times \text{Base area} \times \text{Height}$$

becomes

$$\text{Volume} = \frac{1}{3} \times 0.36 \times 0.9055\ldots$$
$$= 0.11 \text{ m}^3.$$

Remember the apex, A, of the pyramid is vertically above the centre, B, of the square base.

This symmetry of the pyramid means that all sloping faces are the same. Fig. 16.8 shows just one face, AXY. In Chapter 6 we looked at 3 different formulae for calculating the area of a triangle. Because triangle AXY is a simple, isosceles, triangle we can easily apply any of the formulae. We choose the one using all 3 sides, replacing b with x and c with y in the usual formula,

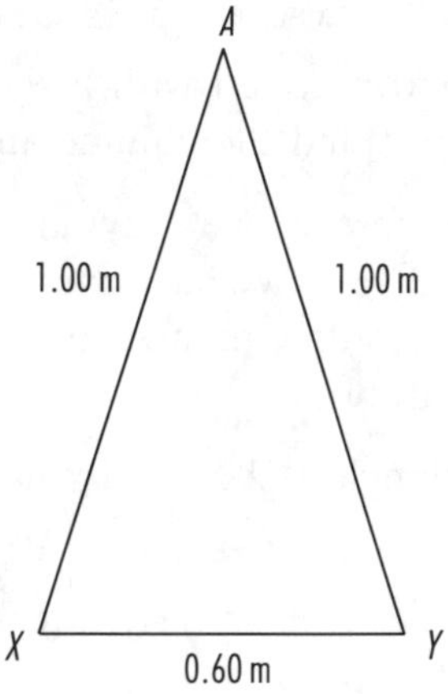

A
1.00 m
1.00 m
X
Y
0.60 m

Fig. 16.8

$$\text{Area} = \sqrt{s(s-a)(s-x)(s-y)}$$

$$\text{where} \quad s = \frac{1}{2}(a+x+y).$$

Using $a = 0.60$, $x = 1.00$ and $y = 1.00$ we get

$$s = \frac{1}{2}(0.60 + 1.00 + 1.00)$$
$$= 1.30 \text{ m} \qquad \text{as half the perimeter.}$$

Applying $\quad s = 1.30$

we get $\quad s - a = 1.30 - 0.60 = 0.70,$

$$s - x = 1.30 - 1.00 = 0.30$$

and $\quad s - y = 1.30 - 1.00 = 0.30.$

These give our area formula as

$$\text{Area} = \sqrt{1.30 \times 0.70 \times 0.30 \times 0.30}$$
$$= \sqrt{0.0819}.$$

$$\text{Total surface area} = \text{Area of base} + 4 \times \text{Area of face } AXY$$
$$= 0.36 + 4 \times \sqrt{0.0819}$$
$$= 0.36 + 1.14$$
$$= 1.50 \text{ m}^2.$$

5. Frustum of a cone

The plural of frustum is frusta.

Suppose we have a cone and remove the top section (i.e. a smaller cone) with a cut parallel to the base. The remaining section is the frustum of the cone. The perpendicular distance between the circular ends is d and the slant height is l. The radius of the top circle is r and of the base circle is R.

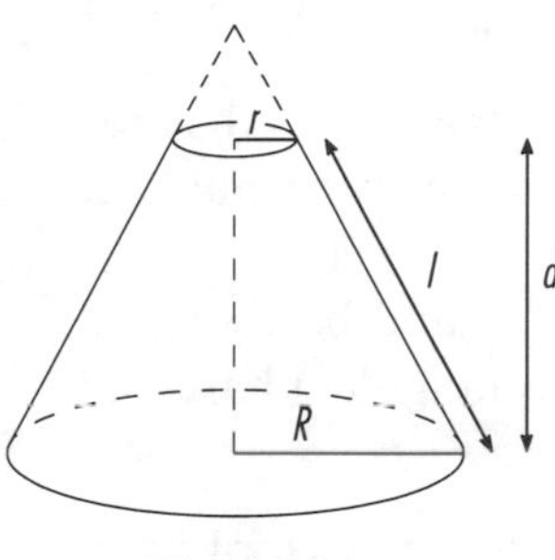

Fig. 16.9

$$\text{Area of top} = a.$$

$$\text{Area of base} = A.$$

$$\text{Volume of frustum} = \frac{1}{3}d(a + \sqrt{aA} + A).$$

$$\text{Curved surface area} = \pi(r + R)l.$$

For the total surface area we add the circular top ($a = \pi r^2$) and base ($A = \pi R^2$) to the curved surface area.

Example 16.6

A drawing office has several waste paper bins. Fig. 16.10 with internal dimensions shows a typical one, the shape being the frustum of a cone. The top radius of 0.175 m and the base radius of 0.125 m are separated by 0.280 m. What is the volume of the bin?

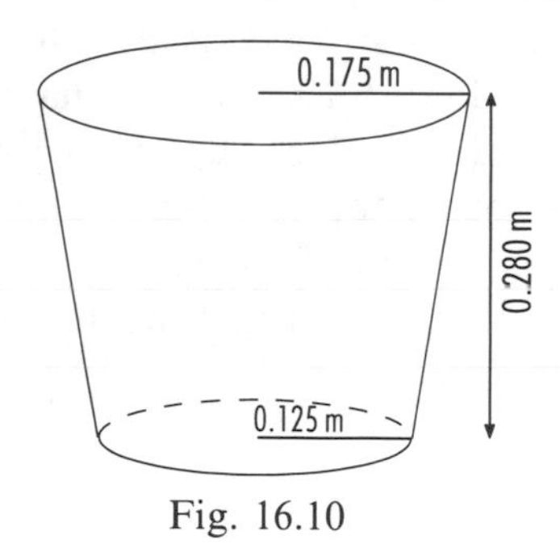

Fig. 16.10

The areas of the top and base are both circular. Remember the area of a circle is πr^2. This means we have

$$\text{Area of top} = \pi(0.175)^2 = 0.096 \text{ m}^2.$$

$$\text{Area of base} = \pi(0.125)^2 = 0.049 \text{ m}^2.$$

In our volume formula for the frustum of a cone we have $a = 0.096$, $A = 0.049$ and $d = 0.280$ so that

$$\begin{aligned}\text{Volume} &= \frac{1}{3}(0.280)(0.096 + \sqrt{(0.096)(0.049)} + 0.049) \\ &= \frac{1}{3}(0.280)(0.096 + 0.069 + 0.049) \\ &= \frac{1}{3} \times 0.280 \times 0.214 \\ &= 0.020 \text{ m}^3 \text{ is the volume of the bin.}\end{aligned}$$

6. Frustum of a pyramid

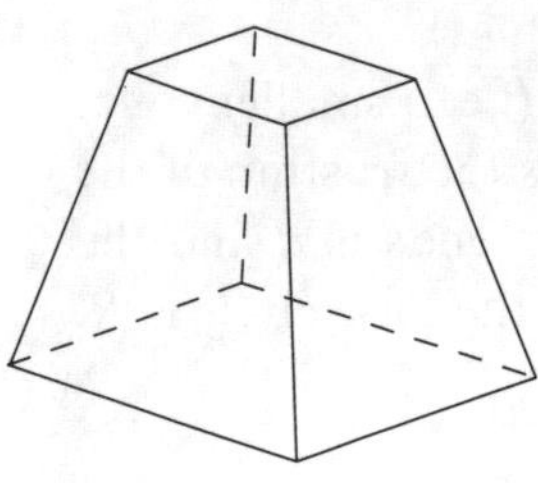
Fig. 16.11

Our diagram (Fig. 16.11) shows a square cross-section for this frustum. The same principle as for a cone applies when creating a frustum of a pyramid. We remove the top section with a cut parallel to the base.

$$\text{Volume of frustum} = \frac{1}{3}\,d\,(a + \sqrt{aA} + A)$$

where a is the area of the top and A is the area of the base, separated by a perpendicular distance, d.

We can find the surface area. To the areas of the top and base we add the sloping faces. For a pyramid with a square (or rectangular) base and top the sloping faces are trapezia.

Total surface area = Area of top + Areas of sloping faces + Area of base.

Example 16.7a

Compared with the waste paper bin of Example 16.6 there is an alternative design. Fig. 16.12, with internal dimensions, shows a frustum of a pyramid with square top and base. The length of the top side is 0.300 m and of the base side is 0.200 m. Find the volume of the bin.

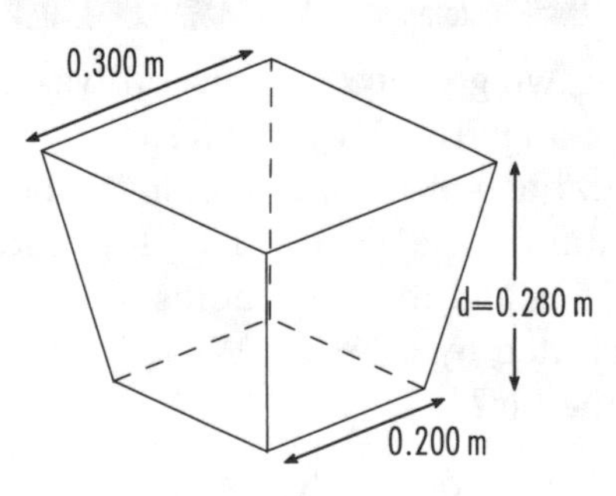

Fig. 16.12

For sides of length l the areas of the squares are l^2. This means we have

$$\text{Area of top} = 0.300^2 = 0.090\ \text{m}^2.$$

$$\text{Area of base} = 0.200^2 = 0.040\ \text{m}^2.$$

In our volume formula for the frustum we have $a = 0.090$, $A = 0.040$ and $d = 0.280$ so that

$$\begin{aligned}
\text{Volume} &= \frac{1}{3}\,(0.280)\,(0.090 + \sqrt{(0.090)\,(0.040)} + 0.040)\\
&= \frac{1}{3}\,(0.280)\,(0.090 + 0.060 + 0.040)\\
&= \frac{1}{3} \times 0.280 \times 0.190\\
&= 0.018\ \text{m}^3 \text{ is the volume of the bin.}
\end{aligned}$$

Example 16.7b

Using the waste paper bin of Example 16.7a find the surface area.

Referring to Figs. 16.13 the left-hand diagram shows the original pyramid from which the frustum was formed. The right-hand diagram shows a vertical cross-section through that original figure. We concentrate on triangle XYZ. For a sloping face in the shape of a trapezium, XY is the distance between the top and base parallel edges.

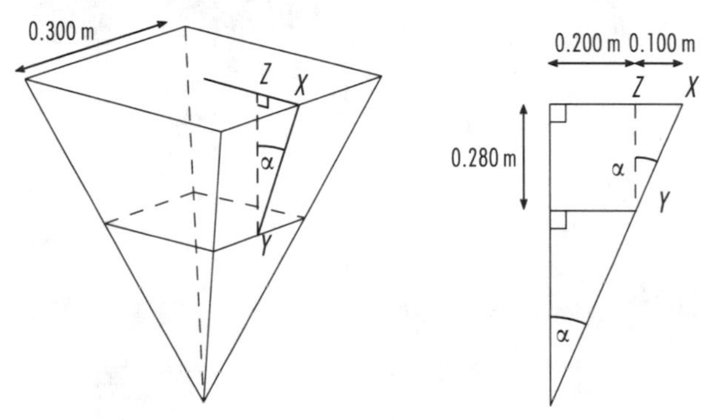

Figs. 16.13

In triangle XYZ, applying Pythagoras' theorem, we get

$$XY^2 = YZ^2 + ZX^2$$

i.e. $$XY^2 = 0.280^2 + 0.100^2$$

i.e. $$XY^2 = 0.0884$$

$\therefore$ $$XY = \sqrt{0.0884} = 0.297 \text{ m.}$$

Calculator value is 0.2973...

Each of the 4 sloping faces of the frustum is a trapezium with parallel sides of 0.300 m and 0.200 m, separated by 0.297 m. Combining these with the square base we get

$$\begin{aligned}\text{Total area} &= \text{Base area} + 4 \times \text{Area of trapezium}\\ &= 0.200^2 + 4 \times \frac{1}{2}(0.300 + 0.200)\,0.297\\ &= 0.34 \text{ m}^2.\end{aligned}$$

7. The prismoidal rule

This is a general formula for calculating the volume of a prism, pyramid, cone or any type of frustum. The volume, V, is

$$V = \frac{h}{6}(A_1 + 4A_m + A_2)$$

where A_1 and A_2 are the areas of the ends a distance h apart. A_m is the cross-sectional area mid-way between the ends.

Obviously for a pyramid or a cone one end has no area, being a vertex. We simply apply the formula, substituting 0 as appropriate.

The next 2 examples are repeats of earlier examples, but using this formula.

Example 16.8

A cone has an altitude of 2.60 m and a base radius of 0.50 m. Find its volume.

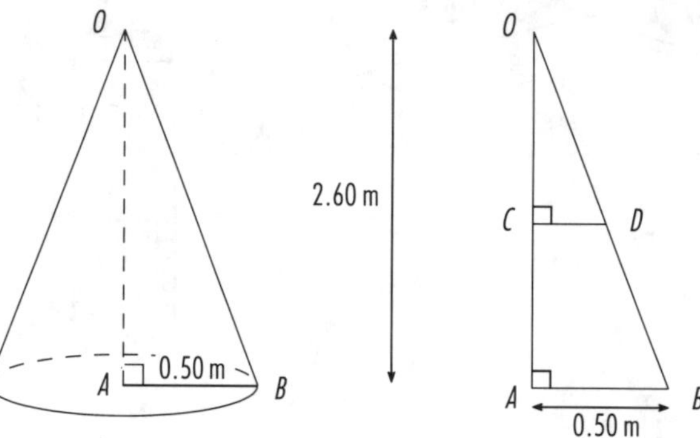

Figs. 16.14

The left-hand diagram in Figs. 16.14 shows the complete cone. The right-hand diagram shows a vertical cross-section through the cone. Triangles *OAB* and *OCD* are similar, i.e. their sides are in proportion. We calculate *CD* because the prismoidal rule requires the cross-sectional area at the middle of the cone. Because *OC* is half of *OA* then *CD* is half of *AB*, i.e. $CD = \frac{1}{2} \times 0.50 = 0.25$ m.

For the prismoidal rule we need to calculate the circular cross-sectional areas (πr^2) that are perpendicular to *OCA* ($h = 2.60$).

At the top, $r = 0$; $A_1 = \pi 0^2 = 0\,\text{m}^2$.
At the middle, $r = 0.25$; $4A_m = 4 \times \pi(0.25)^2 = 0.785\,\text{m}^2$.
At the base, $r = 0.50$; $A_2 = \pi(0.50)^2 = 0.785\,\text{m}^2$.

Now we substitute these values into our formula

$$V = \frac{h}{6}(A_1 + 4A_m + A_2)$$

so that $$\text{Volume} = \frac{2.60}{6}(0 + 0.785 + 0.785)$$

i.e. $$\text{Volume} = \frac{2.60}{6} \times 1.570$$

$$= 0.68\ \text{m}^3.$$

Example 16.9

A frustum of a pyramid has a square top and base (see Fig. 16.15). The length of the top side is 0.300 m and of the base side is 0.200 m. They are separated by 0.28 m. Find the volume of the frustum.

Again using similar triangles we can calculate the side of the middle cross-section. This works out at 0.25 m. For sides of length l the areas of the squares are l^2. This means we have

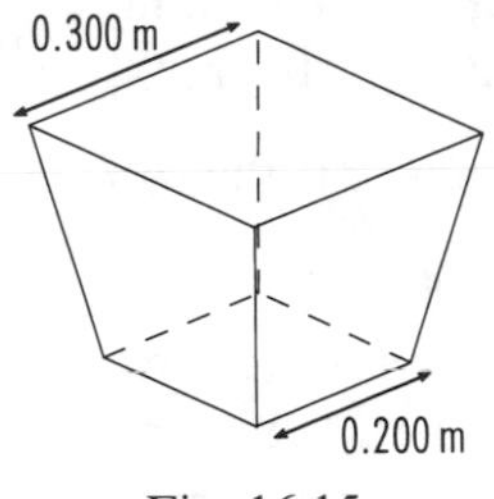

Fig. 16.15

At the top,

$$l = 0.20; \quad A_1 = 0.20^2 \quad = 0.04 \text{ m}^2.$$

At the middle,

$$l = 0.25; \quad 4A_m = 4 \times 0.25^2 = 0.25 \text{ m}^2.$$

At the base,

$$l = 0.30; \quad A_2 = 0.30^2 \quad = 0.09 \text{ m}^2.$$

Now we substitute these values into our formula

$$V = \frac{h}{6}(A_1 + 4A_m + A_2)$$

so that $$\text{Volume} = \frac{0.28}{6}(0.04 + 0.25 + 0.09)$$

i.e. $$\text{Volume} = \frac{0.28}{6} \times 0.38$$

$$= 0.018 \text{ m}^3.$$

EXERCISE 16.1

1. For a sphere of radius 0.275 m calculate the volume and surface area.
2. A solid sphere of radius 0.25 m is placed in a hollow cone. The cone has a base radius of 0.50 m and a vertical height of 1.20 m. What is the volume of air in the cone?
3. All the surfaces of the waste paper bin in Example 16.6 are painted. What is the total surface area covered in paint?
4. A cylindrical piece of alloy has a radius of 0.150 m and length 2.450 m. Calculate its volume.
   The cylinder is melted down. How many spheres of radius 0.125 m can be formed from the molten metal?
   If there is 7.5% waste due to spillage during melting and re-casting how many spheres would now be made?
5. There is a solid pyramid with a square base of side 0.60 m and vertical height 1.80 m. Find its volume and surface area.
   A cylinder has the same height and volume as the pyramid. What is the radius of the cylinder?
6. A cone has a vertical height of 1.50 m and base radius 0.48 m. Draw the isosceles triangle representing the vertical cross-section through the cone. Calculate the volume of the cone.

The small top conical section of height 0.50 m is removed. By calculation, with the appropriate explanation, show that its base radius is 0.16 m. Calculate the volume of this small cone.
Using your two volume answers find the volume of the remaining frustum. Check your answer using the usual frustum formula, $V = \frac{1}{3} d(a + \sqrt{aA} + A)$, mentioned earlier in this chapter. a is the top area, A is the base area and d is the separation of these areas.

7 Calculate the volume and surface area of a sphere of radius 0.150 m. How many such spheres can be cast from $1\,m^3$ of metal? If each sphere is to be spray painted what will be the total surface area needing treatment?
To improve the finished quality each sphere is to have 3 coats of paint. If 1 l of paint will cover $12\,m^2$ how much paint is required?

8 A solid frustum has a rectangular base and top separated by 0.85 m. The dimensions of the base are 1.25 m and 0.75 m. The dimensions of the top are 1.00 m and 0.60 m. Calculate the total surface area of the frustum.

9 The diagram shows a vertical cross-section through a sphere containing a cone. The volume of the cone is $72\pi\,m^3$. Using some simple trigonometry find the dimensions of the cone and sphere. Hence find the volume of the sphere. Show that the surface area of the sphere is $\frac{400}{3}\pi\,m^2$.
The radius of the cone is r and of the sphere is R. The height of the cone is h where $\frac{2}{3}h = R$.

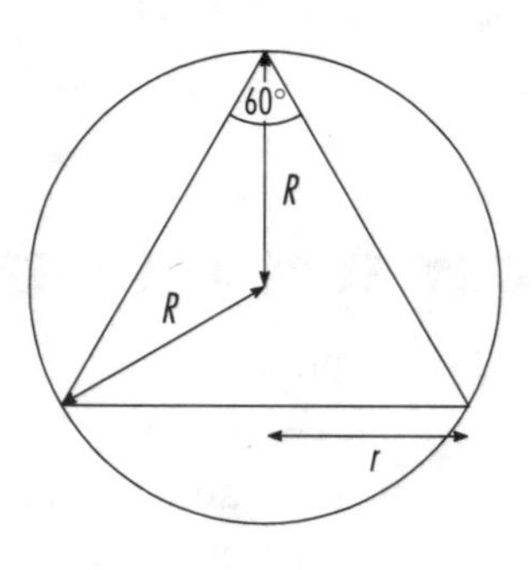

10 Find the volume and total surface area of a hemisphere of radius 0.50 m.

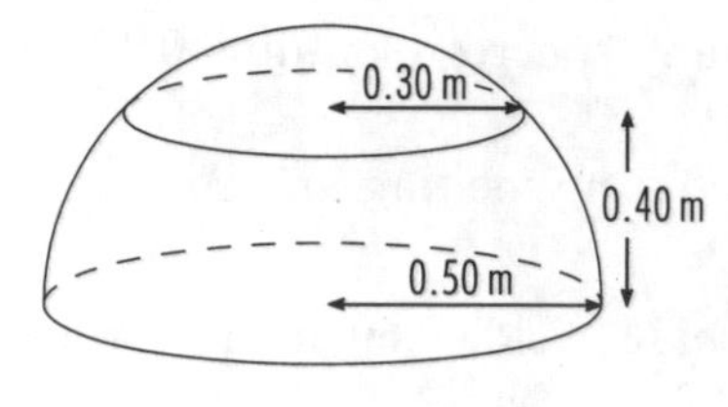

A spherical frustum can be formed by slicing the top cap from a hemisphere as shown in the diagram. The circular top of the frustum has a radius of 0.30 m and is 0.40 m (i.e. h) from the circular base.

Find the volume, V, of the frustum where $V = \frac{\pi h}{6}(h^2 + 3r^2 + 3R^2)$. r is the radius of the top circle and R is the radius of the base circle. By subtraction also find the volume of the cap that has been removed.

Composite bodies

When we looked at quadrilaterals in Chapter 6 we attempted to split them into sections. The aim was to make each section a regular figure (e.g. triangle, square, rectangle, trapezium) so that we could apply standard formulae.

For composite bodies we aim to apply the same principle, just using volumes because we are in 3 dimensions.

Example 16.10

Our diagram (Fig. 16.16) shows a solid cone of height 0.750 m and base radius 0.250 m above a solid hemisphere. The flat surfaces coincide exactly. What is the volume of the composite body?

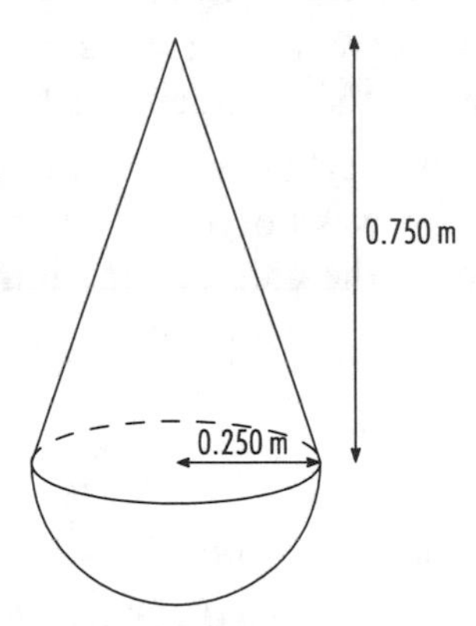

Fig. 16.16

Immediately we split our body into the separate cone and hemisphere (see Figs. 16.17).

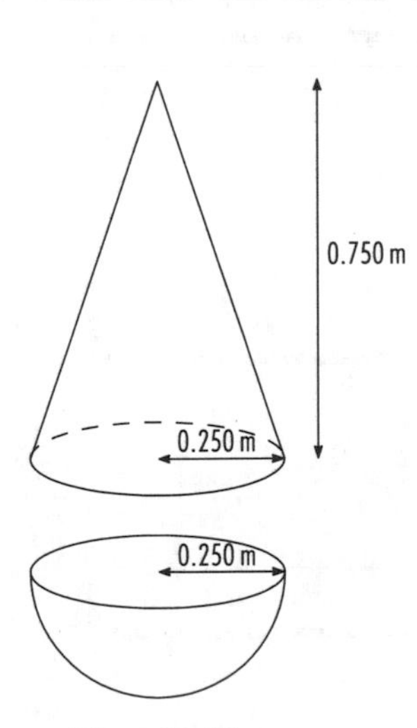

Figs. 16.17

For the cone

$$\text{Volume} = \frac{1}{3}\pi r^2 h$$

becomes

$$\text{Volume} = \frac{1}{3}\pi (0.250)^2(0.750)$$
$$= 0.049 \text{ m}^3.$$

Calculator value is 0.0490...

For the hemisphere

$$\text{Volume} = \frac{1}{2} \times \text{Volume of sphere}$$
$$= \frac{1}{2} \times \frac{4}{3}\pi r^3$$
$$= \frac{2}{3}\pi r^3$$

becomes

$$\text{Volume} = \frac{2}{3}\pi(0.250)^3$$
$$= 0.033 \text{ m}^3.$$

Calculator value is 0.0327...

$$\text{Total volume} = \text{Volume of cone} + \text{Volume of hemisphere}$$
$$= 0.0490\ldots \quad + \quad 0.0327\ldots$$
$$= 0.082\ \text{m}^3.$$

Example 16.11a

Our diagram in Fig. 16.18 started as a solid cylinder of height 1.260 m and radius 0.340 m. A hemisphere of radius 0.400 m was bonded to one end of the cylinder.

Now there is a constant overlap of 0.06 m between the cylinder and hemisphere.

Finally a cylinder of radius 0.170 m and length 1.280 m was bored out of the composite body. What is the volume remaining?

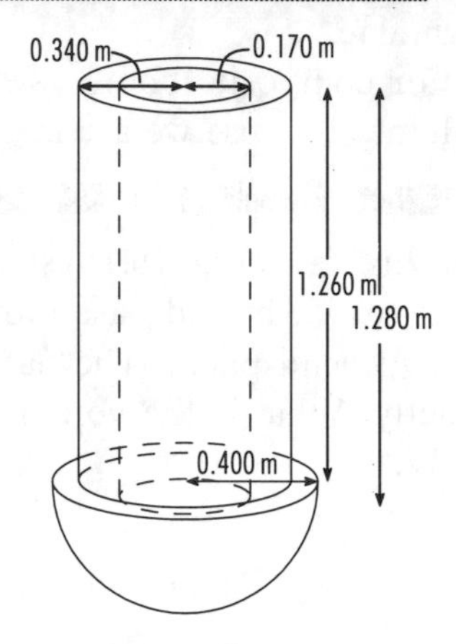

Fig. 16.18

We use our standard results in turn.

For the **original cylinder** $h = 1.260$ and $r = 0.340$ so that

$$\text{Volume} = \pi r^2 h$$

becomes $\text{Volume} = \pi(0.340)^2 1.260 = 0.458\ \text{m}^3.$

Calculator value is 0.4575...

For the **hemisphere** $r = 0.400$ so that

$$\text{Volume} = \frac{2}{3}\pi r^3$$

becomes $\text{Volume} = \frac{2}{3}\pi(0.400)^3 = 0.134\ \text{m}^3.$

Calculator value is 0.1340...

For the **cylindrical bore** $h = 1.280$ and $r = 0.170$ so that

$$\text{Volume} = \pi r^2 h$$

becomes $\text{Volume} = \pi(0.170)^2 1.280 = 0.116\ \text{m}^3.$

Calculator value is 0.1162...

Now we gather together all our results to get

$$\text{Remaining volume} = \text{Volume of cylinder} + \text{Volume of hemisphere} - \text{Volume of cylindrical bore}$$
$$= 0.4575\ldots + 0.1340\ldots - 0.1162\ldots$$
$$= 0.475\ \text{m}^3.$$

Example 16.11b

We use the same body and dimensions of Example 16.11a. This time we calculate the complete surface area.

We work through our calculation in stages.

The top of the body is an annulus (ring).

$$\textbf{Area of annulus} = \text{Area of large circle} - \text{Area of small circle}$$
$$= \pi(0.340)^2 - \pi(0.170)^2 = 0.2724 \text{ m}^2.$$

The inner bored base is a circle.

$$\textbf{Area} = \pi(0.170)^2 = 0.0908 \text{ m}^2.$$

For the original cylinder we need the curved surface area.

$$\textbf{Outer area} = 2\pi(0.340)(1.260) = 2.6917 \text{ m}^2.$$
$$\textbf{Inner area} = 2\pi(0.170)(1.280) = 1.3672 \text{ m}^2.$$

At the join of the cylinder and hemisphere we have another annulus.

$$\textbf{Area of annulus} = \pi(0.400)^2 - \pi(0.340)^2 = 0.1395 \text{ m}^2.$$

The curved surface area of the hemisphere is half that of a sphere.

$$\textbf{Area} = \frac{1}{2} \times 4\pi(0.400)^2 = 1.0053 \text{ m}^2.$$

We add all our results to get the complete surface area, i.e.

$$\text{Surface area} = 0.2724 + 0.0908 + 2.6917 + 1.3672 + 0.1395 + 1.0053$$
$$= 5.57 \text{ m}^2.$$

ASSIGNMENT

We will look at the bollard in 2 sections, the hollow hexagonal frustum and the solid sphere. The base hexagon has a side of length 0.20 m and the top hexagon has a side of length 0.12 m. The height of the bollard is 1.20 m. On top of the frustum is a sphere of diameter 0.24 m.

For the frustum we use our standard volume formula for V,

$$V = \frac{1}{3} d(a + \sqrt{aA} + A)$$

where a is the area of the top and A is the area of the base. d is the height of the frustum. We have calculated the area of a hexagon before. A hexagon is a series of 6 adjacent equilateral triangles, which we have drawn in Fig. 16.19.

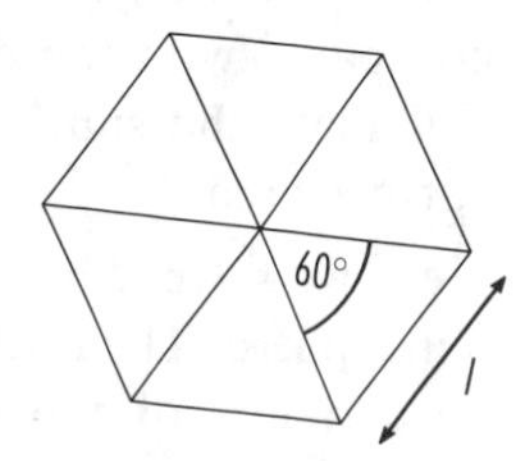

Fig. 16.19

For a triangle of side l,

$$\text{Area of triangle} = \frac{1}{2} l^2 \sin 60°.$$

$\therefore$ Area of hexagon $= 6 \times \frac{1}{2} l^2 \sin 60°$

$= 3l^2 \sin 60°.$

For the top $l = 0.12$ m so that

$$\text{Top area} = 3\,(0.12)^2(0.866) = 0.037 \text{ m}^2.$$

For the base, $l = 0.20$ m so that

$$\text{Base area} = 3(0.20)^2(0.866) = 0.104 \text{ m}^2.$$

In our volume formula $a = 0.037$, $A = 0.104$ and $d = 1.20$ to give

$$\begin{aligned} \text{Volume} &= \frac{1}{3}\,(1.20)\,(0.037 + \sqrt{(0.037)\,(0.104)} + 0.104) \\ &= \frac{1}{3} \times 1.20 \times 0.203 \\ &= 0.081 \text{ m}^3. \end{aligned}$$

Now let us calculate the volume of the solid sphere. If the diameter is 0.24 m this means the radius, r, is 0.12 m.

Using $\qquad \text{Volume} = \frac{4}{3}\pi r^3$

we get $\qquad \text{Volume} = \frac{4}{3}\pi(0.12)^3$

$= 0.007 \text{ m}^3.$

We extend this Assignment in the next set of exercises, asking you to calculate the volume of metal needed.

EXERCISE 16.2

1 We have a metal cylinder of diameter 1.00 m and length 3.60 m. Five separate cylindrical holes of diameter 0.25 m are drilled along the original cylinder. The diagram shows a cross-section through the cylinder at right angles to its length.
What is the volume of the finished metal?
What percentage of metal does this contain compared with the original cylinder?
Calculate the sum of the curved surface areas and hence the total surface area.

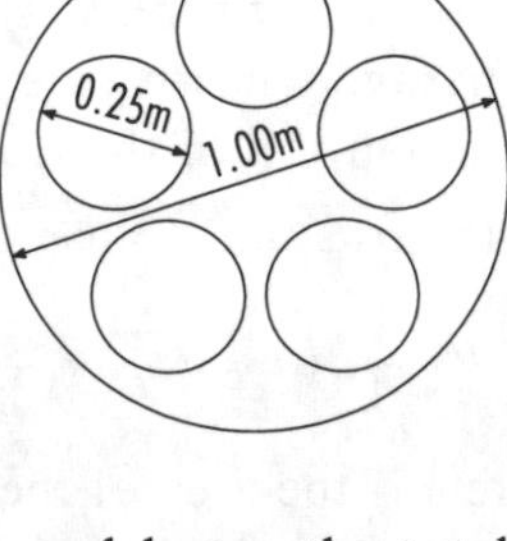

2 Here we have a hollow cone welded to a hollow hemisphere. The radius of the hemisphere and the base radius of the cone are both 0.65 m. The vertical height of the cone is twice its base radius. The internal and external dimensions are almost the same because the metal shell is so thin.

Calculate the volume of this composite shell.
What is the curved surface area?
This shell is to be decorated with 9 circular self-adhesive advertising stickers. If a typical circle has a radius of 100 mm what area remains uncovered by advertisements?

3 The radius of a hemisphere is 0.40 m. The diagram includes a smaller hemispherical hollow of radius 0.20 m. What is the volume of the remaining body?

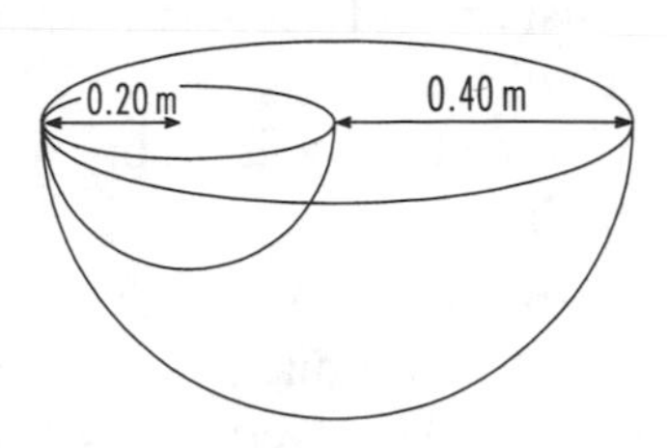

4 Our diagram shows a cross-section through a 5 m length of reinforced plastic channelling. Calculate the shaded area now the semicircle has been moulded within the trapezium. What is the volume of plastic in a 5 m length?

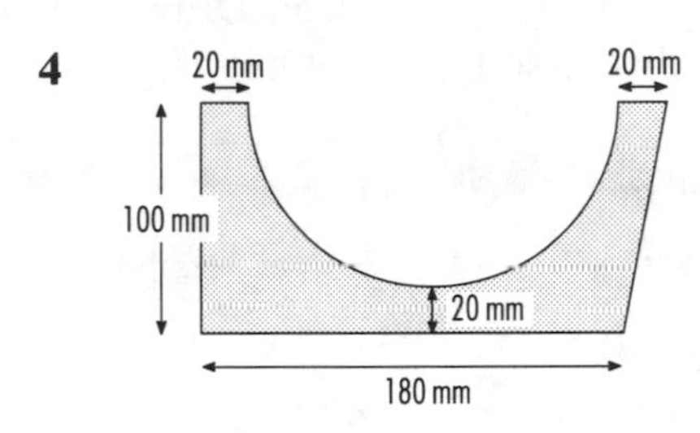

5 A frustum of a pyramid has a square base of side 0.35 m and a square top of side 0.25 m. Its height is 0.64 m. What volume remains once a drill has bored out a circular hole of radius 0.08 m from top to base?

6 A solid cone of height 0.75 m has a base radius of 0.25 m. It is bonded onto a solid hemisphere of radius 0.20 m, creating a constant overlap of 0.05 m. Calculate the total surface area in need of painting.

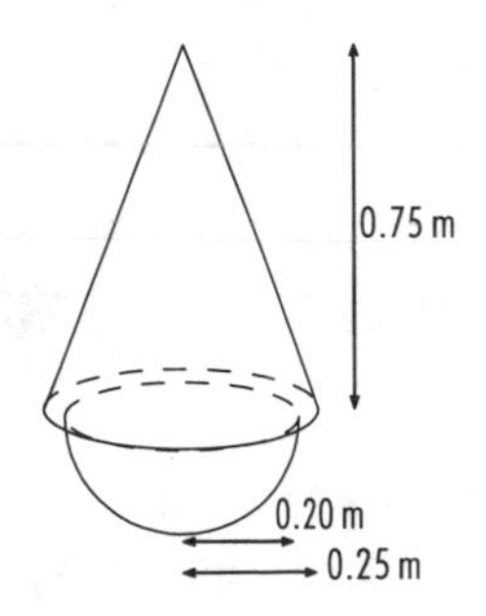

7 A hemisphere has a diameter of 1.00 m. Four smaller hemispheres of diameter 0.25 m are bored out of the larger one. One diagram shows a plan whilst the other shows a cross-section through the hemispheres. Calculate the remaining volume and total surface area.

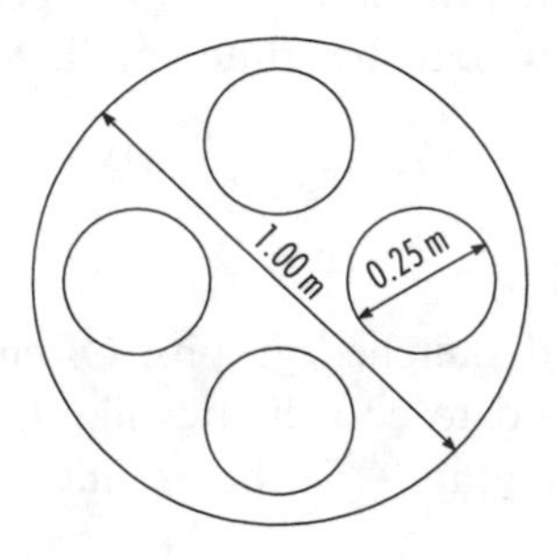

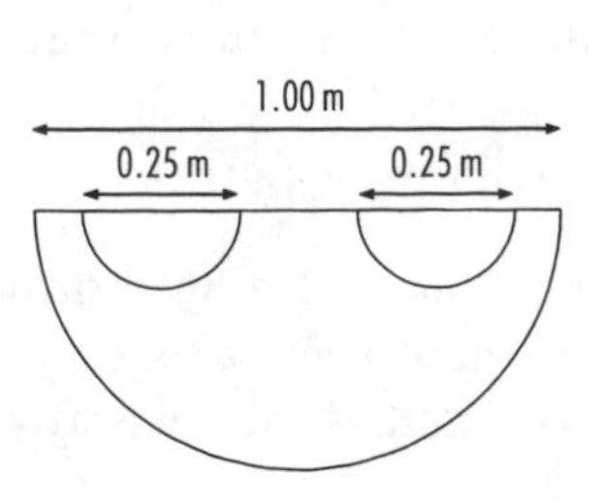

8 This question is based on the Assignment.
We know that the frustum is hollow and the sphere is solid. The inside of the hexagon is a frustum of a cone. The diameters of this base and top are 0.36 m and 0.20 m respectively. The height of the conical frustum is 1.10 m. By calculating the volume of the conical frustum and using subtraction find the total volume of metal used in a bollard.
If the density is approximately 8000 kgm^{-3} find the mass of 60 bollards.

9 A cylinder has a radius of 0.30 m and a height of 0.90 m. Fitting exactly over one flat end is a hemisphere and over the opposite end is a cone. The overall length of this composite body is 1.50 m. Draw a cross-sectional diagram to represent this description. What is its volume?

10 Find the volume of a cone of height 0.90 m and base radius 0.40 m. The cross-sectional diagram shows a smaller cone of half this height and of the same base radius machined from the original one. Calculate the remaining volume and the new surface area.
Explain why the volume of the smaller cone is half the volume of the larger one.

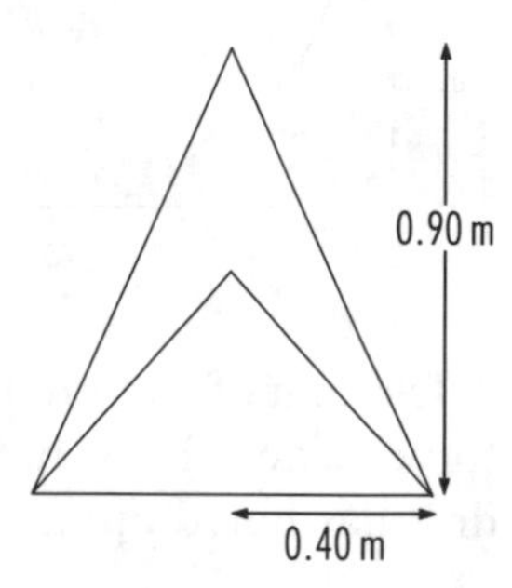

Irregular shaped areas

In this section we move away from standard shapes. This is when we are unable to split up a composite body into standard component shapes. Hence we are unable to apply any of our earlier standard formulae.

We have 3 methods for finding areas of irregular shapes. They are

i) the mid-ordinate rule,
ii) the trapezoidal rule
and iii) Simpson's rule.

These are numerical methods for finding areas. In many cases you get an approximate solution, though often the error can be quite small. With standard formulae you get an accurate answer.

1. The mid-ordinate rule

Many times in this book we have drawn and sketched graphs. Often we have referred to the vertical axis or y-axis and to coordinates like (x, y), e.g. (4, 7). The ordinate is the vertical value, e.g. in (4, 7) the ordinate is 7.

When we apply the mid-ordinate rule we usually split the area into vertical (associated with ordinate) strips. All the strips must be of **equal width**. Let us look at an irregular area, Fig. 16.20. It may be the plan of a small lake. We split it into strips of equal width, w.

> The method works just as well for horizontal strips.

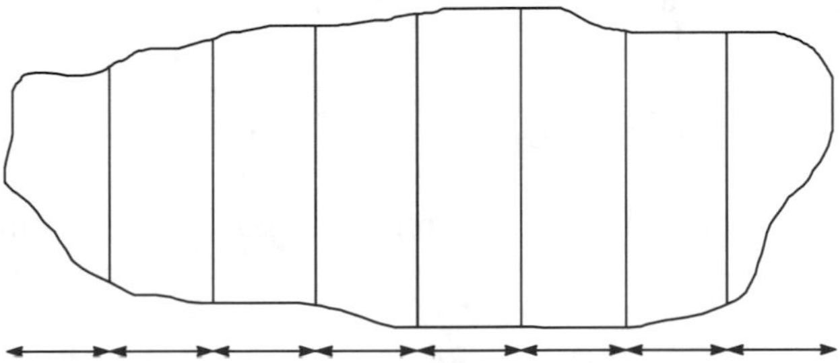

Fig. 16.20

We can look more closely at one of these strips, perhaps choosing the second one as a specimen. This is still irregularly shaped (Fig. 16.21). Suppose we draw a vertical mid-placed line through AB and then horizontal lines through A and B. We suggest that the rectangle $PQRS$ is approximately equal in area to the original strip (Fig. 16.22). Some small sections of the strip lie outside the rectangle. These will approximately fill in the gaps where the rectangle is greater than the strip. We repeat this technique for all our strips. Once complete we will have approximated our irregular shape to a series of rectangles. All the rectangles will have the same width. Their heights will be based on the mid-strip heights (i.e. mid-ordinates).

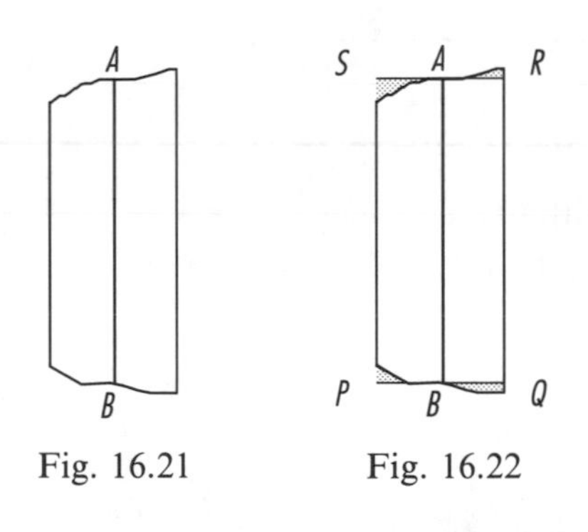

Fig. 16.21 Fig. 16.22

$$\begin{aligned}\text{Area} &= \text{Sum of all rectangular areas}\\ &= (w \times \text{mid-ordinate}_1) + (w \times \text{mid-ordinate}_2) +\\ &\qquad (w \times \text{mid-ordinate}_3) + (w \times \text{mid-ordinate}_4) + \ldots\\ &= w \times (\text{mid-ordinate}_1 + \text{mid-ordinate}_2 + \text{mid-ordinate}_3 \ldots)\end{aligned}$$

We write this as

$$\text{Area} = \text{Strip width} \times \text{Sum of mid-ordinates.}$$

The mid-ordinate rule will work for any area, including standard ones. Generally the more strips we use the more closely each strip resembles a rectangle. This leads to improved accuracy. We look at accuracy and the number of strip widths in Example 16.12b. In the next example we apply the rule to the area under a curve.

Example 16.12a

Use the mid-ordinate rule to find the area between the curve $y = 4 - x^2$ and the horizontal axis from $x = 1.00$ to 2.00.

How many strips to use is our first decision. Suppose we use 5 strips each of width 0.20. These will be 1.00 to 1.20, 1.20 to 1.40, ... and 1.80 to 2.00. We need to find the values of y (ordinates) at the mid-points of these strips, i.e. at $x = 1.10$, 1.30, ... and 1.90. The table of values is

x	1.10	1.30	1.50	1.70	1.90
y	2.79	2.31	1.75	1.11	0.39

We have sketched the curve in Fig. 16.23, but it is not vital for our example.

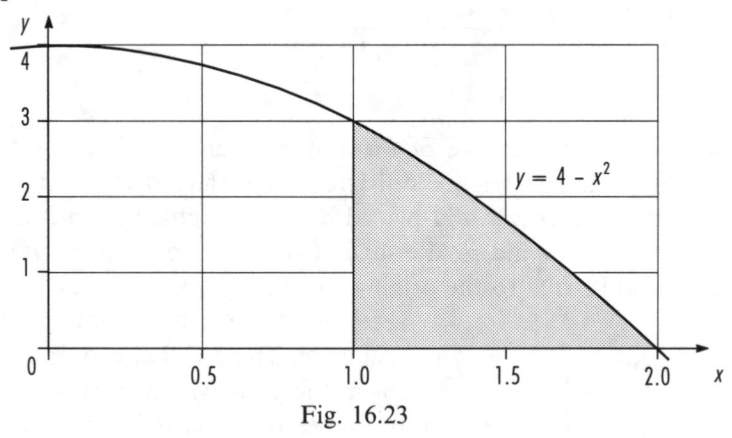

Fig. 16.23

Using $\quad$ Area = Strip width × Sum of ordinates

we get

$$\begin{aligned} \text{Area} &= 0.20(2.79 + 2.31 + 1.75 + 1.11 + 0.39) \\ &= 0.20 \times 8.35 \\ &= 1.67 \text{ unit}^2. \end{aligned}$$

2. The trapezoidal rule

As the name suggests this method uses a series of trapezia (remember **trapezia** is the plural of **trapezium**) as approximations. We split the area into strips (vertically or horizontally) of equal width as in Fig. 16.24.

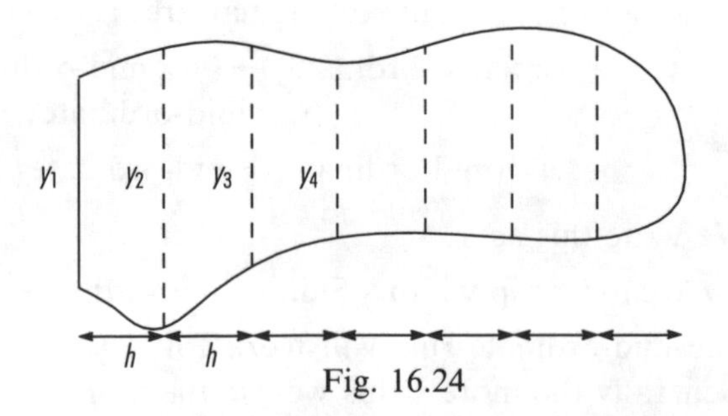

Fig. 16.24

Each strip is approximated to a trapezium and its area calculated using the formula, Area $=\frac{1}{2}(a+b)h$, in the usual way. The complete area of the irregular shape is the sum of all these trapezia. You should notice as we develop the formula that most of the y values appear twice. The exceptions are the first and last y values. All the others appear as the right parallel side of one trapezium and the left parallel side of the next trapezium. For example in the formula you will see $\frac{1}{2}y_3+\frac{1}{2}y_3$ simplify to y_3. Now we can write

$$\begin{aligned}\text{Area} &= \text{Sum of all trapezia}\\ &= \tfrac{1}{2}(y_1+y_2)h+\tfrac{1}{2}(y_2+y_3)h+\tfrac{1}{2}(y_3+y_4)h+\ldots\\ &= h\{\tfrac{1}{2}y_1+\tfrac{1}{2}y_2+\tfrac{1}{2}y_2+\tfrac{1}{2}y_3+\tfrac{1}{2}y_3+\ldots+\tfrac{1}{2}y_n\}\\ &= h\{\tfrac{1}{2}y_1+\tfrac{1}{2}y_n+y_2+y_3+\ldots\}\\ &= h\{\tfrac{1}{2}(y_1+y_n)+y_2+y_3+\ldots\}.\end{aligned}$$

We can think of this formula as

$$\text{Area} = \text{Strip width}\times\{\tfrac{1}{2}(\text{Sum of first and last ordinates}) + \text{Sum of remaining ordinates}\}.$$

When you apply this formula make sure you use each value of y once only in the correct position.

Example 16.12b

Use the trapezoidal rule to find the area between the curve $y=4-x^2$ and the horizontal axis from $x=1.00$ to 2.00. We are going to attempt this example twice, with different strip widths. The larger the strip width the less the strip resembles a trapezium and so the less accurate the answer.

i) As an extreme case suppose we use just 2 strips. From $x=1.00$ to 2.00 this means the width is 0.5.
Our table of values is

x	1.00	1.50	2.00
y	3.00	1.75	0

Just because the last value of y is 0 do not worry. It is as acceptable as any other value.
Using our area formula we get

$$\begin{aligned}\text{Area} &= 0.50\{\tfrac{1}{2}(3.00+0)+1.75\}\\ &= 1.625\ \text{unit}^2.\end{aligned}$$

You can round this area answer to either 1.62 or 1.63.

ii) For a more likely case we use 4 strips, each of width 0.25, with the table

x	1.00	1.25	1.50	1.75	2.00
y	3.0000	2.4375	1.7500	0.9375	0

You might find it useful to label these y values now there are a few more of them as

$$y_1 \qquad y_2 \qquad y_3 \qquad y_4 \qquad y_n$$

Notice the pattern of subscripts, the last one being n and there being 4 strips.
Using our area formula with $h = 0.25$ we get

$$\begin{aligned} \text{Area} &= 0.25\{\tfrac{1}{2}(3.0000 + 0) + 2.4375 + 1.7500 + 0.9375\} \\ &= 0.25\{1.500 + 5.125\} \\ &= 0.25 \times 6.625 \\ &= 1.66 \text{ unit}^2. \end{aligned}$$

These calculations have provided us with 2 answers. At this stage you do not know which is the more accurate. However we can write that the true value is $1.\bar{6}$. We expect a more accurate answer when we split the area into greater rather than fewer strips. This means each strip should be quite narrow for accuracy.

3. Simpson's rule

Again we split the area into a series of equal width strips. It is vital that we have an **even number of strips**, and hence an odd number of ordinates. To help you remember this you can think of an **even** number of wooden fencing panels (i.e. strips) between an odd number of concrete posts (i.e. ordinates).

Fig. 16.25 shows an irregular area labelled with subscripted ys and equal widths h.

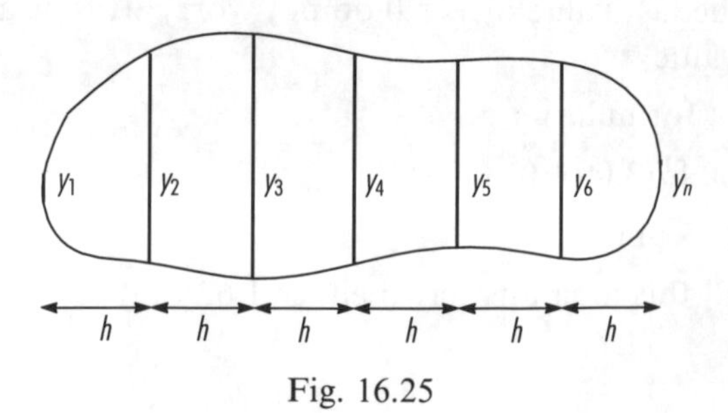

Fig. 16.25

The formula for area using Simpson's rule is

$$\text{Area} = \frac{h}{3}\left\{y_1 + y_n + 4(y_2 + y_4 + y_6 + \ldots) + 2(y_3 + y_5 + \ldots)\right\}$$

We can write this as

$$\text{Area} = \frac{1}{3} \times \text{Strip width} \times \left\{\begin{matrix}\text{sum of first and}\\ \text{last ordinates}\end{matrix} + 4\begin{pmatrix}\text{sum of even}\\ \text{ordinates}\end{pmatrix} + 2\begin{pmatrix}\text{sum of remaining}\\ \text{odd ordinates}\end{pmatrix}\right\}$$

In this formula the *even* of *even ordinates* and the *odd* of *odd ordinates* refer to the subscripts of y.

Example 16.12c

Use Simpson's rule to find the area between the curve $y = 4 - x^2$ and the horizontal axis from $x = 1.00$ to 2.00. Remember that we use an **even** number of strips for Simpson's rule. In this calculation we are going to use 8 strips. We could have used 4 strips again or possibly 6. We avoid using 6 strips in this case. 6 does not readily nor easily divide into our overall width from $x = 1.00$ to 2.00. The resulting decimal of $0.1\bar{6}$ would be awkward.

Using 8 strips means we have a strip width of 0.125.

We construct a table of values for our equation.

x	1.000	1.125	1.250	1.375	1.500	1.625	1.750	1.875	2.000
y	3.0000	2.7344	2.4375	2.1094	1.7500	1.3594	0.9375	0.4844	0

Again we can label our values of y. With so many of them it is very useful during substitution into our formula.

y_1 y_2 y_3 y_4 y_5 y_6 y_7 y_8 y_n

We substitute into our formula to get

$$\begin{aligned}\text{Area} &= \frac{1}{3}(0.125)\{(3.0000 + 0) + 4(2.7344 + 2.1094 + 1.3594 + 0.4844) \\ &\qquad + 2(2.4375 + 1.7500 + 0.9375)\} \\ &= \frac{1}{3}(0.125)\{3.0000 + 4(6.6876) + 2(5.1250)\} \\ &= \frac{1}{3}(0.125)\{3.0000 + 26.7504 + 10.2500\} \\ &= \frac{1}{3} \times 0.125 \times 40.0004 \\ &= 1.67 \text{ unit}^2.\end{aligned}$$

Example 16.13

Find the area bounded by the curve $y=5x^2+3$, the vertical axis and the line $y=10.2$.

We sketch this curve in Fig. 16.26. We are interested in an area different to that of the previous examples. This is the shaded area, for which we have 2 methods of calculation.

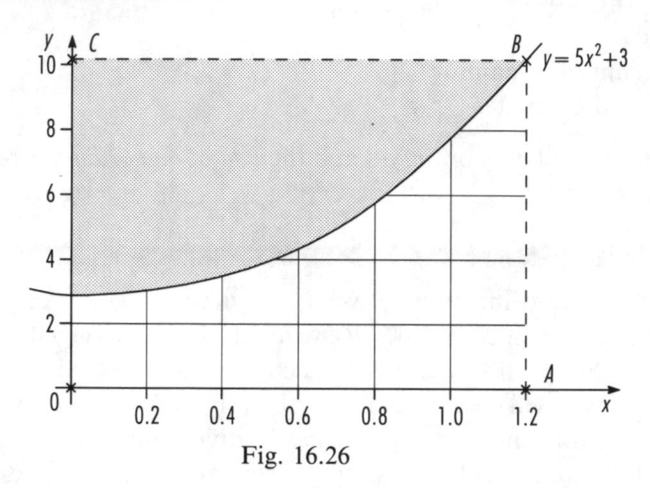

Fig. 16.26

We can find the area of the rectangle $OABC$ and subtract the area under the curve. Alternatively we can split the shaded area into a series of horizontal strips of equal width. Let us look at both of these methods.

i) Firstly we can work out the coordinates $OABC$ by simple substitution into our original equation. These are $O(0,0)$, $A(1.2,0)$, $B(1.2,10.2)$ and $C(0,10.2)$. Hence rectangle $OABC$ has dimensions of 1.2 and 10.2.

$$\begin{aligned}\text{Area of } OABC &= 1.2\times 10.2\\ &= 12.24 \text{ unit}^2.\end{aligned}$$

For the area under the curve we use 6 strips each of width 0.20. This is because 1.2 is easily divided by 6. Then we construct a table of values in the usual way.

x	0	0.20	0.40	0.60	0.80	1.00	1.20
y	3.00	3.20	3.80	4.80	6.20	8.00	10.20
	y_1	y_2	y_3	y_4	y_5	y_6	y_n

We apply Simpson's rule to calculate the area, but alternatively could have used either the mid-ordinate or trapezoidal rules.

$$\begin{aligned}\text{Area} &= \frac{1}{3}(0.20)\{(3.00+10.20)+4(3.20+4.80+8.00)\\ &\qquad +2(3.80+6.20)\}\end{aligned}$$

$$= \frac{1}{3}(0.20)\{13.20 + 4(16.00) + 2(10.00)\}$$

$$= \frac{1}{3}(0.20)\{13.20 + 64.00 + 20.00\}$$

$$= \frac{1}{3} \times 0.20 \times 97.20$$

$$= 6.48 \text{ unit}^2.$$

We substitute our answers to give

Shaded area = Area of rectangle $OABC$ − Area under curve

as Shaded area = 12.24 − 6.48

= 5.76 unit².

ii) Now for the method directly working out the shaded area. Our values of y range from 3.00 to 10.20. It is convenient to split this into 6 equal strips of width 1.20. However we could use more strips if we so wished. To find the values of x corresponding to the y values we need to transpose our original equation. We looked at transposition earlier in Chapter 14 but include the necessary steps as revision. We omit the reasons so that you think carefully about each algebraic re-arrangement.

$$5x^2 + 3 = y$$

becomes $$5x^2 = y - 3$$

i.e. $$x^2 = \frac{y-3}{5}$$

$\therefore$ $$x = \sqrt{\frac{y-3}{5}}.$$

The positive square root is consistent with Fig. 16.26. Now we can construct our table of values.

y	3.00	4.20	5.40	6.60	7.80	9.00	10.20
x	0	0.490	0.693	0.849	0.980	1.095	1.200
	x_1	x_2	x_3	x_4	x_5	x_6	x_n

Again using Simpson's rule we get

$$\text{Area} = \frac{1}{3}(1.20)\{(0 + 1.200) + 4(0.490 + 0.849 + 1.095) + 2(0.693 + 0.980)\}$$

$$= \frac{1}{3}(1.20)\{1.200 + 9.736 + 3.346\}$$

$$= \frac{1}{3} \times 1.20 \times 14.282$$

$$= 5.71 \text{ unit}^2.$$

You can see we have 2 slightly different answers. The question is, which answer is correct? It is the first value. The second method makes an approximation when using the square root and so creates this slight change.

From many of the graphs we have plotted we know the values of y below the horizontal axis are negative. This means that any area below the horizontal axis involves the multiplication of a positive strip width and negative ordinates. Hence any **area below the horizontal axis is negative**. We look at this in the next example.

$(+)(-)=(-)$.

Example 16.14

The area under a velocity-time curve represents displacement. The equation $v=(2-t)^3$ relates a vehicle's velocity, $v\,\mathrm{ms}^{-1}$, to time, t s. Use the trapezoidal rule to find the displacement after 3 seconds and the distance travelled during this time.

Let us look at this curve in Fig. 16.27.

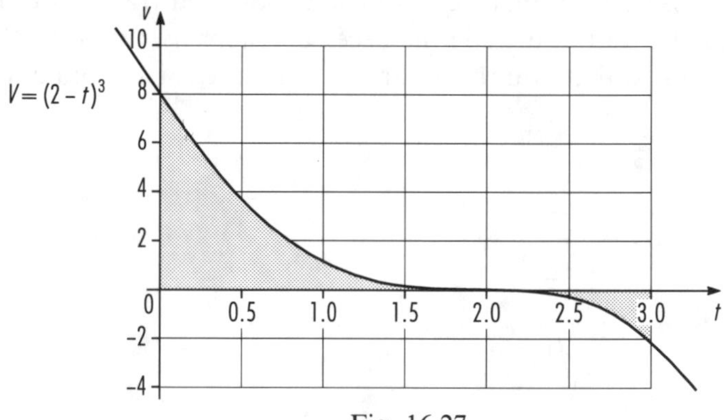

Fig. 16.27

Notice that the curve crosses the horizontal axis at $t=2$. We need to be careful, splitting our calculation to deal separately with positve and negative areas. We construct separate tables from $t=0$ to 2.00 and from $t=2.00$ to 3.00, using a strip width of 0.20 in each case.

t	0	0.20	0.40	0.60	0.80	1.00	1.2
v	8.000	5.832	4.096	2.744	1.728	1.000	0.512

	1.40	1.60	1.80	2.00
	0.216	0.064	0.008	0

Applying the formula we get

$$\begin{aligned}\text{Positive area} &= 0.20\left\{\tfrac{1}{2}(8.00+0)+5.832+4.096+\ \dots\ +0.008\right\}\\ &= 0.20\times 20.200\\ &= 4.04\text{ m.}\end{aligned}$$

t	2.00	2.20	2.40	2.60	2.80	3.00
v	0	−0.008	−0.064	−0.216	−0.512	−1.000

Again applying the formula we get

$$\text{Negative area} = 0.20\left\{\tfrac{1}{2}(0+-1.000)+-0.008+-0.064 +-0.216+-0.512\right\}$$

$$= 0.20 \times -1.300$$
$$= -0.26 \text{ m.}$$

We need to interpret our answers. Remember velocity and displacement are vectors, whereas distance is a scalar. This means that 4.04 m is in one direction and −0.26 m is in the opposite direction, indicated by the negative sign. The vehicle travels for 4.04 m in one direction, suddenly stops (i.e. $v = 0$) and reverses 0.26 m. Displacement, being a vector, includes the negative sign; distance as a scalar does not. This gives us

Displacement, from start to finish $= 4.04 + -0.26 = 3.78$ m

and Distance travelled $= 4.04 + 0.26 = 4.30$ m.

ASSIGNMENT

We are going to find the area under the sine curve of Fig. 16.28. Remember the curve's shape similarities for each of the quadrants.

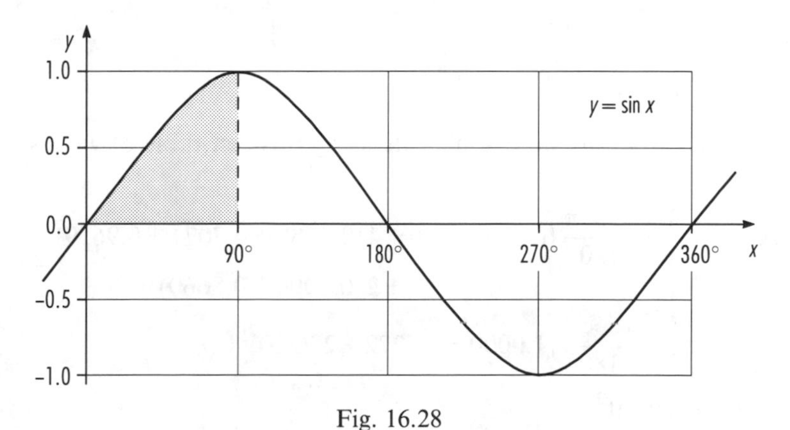

Fig. 16.28

We can simplify the calculation by finding the area for just the first quadrant and extending our answer as necessary. We must mention an important piece of information here. **All area calculations involving trigonometrical curves must use radians,** *not* degrees. So that you can check the table values more easily we have quoted degrees. However, notice that radians are included in the formulae strip widths.

We use a strip width of 15°, i.e. $15 \times \frac{\pi}{180}$ radians.

The strips are 0° to 15°, 15° to 30°, ... and 75° to 90°. For practice we have included all 3 numerical methods in turn.

i) Using the mid-ordinate rule our table of values is

x	7.5°	22.5°	37.5°	52.5°	67.5°	82.5°
$\sin x$	0.1305	0.3827	0.6088	0.7934	0.9239	0.9914

The formula gives

$$\begin{aligned}\text{Area} &= \frac{15\times\pi}{180}(0.1305+0.3827+\ \dots\ +0.9914)\\ &= \frac{15\times\pi}{180}\times 3.8307\\ &= 1.003\ \text{unit}^2.\end{aligned}$$

ii) Using the trapezoidal rule our table of values is

x	0°	15°	30°	45°	60°	75°	90°
$\sin x$	0	0.2588	0.5000	0.7071	0.8660	0.9659	1.0000

For reference we label these values as

	y_1	y_2	y_3	y_4	y_5	y_6	y_n

$$\begin{aligned}\text{Area} &= \frac{15\times\pi}{180}\{\tfrac{1}{2}(0+1.0000)+0.2588+\ \dots +0.9659\}\\ &= \frac{15\pi}{180}\times 3.7978\\ &= 0.994\ \text{unit}^2.\end{aligned}$$

iii) Finally we can apply Simpson's rule using this last table of values to get

$$\begin{aligned}\text{Area} &= \frac{1}{3}\times\frac{15\times\pi}{180}\{(0+1.0000)+4\,(0.2588+0.7071+0.9659)\\ &\qquad +2\,(0.5000+0.8660)\}\\ &= \frac{1}{3}\times\frac{15\times\pi}{180}\{1.0000+7.7272+2.7320\}\\ &= 1\ \text{unit}^2.\end{aligned}$$

Irregular volumes

When we looked at irregular areas our calculations were 2 dimensional with a constant strip width. For irregular volumes we use a constant strip width again. Compared with irregular areas the change occurs with the ordinates, or y values, being replaced by cross-sectional areas. We make this replacement in each of our numerical methods formulae,

i) the mid-ordinate rule,
ii) the trapezoidal rule
and iii) Simpson's rule.

The solution techniques for irregular areas and volumes are very similar, as you can see in this next example. This constant strip width between ordinates becomes a constant strip width between cross-sectional areas.

Example 16.15

Fig. 16.29 shows a small lake. Some cross-sectional areas are labelled, each separated by a constant 12 m (i.e. $h = 12$). The first area, A_1, and the last area, A_n, are both zero. These ends are where the lake tapers to nothing. For this example the areas are

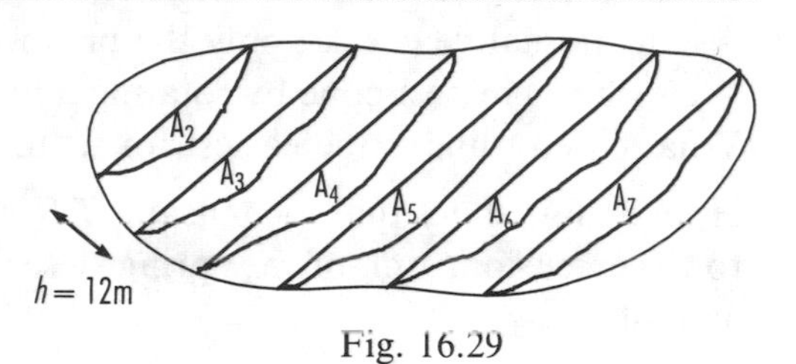

Fig. 16.29

$A_1 = 0\,\text{m}^2$, $A_2 = 15\,\text{m}^2$, $A_3 = 22.3\,\text{m}^2$, $A_4 = 28.7\,\text{m}^2$,
$A_5 = 26.9\,\text{m}^2$, $A_6 = 16.2\,\text{m}^2$, $A_7 = 9.6\,\text{m}^2$, $A_n = 0\,\text{m}^2$.

We use the trapezoidal rule to demonstrate the technique. Replacing the y values in our original formula with A values we get

$$\text{Volume} = h\{\tfrac{1}{2}(A_1 + A_n) + A_2 + A_3 + \ldots + A_7\}$$

Substituting our values gives

$$\begin{aligned}\text{Volume} &= 12\{\tfrac{1}{2}(0+0) + 15 + 22.3 + 28.7 + 26.9 + 16.2 + 9.6\}\\ &= 12 \times 118.7\\ &= 1424.4\ \text{m}^3.\end{aligned}$$

To our degree of accuracy the volume of the small lake is $1400\,\text{m}^3$.

We may also apply the prismoidal rule to irregular volumes. However because it uses only 3 cross-sectional areas it gives only an approximate solution.

EXERCISE 16.3

You have seen 3 different numerical methods for areas and the idea of their extensions to volumes, i.e.

i) the mid-ordinate rule,
ii) the trapezoidal rule
and iii) Simpson's rule.

In this exercise you should practise with more than one method.

1 Calculate the area between the line $y=2x+4$ and the horizontal axis from $x=0$ to $x=3$. Use strip widths of 0.5. By applying the standard formula for the area of a trapezium check the accuracy of your answer.

2 The line $y=2x+4$ between $x=0$ and $x=3$ is rotated about the horizontal axis. The cross-sectional areas perpendicular to this axis are circles. Using strip widths of 0.5 find the volume generated by this rotation.
The volume generated is the frustum of a cone. Using the standard formula for the frustum of a cone check the accuracy of your answer.
As an alternative check apply the prismoidal rule.
We wish to create a cone by rotating a line about the horizontal axis. What type of line is the easiest for us to use?

3 Hooke's law states that the tension, T N, in an elastic spring is related to the extension, x m, of the spring. Use the table of values to plot T against x.

x	1.5	2.5	4.0	5.0	7.5	10.0
T	20.2	33.8	53.6	67.9	101.3	135.1

Tension is a force and extension is a distance. The area under this type of graph represents the work done in stretching the spring. Choose your own constant strip width. Calculate the work done in stretching a spring from an extension of 1.5 m to 10.0 m.

4 Split the range of $x=0$ to $x=\frac{1}{2}\pi$ into 6 strips of equal width. Over this range of values calculate the area under the curve $y=\cos x$. What is the size of the area between the curve and horizontal axis for one cycle?

5 The diagram shows a piece of land bought for development. Calculate its area. Before building begins the contractors will skim off the topsoil for sale elsewhere. They intend to remove it to a depth of about 300 mm over the whole site. Estimate this volume of topsoil.

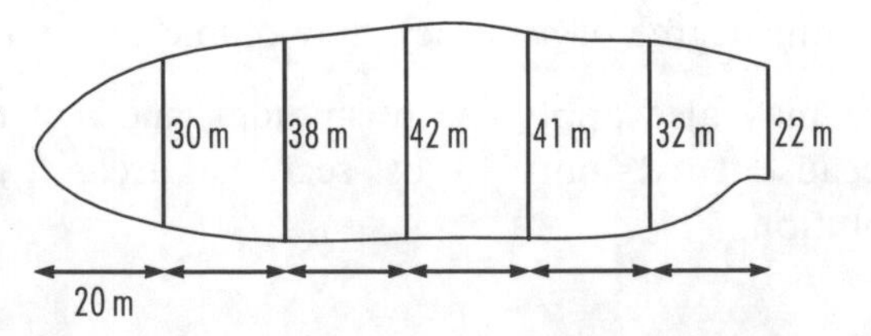

6 A new relief road is to cut through a greenfield site. The following table shows the approximate cross-sectional areas (m^2) through the cutting. These are given at intervals of 50 m from one end.

Distance from end	0	50	100	150	200	250	300
Area	0	1900	2700	4250	4500	3300	0

What volume of earth and rock needs to be removed?

7 A variable force, F N, acts on a body. The table shows the size of the force at various horizontal distances, s m, from the starting point. Plot F against s.

s (m)	0	2	4	6	8	10	12
F (N)	33	40	48	49	57	60	64

The area under such a graph represents work done. Work done by a force equals the change in mechanical energy. In this case of horizontal distances, that change is a change in kinetic energy. By calculating the area under the graph find the change in kinetic energy.

8 a) Plot the graph of $y = \log x$ from $x = 1$ to 10. Find the area under this curve. Repeat the process for $y = \ln x$.

b) Using a large scale, plot the graphs of $y = e^x$ and $y = e^{-x}$ from $x = -1.00$ to $x = 1.00$. Choosing strip widths of 0.20 find the area under each curve. What do you notice about your answers?

9 The following depth soundings (m) are taken across a river at intervals of 2.0 m.

Distance from bank	0.0	2.0	4.0	6.0	8.0	10.0	12.0
Depth sounding	2.8	3.1	3.7	2.2	2.9	1.5	0.5

Calculate the cross-sectional area of the river at this point.

10 We have some information on a vehicle relating its velocity to the distance travelled.

Using the table plot a graph of $\frac{1}{v}$ against s where v is the velocity (ms^{-1}) and s is the distance (m).

s	0	10	20	30	40	50	60	70	80	90	100
v	7.0	7.5	8.1	8.9	10.1	11.9	13.8	16.2	18.6	22.7	24.9

The area under this curve represents the time taken to travel 100 m. Find this time.

The area and perimeter of the ellipse

Remember the ellipse?

We have looked at its shape and formula before, though have sketched it again in Fig. 16.30 for revision.

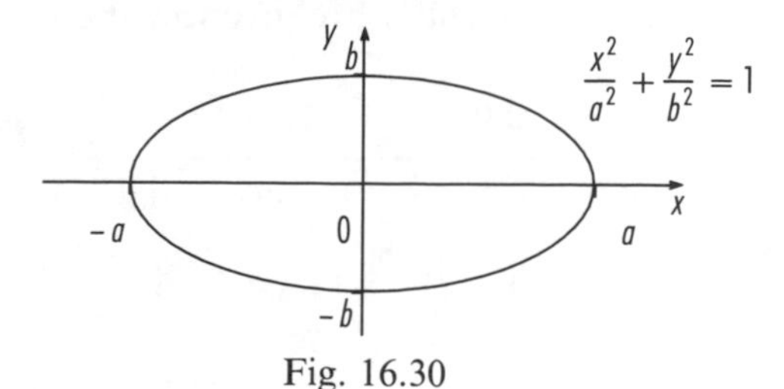

Fig. 16.30

The general equation is $\frac{x^2}{a^2}+\frac{y^2}{b^2}=1$. The length of the major and minor axes are $2a$ and $2b$. Which is the major axis and which is the minor axis depends on the sizes of a and b.

There are 2 simple formulae associated with the ellipse.

$$\text{Area} = \pi ab.$$

$$\text{Approximate perimeter} = \pi(a+b).$$

We demonstrate these formulae in the next example.

Example 16.16

Find the area and perimeter of an ellipse with a major axis of 0.72 m and a minor axis of 0.50 m.

The values for our formulae come from $2a=0.72$, i.e. $a=0.36$, and $2b=0.50$, i.e. $b=0.25$. We substitute these values into the formula

$$\text{Area} = \pi ab$$

to get

$$\begin{aligned}\text{Area} &= \pi \times 0.36 \times 0.25\\ &= 0.28 \text{ m}^2.\end{aligned}$$

Also using

$$\text{Perimeter} = \pi(a+b)$$

we get

$$\begin{aligned}\text{Perimeter} &= \pi(0.36+0.25)\\ &= 1.92 \text{ m}.\end{aligned}$$

We can check the area formula by applying Simpson's rule. The symmetry of the ellipse means we need calculate only the positive quadrant and then scale our answer by a factor of 4.

We re-arrange the general formula

$$\frac{x^2}{a^2}+\frac{y^2}{b^2}=1$$

to get

$$y = b\left(1-\frac{x^2}{a^2}\right)^{\frac{1}{2}}.$$

Using our values for a and b this becomes

$$y = 0.25\left(1-\frac{x^2}{0.36^2}\right)^{\frac{1}{2}}.$$

We split the semi-major axis of 0.36 into 6 strips of constant width 0.06. The values of both x and y are given in the table.

x	0	0.06	0.12	0.18	0.24	0.30	0.36
y	0.2500	0.2465	0.2357	0.2165	0.1863	0.1382	0
	y_1	y_2	y_3	y_4	y_5	y_6	y_n

Substitution into Simpson's rule gives us

$$\text{Area} = \frac{0.06}{3}\Big\{(0.2500+0)+4(0.2465+0.2165+0.1382) + 2(0.2357+0.1863)\Big\}$$

$$= \frac{0.06}{3} \times 3.4988$$

$$= 0.07 \text{ m}^2 \qquad \text{is the area of a quadrant of the ellipse.}$$

For all 4 quadrants we get

$$\text{Total area} = 4 \times 0.07 = 0.28 \text{ m}^2$$

which agrees with our earlier calculation correct to 2 decimal places.

EXERCISE 16.4

1 Find the areas of the ellipses i) $\frac{x^2}{4^2}+\frac{y^2}{9^2}=1$,

ii) $\frac{x^2}{9^2}+\frac{y^2}{4^2}=1$.

Why are these areas the same?
Also calculate both perimeters and explain why they are the same.

2 Find the perimeter of the ellipse $\frac{x^2}{36^2}+\frac{y^2}{9^2}=1$.
Sketch this ellipse.
The ellipse is cut in half horizontally along the major axis. What is the perimeter of the remaining plane shape?
Alternatively the ellipse is cut in half vertically along the minor axis. What is the perimeter of this remaining plane shape?

3 Sketch the circle $x^2+y^2=9$ and the ellipse $\frac{x^2}{9}+y^2=1$. From your sketch decide which graph encloses the larger area.
Calculate the separate areas enclosed by each curve and hence the area between them.

4 Find the area enclosed by the ellipse $\frac{x^2}{81}+\frac{y^2}{16}=1$.
What is the radius of a circle with the same area as this ellipse?

5 An ellipse, $\frac{x^2}{25}+\frac{y^2}{36}=1$, is cut from a rectangle measuring 10 by 12 units. Calculate the remaining area.

The theorems of Pappus

There are 2 theorems. The first one calculates volume and the second calculates surface area. Both theorems are similar in their use of the path of a centroid.

The centroid is the geometric centre of a body. For a uniform body the centroid and the centre of gravity are the same point.

For the first theorem we rotate a plane area (i.e. a cross-sectional area) and need to know the centroid of that area.

1. If we rotate a plane area about a line then the volume generated is given by

Volume = Area × Distance travelled by the centroid.

There is an important feature for the volume we generate in this rotation. The line about which we rotate must *not* cut through the cross-sectional area.

Example 16.17

Figure 16.31 shows a cylindrical sewer pipe of length 3 m and thickness 50 mm. The centroid follows a circular path of radius 0.90 m. Calculate the volume of concrete in the pipe.

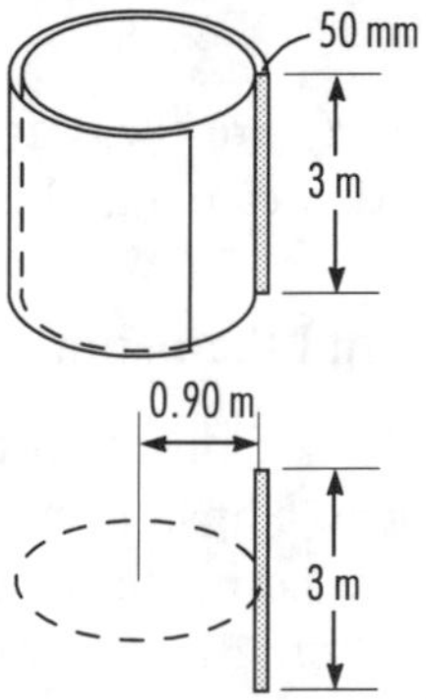

Fig. 16.31

We work out the separate parts of the overall volume formula.

$$\text{Cross-sectional area} = 3 \times 0.050 = 0.15\ \text{m}^2$$

The centroid travels along the circumference of a circle. We substitute $r = 0.90$ into the formula

$$\text{Circumference} = 2\pi r$$

to get $\text{Circumference} = 2\pi(0.90) = 5.65\ldots\ \text{m}.$

Using $\text{Volume} = \text{Area} \times \text{Distance travelled by the centroid}$

we have $\text{Volume} = 0.15 \times 5.65\ldots$

$$= 0.85\ \text{m}^3.$$

Calculator value is 0.84823.

We can re-calculate this volume of concrete, as a check, using a formula from earlier in the chapter. The concrete lies between the outer and inner cylinders. We calculate these cylindrical volumes ($\pi r^2 h$) and subtract to find that volume of concrete.

Larger cylinder, $r = 0.925$ and $h = 3$,

$$\text{Volume} = \pi(0.925)^2 3 = 8.064\ldots\ \text{m}^3$$

Smaller cylinder, $r = 0.875$ and $h = 3$,

$$\text{Volume} = \pi(0.875)^2 3 = 7.215\ldots\ \text{m}^3$$

$\therefore$ Vol. of concrete $= 8.064\ldots - 7.215\ldots$

$= 0.84\ \text{m}^3$ correct to 2 decimal places.

This calculation was easy to check. This may *not* always be the case. Pappus' theorem is useful when there is no obvious standard volume formula, as we see in this next example.

Example 16.18

Calculate the volume of a circular ring of hexagonal cross-section. Fig. 16.32 shows a hexagon of side 0.10 m. Its centroid traces out the circumference of a circle of radius 1.45 m.

We have met the hexagon several times before and can find its area in the usual way.

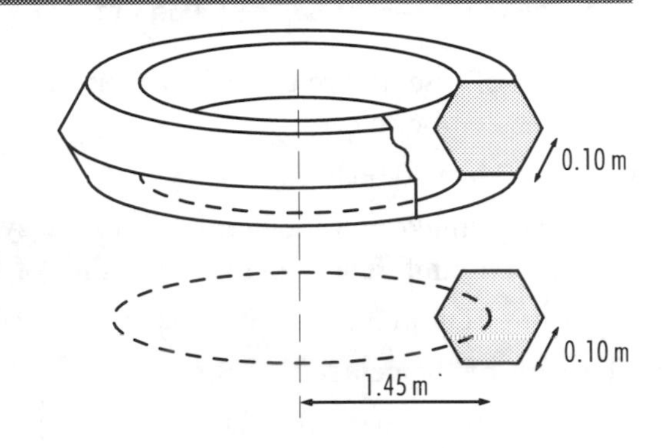

Fig. 16.32

$$\text{Area of hexagon} = 6 \times \text{Area of equilateral triangle}$$
$$= 6 \times \frac{1}{2}(0.10)^2 \sin 60°$$
$$= 0.026\ \text{m}^2.$$

For the path of the centroid we use the formula

$$\text{Circumference} = 2\pi r$$
$$= 2\pi(1.45)$$
$$= 9.111\ \text{m}.$$

Using Volume = Area × Distance travelled by the centroid

we get Volume $= 0.026 \times 9.111$

$= 0.24\ \text{m}^3.$

Example 16.19

We have a cylinder of alloy with a radius of 0.50 m and a length of 0.80 m. Around the circumference is machined a semi-circular groove of radius 0.15 m. Calculate the volume of metal removed.

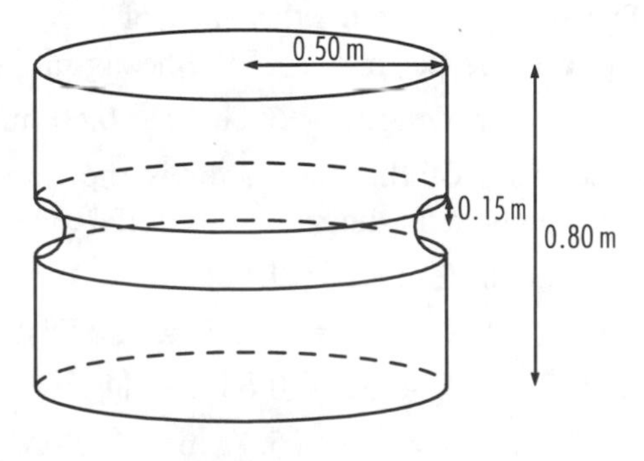

Fig. 16.33

Fig. 16.33 shows the original cylinder together with the groove. We need 2 separate values to substitute into our volume formula.

The area of the semi-circle given by

$$\text{Area} = \frac{1}{2}\pi r^2$$

becomes $\text{Area} = \frac{1}{2}\pi(0.15)^2 = 0.0353 \text{ m}^2.$

The centroid of a semi-circle is $0.424r$ from the straight side where r is the radius of the semi-circle.

Now $\quad 0.424r = 0.0636$ m.

The semi-circle makes an indentation into the cylinder. Hence its centroid traces out a circumference smaller than that of the cylinder.

$$\text{Radius of circumference} = 0.50 - 0.424r = 0.4364 \text{ m}.$$

For the path of the centroid

$$\text{Circumference} = 2\pi R$$

becomes $\text{Circumference} = 2\pi(0.4364) = 2.7420$ m.

Applying Pappus' theorem gives us

$$\begin{aligned}\text{Volume of groove} &= 0.0353 \times 2.7420\\ &= 0.10 \text{ m}^3.\end{aligned}$$

For the second theorem we rotate an arc and need to know the centroid of that arc.

2. If we rotate an arc about a line then the surface area generated is given by

Area = Arc length × Distance travelled by the centroid.

There is an important feature for the surface area we generate in this rotation. The line about which we rotate must *not* cut through the arc.

Example 16.20

Find the surface area generated by rotating the hexagon of Example 16.18.

The hexagon has 6 sides each of length 0.10 m. The arc length in our formula is the perimeter of the hexagon, i.e.

$$\text{Arc length} = 6 \times 0.10 = 0.60 \text{ m}.$$

The centroid of this perimeter is the same as the centroid of the area. Hence the path is the same circumference, i.e.

$$\text{Circumference} = 9.111 \text{ m}.$$

Using $\quad \text{Area} = \text{Arc length} \times \text{Distance travelled by the centroid}$

we get $\quad \text{Area} = 0.60 \times 9.111 = 5.47 \text{ m}^2$

is the surface area of this ring with hexagonal cross-section.

ASSIGNMENT

We return to the bollard Assignment to calculate the volume of a typical link in a chain. The original specification for a link stated a circular cross-section in the shape of an ellipse.

Fig. 16.34 shows the cross-sectional area (πr^2) and the path traced out by the centroid of that area ($\pi(a+b)$). The dimensions are in mm.

The radius of the circular cross-section is 10 mm so that

$$\text{Area} = \pi r^2$$

becomes $\text{Area} = \pi 10^2 = 314.2 \text{ mm}^2.$

Fig. 16.34

The major axis of the elliptical path is 100 mm, i.e. $2a = 100$ so that $a = 50$. The minor axis is 50 mm, i.e. $2b = 50$ so that $b = 25$.

$$\text{Path length} = \pi(a+b)$$

becomes $\text{Path length} = \pi(50+25) = 235.6 \text{ mm}.$

Hence $\text{Volume} = \text{Area} \times \text{Distance travelled by centroid}$

becomes $\text{Volume} = 314.2 \times 235.6$

$= 74\,000 \text{ mm}^3$ correct to 2 significant figures,

is the volume of a typical chain link.

EXERCISE 16.5

1 A small metal tube of large radius is formed by rotating a square of side 0.075 m about a line. The line is 0.80 m from the nearest side of the square. Using the theorems of Pappus find the volume and surface are as generated.

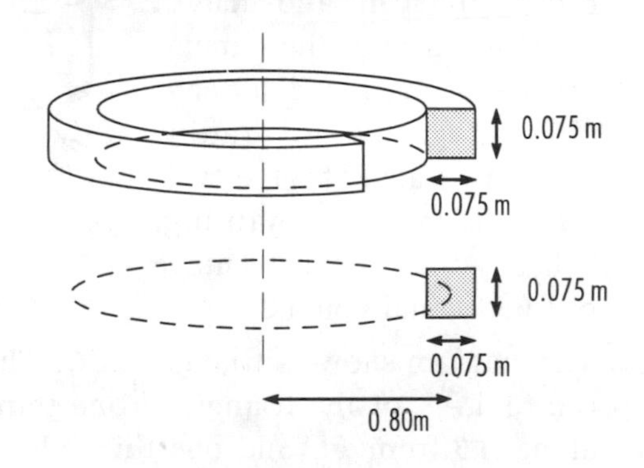

2 An ellipse of high tensile steel has a major axis of length 0.25 m and a minor axis of length 0.15 m. It is rotated about a line. The line is 1.2 m from the centroid of the ellipse. Calculate the volume and surface area of the steel.

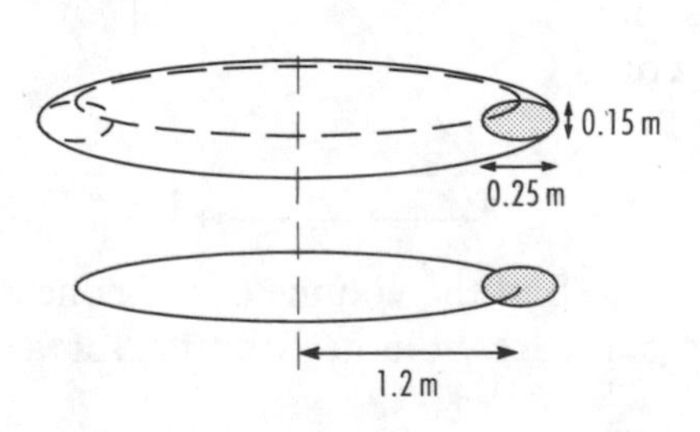

3 A cylinder is generated by rotating the rectangle about the vertical line *AB*. Use Pappus' first theorem to find the volume of the cylinder. Check your answer by applying the standard volume formula for a cylinder.

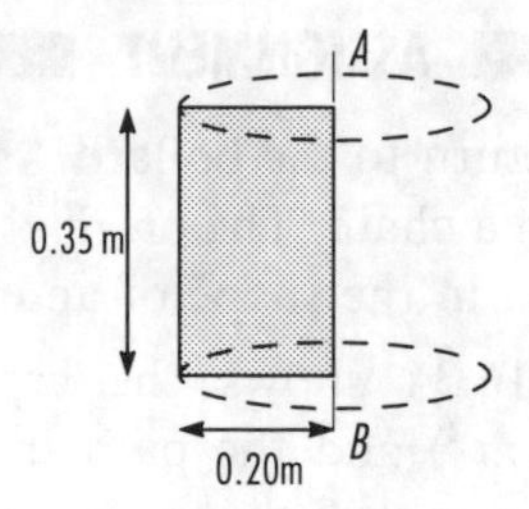

4 A semi-circle of radius 0.30 m is rotated about a vertical line 1.50 m from its straight side. Calculate the volume generated.

5 A hexagon of side 0.24 m turns about a vertical line. Its centroid traces out an ellipse with a major axis of 2.50 m and minor axis of 1.50 m. Calculate the volume and surface area generated.

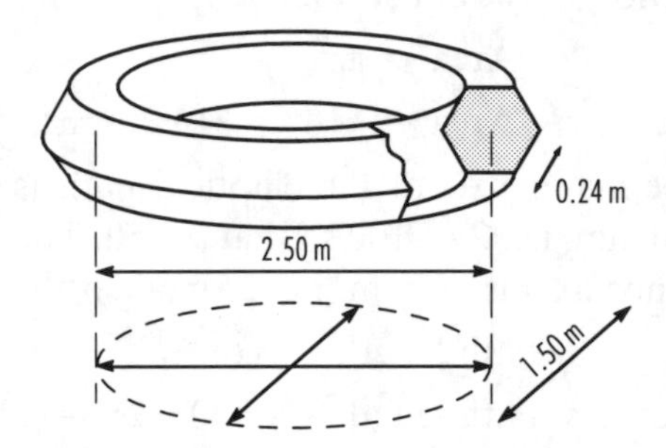

6 A semi-circle of radius 0.30 m is rotated about a vertical line 1.50 m from its straight side. Calculate the volume generated. Why is this answer different from your answer to Question **4**?

7 We have an alloy cylinder of radius 0.650 m and length 0.800 m. Around the circumference is machined a semi-circular groove of radius 0.200 m. Calculate the volume of metal remaining. What is this as a percentage of the original volume?

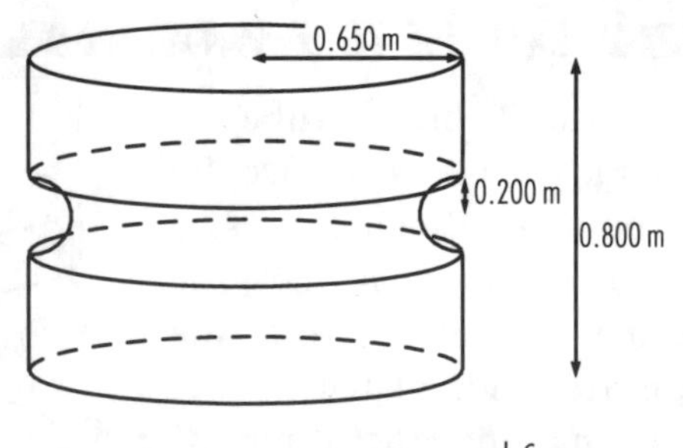

8 The diagram shows a triangle *ABC*. The centroid, *G*, of the triangle is one third along *AB* from *BC* and one third along *BC* from *AB*. Calculate these simple distances. We generate 2 different cones by rotating triangle ABC about

i) AB

and ii) BC.

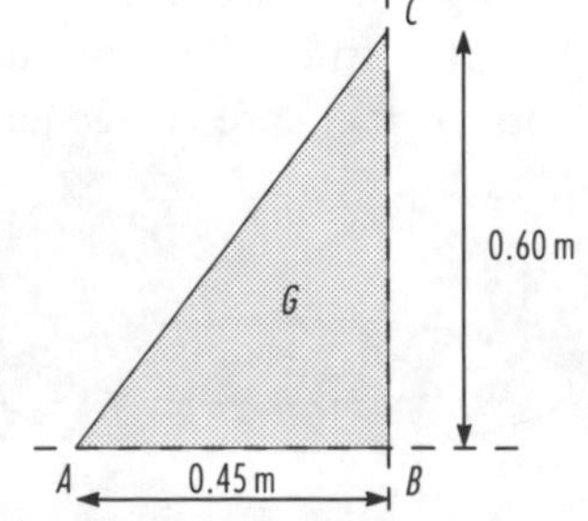

For each case apply Pappus' theorem to find the volume of the cone. Check your answers, applying the standard volume formula for a cone.

9 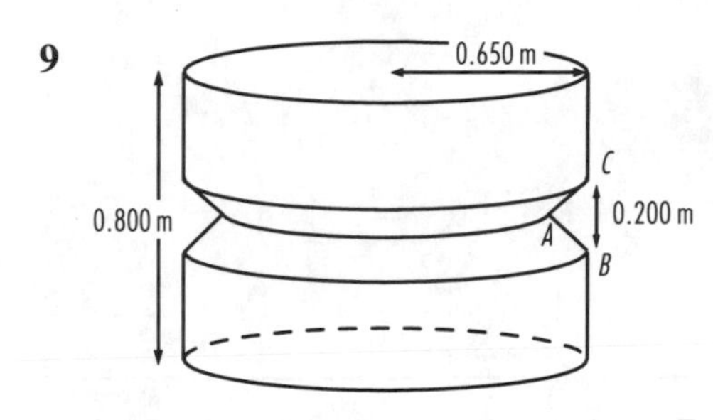

We have an alloy cylinder of radius 0.650 m and length 0.800 m. Around the circumference is machined an equilateral triangular groove of side 0.200 m. Re-draw triangle *ABC* for yourself, labelling the mid-point of *BC* as *D*. Calculate the length of *AD*. The centroid of this triangle is one third along *AD* from *D*. What is this distance? Hence calculate the volume of metal machined to form the groove. What volume of alloy remains?

10 The plane figure shows a square joined to a semi-circle. The complete area is rotated about the vertical line *XY* to generate a volume. By considering the square and semi-circle separately find this volume.

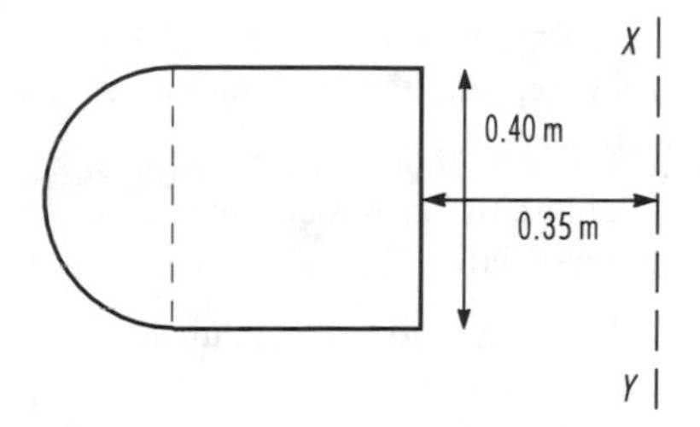

17 Statistics

The objectives of this chapter are to:

1. Define the arithmetic mean, median and mode.
2. Calculate the arithmetic mean for ungrouped data.
3. Place ungrouped data in rank order and determine the median and modal values.
4. Calculate the arithmetic mean for grouped data.
5. Estimate the mode of grouped data using a histogram.
6. Determine the median and other values from cumulative frequency data.
7. Discuss the relevance of each average as a measure of central tendency.
8. Describe the concepts of variability of data and the need to measure scatter.
9. Define the standard deviation and variance.
10. Calculate the standard deviation for ungrouped data.
11. Calculate the standard deviation for grouped data.
12. Use the statistical functions on calculators to obtain mean and standard deviation.

Introduction

Statistics is an area of mathematics widely used throughout engineering, science and business. It involves the **collection, representation, analysis and interpretation of data (i.e. information)**. In this chapter we are going to look at the simpler aspects of Statistics. We will examine the way we compare different sets of data. This comparison uses an average which fairly represents the information and a spread of the data about an average.

Once we have introduced the Assignment for this chapter we need to discuss some basic ideas.

ASSIGNMENT

A cement company wishes to test the accuracy of the machines it uses during the bagging process. The nominal weight of each bag is 50 kg.

The production supervisor checked a test run on one machine and recorded the following weights correct to 2 decimal places:

49.95	50.01	50.23	50.11	50.09	50.19	50.03	50.31	50.07	50.10
50.15	49.95	50.10	50.17	50.04	50.22	49.92	50.25	50.18	50.16
50.08	50.08	50.12	50.40	49.81	50.25	50.16	50.15	49.93	50.12
50.15	50.23	50.41	50.16	50.32	50.14	50.47	50.26	50.09	50.18
50.28	50.17	49.97	50.09	50.26	50.35	50.11	50.33	50.30	50.11
50.13	50.14	50.21	50.13	50.04	50.22	50.52	49.84	50.49	50.24
50.27	50.22	50.18	50.38	50.29	50.27	50.44	50.10	50.26	50.08
49.64	50.15	50.29	49.99	50.15	50.09	50.10	50.12	50.20	50.30
49.87	50.06	50.00	50.03	50.56	50.37	50.66	50.07	50.21	49.86
50.20	50.05	49.76	50.08	50.16	49.90	50.22	50.10	50.14	50.02

During this chapter we will look at ways of presenting and analysing these figures.

Data

Data is the information which we collect or measure in some way. It can be split broadly into **discrete data** or **continuous data. Discrete data can be counted, taking only limited values. Continuous data can take any value in a given range** and is usually associated with measuring processes.

In the following example we look at the difference between these data types.

Example 17.1

Consider a box of assorted screws.

Suppose we are interested in the number of screws of each gauge (6, 8, 10, 12 etc). This type of data can be **counted** and so is **discrete**.

Suppose we are interested in the "exact" lengths of all screws classed as 2″, correct to the nearest 0.01 inch, (they could be 1.99″, 2.01″, etc.). This data must be obtained by **suitable measurement** and so is **continuous**.

You must be aware of this distinction. In later sections of the chapter we refer to these different data types.

EXERCISE 17.1

Decide whether the following data is discrete or continuous.

1 The number of apprentices in each department of a large engineering firm.

2 The life-time of each of a set of electrical components

3 The car mileage recorded by sales representatives on travel claim forms.

4 The quantity of facing bricks in each type of house on a building site.

5 The number of accidents entered each week in the accident book at a factory.

6 The temperature of float glass during different stages of manufacture.

7 The maximum loads supported by a various cables.

8 The number of cars coming off a production line each day.

9 The height of saplings in a commercial tree nursery.

10 The depth of a tidal river.

Frequency distributions

When we collect a large amount of data the unsorted figures are difficult to analyse. In this form the information lacks order. It is difficult to interpret and needs to be **organised** in some way. A first attempt might be to place it in **rank order**. Rank order starts with the smallest value and moves numerically through to the largest value . Alternatively it may start at the largest value and move down to the smallest value. Many of the values occur more than once. **Frequency** is the number telling us how often a particular value occurs.

Let us consider the following example.

Example 17.2

A person from the organisation and management section recorded the time taken by 100 operatives to produce a small electrical component. The readings, correct to the nearest second, were:

76	52	56	56	79	64	87	78	95	97
89	69	69	61	60	91	84	60	59	82
75	90	60	81	80	77	77	83	67	68
71	93	78	72	72	70	66	66	66	65
71	74	94	57	78	82	65	88	85	52
76	81	93	93	65	61	55	82	79	84
62	74	66	68	72	71	85	75	75	56
61	63	63	71	86	88	91	77	57	78
59	88	65	60	59	67	95	58	66	67
60	94	70	97	96	70	72	68	73	73

Represent this information in a frequency distribution table.

The information lacks order and so needs to be organised. Our first attempt places it in rank order, including the frequency for each time value, as in Table 17.1.

Time (s)	Frequency	Time (s)	Frequency	Time (s)	Frequency
52	2	68	3	84	2
53	0	69	2	85	2
54	0	70	3	86	1
55	1	71	4	87	1
56	3	72	4	88	3
57	2	73	2	89	1
58	1	74	2	90	1
59	3	75	3	91	2
60	5	76	2	92	0
61	3	77	3	93	3
62	1	78	4	94	2
63	2	79	2	95	2
64	1	80	1	96	1
65	4	81	2	97	2
66	5	82	3		
67	3	83	1		

Table 17.1

This is a **frequency distribution**. Although it is much better than the unsorted data, it remains difficult for us to see any pattern amongst the figures. There is still too much information for us to assess. We can obtain a clearer overall picture by **grouping** the data. In Table 17.2 we have grouped 5 timings together in each case, e.g. 50, 51, 52, 53, and 54 as 50–54.

Time (s)	Frequency
50–54	2
55–59	10
60–64	12
65–69	17
70–74	15
75–79	14
80–84	9
85–89	8
90–94	8
95–99	5
TOTAL FREQUENCY	= 100

Table 17.2

By grouping the data we have lost the initial individual information from the distribution. For example the grouped table tells us there are two values in the group 50 – 54 but not what they are. The accuracy of any analysis using grouped data is less than if we use ungrouped data. Obviously this is a "trade-off" situation. There is no hard and fast rule for grouping. It is accepted practice to organise the data into between about 6 and 15 groups. An optimum distribution consists of about 10 groups. If we use too few groups the loss of information is too great. If we retain too many groups there is no benefit from the grouping.

Let us look more closely at our grouped data in the previous example. We can consider the group 60 – 64. What is the width of this group? We avoid a simple subtraction of 60 from 64. The group or **class** size is actually five because the group records times from 60 to 64 seconds inclusive, (i.e. 60, 61, 62, 63 and 64). Also, because the data is recorded to the nearest second, it is continuous. We must allow for measurements like 59.7, 64.2 etc. This means the 60 – 64 class must include any value from 59.5 up to 64.5. These become the **true** limits for the class. The grouped table showing the **true class limits** is shown in Table 17.3.

Time (s)	Frequency
49.5 – 54.5	2
54.5 – 59.5	10
59.5 – 64.5	12
64.5 – 69.5	17
69.5 – 74.5	15
74.5 – 79.5	14
79.5 – 84.5	9
84.5 – 89.5	8
89.5 – 94.5	8
94.5 – 99.5	5
	100

Table 17.3

We can look at the ungrouped data and its class limits. For example 52 to the nearest second has values from 51.5 to 52.5. We discuss a reading of exactly 52.5 a little later.

Our original table would start as

51.5 – 52.5	2
52.5 – 53.5	0, etc.

The original data is continuous and so the 'merging' of classes is acceptable.

For discrete data with classes of 0–4, 5–9, 10–14 etc, values like 4.6 or 9.3 may not exist. Then the need to merge classes is less obvious. Later we will see that there can be no gaps between adjacent classes. In the case of discrete data there is a convention. Class limits should extend half a unit on either side of the stated limits. Hence the 5–9 class would use a class interval of 4.5–9.5.

We have mentioned already 'the 0.5 problem' in Example 17.2. Where would you place a value of 54.5 in the grouped data? It could be included in either the 49.5–54.5 class or in the 54.5–59.5 class. If the data can be recorded to a greater degree of accuracy then the problem may be resolved easily. The 54.5 may have arisen from a reading of 54.48 or 54.53 and the correct group is easily identified. 54.48 goes into the 49.5–54.5 group whilst 54.53 goes into the 54.5–59.5 group.

In other cases a consistent convention must be adopted. One convention includes thc 0.5 in the group above. Then 54.5 would then go into the 54.5–59.5 class.

Histograms

A **histogram** is the most commonly used graphical representation of a frequency distribution. It consists of a set of connected vertical rectangles. The width of a rectangle on the scaled horizontal axis is the class interval. Strictly, it is the area of each rectangle which is proportional to the frequency. We can draw a histogram for the data of Example 17.2.

Example 17.3

Draw a histogram for the grouped data tabulated for Example 17.2.

The ⩘ symbol shows a break in the axis and its scale. 0–50 is *not* to scale, whilst the remainder of the axis is to scale. In our example all classes

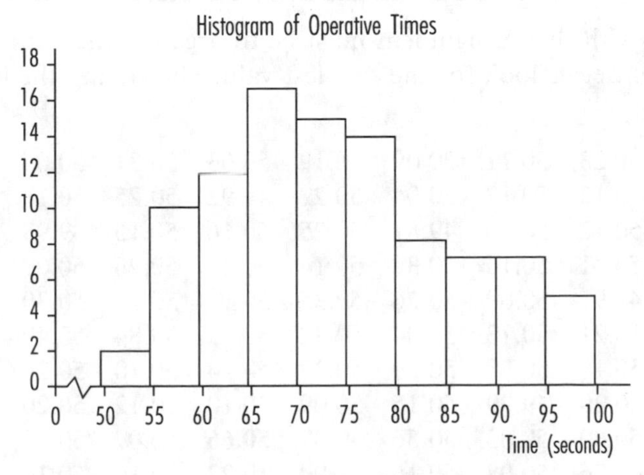

Fig. 17.1

are of equal width. Hence the height is proportional to the area of each rectangle. The height can be used to represent the frequency of each group. It would be technically *incorrect* to label the vertical axis as a frequency scale. It is better that we leave it without any label at this level.

This problem of areas and heights can be illustrated with our own data. Suppose we combine the 75–79 and 80–84 groups as one single group. This would become a group of 75–84 with a frequency of 23. If heights were used to represent frequency we would get the incorrect histogram on the left of Figs. 17.2. This has changed the shape of the distribution and now *cannot* be correct. The solution to the problem lies in the areas of the rectangles. Since the 75–84 rectangle is now twice the width of the remaining rectangles we must halve its height to represent the frequency correctly. The correct histogram with this unequal class is on the right of Figs. 17.2

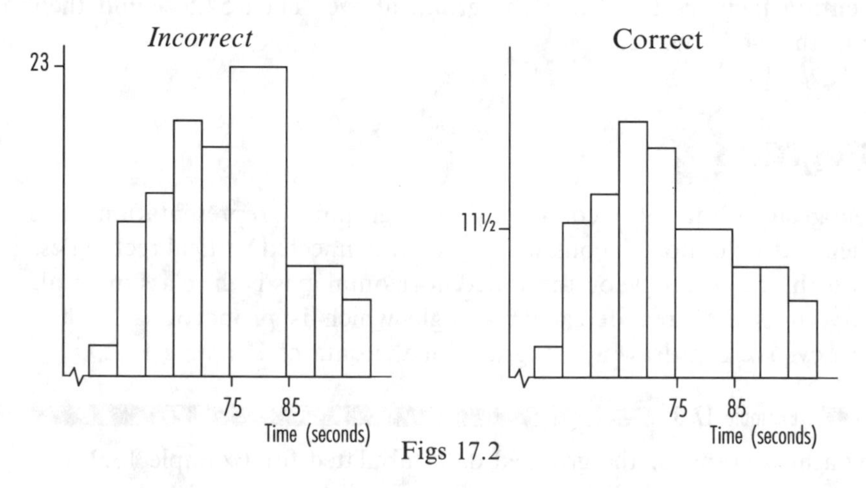

Figs 17.2

ASSIGNMENT

Our first move with this Assignment must be to organise the data into a frequency table. Let us look for the smallest value, 49.64, and the largest value, 50.66.

49.95	50.01	50.23	50.11	50.09	50.19	50.03	50.31	50.07	50.10
50.15	49.95	50.10	50.17	50.04	50.22	49.92	50.25	50.18	50.16
50.08	50.08	50.12	50.40	49.81	50.25	50.16	50.15	49.93	50.12
50.15	50.23	50.41	50.16	50.32	50.14	50.47	50.26	50.09	50.18
50.28	50.17	49.97	50.09	50.26	50.35	50.11	50.33	50.30	50.11
50.13	50.14	50.21	50.13	50.04	50.22	50.52	49.84	50.49	50.24
50.27	50.22	50.18	50.38	50.29	50.27	50.44	50.10	50.26	50.08
49.64	50.15	50.29	49.99	50.15	50.09	50.10	50.12	50.20	50.30
49.87	50.06	50.00	50.03	50.56	50.37	**50.66**	50.07	50.21	49.86
50.20	50.05	49.76	50.08	50.16	49.90	50.22	50.10	50.14	50.02

There are far too many values between 49.64 and 50.66 to list individually. Let us group them into classes of equal width, 0.10. Where do we start and finish? For ease we group them with an eye on pattern, as in Table 17.4.

Weight (kg)	Frequency
49.60 – 49.69	1
49.70 – 49.79	1
49.80 – 49.89	4
49.90 – 49.99	7
50.00 – 50.09	19
50.10 – 50.19	31
50.20 – 50.29	21
50.30 – 50.39	8
50.40 – 50.49	5
50.50 – 50.59	2
50.60 – 50.69	1
	100

Table 17.4

In each group we have gathered those weights where the first three digits are the same. For example 50.40 – 50.49 all have 5, 0 and 4 as the first three digits. Also we could have used the true class limits for the weight column.

Now concentrate on the frequency column. You can see the higher frequencies are towards the middle of the distribution. We will see this more readily with the aid of a diagram. Let us draw a histogram, as in Fig. 17.3.

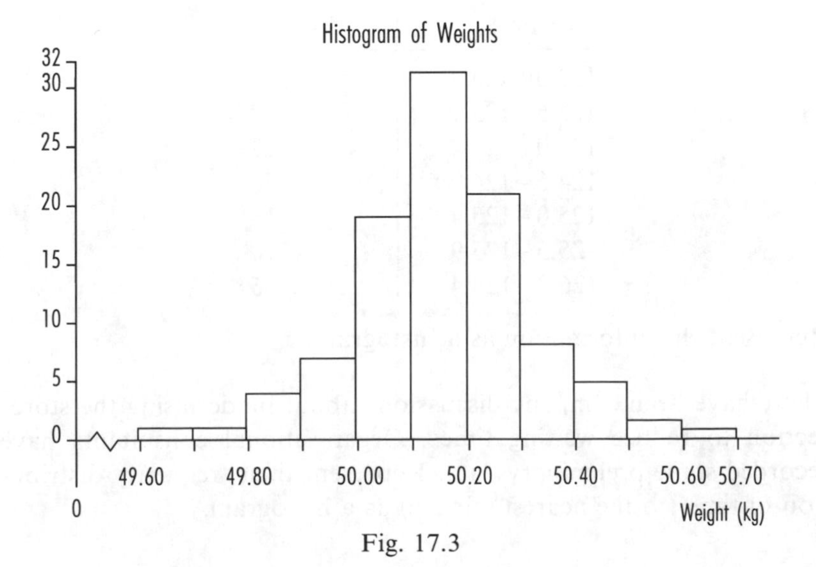

Fig. 17.3

EXERCISE 17.2

1 The personnel department of a local engineering company recorded the number of missed working days per month due to sickness.

68	29	102	58	27	79	38	62	115	80
114	71	7	89	113	75	19	70	45	91
57	101	88	63	45	66	99	37	25	68
48	79	82	93	91	59	9	72	92	51
21	74	41	83	100	89	87	55	36	30

Using class widths of 10 draw up a frequency distribution table. Hence draw a fully labelled histogram.

2 The production manager of a continuous process company has been given his weekly production targets. The following data shows his actual weekly production as a percentage of his target. 50 weekly figures are consistent with a fortnight's Christmas holiday. A value below 100 shows a production shortfall. A value above 100 shows the week's target has been exceeded. With the aid of a grouped frequency distribution table draw a histogram of this data.

94	96	117	83	106	91	110	114	97	110
86	109	101	100	91	99	102	103	108	99
94	111	101	88	96	95	99	101	98	104
105	99	98	103	102	98	102	104	106	103
95	97	102	95	96	100	104	101	93	99

3 Quality control information is contained in the following table. It relates to the length of bolts in today's sample batch.

Length (mm)	Frequency
123.0–123.4	6
123.5–123.9	13
124.0–124.4	22
124.5–124.9	26
125.0–125.4	18
125.5–125.9	10
126.0–126.4	5

Represent this information as a histogram.

4 There have been company discussions about modernising the stores section to reduce waiting times. Organisational consultants have recorded some preliminary data. Represent their frequency distribution of time (to the nearest minute) as a histogram.

Waiting Time (min)	Frequency
2	7
3	9
4	10
5	16
6	24
7	17
8	11
9	6

5 A new electrical product is being given a pre-production trial run. There is some doubt about the consistency of its power rating within the sample batch. The company decides to gather some data and present it in the tablc below. For the following grouped frequency distribution of power ratings draw a histogram.

Power (W)	Frequency
393 – 395	3
396 – 398	7
399 – 401	15
402 – 404	23
405 – 407	26
408 – 410	18
411 – 413	8

Cumulative frequency

With the frequency distribution and histogram we have good representations of the data. We can discuss the results in broad terms without too much detail.

Let us return to our example of operatives producing a small electrical component. Suppose thc company decides to give a bonus to those who make the component in 63 seconds or less. Fearing a decline in production they issue a warning to those who make the component in more than 80 seconds. We need to calculate how many operatives fall into these categories. However it would be better to have them at our finger tips. This is possible by keeping a running total of the frequencies using a **cumulative frequency** table. In a cumulative frequency table we start with the frequency of the first class (or smallest value for ungrouped data). As we progress to the next class we add on its frequency. The sum of these frequencies is the cumulative frequency up to and including the second class. For the third class we add on its frequency as well. The sum of these frequencies is the cumulative frequency up to and including the third class.

The cumulative frequency of any class is the total frequency up to and including the upper limit in that class.

Example 17.4

Construct a cumulative frequency table for the frequency distribution in Table 17.5.

Time (s)	Frequency
49.5 – 54.5	2
54.5 – 59.5	10
59.5 – 64.5	12
64.5 – 69.5	17
69.5 – 74.5	15
74.5 – 79.5	14
79.5 – 84.5	9
84.5 – 89.5	8
89.5 – 94.5	8
94.5 – 99.5	5
	100

Table 17.5

We know to look closely at the upper limits in each class. For the frequencies we keep a running total, as in Table 17.6.

	Upper Class Limit (s)	Frequency	Cumulative Frequency
not more than	54.5	2	2
	59.5	10	10 + 2 = 12
	64.5	12	12 + 12 = 24
	69.5	17	17 + 24 = 41
	74.5	15	15 + 41 = 56
	79.5	14	14 + 56 = 70
	84.5	9	9 + 70 = 79
	89.5	8	8 + 79 = 87
	94.5	8	8 + 87 = 95
	99.5	5	5 + 95 = 100

Table 17.6

We can use the table to obtain some information directly. For example, 70 operatives made the component in 79.5 seconds or less.

More general information can be found from a graphical representation of the cumulative frequency data. We label our horizontal axis with our values (variable or variate) and our vertical axis with the cumulative frequencies. We plot the points remembering to use upper class limits. Finally we draw a smooth curve through them. This curve is called an ogive (pronounced "oh-jive"). It is usually a lazy S shape and can never decrease.

Example 17.5

Using the cumulative frequency table, Table 17.6, draw an ogive.
How many operatives made the component in 63 seconds or less?
How many made the component in more than 80 seconds?

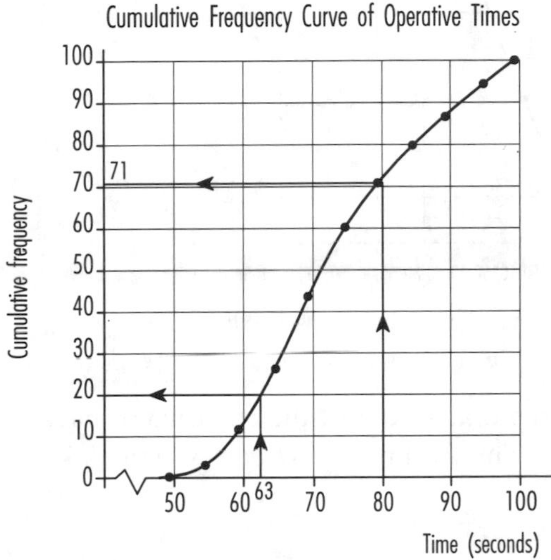

Fig. 17.4

From the graph we see that 20 operatives made the component in 63 seconds or less.

Also we see that 71 operatives made the component in 80 seconds or less. Since there are 100 operatives in total, 29 (100 − 71) made the component in more than 80 seconds.

ASSIGNMENT

We can apply these techniques to the cement bags in our Assignment. In each class we write down the upper limit. Then we keep a running total of the frequency to give the cumulative frequency column, as in Table 17.7.

Weight (kg)	Upper Class Limit	Cumulative Frequency
49.60 – 49.69	49.695	1
49.70 – 49.79	49.795	2
49.80 – 49.89	49.895	6
49.90 – 49.99	49.995	13
50.00 – 50.09	50.095	32
50.10 – 50.19	50.195	63
50.20 – 50.29	50.295	84
50.30 – 50.39	50.395	92
50.40 – 50.49	50.495	97
50.50 – 50.59	50.595	99
50.60 – 50.69	50.695	100

Table 17.7

Now we can use Table 17.7 to draw the ogive in Fig. 17.5.

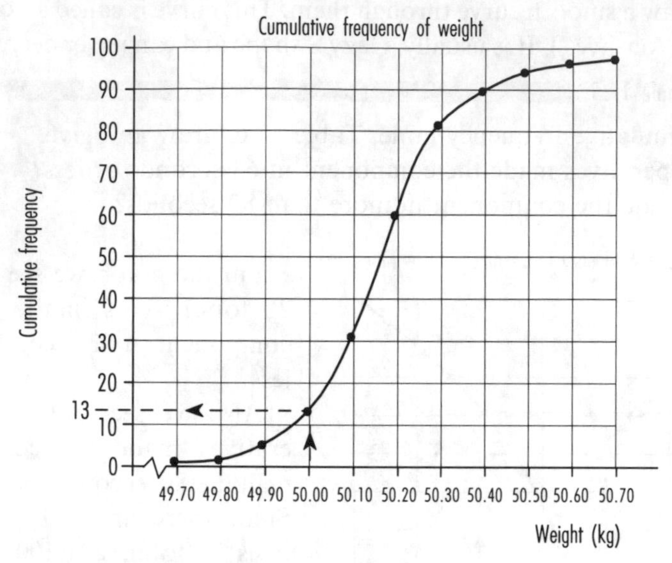

Fig. 17.5

From both the table and ogive we see that 13 out of the 100 cement bags fall below the nominal 50 kg weight. This means that 87 bags weigh 50 kg or more.

EXERCISE 17.3

1 The personnel department of a local engineering company recorded the number of missed working days per month due to sickness.

68	29	102	58	27	79	38	62	115	80
114	71	7	89	113	75	19	70	45	91
57	101	88	63	45	66	99	37	25	68
48	79	82	93	91	59	9	72	92	51
21	74	41	83	100	89	87	55	36	30

Draw a cumulative frequency table based on class widths 0#–#9, 10–19, . . . Hence draw a fully labelled ogive.

2 The production manager of a continuous process company has been given his weekly production targets. The following data shows his actual weekly production as a percentage of his target. 50 weekly figures are consistent with a fortnight's Christmas holiday. A value below 100 shows a production shortfall. A value above 100 shows the week's target has been exceeded.

94	96	117	83	106	91	110	114	97	110
86	109	101	100	91	99	102	103	108	99
94	111	101	88	96	95	99	101	98	104
105	99	98	103	102	98	102	104	106	103
95	97	102	95	96	100	104	101	93	99

With the aid of a grouped cumulative frequency table draw an ogive of this data.

What percentage of the weekly production figures exceed their target?

3 Quality control information is contained in the following table. It relates to the length of bolts in today's sample batch.

Length (mm)	Frequency
123.0 – 123.4	6
123.5 – 123.9	13
124.0 – 124.4	22
124.5 – 124.9	26
125.0 – 125.4	18
125.5 – 125.9	10
126.0 – 126.4	5

Represent this information in an ogive.

What percentage of bolts exceed 125.25 mm?

4 There have been company discussions about modernising the stores section to reduce waiting times. Organisational consultants have recorded some preliminary data. Represent their frequency distribution of time (to the nearest minute) in a cumulative frequency table.

Waiting Time (min)	Frequency
2	7
3	9
4	10
5	16
6	24
7	17
8	11
9	6

Draw an ogive of this information. You are not prepared to wait longer than $4\frac{1}{2}$ minutes. What percentage of waiting times are acceptable to you?

5 A new electrical product is being given a pre-production trial run. There is some doubt about the consistency of its power rating in the sample batch. The company decides to gather some data and present it in the table below. From the grouped frequency distribution of power ratings calculate the cumulative frequencies. Hence draw an ogive.

Power (W)	Frequency
393–395	3
396–398	7
399–401	15
402–404	23
405–407	26
408–410	18
411–413	8

What percentage of ratings are within 5 W of a nominal 400 W?

Now we have enough background information to look in detail at average values and the spread of data around such averages.

Averages

In this section we discover different types of "average".

Suppose we carry out a quality control check on five different electrical components. We examine one box of each type and record the number of faulty components per box. The results are 1, 7, 4, 1 and 2. Obviously we are interested in reducing the number of faulty components, provided it is economically possible. We might decide to present one figure which we feel best represents the data. We might do this by finding the average number of faulty components.

What is an average?

An average is a single value which is representative of the data.

We could ask the majority of people to calculate the average of 1, 7, 4, 1 and 2. They are likely to give the answer 3. This is only one possible answer. Others are 1 and 2 because there are several other kinds of average. The choice depends on how we wish to represent the data. Now we look at three commonly used averages.

1. Mean

The mean (arithmetic mean) is found by adding together all the items of data and dividing by the total number present.

For our faulty components the mean number is

$$\frac{1+7+4+1+2}{5}=3.$$

It is common practice to represent the mean by $\bar{x}$ (x bar).

If we have n items of data, x_1, x_2, x_3, x_4, ... x_n, then

$$\bar{\boldsymbol{x}}=\frac{\boldsymbol{x_1}+\boldsymbol{x_2}+\boldsymbol{x_3}+\boldsymbol{x_4}+\ldots+\boldsymbol{x_n}}{\boldsymbol{n}}$$

This is usually written as

$$\bar{\boldsymbol{x}}=\frac{\sum \boldsymbol{x}}{\boldsymbol{n}}$$

where the Greek symbol $\sum$ (capital sigma) is used for "the sum of".

EXERCISE 17.4

Find the mean of the following sets of data:

1 78, 31, 9, 16, 27, 27, 41.

2 3 N, 5 N, 2 N, 0 N, 6 N.

3 6.2 Ω, 2.4 Ω, 1.7 Ω, 8.3 Ω, 2.9 Ω.

4 65 cm, 2m, 27mm.

5 −6°C, 4°C, 1°C, −2°C, 3°C.

2. Mode

The mode is the most commonly occurring item of data.

We easily identify it when the data is arranged in ascending or descending order. For the faulty component data this means

1, 1, 2, 4, 7 or 7, 4, 2, 1, 1.

The modal number of faulty components is 1 because it occurs more often than any other.

It is possible to have **more than one mode** in a distribution. However, if there are too many modes we do wonder if they are worthwhile.

Example 17.6

i) The data 1, 1, 2, 3, 4, 4, 5 has two modes, 1 and 4. This is because both 1 and 4 appear the most number of times.
It is a **bi-modal** distribution.
(*bi is a prefix meaning two.*)

ii) The data 1, 1, 2, 3, 4, 5, 5, 6, 6 has three modes, 1, 5 and 6.
It is a **tri-modal** distribution.
(*tri is a prefix meaning three.*)

EXERCISE 17.5

Find the mode of the following sets of data:

1 6, 2, 3, 5, 3, 1, 7.

2 7 kg, 2 kg, 1 kg, 3 kg, 5 kg, 3 kg, 2 kg.

3 6.2 V, 4.6 V, 1.7 V, 8.3 V 2.5 V.

4 £8.51, £8.53, £8.50, £8.54, £8.51, £8.52, £8.51.

5 −0.2 J, −0.4 J, 0 J, 0.1 J, 0.2 J, −0.1 J, −0.3 J, −0.2 J.

3. Median

The median is the "middle" value.

To find the mode we must arrange the data in either ascending or descending order, i.e.

1, 1, 2, 4, 7 or 7, 4, 2, 1, 1.

In this example the middle value is the third value, i.e. the median is 2.

In general if there are n items of data, the median is the $\frac{n+1}{2}$th value when the data is arranged in order.

Example 17.7

i) If n is 37 then the median value is the $\frac{(37+1)}{2}$th value, i.e. the 19th value.

ii) If n is an even number then the median is taken to be half-way between the two middle values.
If n is 24 then the median value is the $\frac{(24+1)}{2}$th value, i.e. the $12\frac{1}{2}$th value. This means we need to be half-way between the 12th and 13th values.

iii) Let us consider the numbers 1, 1, 2, 5, 7, 8. The middle numbers are 2 and 5 and so the median is $\frac{2+5}{2}$, i.e. 3.5.

EXERCISE 17.6

Find the median of the following sets of data:

1 2, 3, 4, 5, 6.

2 13 W, 10 W, 12 W, 16 W, 14 W, 12 W.

3 65%, 82%, 71%, 36%, 41%, 22%, 18%, 75%.

4 0.3μF, 0.3μF, 0.4μF, 0.1μF, 0.3μF, 0.5μF.

5 The whole numbers (integers) 1–100 inclusive.

Now we know how to calculate three different figures which we could use as average. Later we will consider which one is best suited to our problem.

The component fault data is simple 'raw' data. This means we have not altered it in any way. Already we have distinguished between ungrouped and grouped data. We must consider how to calculate the three averages for these types of data.

Example 17.8

This example looks at ungrouped data. An assembly plant has 10 production teams producing machine tools. The following results are last week's production figures (to the nearest 10).

Tools Produced (x)	Number of Teams (f)
40	1
50	0
60	1
70	2
80	4
90	2

Table 17.8

x is used to represent the machine tools produced (the variable) and f to represent the number of teams (the frequency).

The mode is easy to identify. The greatest frequency is 4 which corresponds to 80 machine tools produced. The mode (or modal value) is 80.

The table is arranged in ascending order and so the median is easy to find. There are 10 teams (an even number) so the median must lie half-way between the fifth and sixth values (either side of $\frac{n+1}{2}$). This lies amongst the 4 teams who produced 80 machine tools. The median is also 80 machine tools.

With more involved data you may wish to produce a **cumulative frequency column**. This will help you to identify the middle position for the median.

For the mean we could use a longhand method with $\bar{x} = \frac{\sum x}{n}$.

i.e. $$x = \frac{40 + 60 + 70 + 70 + 80 + 80 + 80 + 80 + 90 + 90}{10}$$

$$= 74$$

$= 70$ when rounded off to the accuracy of the original data.

The machine tool production is quoted in multiples of 10.

This method works well for small values of n but can be very laborious for larger distributions. A better method is to use tabulation.

x	f	fx
40	1	40
50	0	0
60	1	60
70	2	140
80	4	320
90	2	180
	$\sum f = 10$	$\sum fx = 740$

Table 17.9

We get the fx column by multiplying corresponding values in the f and x columns, e.g. the 4 teams producing 80 machine tools produce a total of 320.

$$\text{Now}\quad \bar{x} = \frac{\sum fx}{\sum f}$$

$$\text{i.e.}\quad \bar{x} = \frac{740}{10}$$

$$= 74$$

$$= 70$$ as before when rounded off to the accuracy of the original data.

This example demonstrates an amended definition for the mean. **For frequency distributions** $\bar{x} = \dfrac{\sum fx}{\sum f}$.

Remember that answers should be given to the same degree of accuracy as the original information.

EXERCISE 17.7

1 The number of BTEC units awarded to part time students at an FE college in one year is shown in the table below

Units	0	1	2	3	4	5	6	7
Number of students	5	12	23	46	32	20	8	4

Find the mean, median and modal number of units awarded.

2 To remain competitive and retain valuable personnel a local engineering company supports its workers' part-time education and training. The number of hours of study of all workers in one week were

Hours of study	5	6	7	8	9	10	11	12	13	14
Number of all workers	20	72	110	175	240	337	284	71	14	8

Find the mean, median and modal hours of study.

3 There have been company discussions about modernising the stores section to reduce waiting times. Organisational consultants have recorded some preliminary data. Represent their frequency distribution of time (to the nearest minute) in a cumulative frequency table.

Waiting Time (min)	Frequency
2	7
3	9
4	10
5	16
6	24
7	17
8	11
9	6

Find the mean, median and mode.

4 Screws are packaged in 25s. A sample of 100 packs revealed the actual number in each one. Using the data find the mean number of screws per pack.

Number of screws	23	24	25	26	27	28
Number of packs	4	12	63	14	5	2

5 The maintenance section records show last year's reported major faults allowing for a fortnight's total plant shut down. They are collected together to show how frequently each number has been reported.

Faults	0	1	2	3	4	5
Number of weeks	14	19	10	4	2	1

Find the mean number of faults per week for last year.

So far we have looked at **raw, ungrouped, data**. We continue with similar calculations, now using data that has been grouped.

Example 17.9

For this example we look again in Table 17.10 at the timed operatives producing a small electrical component. Remember the timings have been grouped.

Time (secs)	Frequency
49.5 – 54.5	2
54.5 – 59.5	10
59.5 – 64.5	12
64.5 – 69.5	17
69.5 – 74.5	15
74.5 – 79.5	14
79.5 – 84.5	9
84.5 – 89.5	8
89.5 – 94.5	8
94.5 – 99.5	5
	100

Table 17.10

We *cannot* find these averages as easily as before.

For the mode we see there is a **modal group** or **modal class**, the 64.5 – 69.5 class, containing the most operatives. This modal class indicates how the data is clustered around a group of values. We *cannot* determine the exact location of the mode within the group. Although there are 17 values in the 64.5 – 69.5 class we do *not* know how many times any individual value occurred. By grouping the data we have destroyed original individual information.

From grouped data we can only estimate a single value for the mode.

An accepted method of doing this involves a simple construction on the histogram. We show this in Fig. 17.6.

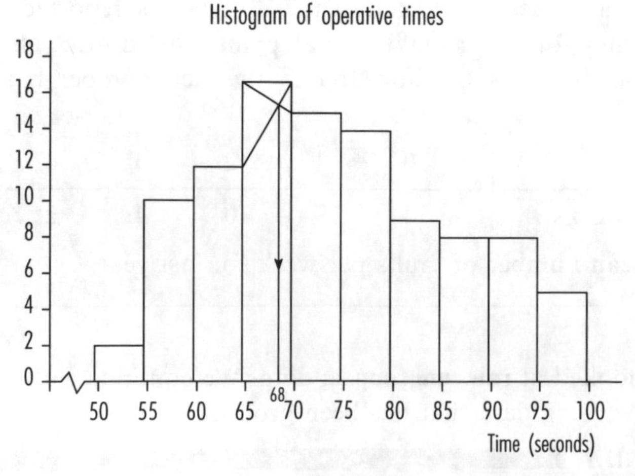

Fig. 17.6

We read the **estimate** for the mode directly from the **horizontal axis** as 68 seconds.

For the median we need to produce the cumulative frequency table in Table 17.11.

	Time (secs)	Cumulative Frequency
Not more than	54.5	2
	59.5	12
	64.5	24
	69.5	41
	74.5	56
	79.5	70
	84.5	79
	89.5	87
	94.5	95
	99.5	100

Table 17.11

One way of estimating the median is to make use of the cumulative frequency curve. We should find the $\frac{(n+1)}{2}$th value. However, it is conventional with grouped data to find the $\left(\frac{n}{2}\right)$ th value. There is little practical difference between these values for large frequencies already distorted by grouping. We mark the "half-way" value on the ogive in Fig. 17.7.

For the median we read our estimate directly from the **horizontal axis** as 72 seconds.

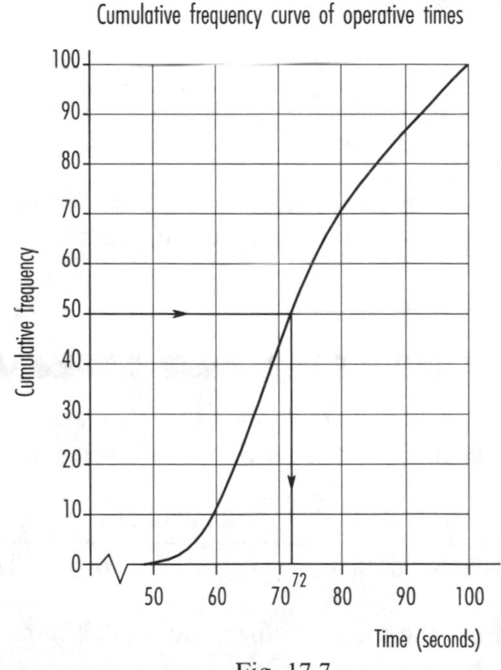

Fig. 17.7

We calculate the mean by extending the frequency table as before to produce an fx column. But what can we use as x when the data has been grouped?. We have to assume the data is evenly distributed in the group and use the **mid-value of each group** as x. This is the best approximation in a frequency distribution already distorted by grouping. The mid-value is $\frac{(\text{lower limit}+\text{upper limit})}{2}$.

Time (secs)	x	f	fx
49.5 – 54.5	52	2	104
54.5 – 59.5	57	10	570
59.5 – 64.5	62	12	744
64.5 – 69.5	67	17	1139
69.5 – 74.5	72	15	1080
74.5 – 79.5	77	14	1078
79.5 – 84.5	82	9	738
84.5 – 89.5	87	8	696
89.5 – 94.5	92	8	736
94.5 – 99.5	97	5	485
		$\sum f = 100$	$\sum fx = 7370$

Table 17.12

Using $\bar{x} = \dfrac{\sum fx}{\sum f}$

we get $\bar{x} = \dfrac{7370}{100}$

$= 73.7$

$= 74$ seconds, when rounded to the same accuracy as the original data.

EXERCISE 17.8

1 The following table shows the % marks awarded to a group of BTEC students in an end-of-unit examination.

Marks (%)	21 – 30	31 – 40	41 – 50	51 – 60	61 – 70	71 – 80
Number of Students	2	6	11	20	7	4

Find the mean, median and modal marks.

2 The number of hours' overtime worked by employees at an auto-engineering firm in one month is recorded below:

Hours	1 – 5	6 – 10	11 – 15	16 – 20	21 – 25	26 – 30	31 – 35	36 – 40
Number of Employees	6	7	18	9	5	4	3	2

Find the mean, median and modal number of hours of overtime.

3 A batch of resistors has a nominal resistance of 15 Ω. Quality control checks found the following sample

Resistance	14.92–14.95	14.96–14.99	15.00–15.03	15.04-15.07	15.08–5.11
Resistors per group	3	9	33	11	4

What are the mean, median and modal values of these resistances?

4 Quality control information is contained in the following table. It relates to the length of bolts in today's sample batch.

Length (mm)	Frequency
123.0 – 123.4	6
123.5 – 123.9	13
124.0 – 124.4	22
124.5 – 124.9	26
125.0 – 125.4	18
125.5 – 125.9	10
126.0 – 126.4	5

Find the mean length of these bolts.

5 A new electrical product is being given a pre-production trial run. There is some doubt about the consistency of its power rating within the sample batch. The company decides to gather some data and present it in the table below. For the following grouped frequency distribution of power ratings calculate the mean power rating.

Power (W)	Frequency
393 – 395	3
396 – 398	7
399 – 401	15
402 – 404	23
405 – 407	26
408 – 410	18
411 – 413	8

Which average to use?

Now we have looked at the three average calculations we consider their advantages, [+], and disadvantages, [–]. Then we can decide which is the most applicable in any particular case. We list the major features of each average.

1. Mean

[+] Because of its common usage the mean is a well accepted average. It is obtained by a fairly simple calculation using all of the data. It is mathematically sound and can be used in further statistical work. When the only information available is the sum total of the data and the number of items of data we can still calculate it.

[−] It can be influenced by extreme values which are not representative of the data. Also the calculation often produces a result which is not an actual item of data. It may make the answer appear unrealistic.

2. Median

[+] The median is a simple idea and is usually an actual item of data. It is not affected by extreme values and can be found in some cases when the data is incomplete.

[−] It only uses the data to determine its position and so is not a suitable value for further mathematical work. In addition, if the data is erratically distributed it may not be a representative item.

3. Mode

[+] The mode is a commonly used value which is easily understood. It is a typical item of data and, like the median, is unaffected by extreme values. It has many practical applications in stock control and can often be supplied without data by inexperienced personnel.

[−] There can be more than one mode, and perhaps none at all, in some distributions. It does not use all the data and cannot be used as a mathematical basis for further work. With grouped data it is only a measure of clustering rather than an actual modal item.

Each of the three averages is an important measure in its own right, particularly with simple applications. If the values are evenly distributed then the mean is the most suitable. If the half-way value is required then the median should be chosen. 50% of the values are below and 50% are above the median. If the most typical value is required then the mode is the most appropriate.

ASSIGNMENT

Let us look again at our assignment to find the 3 types of average. We can use some of our earlier work. Fig. 17.8 shows the histogram we drew earlier together with our simple construction lines. We read directly from the horizontal axis that the mode is 50.15 kg.

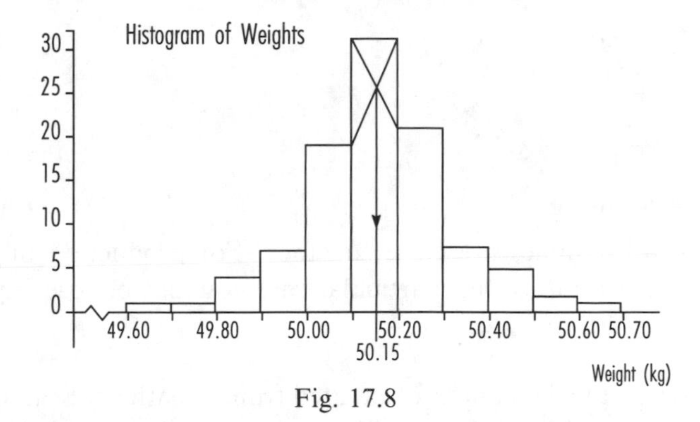

Fig. 17.8

Fig. 17.9 shows our earlier ogive. Remember the median is the middle value. Its position occurs half-way along the cumulative frequency axis. Again we read directly from the horizontal axis that the median is 50.13 kg.

To find the mean value we need to construct a table rather than use earlier diagrams. Our table is based on Table 17.4. The weights are grouped so we need to use the mid-point of each class for x. Also we need to multiply corresponding values from the x and f columns to obtain the fx values.

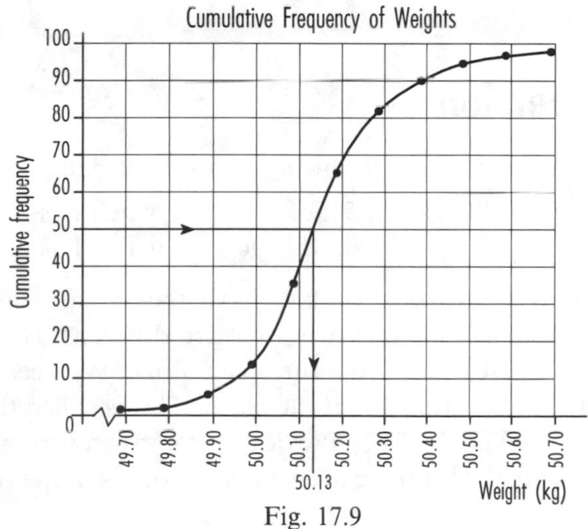

Fig. 17.9

Weight (kg)	x	f	fx
49.60 – 49.69	49.645	1	49.645
49.70 – 49.79	49.745	1	49.745
49.80 – 49.89	49.845	4	199.380
49.90 – 49.99	49.945	7	349.615
50.00 – 50.09	50.045	19	950.855
50.10 – 50.19	50.145	31	1554.495
50.20 – 50.29	50.245	21	1055.145
50.30 – 50.39	50.345	8	402.760
50.40 – 50.49	50.445	5	252.225
50.50 – 50.59	50.545	2	101.090
50.60 – 50.69	50.645	1	50.645
		100	5015.600

Table 17.13

Using $\bar{x} = \dfrac{\sum fx}{\sum f}$

we get $\bar{x} = \dfrac{5015.600}{100}$

$= 50.16$ kg.

You see that all 3 values are close together. For production of quality goods this is important. In this particular case you may choose any of the averages to support your decision.

The mean uses all the data and is calculated from a mathematical formula. It is the average used in more advanced analysis. In any statistical work you meet in later units of your BTEC course the mean will almost certainly be the average you use. Now let us look at some of that further analysis.

Dispersion

Let us consider the following two sets of data:

Set A:	48, 49, 50, 51, 52	which has a mean of 50.
Set B:	0, 20, 45, 50, 85, 100	which has a mean of 50.

Although both sets of data have the same mean and hence the same representative item, clearly they are different. Hence proper comparison of the figures *cannot* use the mean alone. We need to consider the spread of the information about the mean. This is called the **scatter (variability or dispersion)** of the data. Previously we saw several methods of quantifying an average. Also there are several methods of measuring the dispersion of data.

1. Range

The range is the **difference between the highest and lowest values in the distribution.**

For set A the range is $52-48 = 4$.
For set B the range is $100-0 = 100$.

It is a simple quantity to understand and calculate, but has only limited application. It fails to use all the data. The two values it does use are extreme, and could be abnormal, items.

2. Interquartile range

The interquartile range overcomes the problem of using possible abnormal items. It states the **range over the middle 50% of the data**. In most distributions this is where the majority of the data will lie. It is an improvement on the range but still does not use all the data.

3. Mean deviation

The mean deviation **involves the variation of each item from the mean**. It finds the mean value of these variations.

The deviations from the mean in each case are

For set A:	$48-50$,	$49-50$,	$50-50$,	$51-50$,	$52-50$	
i.e	-2,	-1,	0,	1,	2	
For set B:	$0-50$,	$20-50$,	$45-50$,	$50-50$,	$85-50$,	$100-50$
i.e.	-50,	-30,	-5,	0,	35,	50

Obviously the deviations are greater for set B. We can represent these deviations by a single value. From our earlier work on averages we do this by calculating the mean of the deviations.

For set A this gives $\dfrac{-2+-1+0+1+2}{5} = \dfrac{0}{5} = 0.$

The positive and negative deviations from the mean will always **cancel out**. We overcome this problem by ignoring the signs and considering only the **magnitude** (or **modulus** or **size**) of each deviation.

For set A this gives $\dfrac{2+1+0+1+2}{5} = 1.2.$

For set B this gives $\dfrac{50+30+5+0+35+50}{6} = 28.\bar{3}.$

By comparing the mean deviations we see the different types of spreads of the two data sets, i.e. set B is about 24 times more variable than set A ($28.\bar{3}$ is approximately 24×1.2). This measure is a great improvement on the range and interquartile range. It makes use of every item of data, simply comparing sets of data. However, the value cannot be used for more advanced analysis. This is because it ignores signs and considers only the magnitudes of the deviations.

4. Standard deviation

This is an important measure of spread.

The standard deviation rectifies the mathematical flaw of the mean deviation by squaring the deviations from the mean.

For our 2 data sets, A and B, we calculate these squared deviations.

For set A this gives 4, 1, 0, 1, 4.
For set B this gives 2500, 900, 25, 0, 1225, 2500.

Once again we need a single representative for each set and so we calculate the mean of these squared deviations.

For set A this gives $\dfrac{4+1+0+1+4}{5} = \dfrac{10}{5} = 2.$

For set B this gives $\dfrac{2500+900+25+0+1225+2500}{6} = \dfrac{7150}{6} = 1191.\dot{6}.$

The mean of the squared deviations is called the **variance** of the data.

The squared deviations create a problem with the units, i.e. units squared. Suppose the original data was in £. Because we have squared the deviations, the variance would be in £^{2}. This is *not* consistent with the original data. To resolve this problem we find the square root of the variance to give the **standard deviation**.

It is common practice to use the lower case Greek symbol σ (sigma) for standard deviation.

For set A $\sigma = \sqrt{2} = 1.4.$

For set B $\sigma = \sqrt{1191.\dot{6}} = 34.5.$

We can compare these results in the same way as for mean deviation. This time set B is about 25 times more variable than set A. Thus the mean deviation and the standard deviation give similar comparisons for the two sets of data.

You should be wary of implying too much from the standard deviation at this level. It is used here only to compare data. For "raw" data (as above) the standard deviation formula is

$$\sigma = \sqrt{\frac{\sum (x - \bar{x})^2}{n}}$$

where $(x - \bar{x})$ is the deviation of each item from the mean.

Standard deviation is the most difficult, but most useful, measure of dispersion to calculate. Good tabulation technique reduces this difficulty as we see in the next example.

Example 17.10

In this example of simple, ungrouped, data we calculate standard deviation.

We use our data set A to demonstrate the required calculation steps. We start by listing the data in a column.

x
48
49
50
51
52
$\sum x = 250$

Next we calculate $\bar{x}$ using $\bar{x} = \frac{\sum x}{n}$

$$= \frac{250}{5}$$

$$= 50.$$

Thirdly we extend the table to include columns for $(x - \bar{x})$ and $(x - \bar{x})^2$, as in Table 17.14.

x	$x - \bar{x}$	$(x - \bar{x})^2$
48	−2	4
49	−1	1
50	0	0
51	1	1
52	2	4
	$\sum(x - \bar{x})^2 =$	10

Table 17.14

Finally we calculate σ.

$$\sigma = \sqrt{\frac{\sum(x - \bar{x})^2}{n}}$$

$$= \sqrt{\frac{10}{5}}$$

$$= \sqrt{2}$$

$= 1.41$ is the standard deviation correct to 3 significant figures.

EXERCISE 17.9

Find the standard deviation of the data in Questions **1**–**5** of Exercise 17.4.

We can extend our theory to find the standard deviation for a frequency distribution. It does not matter whether we consider a grouped or ungrouped distribution. The only difference between them is the value of x we use. This is the group mid-point for grouped distributions. For these frequency distributions our formula needs to be modified to include frequency, f,

$$\sigma = \sqrt{\frac{\sum f(x - \bar{x})^2}{\sum f}}$$

Example 17.11

Let us return to our grouped "time" data to find both the mean and standard deviation, as in Table 17.15.

Time (secs)	Frequency
49.5 – 54.5	2
54.5 – 59.5	10
59.5 – 64.5	12
64.5 – 69.5	17
69.5 – 74.5	15
74.5 – 79.5	14
79.5 – 84.5	9
84.5 – 89.5	8
89.5 – 94.5	8
94.5 – 99.5	5
	100

Table 17.15

For the simple data in Example 17.10 we progressively built up the table. Similarly we start in Table 17.16 with the x, f and fx columns to calculate the mean. Remember the x values are the class mid-points.

Time (secs)	x	f	fx
49.5 – 54.5	52	2	104
54.5 – 59.5	57	10	570
59.5 – 64.5	62	12	744
64.5 – 69.5	67	17	1139
69.5 – 74.5	72	15	1080
74.5 – 79.5	77	14	1078
79.5 – 84.5	82	9	738
84.5 – 89.5	87	8	696
89.5 – 94.5	92	8	736
94.5 – 99.5	97	5	485
		100	7370

Table 17.16

Using $\bar{x} = \dfrac{\sum fx}{\sum f}$

we get $\bar{x} = \dfrac{7370}{100}$

$= 73.7$

$= 74$ seconds when rounded to the accuracy of the original data.

We extend the table to include $x - \bar{x}$, $(x - \bar{x})^2$ and $f(x - \bar{x})^2$ columns. Multiplying corresponding values of f and $(x - \bar{x})^2$ forms the $f(x - \bar{x})^2$ column.

Time (secs)	x	f	fx	$x - \bar{x}$	$(x - \bar{x})^2$	$f(x - \bar{x})^2$
49.5–54.5	52	2	104	−22	484	968
54.5–59.5	57	10	570	−17	289	2890
59.5–64.5	62	12	744	−12	144	1728
64.5–69.5	67	17	1139	−7	49	833
69.5–74.5	72	15	1080	−2	4	60
74.5–79.5	77	14	1078	3	9	126
79.5–84.5	82	9	738	8	64	576
84.5–89.5	87	8	696	13	169	1352
89.5–94.5	92	8	736	18	324	2592
94.5–99.5	97	5	485	23	529	2645
		100	7370			13770

Table 17.17

Using $\sigma = \sqrt{\dfrac{\sum f(x - \bar{x})^2}{\sum f}}$

we get $\sigma = \sqrt{\dfrac{13770}{100}}$

$= \sqrt{137.7}$

$= 11.7$

$= 12$ seconds when rounded to the accuracy of the original data,

i.e. the grouped data has a mean of 74 seconds and a standard deviation of 12 seconds. These values are rounded to the same degree of accuracy as the original data.

ASSIGNMENT

We now take our final look at the Assignment. This time we aim to find the standard deviation from the mean weight of 50.156 kg. We extend our previous table, Table 17.11, with the necessary columns of $(x - \bar{x})$, $(x - \bar{x})^2$ and $f(x - \bar{x})^2$, as in Table 17.18.

Weight (kg)	x	f	$(x - \bar{x})$	$(x - \bar{x})^2$	$f(x - \bar{x})^2$
49.60–49.69	49.645	1	−0.511	0.261	0.261
49.70–49.79	49.745	1	−0.411	0.169	0.169
49.80–49.89	49.845	4	−0.311	0.097	0.387
49.90–49.99	49.945	7	−0.211	0.045	0.312
50.00–50.09	50.045	19	−0.111	0.012	0.234
50.10–50.19	50.145	31	−0.011	small	0.004
50.20–50.29	50.245	21	0.089	0.008	0.166
50.30–50.39	50.345	8	0.189	0.036	0.286
50.40–50.49	50.445	5	0.289	0.084	0.418
50.50–50.59	50.545	2	0.389	0.151	0.303
50.60–50.69	50.645	1	0.489	0.239	0.239
		100			2.779

Table 17.18

Using $\sigma = \sqrt{\dfrac{\sum f(x - \bar{x})^2}{\sum f}}$

we get $\sigma = \sqrt{\dfrac{2.779}{100}}$

$= \sqrt{0.02779}$

$= 0.17$ kg correct to 2 decimal places.

To our accuracy we have a mean cement bag weight of 50.16 kg with a standard deviation of 0.17 kg.

Now we are able to compare our results with results from other bagging machines. Suppose another machine bagged cement with a mean weight of 50.28 kg and standard deviation of 0.20 kg. Firstly, comparing the means we see the other machine consistently uses more cement. The other standard deviation is also greater than our standard deviation. This means it is more variable than our machine. However, the differences are not great, particularly compared with the expected wastage between source and customer.

EXERCISE 17.10

1 Find the standard deviation of the data using the questions of Exercise 17.7.

2 Find the standard deviation of the data using the questions of Exercise 17.8.

Calculator methods

Many calculators now have standard statistics functions which allow rapid calculation of the mean and standard deviation. These may be either raw figures or figures presented in a frequency distribution. You should read the calculator instruction manual in conjunction with our explanation of the statistics. They are useful as a means of rapidly cross-checking answers. There are a variety of calculators on the market so it is impossible to explain the method for every model. For illustration purposes two of the above examples are repeated using a CASIO fx85M calculator. Most of the CASIO models will use the same steps.

Example 17.12

This example uses the raw data of our earlier set B,

i.e. 0, 20, 45, 50, 85 and 100.

We use this calculator technique to find the mean and standard deviation.

Firstly use the mode function to select SD.

Press INV AC to clear the memory.

Insert the raw data 0, 20, 45, 50, 85, 100 by:

0 followed by M+ (the data entry or x key in SD mode).
20 followed by M+
45 followed by M+
50 followed by M+
85 followed by M+
100 followed by M+

Now press INV 6 (n in SD mode) to check that you have entered the correct number of items.

The answer should be 6.

Press INV 7 for the mean ($\bar{x}$ in SD mode).

The answer should be 50.

Press INV 8 for the standard deviation (σ_n in SD mode)

The answer should be 34.5.

EXERCISE 17.11

Check your calculations of Exercise 17.9 using the statistics functions on your calculator.

Example 17.13

This example uses in Table 17.19 a frequency distribution of grouped data we have seen often in this chapter.

Time (secs)	x	f
49.5–54.5	52	2
54.5–59.5	57	10
59.5–64.5	62	12
64.5–69.5	67	17
69.5–74.5	72	15
74.5–79.5	77	14
79.5–84.5	82	9
84.5–89.5	87	8
89.5–94.5	92	8
94.5–99.5	97	5
		100

Table 17.19

For frequency distributions we need to amend the data inputs slightly. The sequence is:

Use the mode function to select SD.

Press inv AC to clear the memory.

Now enter the data by

52 × 2 followed by M+

57 × 10 followed by M+

62 × 12 followed by M+

67 × 17 followed by M+

72 × 15 followed by M+

77 × 14 followed by M+

82 × 9 followed by M+

87 × 8 followed by M+

92 × 8 followed by M+

97 × 5 followed by M+

There is no = sign.

Now press inv| 6| (n in SD mode) to check that you have entered the correct number of items.

The answer should be 100.

Press inv| 7| for the mean (x in SD mode).

The answer should be 73.7.

Press inv| 8| for the standard deviation (σ_n in SD mode).

The answer should be 11.7.

EXERCISE 17.12

Check the calculations of Exercise 17.10 using the statistics functions on your calculator.

Compare this calculator method with the methods we have used earlier. You should be favourably impressed.

Using your calculator in this way you need some appreciation of the size of the answer before you start. Remember the calculator will just give an answer. You have no way of checking that your data entry was accurate.

Our final set of exercises lets you practise the techniques from throughout the chapter.

EXERCISE 17.13

1 The figures below are last month's assessment marks (out of 160) for second year apprentices at a local company

136, 150, 105, 96, 108, 116, 120, 98, 105, 113.

i) Find the mean, median and modal marks, correct to the nearest whole number (integer).

ii) In the current round of appraisals the personnel manager argues that the average assessment mark is 115.

Do you think that this is a fair statement to make? Give reasons for your answer.

2 A neighbouring company and its rival quote their assessment marks (out of 160) for last month as

112, 115, 111, 118, 120, 116, 115, 114, 113, 116.

i) Find the mean and standard deviation of this data.

ii) Find the standard deviation for the corresponding data in Question **1**.

iii) Compare the two sets of figures using their means and standard deviations. What does this comparison tell you about the apprentices of the two firms?

3 The marks, out of 40, of 15 students in a BTEC timed assignment were as follows

34 29 25 34 32 38 26 26 32 24 35 34 27 37 33

i) Calculate the mean, median and modal marks.
ii) Two students who scored 38 and 27 respectively were later disqualified for copying and each awarded 0. Calculate the mean, median and mode for the new set of marks.
iii) Compare the two sets of results. What feature of the averages do the new results illustrate?

4 In the quest for alternative technologies, the Green Electric company is considering the possibility of building a hydro-electric generating plant. Three suitable fast flowing rivers have been suggested. The final choice will depend on an analysis of the rainfall in each area. The following table gives the rainfall (correct to the nearest 0.1 inch) during each of the preceding 12 months in those areas.

Month	Area A	Area B	Area C
Jan	2.3	1.8	3.1
Feb	1.9	1.5	2.9
Mar	1.9	1.6	2.8
Apr	1.9	1.6	2.6
May	1.7	1.5	1.9
Jun	1.7	1.4	1.4
Jul	1.9	1.7	1.1
Aug	1.9	1.7	1.1
Sep	2.1	1.6	1.2
Oct	2.2	1.9	1.4
Nov	2.2	1.8	2.0
Dec	2.2	1.8	2.6

i) Find the mean monthly rainfall and standard deviation for each area.
ii) Use these results to formulate your recommendations to the Green Electric Company.

5 A firm of design and build consultants is developing a new family timeshare holiday complex. It is trying to plan the optimum number of bedrooms per holiday home. To do this it commissions a survey to find the average number of children per family in the UK.
The results indicate that there are an average of 2.4 children in each family.

i) Which average has been used to obtain this figure?
ii) Which would be the most suitable average to use for this purpose? Give reasons for your answer.

6 The frequency table below shows the attendance at a new technology training course over an 8 week period.

Number of attendances	0	1	2	3	4	5	6	7	8
Number of employees	1	4	7	18	23	22	11	8	6

i) Calculate the mean, median and modal number of attendances.
ii) Calculate the standard deviation for this data.
iii) Previous experience of similar courses has been monitored by the organisers. They expect the attendance to have a mean of 6 attendances and a standard deviation of 2.5 attendances per 100 trainees. What do your results suggest about these trainees?

7 The following table shows the frequency distribution of the examination marks for 700 candidates in a BTEC end-of-unit Maths II examination.

Marks	10–19	20–29	30–39	40–49	50–59	60–69	70–79	80–89	90–99
Number of students	10	32	45	87	117	148	127	90	44

Grades are awarded according to

Less than 40 marks	Fail
40 marks or more	Borderline Fail/Pass
50 marks or more	Pass
65 marks or more	Merit
85 marks or more	Distinction.

Construct a cumulative frequency table and draw a cumulative frequency curve (ogive).
Use your curve to
i) estimate the median mark,
ii) estimate the percentage of students in each award category.

8 Design engineers are testing their new long-life battery. The data records the life of each battery (hours) in the sample batch of 100.

Hours	100–109	110–119	120–129	130–139	140–149	150–159	160–169	170–179	180–189	190–199	200–209	210–219
Number of batteries	2	6	23	16	15	12	10	8	3	2	2	1

i) Calculate the mean, median and mode for this distribution.
ii) In the publicity material, which average should the company quote for the average life of a battery?
iii) Do you think that the customers will, in general, feel satisfied with the company's publicity claim?

9 A machine produces rods with a nominal diameter of 10 mm. The distribution of a sample of 150 rods is given below. The diameters are correct to 0.01 mm.

Diameter (mm)	9.76–9.81	9.82–9.87	9.88–9.93	9.94–9.99	10.00–10.05	10.06–10.11	10.12–10.17
Number of rods	1	22	30	45	34	16	2

i) Calculate the mean and standard deviation for this sample.
ii) When the machine is functioning correctly the company expects a mean and standard deviation of 10.00 mm and 0.05 mm repectively. The machine is overhauled when the variablity exceeds the norm by 50% or more. What do your results suggest to the company?

10 This final question is more of a data gathering exercise followed by analysis. Apply the statistical techniques to this textbook to find the mean and standard deviation related to

i) number of pages/chapter,
ii) number of questions/chapter,
iii) number of figures/ chapter.

From your results you should suggest values for a typical chapter.

18 Calculus: I – Differentiation

The objectives of this chapter are to:

1 Determine average and instantaneous gradients to a simple curve, e.g. $y = ax^2$ where a is a constant.

2 Deduce the process of moving a point on a curve towards a fixed point on the curve causes the gradient of the chord to approach that of the tangent to the curve at the fixed point.

3 Identify incremental changes in the x and y directions as δx and δy.

4 Determine the ratio $\frac{\delta y}{\delta x}$ in terms of $x$and δx.

5 Derive the limit of $\frac{\delta y}{\delta x}$ as δx tends to zero and define it as $\frac{dy}{dx}$.

6 Derive $\frac{dy}{dx}$ for the functions $y = ax^n$, $n = 0, 1, 2, \ldots$ from first principles.

7 Differentiate simple algebraic functions.

8 Demonstrate graphically results of the derivatives of $\sin\theta$ and $\cos\theta$.

9 Differentiate functions of the form $y = a\cos\theta + b\sin\theta$ where a and b are constants.

10 Define the differential properties of the exponential and logarithmic functions.

11 Use the derivatives of the functions $\sin x$, $\cos x$, $\ln x$, e^x and x^n.

12 State the rate of change at a maximum or minimum point of a curve is zero.

13 Evaluate derivatives at given points.

Introduction

In this chapter we are going to look at change. Many things change continually, though some do remain constant. On page 47 in Chapter 2 we looked at change when we calculated the gradients of straight line graphs. You will remember the gradient is the same all along a straight line. Later in this volume we looked at various curves where the gradient changed along the curve. Mathematics has a set of rules for these changes. We will briefly introduce these rules now, and extend them in Volume 2.

ASSIGNMENT

The Assignment for this chapter links together those for Chapters 1 and 2. We have a vehicle travelling at a speed of $30\,\text{ms}^{-1}$ when the brakes are applied to produce a deceleration of $4\,\text{ms}^{-2}$. Allowing for a minor measurement error we have the distance travelled given by

$$s = 1 + 30t - 2t^2$$

where s is the distance travelled
and t is the time from when the brakes are applied.

Deceleration and retardation, being negative acceleration, mean a slowing down.

Also the speed, v, is linked to time, t, by the formula $v = 30 - 4t$.

The gradient of a graph

When we looked at a gradient before we considered the change in y due to a change in x. Remember x is the independent variable and y is the dependent variable. This means the change in x automatically causes a change in y. For many things this change takes place continually, though obviously a constant **never changes**. We may wish to look at an average change or an instantaneous change, as in this first example.

Example 18.1

Let us look at the graph of $y = 3x^2$.

Firstly we can look at the change over a portion of the curve. We may start anywhere on the graph and either increase or decrease the x value. Suppose we start at the point (1, 3) and move along the curve increasing

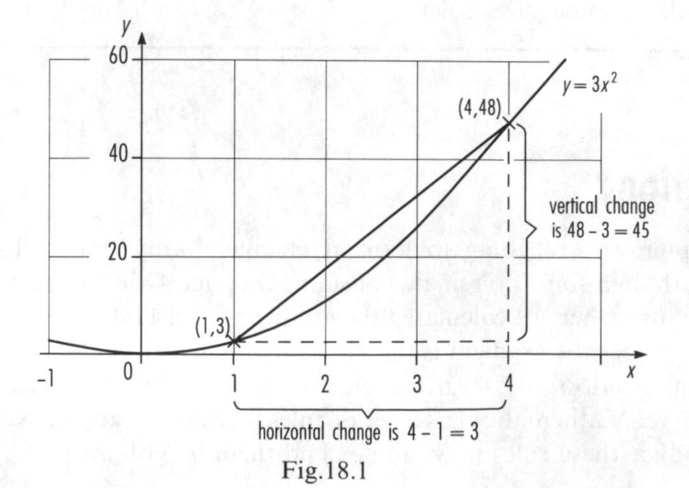

Fig.18.1

the x values. Automatically we change the y values, which in this case also increase. We draw this in Fig. 18.1, stopping at the point (4, 48). From our start to finish the curve is getting steeper. We can look at the average change by drawing a straight line from (1, 3) to (4, 48). This is a **chord**. We know from Chapter 2 that

$$\text{gradient} = \frac{\text{vertical change}}{\text{horizontal change}}$$

$$\text{so that gradient} = \frac{48-3}{4-1} = \frac{45}{3} = 15.$$

This gradient calculation only tells us about the **average value**. It tells us nothing about what happens to the gradient of this portion of the curve at particular points.

Alternatively we may find the change at a specific, instantaneous, point on the curve. In this case we draw a tangent to the curve at that point. In Fig. 18.2 we are interested in the gradient of the curve $y=3x^2$ at the points where $x=2.5$ and $x=-2$.

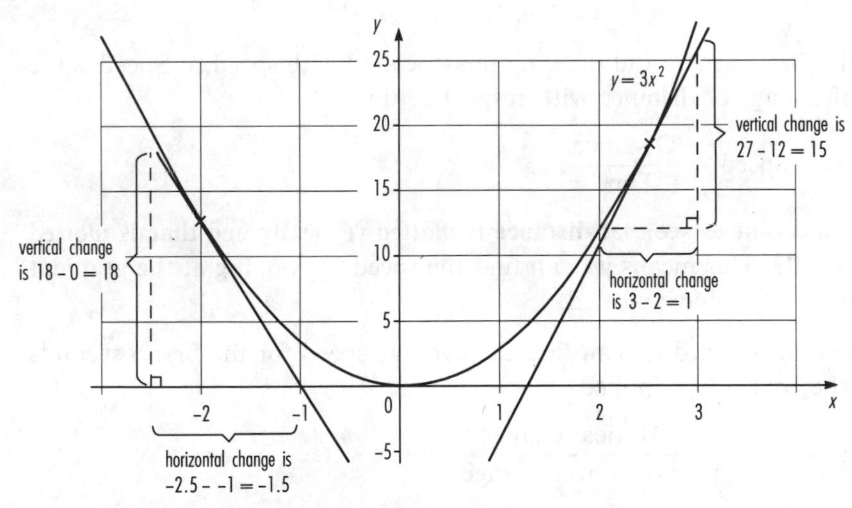

Fig.18.2

Figure 18.2 shows the tangents together with their associated right-angled triangles.

$$\text{Using} \quad \text{gradient} = \frac{\text{vertical change}}{\text{horizontal change}}$$

at $x=-2$

$$\text{gradient} = \frac{18}{-1.5} = -12.$$

at $x=2.5$

$$\text{gradient} = \frac{15}{1} = 15.$$

ASSIGNMENT

We have used the techniques we learned in Chapter 1 to construct a table of $s = 1 + 30t - 2t^2$ and plot its graph. The table from that chapter is not repeated but the graph is given in Fig. 18.3.

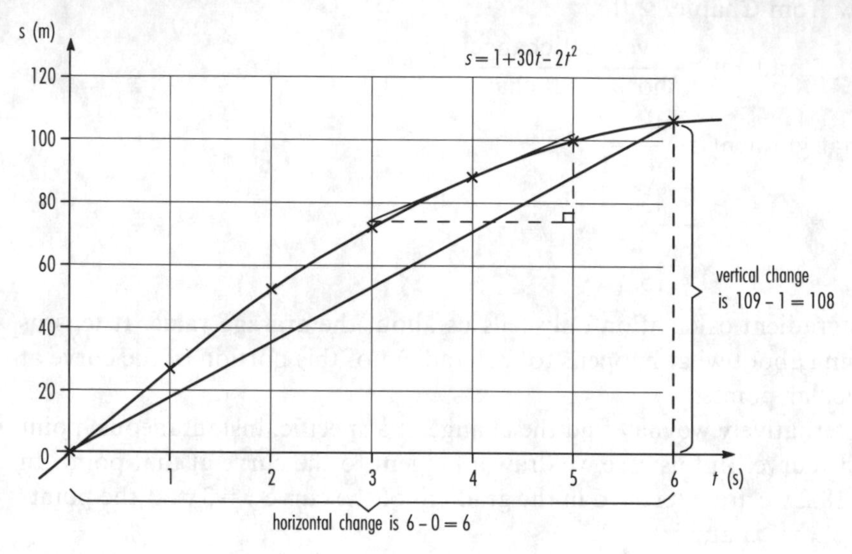

Fig. 18.3

Distance travelled, *s*, and time, *t*, are associated with speed, *v*. Speed is the rate of change of distance with respect to time,

i.e. $$\text{Speed} = \frac{\text{Distance}}{\text{Time}}.$$

It is important to see that **distance is plotted vertically and time is plotted horizontally**. This means we can find the speed by looking at the gradient of the curve we have plotted.

By drawing a chord we can find the average speed for the first 6 seconds after the brakes are applied.

Using $$\text{gradient} = \frac{\text{vertical change}}{\text{horizontal change}}$$

i.e. $$\text{speed} = \frac{\text{distance travelled}}{\text{time taken}}$$

we get $$\text{speed} = \frac{108\,\text{m}}{6\,\text{s}}$$

$$= 18\ \text{ms}^{-1}.$$

Also we can find the speed at a specific time, say 4 seconds after the brakes are applied. We draw a tangent at the point on the curve where $t = 4$. Using the usual gradient formula we get

$$\text{speed} = \frac{103 - 75}{5 - 3} = \frac{28}{2} = 14\ \text{ms}^{-1}.$$

This speed is correct only after 4 seconds because we drew the tangent only at this point.

EXERCISE 18.1

1 Plot the graph of $y = x^2 - 2x$. In your table use values of x from $x = -1$ to $x = 5$ at intervals of 1.
From your graph find the average change of y over this range. Also draw a tangent to your graph at the point where $x = 3.25$. What is the gradient at this point?

2 From $x = -2.0$ to $x = 2.0$ at intervals of 0.5 construct a table of values
for i) $y = 2e^x$
and ii) $y = 2e^{-x}$.
On one set of axes plot these graphs. For each graph find the average change in y between $x = -0.5$ and $x = 0.5$.
Also find the gradient at the point where $x = 0.2$ in each case.
Do you notice any similarities/differences between your answers for each pair of calculations?

3 For values of x from $x = 0$ to $x = \pi$ radians plot the graph of $y = \sin x$.
At $x = \frac{1}{2}\pi$ what is the gradient of this curve?
Find the average gradients from $x =$ i) 0 to $\frac{1}{2}\pi$,
ii) $\frac{1}{2}\pi$ to π,
and iii) 0 to π.

4 A rigid beam of length 10 m carries a uniformly distributed total load of 10 000 N. The bending moment, M, is related to its distance from one end of the beam, x, according to

$$M = -10\,000 + 5000x - 500x^2.$$

What is the bending moment at either end of the beam? What is the difference between the values?

5 Newton's law of cooling relates temperature of a body to its surroundings over a period of time, t seconds. If the body is hotter than its surrounding environment then its temperature will decrease over time. Let T°C be the difference in temperature between a body and its environment. Suppose that the relation is $T = 400e^{-t/1500}$. Find the average temperature gradient over the first 10 minutes without plotting the graph.

The graphical methods for finding gradients take too much time. Drawing an accurate tangent can prove to be difficult. To overcome this problem we derive a method that avoids the use of graphs. It is **differentiation from first principles**.

Differentiation from first principles

In this section we establish the idea behind differentiation. Once you have seen what is involved all you need to do is obey the rules.

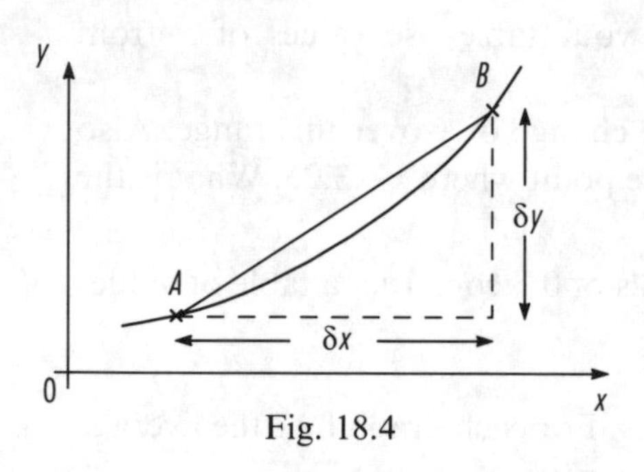

Fig. 18.4

Fig. 18.4 shows a curve, any curve. Think of it as a magnification under a microscope with A and B *very* close together. Let A be a general point (x, y). Now consider a small change in x. This small change is δx where δ is the Greek letter delta. δx is one complete symbol. The original x together with the small change, δx, means we have moved to a new point with the horizontal value $x + \delta x$.

Because y depends on x this causes a small change in y. This change is δy where δy is again one complete symbol. The original y together with the small change, δy, means the new point we mentioned before has a vertical value of $y + \delta y$. Let B be the point $(x + \delta x, y + \delta y)$. In Fig. 18.4 we have labelled A and B and the horizontal and vertical changes.

Using $\quad \text{gradient} = \dfrac{\text{vertical change}}{\text{horizontal change}}$

we have the gradient of the chord AB to be $\dfrac{\delta y}{\delta x}$.

If A is fixed and we move B closer and closer to A the chord gets nearer to the curve. Eventually, as B reaches A, the chord resembles a tangent to the curve at A. We show this in Figs. 18.5. As B approaches A the changes, δx and δy, get smaller and smaller. We write that δx tends to zero, written as $\delta x \to 0$. We are interested in the limit of δy as $\dfrac{\delta y}{\delta x} \to 0$.

Figs. 18.5

$\frac{dy}{dx}$ (spoken as "*dy* by *dx*") represents the gradient of the curve at a general point. This gives us $\frac{dy}{dx} = \lim_{\delta x \to 0} \frac{\delta y}{\delta x}$.

It means that the gradient at a point is the limiting value (Lim) as δx tends to 0 ($\delta x \to 0$) of $\frac{\delta y}{\delta x}$.

$\frac{dy}{dx}$ has various interpretations:

i) the change in y due to the change in x,
ii) the differentiation of y with respect to x,
iii) the first differential,
iv) the first derivative,
v) the gradient,
vi) $\tan \alpha$ where α is the angle of inclination of the tangent to the horizontal.

In the next example we apply this new technique of differentiation from first principles.

Examples 18.2

Differentiate from first principles i) $y = 3x^2$
and ii) $y = \frac{5}{4x}$.

i) Let us look at a general point (x, y) on the curve. Now consider a small change in x, δx, to a new point on the curve $x + \delta x$. In turn this causes a small change in y, δy, to $y + \delta y$. Because the original point (x, y) and the new point $(x + \delta x,\ y + \delta y)$ lie on the curve they both satisfy the equation of that curve.

We write $y + \delta y = 3(x + \delta x)^2$
and $y = 3x^2$.

The original equation.

We have 4 unknowns in this pair of equations; x, y, δx and δy. By subtraction we can attempt to eliminate some of them,

i.e.
$$\delta y = 3(x + \delta x)^2 - 3x^2$$
$$= 3((x + \delta x)^2 - x^2)$$
$$= 3(x^2 + 2x(\delta x) + (\delta x)^2 - x^2)$$

Expanding $(x + \delta x)^2$.

$$= 3(2x(\delta x) + (\delta x)^2)$$

i.e.
$$\delta y = 3(2x + \delta x)(\delta x)$$

Removing a common factor of δx.

so that $\dfrac{\delta y}{\delta x} = 3(2x + \delta x)$.

Dividing by δx.

Using $\dfrac{dy}{dx} = \lim\limits_{\delta x \to 0} \dfrac{\delta y}{\delta x}$

we see that $2x + \delta x \to 2x$

Because $\delta x \to 0$.

to give $\dfrac{dy}{dx} = 3(2x) = 6x$

i.e. the gradient of the curve $y = 3x^2$ is always $6x$.

ii) Let us look at a general point (x, y) on the curve. Now consider a small change in x, δx, to a new point on the curve $x + \delta x$. In turn this causes a small change in y, δy, to $y + \delta y$. Because the original point (x, y) and the new point $(x + \delta x,\ y + \delta y)$ lie on the curve they both satisfy the equation of that curve.

We write $y + \delta y = \dfrac{5}{4(x + \delta x)}$

and $y = \dfrac{5}{4x}$.

The original equation.

We attempt to eliminate some of the variables by subtraction

i.e.
$$\delta y = \frac{5}{4(x + \delta x)} - \frac{5}{4x}$$
$$= \frac{5}{4}\left(\frac{1}{x + \delta x} - \frac{1}{x}\right)$$
$$= \frac{5}{4}\left(\frac{x - (x + \delta x)}{(x + \delta x)x}\right)$$

i.e.
$$\delta y = \frac{-5\delta x}{4(x + \delta x)x}$$

so that $\dfrac{\delta y}{\delta x} = \dfrac{-5}{4(x + \delta x)\,x}$.

Dividing by δx.

Using $\dfrac{dy}{dx} = \lim\limits_{\delta x \to 0} \dfrac{\delta y}{\delta x}$

we see that $x + \delta x \to x$

Because $\delta x \to 0$.

to give $\dfrac{dy}{dx} = \dfrac{-5}{4(x)\,x} = \dfrac{-5}{4x^2}$

i.e. the gradient of the curve $y = \dfrac{5}{4x}$ is always $\dfrac{-5}{4x^2}$.

We can add/subtract gradients according to the usual rules of algebra. Also we can multiply them by a scalar.

Examples 18.3

We use our two previous results to demonstrate some algebraic variations.

For $y=3x^2$, $\frac{dy}{dx}=6x$ and for $y=\frac{5}{4x}$, $\frac{dy}{dx}=\frac{-5}{4x^2}$.

i) If $y=\frac{5}{4x}+3x^2$ then $\frac{dy}{dx}=\frac{-5}{4x^2}+6x$.

ii) If $y=3x^2-\frac{5}{4x}$ then $\frac{dy}{dx}=6x+\frac{5}{4x^2}$.

$(-)(-)=+$.

iii) If $y=21x^2$, i.e. $7(3x^2)$, then $\frac{dy}{dx}=42x$, i.e. $7(6x)$.

ASSIGNMENT

We can move on from our graphical attempts to differentiate $s=1+30t-2t^2$ from first principles.

Let us look at a general point (t, s) on the curve. Now consider a small change in t, δt, to a new point on the curve $t+\delta t$. In turn this causes a small change in s, δs, to $s+\delta s$. Because the original point (t, s) and the new point $(t+\delta t, s+\delta s)$ lie on the curve they both satisfy the equation of that curve.

We write $s+\delta s = 1+30(t+\delta t)-2(t+\delta t)^2$

and $s = 1+30t \qquad -2t^2$.

We subtract these equations to eliminate some of these variables,

i.e.
$$\begin{aligned}\delta s &= 30(t+\delta t-t)-2(t^2+2t(\delta t)+(\delta t)^2-t^2)\\ &= 30\delta t-2(2t(\delta t)+(\delta t)^2)\\ &= 30\delta t-4t(\delta t)-2(\delta t)^2\end{aligned}$$

i.e.
$$\delta s = (30-4t-2(\delta t))(\delta t)$$

so that
$$\frac{\delta s}{\delta t} = 30-4t-2\delta t.$$

Dividing by δt.

Using
$$\frac{ds}{dt} = \lim_{\delta t\to 0}\frac{\delta s}{\delta t}$$

we get
$$\frac{ds}{dt} = 30-4t$$

i.e. the rate of change with repect to time for this distance equation is the speed, $30-4t$.

Generally we write $v=\frac{ds}{dt}$ because speed is the rate of change of distance with respect to time.

EXERCISE 18.2

For the following curves differentiate from first principles to find general expressions for their gradients.

1 $y=x$
2 $y=3x$
3 $y=x^2$
4 $s=2t^2$
5 $s=t^2+t$
6 $y=2x^2-3x+1$
7 $y=x^3$
8 $y=\frac{2}{x}$
9 $s=\frac{4}{t^2}$
10 $y=\frac{2}{x}-3x$

Differentiation of algebraic functions, $y=ax^n$

In the previous exercise you should have obtained some general results including

$x^0=1$ and $1\times1=1$.

For $y=x$, $\frac{dy}{dx}=1$ or $1x^0$.

For $y=x^2$, $\frac{dy}{dx}=2x$ or $2x^1$.

For $y=x^3$, $\frac{dy}{dx}=3x^2$.

Following a similar pattern you would expect that

for $y=x^4$, $\frac{dy}{dx}=4x^3$.

Generally for $y=x^n$ we have $\frac{dy}{dx}=nx^{n-1}$ where n is a constant.

Also we can compare two of our earlier results.

For $y=x^2$ we have $\frac{dy}{dx}=2x$ and

for $y=3x^2$ we have $\frac{dy}{dx}=6x$, i.e. $3(2x)$.

This is a particular case of a more general result:

For $y=ax^n$ we have $\frac{dy}{dx}=anx^{n-1}$ where n and a are constants.

There is an alternative notation of

$$\frac{d}{dx}(ax^n) = anx^{n-1}.$$

It is important to follow the pattern of differentiation. The old power n comes forward to multiply and the power is reduced by 1.

Earlier in this chapter we mentioned that not all things change, i.e. some things are constant. This leads us to the general result that if $y = \text{constant}$ then $\frac{dy}{dx} = 0$.

Examples 18.4

In this set of examples we start with y in terms of x and find a general expression for $\frac{dy}{dx}$.

i) $y = 18.5$, which is a constant, $\quad \frac{dy}{dx} = 0.$

ii) $y = x^7, \quad \frac{dy}{dx} = 7x^{7-1} \quad = 7x^6.$

iii) $y = 4x^7, \quad \frac{dy}{dx} = 4 \times 7x^{7-1} \quad = 28x^6.$

iv) $y = \frac{1}{2}x^{10}, \quad \frac{dy}{dx} = \frac{1}{2} \times 10x^{10-1} \quad = 5x^9.$

v) $y = \frac{1}{2x} = \frac{1}{2}x^{-1}, \quad \frac{dy}{dx} = \frac{1}{2} \times (-1)x^{-1-1} \quad = -\frac{1}{2}x^{-2}$ or $\frac{-1}{2x^2}.$

vi) $y = \frac{7}{2x^4} = \frac{7x^{-4}}{2}, \quad \frac{dy}{dx} = \frac{7}{2} \times (-4)x^{-4-1} \quad = -14x^{-5}$ or $\frac{-14}{x^5}.$

Notice how we move the x term to the numerator before we differentiate it. In Volume 2 we will look at alternative methods of differentiation.

Not all expressions that we differentiate have only one term. In this next example we simplify the algebra first. The next step is to differentiate each term according to our rule.

Example 18.5

For $y = 4x^3 + \frac{(2x^2 - x)}{3x}$ first we simplify the algebra. We put each term in the bracket separately upon the common denominator. This gives

$$y = 4x^3 + \frac{2x^2}{3x} - \frac{x}{3x}$$

i.e. $$y = 4x^3 + \frac{2x}{3} - \frac{1}{3}.$$

Differentiating term by term gives

$$\frac{dy}{dx} = 12x^2 + \frac{2}{3}.$$

ASSIGNMENT

The last time we looked at the Assignment we differentiated from first principles. Now our improved technique using the differentiation rules allows us to write

$$s = 1+30t-2t^2$$

and $\dfrac{ds}{dt} = 30-4t$ immediately.

This is a simpler method of finding the general expression for speed, v, which in this case is $v=30-4t$.

EXERCISE 18.3

Differentiate the following expressions by rule.

1 $y=9$

2 $y=3x$

3 $y=3x-9$

4 $y=4x^2$

5 $s=4t^2-3t+9$

6 $y=9-3x-4x^2$

7 $y=x^2+x$

8 $y=2x^2-3x+1$

9 $s=6t^{-1}+2t^{-2}$

10 $s=\dfrac{4}{t^2}$

11 $y=\dfrac{6}{x^2}+\dfrac{2}{x}$

12 $y=3x-\dfrac{2}{x}$

13 $y=\dfrac{2}{x^6}-3x^2$

14 $y=\dfrac{3x^3+2x^2}{x}$

15 $y=\dfrac{x-x^2}{x^4}$

So far we have concentrated on powers that are whole numbers (integers). Exactly the same ideas apply to fractional or decimal powers.

Examples 18.6

Again in this set of examples we start with y in terms of x and find a general expression for $\dfrac{dy}{dx}$ by rule.

i) $y = \sqrt{x} = x^{\frac{1}{2}}$, $\quad \dfrac{dy}{dx} = \dfrac{1}{2}x^{\frac{1}{2}-1} \quad = \dfrac{1}{2}x^{-\frac{1}{2}}$

or $\dfrac{1}{2x^{\frac{1}{2}}}$ or $\dfrac{1}{2\sqrt{x}}$.

ii) $y = 3\sqrt{x} = 3x^{\frac{1}{2}}$, $\quad \dfrac{dy}{dx} = 3 \times \dfrac{1}{2}\, x^{\frac{1}{2}-1} \quad = \dfrac{3}{2}\, x^{-\frac{1}{2}}$

or $\dfrac{3}{2x^{\frac{1}{2}}}$ or $\dfrac{3}{2\sqrt{x}}$.

iii) $y = 8\sqrt[4]{x^3} = 8x^{\frac{3}{4}}$, $\quad \dfrac{dy}{dx} = 8 \times \dfrac{3}{4} x^{\frac{3}{4}-1} \quad = 6x^{-\frac{1}{4}}$

or $6x^{-0.25}$.

iv) $y = \dfrac{5}{x^{\frac{1}{2}}} = 5x^{-\frac{1}{2}}$, $\quad \dfrac{dy}{dx} = 5 \times \left(-\dfrac{1}{2}\right) x^{-\frac{1}{2}-1} \quad = -\dfrac{5}{2}\, x^{-\frac{3}{2}}$

or $-2.5x^{-1.5}$.

v) $y = \dfrac{2}{3\sqrt[3]{x}} = \dfrac{2}{3x^{\frac{1}{3}}} = \dfrac{2}{3} x^{-\frac{1}{3}}$, $\dfrac{dy}{dx} = \dfrac{2}{3} \times \left(-\dfrac{1}{3}\right) x^{-\frac{1}{3}-1} = \dfrac{-2}{9}\, x^{-\frac{4}{3}}$.

We leave the answer for the final part as a fraction. If you convert it to a decimal make sure you are accurate with the decimal places.

The next set of exercises continues with differentiation by rule for these types of powers.

EXERCISE 18.4

Differentiate the following expressions by rule.

1 $y = 4\sqrt{x}$

2 $y = 3.5x^{0.5}$

3 $y = 8x^{\frac{1}{3}}$

4 $y = \dfrac{1}{3}\, x^{\frac{1}{8}}$

5 $s = 4t^{\frac{1}{2}}$

6 $y = \sqrt{x^3} - 4$

7 $y = \sqrt[3]{x^2} + 3$

8 $s = 4t^{0.2} + 2t^{0.4}$

9 $y = 7x^{0.5} - 2x^{-0.5}$

10 $y = \dfrac{2}{3x^{0.5}}$

11 $y = \dfrac{2}{3}\, x^{-0.5} + \dfrac{1}{3}\, x^{-1.5}$

12 $y = \dfrac{3}{2\sqrt[3]{x}} + \dfrac{1}{\sqrt{x}}$

13 $s = \sqrt{t} - \dfrac{1}{\sqrt{t}}$

14 $y = 4\sqrt{x} - \dfrac{1}{2\sqrt{x}}$

15 $y = \dfrac{\sqrt{x} + 1}{x^2}$

Differentiation of sine and cosine

We know the shape of the graphs of $y = \sin\theta$ and $y = \cos\theta$ from Chapter 5. From our early work in this chapter we can find the gradient of a curve graphically. For each of these curves we can find the instantaneous gradients by drawing a series of tangents. In Example 18.7 we start off the technique and leave you to complete the task.

Example 18.7

For the curve $y = \sin\theta$ we plot the graph over one cycle (i.e. from 0 to 2π radians) at intervals of $\frac{\pi}{6}$ radians. At the same intervals draw tangents to the curve and find their gradients.

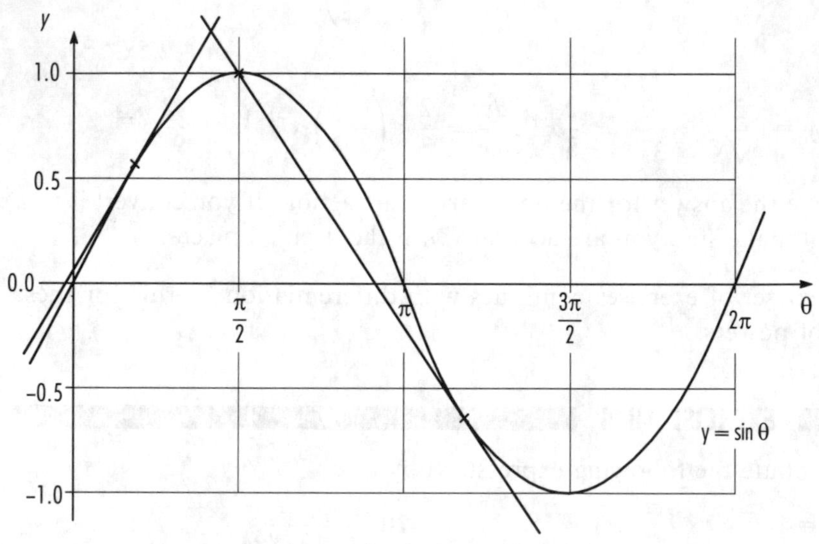

Fig. 18.6

In Fig. 18.6 we have drawn 3 specimen gradients to the sine curve to show the idea. You need to do this for yourself and check your gradient values against those in our table.

θ	0	$\frac{\pi}{6}$	$\frac{\pi}{3}$	$\frac{\pi}{2}$	$\frac{2\pi}{3}$	$\frac{5\pi}{6}$
$\sin\theta$	0	0.500	0.866	1	0.866	0.500
gradient	1	0.866	0.500	0	−0.500	−0.866

θ	π	$\frac{7\pi}{6}$	$\frac{4\pi}{3}$	$\frac{3\pi}{2}$	$\frac{5\pi}{3}$	$\frac{11\pi}{6}$	2π
$\sin\theta$	0	−0.500	−0.866	−1	−0.866	−0.500	0
gradient	−1	−0.866	−0.500	0	0.500	0.866	1

You ought to be able to spot a pattern. Each gradient value is the **value of the cosine at that point**.

For yourself you should repeat this technique using the cosine curve, again over one cycle using radians. In this case the gradient at each point should be the negative of the sine value.

The patterns give us the differentiation rules

for $y = \sin\theta$ and for $y = \cos\theta$

$$\frac{dy}{d\theta} = \cos\theta \qquad \frac{dy}{d\theta} = -\sin\theta$$

or using the alternative notation

$$\frac{d}{d\theta}(\sin\theta) = \cos\theta \quad \text{and} \quad \frac{d}{d\theta}(\cos\theta) = -\sin\theta$$

provided θ is in radians.

When we looked at the gradients of the sine and cosine curves we saw that sometimes the gradients were zero. For many other curves there are particular points where the gradients are zero. In Figs. 18.7 we have sketched examples of curves and highlighted where the gradients are zero. Points ***A*** and ***B*** are called **turning points**. At the turning point ***A***, y has a maximum value. At the turning point ***B***, y has a minimum value. We shall explore points of these types further in Volume 2.

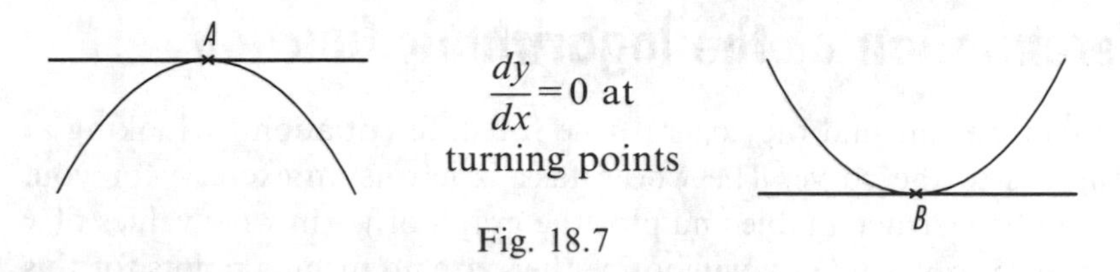

Fig. 18.7

Examples 18.8

We differentiate the trigonometric expressions by rule.

i) For $y = 3\sin\theta$

$$\frac{dy}{dx} = 3\cos\theta.$$

ii) For $y = 4\cos\theta$

$$\frac{dy}{dx} = -4\sin\theta.$$

iii) For $y = 2.5\sin\theta + 3\cos\theta$

$$\frac{dy}{dx} = 2.5\cos\theta - 3\sin\theta.$$

Differentiation of the exponential function

Again we can find a general rule for differentiation by looking at the tangents to the curve. This time the whole task is left as an exercise for you. You should construct a table and plot the graph of $y = e^x$ for values of x from $x = -2$ to $x = 2$. (You will notice there are no turning points for this curve.) Within this range you should draw tangents to the curve at intervals of 0.25. Calculate the gradient of each tangent and include it in your table as we did for $y = \sin\theta$. With accuracy you should find each gradient value is the same as the y value at that point. This gives us the general rule for $y = e^x$

$$\frac{dy}{dx} = e^x$$

or using the alternative notation $\dfrac{d}{dx}(e^x) = e^x$.

Example 18.9

By rule if $y = 2.75e^x$

then $\dfrac{dy}{dx} = 2.75e^x$.

Differentiation of the logarithmic function

Once again we can find the general rule for differentiation by looking at the tangents to the curve. The whole task is left as an exercise for you. You should construct a table and plot the graph of $y = \ln x$ for values of x from $x = 0.25$ to $x = 3$. (You will notice there are no turning points for this curve.) Within this range you should draw tangents to the curve at intervals of 0.25. Calculate the gradient of each tangent and include it in your table as we did for the exponential function. With accuracy you should find each gradient value is the same as $\frac{1}{x}$ value at that point. This gives us the general rule for $y = \ln x$

$$\frac{dy}{dx} = \frac{1}{x}$$

or using the alternative notation $\dfrac{d}{dx}(\ln x) = \dfrac{1}{x}$.

Example 18.10

Differentiate the following logarithmic functions by rule

i) For
$$y = 3\ln x$$
$$\frac{dy}{dx} = 3 \times \frac{1}{x} = \frac{3}{x}.$$

ii) For
$$y = \ln(3x)$$
i.e.
$$y = \ln 3 + \ln x$$
so that
$$\frac{dy}{dx} = 0 + \frac{1}{x} = \frac{1}{x}.$$

First log law.

ln 3 is a constant.

iii) For $y = \ln\left(\frac{x}{4}\right)$

i.e. $y = \ln x - \ln 4$

so that $\frac{dy}{dx} = \frac{1}{x} - 0 = \frac{1}{x}$.

Second log law.

ln 4 is a constant.

iv) For $y = \ln\left(\frac{1}{x}\right)$

i.e. $y = \ln 1 - \ln x$

i.e. $y = -\ln x$

so that $\frac{dy}{dx} = -\frac{1}{x}$.

Second log law.

$\ln 1 = 0$.

v) For $y = \ln(x^2)$

i.e. $y = 2\ln x$

so that $\frac{dy}{dx} = \frac{2}{x}$.

Third log law.

The next set of exercises allows you to practise each of these differentiations by rule.

EXERCISE 18.5

Differentiate the following expressions by rule.

1 $y = 4\cos\theta - 2\sin\theta$

2 $y = 2.4\sin\theta + 3\cos\theta$

3 $s = 3.5e^t$

4 $y = 5\ln x$

5 $y = 2.4\sin\theta - 3\cos\theta$

6 $y = 2e^t + 3\cos t$

7 $y = 5\ln x + \ln(5x)$

8 $y = 3e^x + \ln(2x)$

9 $y = 3\ln x + 2\ln\left(\frac{1}{x}\right)$

10 $y = \ln(2x^3)$

Gradient at a point

We know how to differentiate using a set of rules. Each time we have started with an equation and found a general expression for the gradient, $\frac{dy}{dx}$, or similar.

As we move along a curve the values of x and y change. This means the gradient changes. (Only along a straight line does the gradient remain the same.) Suppose we wish to find the gradient at a particular point. We find the general expression for the gradient *before* substituting any particular numbers.

Example 18.11

Find the gradient of the curve $y=3x^2$ at the points where

i) $x=-2$

and ii) $x=2.5$.

i) For this particular curve we know, differentiating by rule, that

$$\frac{dy}{dx} = 6x.$$

We substitute for $x=-2$ to get

$$\frac{dy}{dx} = 6(-2) = -12$$

i.e. the gradient of the curve where $x=-2$ is -12, or the gradient of the tangent to the curve where $x=-2$ is -12.

ii) Again using $\frac{dy}{dx} = 6x$

we substitute for $x=2.5$ to get

$$\frac{dy}{dx} = 6(2.5) = 15$$

i.e. the gradient of the curve where $x=2.5$ is 15.

Example 18.12

A bacterial growth, B, is related exponentially to time, t hours, by $B=B_0e^t$ where $B_0=1.75\times10^3$. Find the rate of growth of the bacteria after 2 hours.

We differentiate our equation by rule to get the general expression

$$\frac{dB}{dt} = 1.75\times10^3e^t.$$

Now we substitute for our particular value of $t=2$ so that

$$\frac{dB}{dt} = 1.75\times10^3e^2 = 12.9\times10^3,$$

i.e. the bacterial growth at this particular time is 12.9×10^3 per hour.

ASSIGNMENT

Let us take a final look at our vehicle with

$$s = 1+30t-2t^2$$

and $$v = \frac{ds}{dt} = 30-4t.$$

s is distance (m).
t is time (s).
v is speed (ms^{-1}).

Suppose we wish to find the initial speed. Initial means "at the beginning", i.e. at $t=0$ so that

$$v = 30-4(0) = 30\,ms^{-1} \quad \text{is the initial speed.}$$

To find the speed after 3.5 seconds we substitute for $t=3.5$ in

$$v = 30-4t$$

to get $\quad v = 30-4(3.5) = 16\,\text{ms}^{-1}$ as the speed.

EXERCISE 18.6

1 If $y=2x+3e^x$ find $\dfrac{dy}{dx}$ where $x=1.5$.

2 $yx^{3.5}=2$. Make y the subject of this formula. What is the value of $\dfrac{dy}{dx}$ where $x=2.25$?

3 $y=4\sin t-\cos t$. Find the rate of change of y with respect to t when $t=\dfrac{\pi}{3}$.

4 Given that $y=\dfrac{10-\pi x^2}{\pi x}$ find a general expression for $\dfrac{dy}{dx}$. What is the value of this gradient where $x=3.4$?

5 Find the value of $\dfrac{dy}{d\theta}$ where $\theta=\dfrac{\pi}{4}$ for the curve $y=2(\theta-3.2\sin\theta)$.

6 $s=15.6t-1.06t^2$ refers to the motion of a particular vehicle. s is the displacement in metres after t seconds. What is the initial velocity $\left(\text{i.e. } \dfrac{ds}{dt}\right)$ of the vehicle? After what time will the vehicle come to rest?

7 A rigid beam of length 10 m carries a uniformly distributed total load of 10 000 N. The bending moment, M, is related to its distance from one end of the beam, x, according to $M=-10\,000+5000x-500x^2$. What is the rate of change of bending moment $\left(\text{i.e. } \dfrac{dM}{dx}\right)$ at 3 m from either end of the beam?

8 The diagram shows a belt in contact with a pulley. The length of belt in contact with the pulley subtends an angle θ radians with the centre. The coefficient of friction is 0.4. The tension on the taut side is T and on the slack side is 25 N. If $\theta=2.5\ln\left(\dfrac{T}{25}\right)$ find $\dfrac{d\theta}{dT}$ when $T=37.5$ N.

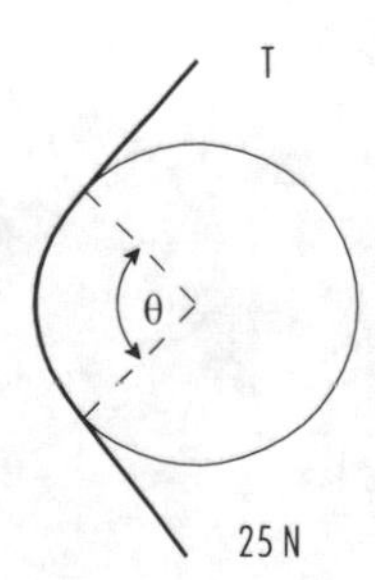

9 A disc is spun from rest. It spins through y° in t seconds according to $y=100t^2-5t^3$. What is the value of $\dfrac{dy}{dt}$ after 2.5 seconds?

10 One particular machine in an engineering workshop costs £C to lease each week according to the formula $C=200+\dfrac{t^3}{20}$. t is the number of hours/week worked by the machine. $\frac{dC}{dt}$ is the rate of increase of cost during the week. Find a general expression for $\frac{dC}{dt}$. When does this rate exceed £150 per hour?

Summary of differentiation rules

y	$\frac{dy}{dx}$	
ax^n	anx^{n-1}	where a and n are numbers.
$\sin x$	$\cos x$	x in radians.
$\cos x$	$-\sin x$	x in radians.
e^x	e^x	
$\ln x$	$\frac{1}{x}$	

19 Calculus: II – Integration

The objectives of this chapter are to:

1. Define indefinite integration as the reverse process to differentiation.
2. Determine indefinite integrals of simple algebraic functions, and functions involving $\cos\theta$, $\sin\theta$ and exponentials.
3. Recognise the need to include an arbitrary constant of integration.
4. Evaluate $\int_a^b y\,dx$ by $\left[F(x)\right]_a^b = F(b) - F(a)$ for simple functions, where $F(x)$ is the indefinite integral of y.
5. Define $\int_a^b y\,dx$ as the area under the curve between the ordinates $x = a$ and $x = b$.
6. Determine areas by applying the definite integral for simple functions.

Introduction

We introduced Calculus in Chapter 18 through differentiation. In this chapter we look at the opposite process to differentiation. **Integration** is that reverse process. Our early examples each start with the gradient of a graph and aim to find that graph's equation. Later we look to apply the technique to find the area under a curve.

ASSIGNMENT

Our Assignment for this chapter looks again at the curve $y = 4 - x^2$. We saw this curve in Example 16.12 where we calculated areas. In that set of 3 examples we calculated the area bounded by the curve, the horizontal axis and the lines $x = 1$ and $x = 2$. Then we used 3 numerical methods (Mid-ordinate rule, Trapezoidal rule and Simpson's rule). This time we will find the area under the curve using integration.

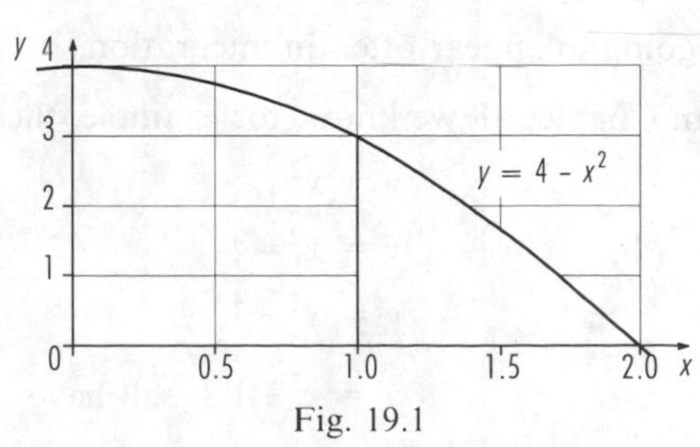

Fig. 19.1

Geometrical interpretation

Suppose we know the gradient of a graph. For example the gradient might be 5. Based on Chapter 2 we know the graph must be a straight line with $m=5$, i.e. $y = 5x+c$.

Remember c is the intercept on the vertical axis of our graph. We know there are many parallel lines with gradients of 5. In each case they cut the vertical axis at a different point. Fig. 19.2 shows a few examples.

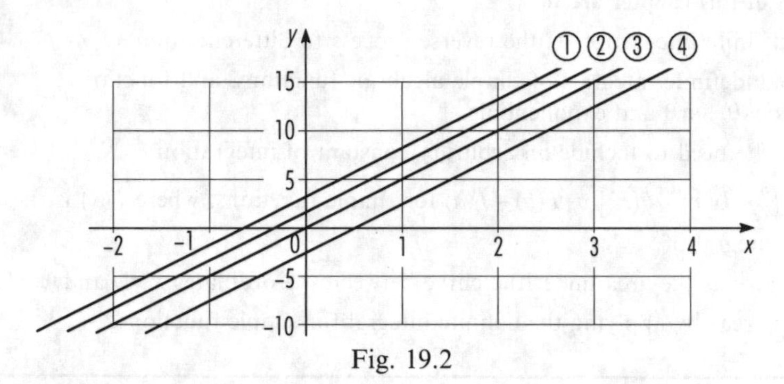

Fig. 19.2

From the many possibilities we can decide upon our particular line if we know a point lying on it.

Example 19.1

Find the equation of the graph with gradient 5 passing through the point (2, 13).

Already we know the graph is a straight line with $y=5x+c$. Because it passes through the point (2, 13) we can substitute for $x=2$ and $y=13$ to get

$$13 = 5(2)+c$$

i.e. $\quad 13-10 = c$

i.e. $\quad 3 = c.$

This means the equation of the straight line is $y=5x+3$.

c is going to appear often in integration. It is the **constant of integration**.

From Chapter 18 we know, for example, the curves

$$y = x^2$$

$$y = x^2-4$$

$$y = x^2+3$$

and $\quad y = x^2+1.8 \quad$ all have $\dfrac{dy}{dx}=2x$.

If we start with $\dfrac{dy}{dx} = 2x$

or $\dfrac{dy}{dx} = 2x+0$

we integrate to get $y = x^2+c$.

c, the constant of integration, should appear because when we differentiate a constant we get 0.

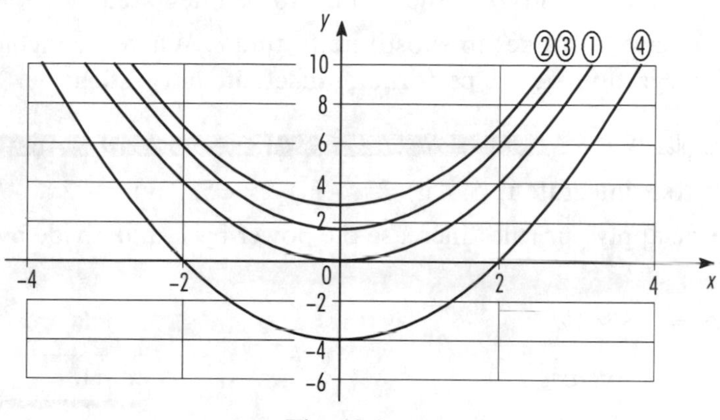

Fig. 19.3

In fact $y=x^2+c$ represents a family of curves (Fig. 19.3) all with gradient $2x$.

Integration by rule

When we differentiated we saw a pattern of movement of numbers. For integration we have the reverse pattern of movement.

Generally if $\dfrac{dy}{dx} = x^n$

then $y = \dfrac{x^{n+1}}{n+1}+c \qquad n\neq -1$

where n is a numerical value.

Notice the important exception to this rule when $n \neq -1$. We cannot allow the possibility of -1 because this would create a divisor of 0. Remember we cannot divide by 0 in Mathematics. We say this integration rule as "increase the power by 1, and divide by the new power".

Remember integration is the reverse process to differentiation. You can check for yourself the differentiation of $y = \dfrac{x^{n+1}}{n+1} + c$ gives the correct answer for $\dfrac{dy}{dx}$ above.

There is an alternative notation,

$$\int x^n\,dx = \frac{x^{n+1}}{n+1} + c \qquad n \neq -1$$

$\int$ is the symbol for integration.

dx means we are going to integrate with respect to x, i.e. x is the variable being processed.

The term(s) between $\int$ and dx is the section to be integrated.

At present we have no values to substitute to find c. Where we include a constant of integration we are performing **indefinite integration**.

Examples 19.2

With respect to x integrate i) 5, ii) $2x$, iii) $2x+5$, iv) $5-2x$.

In each case we apply our rule "increase the power by 1, and divide by the new power".

i)
$$\begin{aligned}\int 5\,dx &= \int 5 \times 1\,dx \\ &= \int 5x^0\,dx \\ &= \frac{5x^{0+1}}{0+1} + c \\ &= 5x + c.\end{aligned}$$

$x^0 = 1.$

$x^1 = x.$

ii)
$$\begin{aligned}\int 2x\,dx &= \int 2x^1\,dx \\ &= \frac{2x^{1+1}}{1+1} + c \\ &= \frac{2x^2}{2} + c \\ &= x^2 + c.\end{aligned}$$

In this example 2 is a scalar multiplier. We can always move this type of multiplier out of the integral. *We can never do it to a variable.*

Let us have another look at this example.

ii) Again
$$\begin{aligned}\int 2x\,dx &= 2\int x\,dx \\ &= 2 \times \frac{x^{1+1}}{1+1} + c \\ &= 2 \times \frac{x^2}{2} + c \\ &= x^2 + c \qquad \text{as before.}\end{aligned}$$

We can add/subtract our integrals, integrating each term individually. In this case we need only include one constant of integration, c.

iii) $\int 2x + 5\,dx = x^2 + 5x + c.$

iv) $\int 5 - 2x\,dx = 5x - x^2 + c.$

Examples 19.3

With respect to x integrate i) $3x^7 + 2x^2 - \dfrac{4}{x^3}$

and ii) $\dfrac{2x^3 - 3x^2}{x}$.

i) We need to re-arrange $\dfrac{4}{x^3}$ before we integrate our expression term by term. $\dfrac{4}{x^3}$ needs to be re-written as $4x^{-3}$.

Then
$$\int 3x^7 + 2x^2 - 4x^{-3}\,dx = \frac{3x^{7+1}}{7+1} + \frac{2x^{2+1}}{2+1} - \frac{4x^{-3+1}}{-3+1} + c$$
$$= \frac{3x^8}{8} + \frac{2x^3}{3} - \frac{4x^{-2}}{-2} + c$$
$$= \frac{3x^8}{8} + \frac{2x^3}{3} + 2x^{-2} + c.$$

ii) We need to simplify our expression before attempting to integrate, again term by term, i.e.
$$\frac{2x^3 - 3x^2}{x} = \frac{2x^3}{x} - \frac{3x^2}{x}$$
$$= 2x^2 - 3x.$$

Then
$$\int \frac{2x^3 - 3x^2}{x}\,dx = \int 2x^2 - 3x\,dx$$
$$= \frac{2x^{2+1}}{2+1} - \frac{3x^{1+1}}{1+1} + c$$
$$= \frac{2x^3}{3} - \frac{3x^2}{2} + c.$$

EXERCISE 19.1

Integrate the following expressions.

1 $\int 7\,dx$

2 $\int 5x\,dx$

3 $\int 7 - 5x\,dx$

4 $\int 3x^2\,dx$

5 $\int 3t^2 + 5t - 7\,dt$

6 $\int 7 - 5x - 3x^2\,dx$

7 $\int x^4 + 2x\,dx$

8 $\int 4x^6 - 2x^3\,dx$

9 $\displaystyle\int 5t^{-2} + 3t^{-4}\,dt$

10 $\displaystyle\int \frac{3}{t^3}\,dt$

11 $\int \frac{5}{x^2} - \frac{3}{x^4}\, dx$

12 $\int 2t^3 - \frac{3}{t^2}\, dt$

13 $\int \frac{3}{2x^5} - 3x^5\, dx$

14 $\int \frac{3x + 4x^3}{x}\, dx$

15 $\int \frac{2x^2 - x}{x^4}\, dx$

So far we have concentrated on whole number (integer) powers. We follow exactly the same rule with fractional and decimal powers, i.e. "increase the power by 1, and divide by the new power".

Examples 19.4

i) $\int \sqrt{x}\, dx = \int x^{\frac{1}{2}}\, dx$

$$= \frac{x^{\frac{1}{2}+1}}{\frac{1}{2}+1} + c$$

$$= \frac{x^{\frac{3}{2}}}{\frac{3}{2}} + c$$

$$= \frac{2}{3} x^{\frac{3}{2}} + c.$$

Dividing by a fraction: invert and multiply.

ii) $\int \frac{2}{5\sqrt{x}}\, dx = \frac{2}{5} \times \int x^{-\frac{1}{2}}\, dx$

$$\frac{1}{\sqrt{x}} = \frac{1}{x^{\frac{1}{2}}} = x^{-\frac{1}{2}}.$$

$$= \frac{2}{5} \times \frac{x^{-\frac{1}{2}+1}}{-\frac{1}{2}+1} + c$$

$$= \frac{2}{5} \times \frac{x^{\frac{1}{2}}}{\frac{1}{2}} + c$$

$$\frac{2}{5} \times \frac{1}{\frac{1}{2}} = \frac{2}{5} \times \frac{2}{1} = \frac{4}{5}.$$

$$= \frac{4}{5} x^{\frac{1}{2}} + c.$$

EXERCISE 19.2

Integrate the following expressions.

1 $\int 6\sqrt{x}\, dx$

2 $\int 5x^{0.5}\, dx$

3 $\int 8t^{\frac{1}{3}}\, dt$

4 $\int \frac{1}{5} x^{\frac{1}{4}}\, dx$

5 $\int 4t^{\frac{1}{2}}\, dt$

6 $\int \sqrt{x^3} - 6\, dx$

7 $\int \sqrt[3]{x^2} - 4\, dx$

8 $\int 4t^{0.2} + 2t^{0.4}\, dt$

9 $\int 9x^{0.5}+4x^{-0.5}\,dx$

10 $\int \frac{3}{2x^{0.5}}\,dx$

11 $\int \frac{2}{5}x^{-0.5}+\frac{1}{5}x^{-1.5}\,dx$

12 $\int \frac{5}{2\sqrt[3]{x}}+\frac{2}{\sqrt{x}}\,dx$

13 $\int 3\sqrt{t}-\frac{4}{\sqrt{t}}\,dt$

14 $\int 8\sqrt{x}+\frac{1}{4\sqrt{x}}\,dx$

15 $\int \frac{\sqrt{x}+4}{2x^2}\,dx$

Integration of sine and cosine

Remembering that integration is the reverse process to differentiation we have

$$\int \cos\theta\, d\theta = \sin\theta + c$$

and $$\int \sin\theta\, d\theta = -\cos\theta + c$$

provided θ is in radians.

You need to be careful with the +/– signs. Do *not* confuse differentiation and integration.

Example 19.5

$$\int 3\cos\theta + 2\sin\theta\, d\theta = 3\sin\theta - 2\cos\theta + c.$$

Integration of the exponential function

The integration is as simple as the differentiation, i.e.

$$\int e^x\, dx = e^x + c.$$

Example 19.6

$$\int 3e^x\, dx \text{ or } 3\int e^x\, dx = 3\,e^x + c.$$

Integration of $\frac{1}{x}$

You will remember our general rule for algebraic functions had the exception of $n \neq -1$. Now we will deal with that case where $n = -1$. Remembering the connection between differentiation and integration we have

$$\int \frac{1}{x}\, dx = \ln x + c.$$

You should check back to the differentiation of $\ln x$ in Chapter 18.

Example 19.7

$$\int \frac{3}{4x}\, dx = \frac{3}{4}\int \frac{1}{x}\, dx = \frac{3}{4}\ln x + c.$$

This next set of exercises allows you to practise a variety of integrations.

EXERCISE 19.3

Integrate the following expressions.

1 $\int 7\cos\theta\, d\theta$

2 $\int 4e^x\, dx$

3 $\int \frac{1}{2x}\, dx$

4 $\int 2\sin\theta\, d\theta$

5 $\int 2e^x + 3\sin x\, dx$

6 $\int 5\cos x - \frac{1}{2}e^x\, dx$

7 $\int 7\cos x - 2\sin x\, dx$

8 $\int e^t + 4\sin t\, dt$

9 $\int \frac{x^2+2}{x^3}\, dx$

10 $\int \frac{x^2 - 3x\cos x}{x}\, dx$

Definite integration

Earlier we saw the indefinite integral $\int 2x\, dx = x^2 + c$, not knowing the value of c.

In contrast, for **definite integration** we get a value for the integral rather than a general expression. This value comes from the numerical substitution of **limits**.

The general rule is

$$\int_a^b x^n\, dx = \left[\frac{x^{n+1}}{n+1}\right]_a^b.$$

a is the **lower limit** of x and b is the **upper limit** of x. We expect $a < b$.

We integrate to get some function of x, say $F(x)$. Next we substitute for $x = a$ and $x = b$, and subtract, according to

$$\Big[F(x)\Big]_a^b = F(b) - F(a).$$

In the following examples we show how to substitute for the values in the a and b positions.

Example 19.8

Find the value of $\int_1^3 5-2x\,dx$.

We know from Examples 19.2 that

$$\int 5-2x\,dx = 5x-x^2+c.$$

Thus $\int_1^3 5-2x\,dx = \left[5x-x^2+c\right]_1^3$

We substitute for $x=3$ and for $x=1$ and subtract to get

$$\begin{aligned}&(5(3)-3^2+c)-(5(1)-1^2+c)\\&=(15-9+c)-(5-1+c)\\&=6+c-4-c\\&=2.\end{aligned}$$

Notice how the cs cancel out. Because this always happens we do *not* need to include them.

Example 19.9

Evaluate $\int_0^{\frac{\pi}{6}} \cos x+2e^x\,dx$.

Because of the trigonometric ratio, $\cos x$, notice the integration limits are in radians. We obey our rules of integration to get

$$\begin{aligned}\int_0^{\frac{\pi}{6}} \cos x+2e^x\,dx &= \left[\sin x+2e^x\right]_0^{\frac{\pi}{6}}\\&=\left(\sin\frac{\pi}{6}+2e^{\frac{\pi}{6}}\right)-\left(\sin 0+2e^0\right)\\&=0.500+3.376-0-2\\&=1.88.\end{aligned}$$

EXERCISE 19.4

Find the values of the following definite integrals.

1. $\int_1^2 2x^4+3\,dx$
2. $\int_{\frac{\pi}{4}}^{\frac{\pi}{2}} 5\cos\theta\,d\theta$
3. $\int_0^2 3e^x\,dx$
4. $\displaystyle\int_1^{2.5} \frac{6}{x}\,dx$
5. $\int_{\frac{\pi}{6}}^{\frac{\pi}{3}} 2\cos x-7\sin x\,dx$
6. $\displaystyle\int_9^{16} 4\sqrt{x}-\frac{2}{\sqrt{x}}\,dx$
7. $\int_{1.5}^{2.5} 1+3x-6x^2\,dx$
8. $2\int_{-\frac{\pi}{6}}^{\frac{\pi}{6}} \cos\theta\,d\theta$
9. $\int_{-1}^{1} 2e^x\,dx$
10. $\displaystyle\int_0^{\frac{\pi}{2}} \frac{3x^3-2x\sin x}{x}\,dx$

Area under a curve

Integration has many applications. Finding the area under a curve is one simple case.

Fig. 19.4 shows a general curve. We wish to find the area under the curve bounded by the x-axis and the lines $x=a$ and $x=b$. To do this we split the area into a series of thin parallel strips each of width δx.

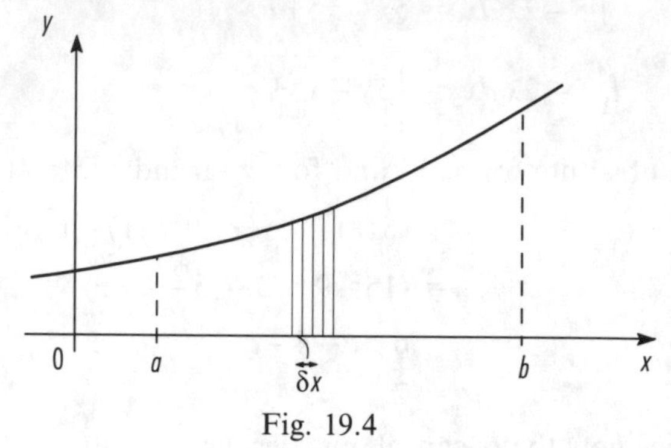

Fig. 19.4

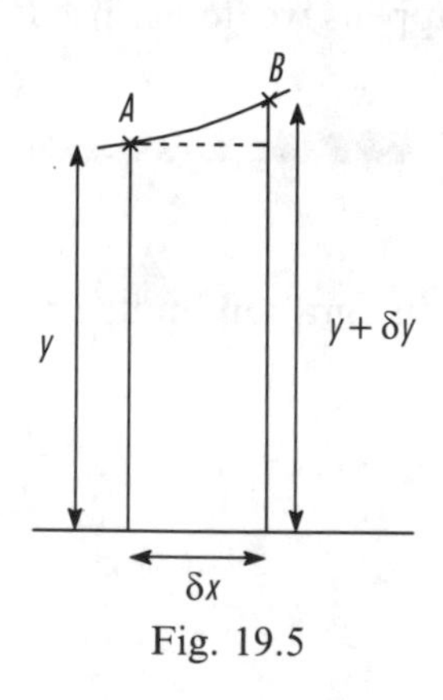

Fig. 19.5

Let us magnify one of those strips. Suppose A is the point (x, y). Remember a small change in x, δx, causes a small change in y, δy. Because B is a small distance away we know it is the point $(x+\delta x, y+\delta y)$. We approximate this strip to a rectangle. As the strip width gets narrower and narrower (i.e. as $\delta x \to 0$) the accuracy of the approximation improves. This means the extra area above the rectangle becomes very small. Then

$$\begin{aligned}\text{Area of thin strip} &\approx \text{Area of rectangle}\\ &\approx y\delta x.\end{aligned}$$

> $\approx$ is "equals approximately".

The area between $x=a$ and $x=b$ is the sum of all such strips, i.e

$$\text{Area} = \lim_{\delta x \to 0} \sum_{x=a}^{x=b} y\delta x.$$

In practice this sum to find the area would be difficult to calculate. We can reach the same result more easily using the integration formula

$$\text{Area} = \int_a^b y\,dx.$$

We easily apply this formula in the next examples.

Example 19.10

Find the area under the curve $y=(1+x)(4-x)$ from $x=0.6$ to $x=3$.

We assume the area under the curve is bounded by the horizontal axis. The shading in Fig. 19.6 shows our area of interest.

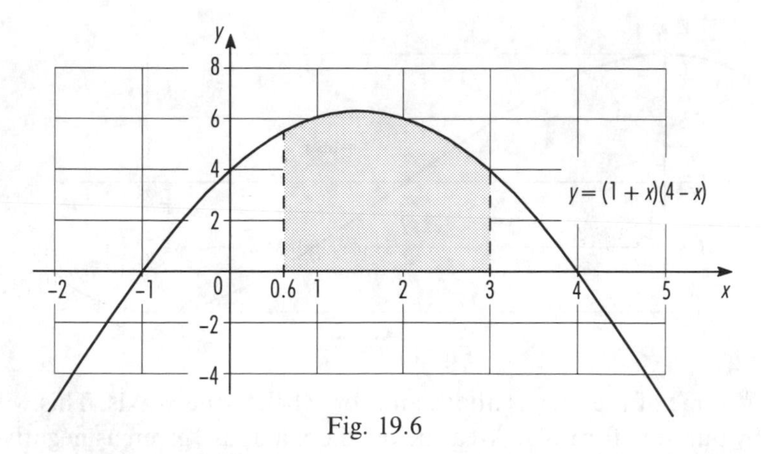

Fig. 19.6

Before attempting the integration we need to multiply out the brackets, i.e.

$$(1+x)(4-x) = 4+3x-x^2$$

Now using our formula

$$\text{Area} = \int_a^b y\,dx$$

we substitute for $a=0.6$, $b=3$ and y to get

$$\text{Area} = \int_{0.6}^{3} 4+3x-x^2\,\mathrm{dx}$$

$$= \left[4x+\frac{3x^2}{2}-\frac{x^3}{3}\right]_{0.6}^{3}$$

$$= \left(4(3)+\frac{3(3)^2}{2}-\frac{3^3}{3}\right)-\left(4(0.6)+\frac{3(0.6)^2}{2}-\frac{0.6^3}{3}\right)$$

$$= 16.500-2.868$$

$$= 13.63 \text{ unit}^2.$$

Because we have been given no units (e.g. mm or m) we use unit^2 (e.g. mm^2 or m^2) for the area.

Examples 19.11

Find the area bounded by the curve $y=\cos x$ and the x-axis between

i) $x=0$ and $x=\dfrac{\pi}{2}$,

ii) $x=\dfrac{\pi}{2}$ and $x=\pi$

and iii) $x=0$ and $x=\pi$.

Fig. 19.7 is a reminder of the cosine curve.

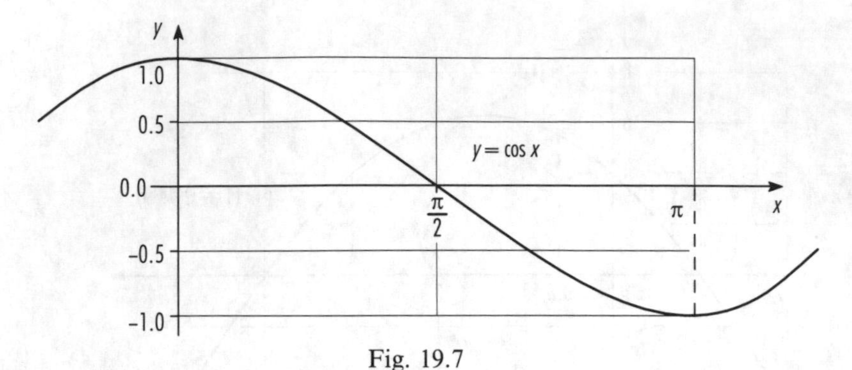

Fig. 19.7

Notice that some of the area is above and some below the x-axis. This will be noted by our area formula. Areas below the x-axis are given as negative and we need to interpret that sign. We apply our area formula to these questions.

i)
$$\begin{aligned} \text{Area} &= \int_0^{\frac{\pi}{2}} \cos x \, dx \\ &= \Big[\sin x\Big]_0^{\frac{\pi}{2}} \\ &= \sin\frac{\pi}{2} - \sin 0 \\ &= 1 \text{ unit}^2. \end{aligned}$$

ii)
$$\begin{aligned} \text{Area} &= \int_{\frac{\pi}{2}}^{\pi} \cos x \, dx \\ &= \Big[\sin x\Big]_{\frac{\pi}{2}}^{\pi} \\ &= \sin \pi - \sin\frac{\pi}{2} \\ &= -1 \end{aligned}$$

i.e. the area is 1 unit^2 *below* the x-axis.

iii) Because the graph crosses the horizontal axis it splits the areas. We need to deal with each section separately. If we fail to do this then the positive and negative areas may cancel out each other.

We consider the area from $x = 0$ to $x = \pi$ as

the area from $x = 0$ to $x = \dfrac{\pi}{2}$

and the area from $x = \dfrac{\pi}{2}$ to $x = \pi$.

Remember we know the meaning of the negative sign and do *not* have to worry about it.

Total area $= 1 + 1 = 2$ unit^2.

ASSIGNMENT

We have needed quite an amount of work before being able to look at our assignment. Fig. 19.8 shows the shaded area we are going to calculate. Using our formula

$$\text{Area} = \int_a^b y\,dx$$

we substitute for $a=1$, $b=2$ and $y=4-x^2$ to get

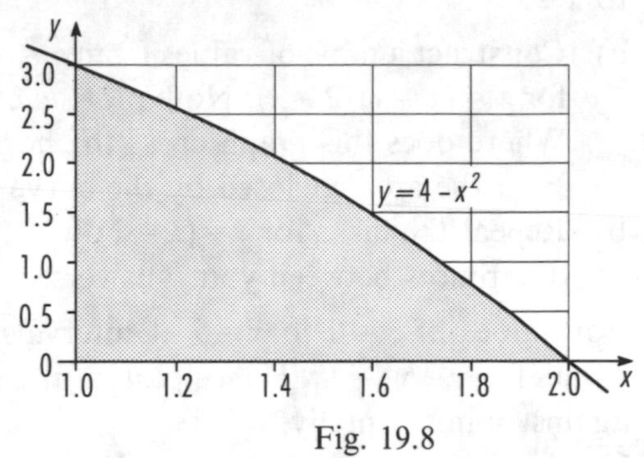

Fig. 19.8

$$\begin{aligned}\text{Area} &= \int_1^2 4-x^2\,dx \\ &= \left[4x-\frac{x^3}{3}\right]_1^2 \\ &= \left(4(2)-\frac{2^3}{3}\right)-\left(4(1)-\frac{1^3}{3}\right) \\ &= 5.\bar{3}-3.\bar{6} \\ &= 1.\bar{6}\text{ unit}^2.\end{aligned}$$

EXERCISE 19.5

1 Find the area bounded by the curve $y=x^3+2$ and the x-axis from $x=0$ to $x=1.5$.

2 $y=\dfrac{x^4+2}{x^2}$. Find the area lying between this curve and the x-axis from $x=1$ to $x=2$.

3 Sketch the curve $y=e^x$. Find the area bounded by both the axes, the curve and the line $x=2.75$.

4 From $x=5$ to $x=10$ find the area between the x-axis and the curve $y=\dfrac{1}{x}$.

5 From $\theta=\pi$ to $\theta=\dfrac{3\pi}{2}$ find the area between the curve $y=\cos\theta$ and the horizontal axis. With a sketch of the graph confirm that your answer should be negative.

6 What is the area between the curve $y=\dfrac{4}{x}$ and the x-axis from $x=\frac{1}{2}$ to $x=1$?

7 Over one cycle find the area bounded by the curve $y=4\sin\theta$ and the horizontal axis.

8 Find the area bounded by the curve $y=x^3$ and the x-axis from $x=-2$ to $x=2$.

9 a) Construct a table of values from $x=-3$ to $x=3$ at intervals of 0.5 for $y=(1-x)(2-x)$. Now plot y vertically and x horizontally. Where does this graph cross the horizontal axis?
Find the area enclosed by the curve and the horizontal axis.
b) Repeat the tasks for $y=(x-1)(x-2)$ and notice any similarities/differences between your answers.

10 For values of $x=0$ to $x=5$ at intervals of 0.5 construct a table of values for $y=x^2-3x$. Hence plot a fully labelled graph of y vertically against x horizontally.
On the same axes plot the straight line $y=x$.
Show that these two graphs intersect at the points (0, 0) and (4, 4).
From $x=0$ to $x=4$ find the area between
i) $y=x^2-3x$ and the x-axis.
ii) $y=x$ and the x-axis,
iii) the two graphs.

Now we can apply our area formula to a further practical problem before you attempt a final short exercise.

Example 19.12

The area under a velocity-time graph represents displacement. Fig. 19.9 shows the curve $v=\frac{1}{10}(4-t)(2+t)$ where v (ms^{-1}) is the velocity at time, t (s). For the first 6 seconds find the

i) distance travelled

and ii) displacement from the starting point.

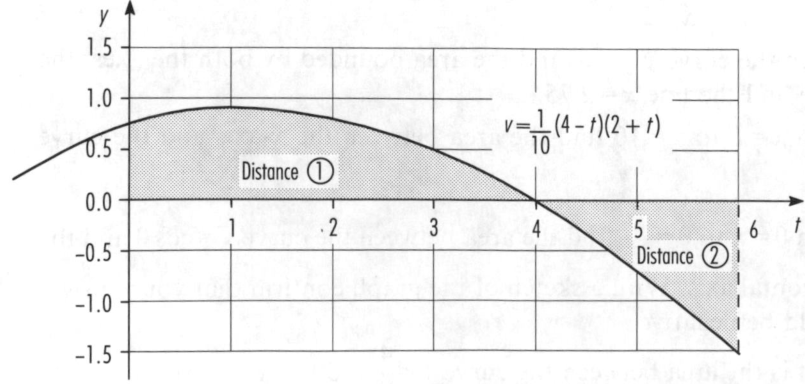

Fig. 19.9

Because our area is split above and below the horizontal axis we need to be careful. When this graph crosses the horizontal axis the velocity is 0. There the velocity changes from being positive to negative. For the area below the horizontal axis the distance is negative, being in the opposite direction, i.e. re-tracing part of the original path.

Before attempting to find the distance using our area formula we need to multiply out the brackets, i.e.

$$\frac{1}{10}(4-t)(2+t) = \frac{1}{10}(8+2t-t^2).$$

Applying our formula of

$$\text{Area} = \int_a^b y\,dx$$

we get

i) Distance ① = Area above the horizontal axis

$$= \frac{1}{10}\int_0^4 8+2t-t^2\,dt$$

$$= 0.1\left[8t+t^2-\frac{t^3}{3}\right]_0^4$$

$$= 0.1\left(\left(8(4)+4^2-\frac{4^3}{3}\right)-(0)\right)$$

$$= 0.1(26.\bar{6})$$

$$= 2.\bar{6}\ \text{m}.$$

Distance ② = Area below the horizontal axis

$$= \frac{1}{10}\int_4^6 8+2t-t^2\,dt$$

$$= 0.1\left[8t+t^2-\frac{t^3}{3}\right]_4^6$$

$$= 0.1\left(8(6)+6^2-\frac{6^3}{3}\right)-\left(8(4)+4^2-\frac{4^3}{3}\right)$$

$$= 0.1(12-26.\bar{6})$$

$$= -1.4\bar{6}\ \text{m}.$$

The negative sign confirms the distance is in the reverse direction.

Total distance $= 2.\bar{6}+1.4\bar{6} = 4.1\bar{3}$ m.

ii) Displacement is a vector quantity (Chapter 13). It accepts that a distance in one direction can cancel a distance in an opposite direction. This means we must include our negative sign in $-1.4\bar{6}$ m. Then we get

Displacement $= 2.\bar{6}-1.4\bar{6} = 1.20$ m.

EXERCISE 19.6

1 $v = 30 - 4t$ relates speed, v (ms^{-1}), to time , t (s). The area under a speed-time graph represents the distance travelled. By integration find the distance travelled in the first 5 seconds.
Plot this straight line graph of v against t for the first 5 seconds. Using the formula for the area of a trapezium check your first answer for the distance travelled.

2 Hooke's law relates the tension in an elastic spring, T (N), according to $T = \frac{\lambda x}{l}$. λ (N) is the coefficient of elasticity, l (m) is the natural length and x (m) is the extension from that natural length. If $\lambda = 30$ N and $l = 0.75$ m show that $T = 40x$. For T plotted vertically and x plotted horizontally the area under the graph represents the work done (J) in stretching the spring. This means

$$\text{Work done} = \int_a^b T\,dx.$$

Find the work done in stretching the spring from its natural length to 1.05 m.

3 In electrical theory we may plot voltage, V, vertically against current, I (A), horizontally. The area under such a curve is associated with power, P (W). This is

$$\text{Power} = 2\int_a^b V\,dI.$$

For a resistor of 470 Ω we have $V = 470I$. Find the power associated with an increase in current from 1×10^{-3} A to 1.08×10^{-3} A.

4 A variable force, F (N), is related to distance, s (m), by $F = 5s^2 + 2s$. The work done (J) by this force is given by

$$\text{Work done} = \int_a^b F\,ds.$$

Calculate the work done by the force as s increases from 3.50 m to 4.75 m.

5 The pressure, p (Nm^{-2}), and volume, V (m^3) of a gas are related by $pV^{1.5} = k$. If $p = 150 \times 10^3$ Nm^{-2} and $V = 0.5$ m^3 evaluate k.
Re-arrange your general formula to make p the subject.
The work done (J) expanding the gas from this volume to 0.75 m^3 is given by the integral formula

$$\text{Work done} = \int_{0.50}^{0.75} p\,dV.$$

Calculate this work done.

Summary of integration rules

y	$\int y\,dx$	
x^n	$\frac{x^{n+1}}{n+1}+c$	where n is a number and $n \neq -1$.
$\cos x$	$\sin x + c$	x in radians.
$\sin x$	$-\cos x + c$	x in radians.
e^x	$e^x + c$	
$\frac{1}{x}$	$\ln x + c$	

Answers

Chapter 1

Exercise 1.1

1	90	**2**	−90	**3**	71	**4**	−71
5	71	**6**	90	**7**	90	**8**	22
9	18	**10**	426	**11**	5	**12**	9
13	−4	**14**	5	**15**	24	**16**	31
17	17	**18**	51	**19**	32	**20**	32

Exercise 1.2

1	81	**2**	243	**3**	243	**4**	243
5	729	**6**	729	**7**	36	**8**	43
9	193	**10**	16	**11**	± 3	**12**	± 9
13	± 9	**14**	± 27	**15**	± 1	**16**	± 2
17	12, 16	**18**	13, 15	**19**	−23, 31	**20**	7, 9
21	1296	**22**	1296	**23**	−1296	**24**	1298
25	−1294	**26**	54	**27**	384	**28**	−33
29	13845	**30**	15				

Exercise 1.3

1	0.99	**2**	0.3885	**3**	17.39
4	5.7	**5**	21.40	**6**	4
7	10	**8**	220 000	**9**	0.00376
10	3200	**11**	62.0	**12**	6.2
13	20	**14**	620.4	**15**	620.5
16	100	**17**	5490	**18**	1 000 000
19	4200	**20**	0.071	**21**	9.92×10^{-1}
22	3.8847×10^{-1}	**23**	1.738847×10	**24**	5.69
25	2.1399×10	**26**	4.32	**27**	1.432×10
28	2.17×10^{5}	**29**	3.7602×10^{-3}	**30**	3.165×10^{3}
31	6.204×10	**32**	6.204	**33**	2.03×10
34	6.20401×10^{2}	**35**	6.20461×10^{2}	**36**	9.98×10
37	5.48997×10^{3}	**38**	1.4×10^{6}	**39**	4.23906×10^{3}
40	7.0629×10^{-2}				

Exercise 1.4

i) Estimated answers:

1	24	**2**	6.5	**3**	14 or 15	**4**	0.5
5	30	**6**	42	**7**	67	**8**	43
9	68	**10**	−25	**11**	25	**12**	$\frac{1}{4}$
13	4000	**14**	4000	**15**	1200	**16**	1.5
17	75	**18**	$\frac{1}{2}$	**19**	$2\frac{1}{2}$	**20**	24

ii) Calculator answers:

1	24.6	(1 dp)	**2**	7	(1 sf)
3	14.5	(1 dp)	**4**	0.41	(2 dp)
5	30	(1 sf)	**6**	48	(2 sf)
7	67	(2 sf)	**8**	46.7	(1 dp)
9	68	(2 sf)	**10**	−50	(1 sf)
11	25.7	(1 dp)	**12**	0.3	(1 sf)
13	4436	(4 sf)	**14**	4436	(4 sf)
15	1157	(4 sf)	**16**	1.5	(2 sf)
17	75.9	(3 sf)	**18**	0.4	(1 dp)
19	3	(1 sf)	**20**	27	(2 sf)

Exercise 1.5

1	37.8	(3 sf)	**2**	112	(3 sf)
3	3.8	(2 sf)	**4**	18.4	(3 sf)
5	36.8	(3 sf)	**6**	±2.9	(2 sf)
7	±2.9	(2 sf)	**8**	3.22	(3 sf)
9	−3.22	(3 sf)	**10**	−0.28, −6.2	(2 sf)
11	6.2, 0.28	(2 sf)	**12**	2	(1 sf)
13	0.159	(3 dp)	**14**	0.1	(1 dp)
15	0.9	(1 dp)	**16**	2	(1 sf)
17	0.5	(1 dp)	**18**	1.3	(1 dp)
19	1.8	(1 dp)	**20**	3.7	(1 dp)

Exercise 1.6

1	0.2 m^2	(1 dp)	**2**	1500 mm^2	(2 sf)
3	12 N, 51 N	(2 sf)	**4**	71.3 m	(1 dp)
5	0.356 m	(3 dp)	**6**	1070	(3 sf)
7	4.7 W	(1 dp)	**8**	140 mm	(2 sf)
9	2 Ω	(1 sf)	**10**	0.2 m	(1 sf)
11	131 m	(3 sf)	**12**	1.1 ms^{-2}	(2 sf)
13	126 000 mm^2	(3 sf)	**14**	2.5 s	(2 sf)
15	22 m	(2 sf)	**16**	67.5 mm, 780 mm^2	
17	7.1, 300.3°K		**18**	7.9, 300 Ω	

19

v	0	2	4	6	8	10	12
Q	0	0.01	0.02	0.03	0.04	0.05	0.06

20

t	0	10	20	30	40	50	60
v	1.00	25.5	50.0	74.5	99.0	123.5	148.0

Chapter 2

Exercise 2.1

1 31 **2** −16 **3** 31 **4** 7
5 −6 **6** 2 **7** −3 **8** 12
9 $\gamma-\alpha$ **10** $\alpha-\beta$

Exercise 2.2

1 0.8 **2** 1.3 **3** −1.8 **4** −10.7
5 28 **6** 0.33 **7** 4.26 **8** $2.2\bar{6}$
9 −2.5 **10** −0.34

Exercise 2.3

1 1.5 **2** 5.3 **3** 20 **4** −20
5 2.8 **6** 9 **7** 18 **8** 13.125
9 −11 **10** 2.4

Exercise 2.4

1 −0.625 **2** $1.8\bar{3}$ **3** $-1.\bar{3}$ **4** $7.\bar{3}$
5 20 **6** −20 **7** 6.6 **8** 15
9 $-0.\bar{0}\bar{9}$ **10** $-1.1\bar{3}\bar{6}$

Exercise 2.5

Questions 1–10 See Fig. A.

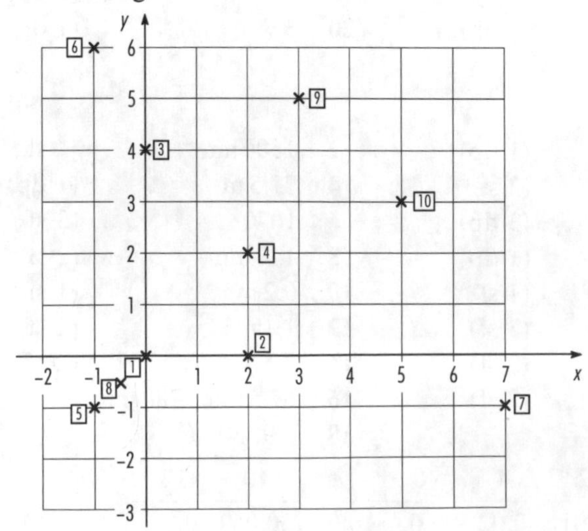

11 Lie on a line at 45° to the horizontal.
12 Lie on a line at 45° to the horizontal.
13 Lie on a horizontal line through $y = 1$.
14 Lie on a vertical line through $x = 3$.
15 No pattern for them all.

Exercise 2.6

	i) Gradient	ii) Intercept	iii) Equation
1	3	15	$y = 3x + 15$
2	−1.25	12.5	$y = -1.25x + 12.5$
3	−2	−10	$y = -2x - 10$
4	5	0	$y = 5x$
5	−25	500	$y = -25x + 500$
6	−1	4	$y = -x + 4$
7	$\frac{2}{3}$	$-\frac{1}{3}$	$y = \frac{2}{3}x - \frac{1}{3}$
8	−1	8	$y = -x + 8$
9	1	3	$y = x + 3$
10	$\frac{3}{13}$	$\frac{-8}{13}$	$y = \frac{3}{13}x - \frac{8}{13}$

Exercise 2.7

2 2, −1; 15, 2.5
3 3, 50; 440 N
4 125 s, 3.19
5 *a*, *b*; 1.8, 32; 20
6 400, −1200; −500 N, 3 m
7 13.5, gradient
8 $v = mt + c$; 0.24, 1.7; 16.1 ms^{-1}, 58.0 kmh^{-1}, 36.2 mph
9 39 Ω
10 0.2, 1.7×10^{-5}; 586°C

Chapter 3

Exercise 3.1

1 $x = -\alpha + \beta$
2 $x = -\alpha + \beta + \gamma$
3 $x = -\alpha + \beta + \gamma$
4 $x = \frac{\beta + \gamma}{\alpha}$
5 $x = \alpha(\delta - \beta)$
6 $x = \left(\frac{\delta}{\gamma} - \beta\right)\alpha$
7 $x = \frac{\gamma}{\beta - \alpha}$
8 $x = \frac{ac}{1 - ab}$
9 $x = \frac{ab(c - d)}{b - a}$
10 $x = \frac{1}{b - a}$
11 $x = \frac{a}{b + c}$
12 $x = \frac{bc}{1 - ab}$
13 $x = \frac{a - b}{d}$
14 $P = \frac{100I}{RT}$
15 $d = \frac{C}{\pi}$
16 $b = \frac{2A}{h}$
17 $F = \frac{9C}{5} + 32$
18 $h = \frac{2A}{a + b}$

19 $x = \frac{S+2}{W}$

20 $t = \frac{2s}{u+v}$

21 $I_1 = -(I_2 + I_3 + I_4)$

22 $r = \frac{C}{2\pi}$

23 $R = \frac{100I}{PT}$

24 $h = \frac{2A}{b}$

25 $c_1 = \frac{Cc_2}{c_2 - C}$

26 $v = \frac{fu}{u-f}$

27 $T = \frac{pV}{mR}$

28 $p_2 = \frac{p_1 V_1}{V_2}$

29 $t = \left(\frac{R}{R_0} - 1\right)\frac{1}{\alpha}$

30 $R = N(1-S)$

31 $a = \frac{2A}{h} - b$

32 $T = \frac{100I}{PR}$

33 $u = \frac{2s}{t} - v$

34 $N = \frac{4V}{\pi D^2 f}$

35 $b = \frac{2A}{h} - a$

36 $v = \frac{Ft}{m} + u$

37 $N = \frac{R}{1-s}$

38 $\alpha = \frac{\Omega - \omega}{t}$

39 $t = \left(\frac{l}{l_0} - 1\right)\frac{1}{\alpha}$

40 $\alpha = \frac{R_1 - R_2}{R_2 t_1 - R_1 t_2}$

Exercise 3.2

1 $r = \pm\sqrt{\frac{A}{\pi}}$

2 $v = \pm\sqrt{\frac{32Pr}{W}}$

3 $b = \pm\sqrt{a^2 - c^2}$

4 $r = \pm\sqrt{\frac{V}{\pi h}}$

5 $x = \pm\sqrt{\frac{2aW}{\lambda}}$

6 $\omega = \pm\frac{v}{r}$

7 $x = \pm\sqrt{2(1-C)}$

8 $x = \pm\sqrt{\frac{y-b}{a}}$

9 $h = \frac{v^2}{2g}$

10 $x = \left(\frac{y-b}{a}\right)^2$

11 $x = \pm\sqrt{2(Ch-1)}$

12 $r = 100\left(1 \pm \sqrt{\frac{A}{P}}\right)$

13 $b = \pm\sqrt{a^2 - c^2}$

14 $d = \pm\sqrt{(A - 320)\,\frac{4}{\pi}}$

15 $V_R = \pm\sqrt{V^2 - (V_L b - V_C)^2}$

16 $r = 200\left(\pm\sqrt{\frac{A}{P}} - 1\right)$

17 $u = \pm\sqrt{v^2 - \frac{2Ck}{m}}$

18 $t = \frac{v^2 - grv}{gr + v^2 v}$

19 $s = c \pm \frac{1}{\sqrt{Lf}}$

20 $t = \pm\sqrt{\frac{1-C}{1+C}}$

Exercise 3.3

1 $x = \sqrt[3]{\dfrac{y-b}{a}}$ **2** $x = \sqrt[5]{\dfrac{y-c}{m}}$

3 $x = \pm\sqrt[4]{\dfrac{-y-c}{m}}$ **4** $x = \left(\dfrac{y-b}{a}\right)^3$

5 $x = \left(\dfrac{y-c}{m}\right)^4$ **6** $x = \left(\dfrac{-y-c}{m}\right)^5$

7 $c = \pm z^{\frac{1}{4}} - ts$ **8** $R = 100\left(\left(\dfrac{A}{P}\right)^{\frac{1}{10}} - 1\right)$

9 $T = \left(\dfrac{c}{S}\right)^7$ **10** $D = \left(\dfrac{32z}{\pi}\right)^{\frac{1}{3}}$

11 $V = \left(\dfrac{c}{p}\right)^{\frac{1}{\gamma}}$ **12** $d = \left(\dfrac{12I}{b}\right)^{\frac{1}{3}}$

13 $T = \left(\dfrac{c}{S}\right)^{10}$ **14** $s = \dfrac{c - z^{\frac{1}{3}}}{t}$

15 $R = 100\left(1 - \left(\dfrac{A}{P}\right)^{\frac{1}{7}}\right)$ **16** $D = \left(\dfrac{T}{kf^{0.75}}\right)^{\frac{5}{9}}$

17 $D = \pm\left(\dfrac{32I}{\pi}\right)^{\frac{1}{4}}$ **18** $f = \left(\dfrac{T}{kD^{1.8}}\right)^{\frac{4}{3}}$

19 $D = \left(\dfrac{12I + bd^3}{B}\right)^{\frac{1}{3}}$ **20** $d = \pm\left(D^4 - \dfrac{32Dz}{\pi}\right)^{\frac{1}{4}}$

Chapter 4

Exercise 4.1

Questions **1–20**: the unit is the radian.

1 $\dfrac{\pi}{18}$ **2** $\dfrac{\pi}{4}$ **3** $\dfrac{\pi}{3}$ **4** $\dfrac{5\pi}{6}$

5 π **6** $\dfrac{7\pi}{6}$ **7** $\dfrac{7\pi}{4}$ **8** 2π

9 3π **10** 6π **11** 0.44 **12** 1.28
13 2.92 **14** 3.93 **15** 4.45 **16** 5.17
17 5.37 **18** 6.02 **19** 8.73 **20** 10.65

Questions **21–30**: the unit is the degree.

21 401.07 **22** 260.70 **23** 297.94 **24** 120.00
25 71.62 **26** 779.22 **27** 225.00 **28** 108.86
29 300.00 **30** 352.37

Exercise 4.2

1a)	0.39 m	**b)**	36.7 mm	**c)**	116 mm	**d)**	1.9 m
2a)	0.86 m	**b)**	0.80 m	**c)**	14.5 mm	**d)**	19 mm
3a)	1.6 rad	**b)**	3 rad	**c)**	0.5 rad	**d)**	3.5 rad
4a)	0.37 m^2	**b)**	280 mm^2	**c)**	8360 mm^2	**d)**	1.4 m^2
5a)	26.8 mm	**b)**	25.8 mm	**c)**	34.8 mm	**d)**	35.7 mm

6a) $0.2\bar{6}$ rad = 15.3° **b)** 0.74 rad = 42.4°
c) 2.64 rad = 151° **d)** 1.33 rad = 76.2°

7	1.25 rad = 71.6°, 1.01 m	**8**	1.57 m, 0.43 m, 1.72 rad = 98.5°
9	0.25 m^2	**10**	785 mm, 157 mm, 5:1, Yes
11	14.5 rad = 831°, 2.3 rev	**12**	$5.2\bar{7}$ rad = 302°
13	1.37 m^2, 26.2°	**14**	75 mm, 135°

15 $l^2 = h^2 + r^2$, l is the radius of the circle from which the sector is created.

Chapter 5

Exercise 5.1

1	0.5446	**2**	0.7986	**3**	−0.8572	**4**	0.7986
5	−7.1154	**6**	2.1445	**7**	−0.8572	**8**	−0.9925
9	−2.1445	**10**	−0.6691	**11**	0	**12**	0.4027
13	1.6977	**14**	−1.6003	**15**	0.9455	**16**	−0.9063
17	0.4877	**18**	−0.8746	**19**	−0.4067	**20**	1.0355
21	−4.3315	**22**	−0.7632	**23**	−0.6494	**24**	−0.7986
25	−0.3256	**26**	1.4281	**27**	0.9397	**28**	0.3256
29	0.5446	**30**	−0.8572				

Exercise 5.2

1	0.9093	**2**	0.7248	**3**	0.9425	**4**	0.2203
5	−1	**6**	−0.9995	**7**	−0.4461	**8**	−0.8660
9	−0.8660	**10**	−1.2602				

Exercise 5.3

1	60°, 120°	**2**	45°, 135°
3	60°, 300°	**4**	76.35°, 283.65°
5	45°, 315°	**6**	210°, 330°
7	240°, 300°	**8**	36.41°, 143.59°
9	0°, 360°	**10**	153.25°, 206.75°
11	82.38°, 277.62°	**12**	90°
13	0°, 180°, 360°	**14**	no solution

15 120°, 240°
16 270°
17 no solution
18 90°, 270°
19 180°
20 225°, 315°

Exercise 5.4

1 26.57°, 206.57°
2 56.31°, 236.31°
3 112.20°, 292.20°
4 80.96°, 260.95°
5 45°, 225°
6 120°, 300°
7 123.69°, 303.69°
8 72.90°, 252.90°
9 107.10°, 287.10°
10 96.83°, 276.83°

Chapter 6

Exercise 6.1

1 $b = 0.37$ m, $c = 0.37$ m, $\angle C = 70°$
2 $a = 0.124$ m, $b = 0.119$ m, $\angle B = 52°$
3 $b = 1.74$ m, $c = 1.46$ m, $\angle A = 118°$
4 $a = 8.90$ m, $b = 5.08$ m, $\angle C = 80°$
5 $b = 0.54$ m, $c = 0.56$ m, $\angle A = 86°$
6 $c = 0.32$ m, $\angle B = 72°$, $\angle C = 58°$;
$c = 0.14$ m, $\angle B = 108°$, $\angle C = 22°$
7 $a = 0.72$ m, $\angle A = 132.8°$, $\angle C = 21.7°$
8 $b = 1.2$ m, $\angle A = 22.5°$, $\angle B = 37.5°$
9 $b = 0.36$ m, $\angle B = 81°$, $\angle C = 58°$;
$b = 0.11$ m, $\angle B = 17°$, $\angle C = 122°$
10 $c = 1.05$ m, $\angle A = 27°$, $\angle C = 105°$

Exercise 6.2

1 $c = 0.83$ m, $\angle A = 53°$, $\angle B = 65°$
2 $a = 0.64$ m, $\angle B = 116°$, $\angle C = 21°$
3 $b = 0.53$ m, $\angle A = 14°$, $\angle C = 95°$
4 $c = 0.63$ m, $\angle A = 106°$, $\angle B = 49°$
5 $b = 0.70$ m, $\angle A = 25.5°$, $\angle C = 45.5°$
6 $\angle A = 15°$, $\angle B = 112°$, $\angle C = 53°$
7 $\angle A = 20°$, $\angle B = 97°$, $\angle C = 63°$
8 $\angle A = 33°$, $\angle B = 38°$, $\angle C = 109°$
9 no triangle possible
10 $\angle A = 126°$, $\angle B = 22°$, $\angle C = 32°$

Exercise 6.3

1 0.096 m^2
2 1.96 m^2
3 0.15 m^2
4 0.032 m^2
5 0.032 m^2
6 0.28 m^2
7 0.10 m^2
8 0.04 m^2

9 $0.34\,m^2$ **10** $0.08\,m^2$ **11** $5.6\,m^2$ **12** $7.6\,m^2$
13 $0.31\,m^2$ **14** no triangle possible **15** $6.46\,m^2$

Exercise 6.4

1 $3\,000\,mm^2$, $16\,000\,mm^2$ **2** $121.4\,m^2$, $17.5\,m^2$, 14.4%
3 $2680\,mm^2$ **4** $0.16\,m^2$
5 $0.52\,m^2$ **6** $0.56\,m^2$
7 $0.50\,m^2$ **8** 13.5 unit^2
9 $4172\,m^2$ **10** $2.43\,m^2$

Exercise 6.5

1 0.97 m, 14° **2** 0.46° **3** 9.8°, 8.2° **4** 10 m
5 24.33 m, 6.13 m

Exercise 6.6

1 i) 69.3°, ii) 75° **2** 7.3°, 30.5°
3 0.91 m, 62.1°, 69.4° **4** $0.42\,m^2$, $0.39\,m^2$, 28°
5 55.2°, 1.62 m **6** $0.82\,m^2$
7 $144\,m^2$, 32 m **8** 8.19°, 8.21°
9 i) 53.1°, ii) 57°; $2.65\,m^2$
10 i) 26.6°, ii) 26.6°, iii) 26.6°, iv) 30°, v) 30°

Chapter 7

Exercise 7.1

1 $x^2+7x+10$ **2** $10x^2+7x+1$ **3** $x^2+7x+10$
4 $10x^2+26x+12$ **5** $3x^2+17x+10$ **6** $12x^2+26x+10$
7 $x^2-7x+10$ **8** $12-7x+x^2$ **9** $x^2-14x+40$
10 $12x^2-17x+6$ **11** $12x^2-29x+14$ **12** $6x^2-5x+1$
13 $x^2-3x-10$ **14** $x^2-5x-36$ **15** $4x^2-2x-12$
16 $2x^2+3x-2$ **17** $49x^2-14x-3$ **18** $66x^2+x-45$
19 $x^2+3x-10$ **20** $9x^2+x-10$ **21** $6x^2+20x-26$
22 $22x^2+86x-8$ **23** $2x^2+x-10$ **24** $6x^2+5x-6$
25 x^2+2x+1 **26** $12x^2-7x-10$ **27** $16-8x+x^2$
28 x^2-2x+1 **29** $x^2-8x+16$ **30** $x^2+10x+25$
31 $4x^2+28x+49$ **32** $49-28x+4x^2$ **33** x^2-4
34 $4x^2-169$ **35** $6x^2-6$ **36** $7x+5x^2$
37 $28x-4x^2$ **38** $-14x^2-63x$ **39** $10x^2+23x+12$
40 $8x^2-10x-33$ **41** $3-16x-12x^2$ **42** $9-82x+9x^2$
43 $2x+x^2$ **44** $9-61x-14x^2$ **45** $6x^2-10.5x^{-3}$
46 $15-t-2t^2$ **47** $6t^2-7t-49$ **48** $8x-12x^2$
49 $6t^2-7\mathrm{t}-49$ **50** $14t^2+61\mathrm{t}-9$

Exercise 7.2

1

x	−3.5	−3.0	−2.5	−2.0	−1.5	−1.0	−0.5	0	0.5	1.0
$2x^2$	24.5	18.0	12.5	8.0	4.5	2.0	0.5	0	0.5	2.0
$4x$	−14.0	−12.0	−10.0	−8.0	−6.0	−4.0	−2.0	0	2.0	4.0
−5	−5.0	−5.0	−5.0	−5.0	−5.0	−5.0	−5.0	−5.0	−5.0	−5.0
y	5.5	1.0	−2.5	−5.0	−6.5	−7.0	−6.5	−5.0	−2.5	1.0

Exercise 7.3

1 −5, 1
2 −2, 3
3 −3 repeated
4 0, 7
5 −5, 5
6 −0.30, 3.30
7 −0.52, 2.32
8 −2.19, 0.69
9 $1, \frac{5}{3}$
10 no real solution

Exercise 7.4

1 −5, −1
2 −3, −2
3 −4, −3
4 −2, −1
5 −7 repeated
6 2, 3
7 3, 9
8 5, 6
9 5 repeated
10 2, 6
11 −5, 1
12 −6, 2
13 −4, 2
14 −9, 2
15 −16, 3
16 −3, 5
17 −2, 5
18 −6, 8
19 −7, 8
20 −7, 9
21 −10, 4
22 −10, 7
23 −10 repeated
24 3 repeated
25 −8, 9
26 −3, 4
27 −12, 6
28 −9, 4
29 −2, 20
30 −4, 15

Exercise 7.5

1 $-1, \frac{2}{3}$
2 $\frac{1}{2}$ repeated
3 $-\frac{1}{2}, \frac{2}{3}$
4 $-\frac{2}{3}, 1$
5 $-2, -\frac{3}{2}$
6 $-5, -\frac{3}{2}$
7 $-3, \frac{1}{3}$
8 $\frac{3}{2}, 2$
9 $-5, -\frac{1}{2}$
10 $-\frac{5}{2}, \frac{4}{3}$
11 $-\frac{1}{2}, \frac{7}{3}$
12 $\frac{2}{3}, \frac{3}{2}$
13 $-\frac{1}{3}, \frac{2}{5}$
14 $-4, -\frac{1}{3}$
15 $\frac{3}{2}$ repeated
16 $\frac{3}{5}$ repeated
17 $-\frac{3}{2}, \frac{5}{4}$
18 $-\frac{3}{2}, -\frac{2}{3}$
19 $-\frac{2}{3}$ repeated
20 $-\frac{3}{8}, 6$

Exercise 7.6

1 −4, 0
2 $0, \frac{8}{3}$
3 0, 4
4 $-\frac{10}{3}, \frac{10}{3}$
5 0, 1
6 $-\frac{10}{3}, 0$
7 0, 3
8 no real solution
9 3
10 −3, 2

Exercise 7.7

1 −21.576, 1.576 **2** −0.618, 1.618 **3** −1.772, 6.772
4 −0.629, 28.629 **5** −3.562, 0.562 **6** −0.464, 6.464
7 6 **8** −4.519, −0.148 **9** no real solution
10 0.147, 0.453

Exercise 7.8

1 0.291 A, 1.376 A **2** 7.42 s
3 2; 6 **4** 18.3%
5 11.22 V **6** $-6.32\,ms^{-1}$, $6.32\,ms^{-1}$
7 6.60 **8** 4 m, 6 m; 2.76 m, 7.24 m
9 0.283 m; 1.595 m **10** 5.38 Ω, 15.62 Ω

Chapter 8

Exercise 8.1

1 −2, 1, 3 **2** −0.73, 1, 2.74
3 −2, 0.5 repeated **4** −0.93
5 −2.60, 0.35, 2.26
7 172.1°, 2.58 seconds **8** $1.128 \times 10^6\,mm^3$, 56.4mm
9 i) £200, ii) $42\frac{1}{2}$ hours, iii) £2800
10 3.47 tonne, £8350

Chapter 9

Exercise 9.1

1 $x = 6, y = 5$ **2** $x = 5, y = 1$
3 $x = 5, y = 2$ **4** $x = \frac{1}{2}, y = 1$
5 $x = -\frac{1}{5}, y = 1$ **6** $x = 3.8, y = 0.6$
7 $x = 0, y = 1$ **8** no solution
9 $x = -\frac{10}{7}, y = \frac{29}{7}$ **10** no solution

Exercise 9.2

1 $x = 10, y = -9$ **2** $x = -2, y = 13$
3 $x = 3, y = 3$ **4** $x = 16, y = 10$
5 $x = 4, y = 10$ **6** $x = -7, y = -2$
7 $x = -1, y = 2.5$ **8** $x = 24, y = 4$
9 $x = 2, y = -1$ **10** $x = 3, y = -0.5$

Exercise 9.3

1 $x = -1$, $y = -2$ and $x = 4$, $y = 18$
2 $x = -1.77$, $y = 8.66$ and $x = 0.94$, $y = 4.59$
3 $x = -2$, $y = -9$ repeated
4 no real solution
5 $x = 0.56$, $y = 2.72$ and $x = 2.69$, $y = 1.65$

Exercise 9.4

1 $x = -1.53$, $y = -3.56$; $x = -0.36$, $y = -0.05$; $x = 1.86$, $y = 6.60$
2 $x = -1.69$, $y = -6.76$; $x = 0$, $y = 0$; $x = 1.19$, $y = 4.76$
3 $x = -2.08$, $y = -9$
4 $x = -2.07$, $y = -1.07$; $x = 0.72$, $y = 1.72$; $x = 3.35$, $y = 4.35$
5 $x = 1.1$, $y = 4.6$

Exercise 9.5

1 $x = -0.83$, $y = -0.57$; $x = 0$, $y = 0$; $x = 4.83$, $y = 113$
2 $x = -0.81$, $y = -0.53$
3 $x = -3.43$, $y = -20.2$
4 $x = -0.89$, $y = -2.11$; $x = 1.27$, $y = 6.15$
5 $x = -1.68$, $y = 9.48$

Exercise 9.6

1 $I_1 = 0.4$ A, $I_2 = 0.2$ A
2 $M_{max} = 200$ Nm, $x = 10$ m; $M_{max} = 225$ Nm, $x = 7.5$ m; $x = 0$ m, 10 m; $M = 0$ Nm, 200 Nm
3 99°C; i) 15.5°C, ii) 36355 Ω
4 i) 1.03 m ii) $V = 2.30\,m^3$
5 13.5 s, $173.\bar{3}$ m, 9.75 s
6 Fixed costs = 0.5, Variable costs = $0.75x$; $x = 0.67$, 2; Greatest profit at $x = 2.4$, Worst loss at $x = 1.45$
7 Supply is $y = 100 + 10x$, Demand is $y = 350 - 8x$; £13.89; £13.20
8 Yes, depending on accuracy required
9 Yes, depending on accuracy required
10 $x = 0.78$ rad $= 44.7°$ is accurate,
$x = -1.89$ rad $= -108.3°$ is inaccurate

Chapter 10

Exercise 10.1

1 0.760 **2** 1.127 **3** 3 **4** 1.428
5 2.301 **6** −0.301 **7** 6 **8** −0.602

9	not defined	**10**	0.246	**11**	17	**12**	0.282
13	4	**14**	1.5	**15**	288	**16**	5
17	0.0562	**18**	45.5	**19**	4900	**20**	88.9

Exercise 10.2

1	2.757	**2**	1.224	**3**	6.908	**4**	1.917
5	5.521	**6**	−1.897	**7**	6	**8**	−1
9	not defined	**10**	5.165	**11**	3.42	**12**	0.577
13	1.83	**14**	1.19	**15**	11.7	**16**	2.01
17	0.287	**18**	5.25	**19**	40	**20**	7.02

Exercise 10.3

1	3	**2**	1.64	**3**	1.21	**4**	4.42
5	4.40	**6**	3.42	**7**	7.13	**8**	−9.81
9	1.16	**10**	1.00, 1.77				

Exercise 10.4

1	1.09	**2**	2.25	**3**	−0.15	**4**	2.00
5	0.49	**6**	5.76	**7**	−1.50	**8**	11.51
9	16.99	**10**	6.91				

Exercise 10.5

1 17.78 dB, 30 W

2 $n = \dfrac{\log c - \log S}{\log T}$

3 0.13

4 $h = \dfrac{1}{a}\ln\left(\dfrac{c}{p}\right)$, 6320 m

5 1.36 rad, 0.42

6 8.50, 6.94, 14.87%

7 0.34, 16.2 mm

8 25%

10 0.025, $t = -T\ln\left(1 - \dfrac{Ri}{E}\right)$, 0.047 s

9 0.34, 16.2 mm

10 25%

Chapter 11

Exercise 11.1

1	5.474	**2**	1636	**3**	0.3679	**4**	1
5	3272	**6**	597.7	**7**	53732	**8**	54.68
9	1.001	**10**	1.001	**11**	54.68	**12**	3.158
13	34.81	**14**	109.9	**15**	109.9	**16**	2749
17	549.7	**18**	3.158	**19**	0.04344	**20**	1.439

Exercise 11.2

17 1.06 **18** 2.12 **19** 1328 **20** 1328

Exercise 11.3

1 4.29×10^{-3}

2 0.02, i) −88.8, ii) 0.014 s

3 2.61×10^3, 1 hr 52 min

4 0.3648

5 0.05 A, 0.8829 As^{-1}

Exercise 11.4

1 $p_0 = 75.8,\ k = -\dfrac{1}{15000}$

2 $E = 5.50,\ T = \dfrac{1}{63},\ 0.011\,\text{s}$

3 $\alpha = 515,\ \gamma = 0.254$

4 −1.95

5 1.06

6 $B_0 = 1.60 \times 10^3,\ a = 0.75$

7 1.525

8 0.3648

9 0.0417 A

10

y	1	2	3	4	5	6
P	150	180	216	259	311	373

yes, $P_1 = 125,\ r = 1.2$

Exercise 11.5

1 $p_0 = 75.8,\ k = -\dfrac{1}{15000}$

2 $E = 5.50,\ T = \dfrac{1}{63},\ 0.011\,\text{s}$

3 $\alpha = 515,\ \gamma = 0.254$

4 −1.95

5 1.06

6 $B_0 = 1.60 \times 10^3,\ a = 0.75$

7 1.525

8 0.3648

9 0.0417 A

10

y	1	2	3	4	5	6
P	150	180	216	259	311	373

yes, $P_1 = 125,\ r = 1.2$

Chapter 12

Exercise 12.1

1 0.4, 0.6 **2** 192 **3** 10, 1.17

4 $7.\overline{27}$ **5** 3.266, k **6** ± 6.36

7 15, no solution **8** 50 **9** 150, 17.36

10 672 m **11** $33.\overline{3}$ N **12** 0.46 m^3

13 1.96 s **14** 17.15 ms^{-1} **15** 0.57 s

Exercise 12.4

1 Centre $(0, 0)$, radius $= 6$
2 Centre $(0, 0)$, radius $= 4.24$
3 Centre $\left(-\frac{1}{2}, 0\right)$, radius $= 6$
4 Centre $\left(\frac{1}{2}, 0\right)$, radius $= 5$
5 Centre $(0, 3)$, radius $= 7$
6 Centre $\left(0, -\frac{1}{4}\right)$ radius $= 2.5$
7 Centre $\left(-\frac{1}{4}, \frac{1}{3}\right)$ radius $= 3$
8 Centre $(2, 2.5)$, radius $= 1$
9 Centre $(-2, -1.7)$, radius $= 2$
10 Centre $\left(\frac{2}{5}, -\frac{3}{4}\right)$, radius $= 1$
11 -2.33, 6.33; -2.08, 7.08
12 Centre of $(2, 2.5)$ is greater than 1 from each axis.

Exercise 12.5

1 minor 8, major 10
2 minor 4, major 4.47
3 minor 2, major 4.47
4 major 2.83, minor 2
5 major 18, minor 12

Exercise 12.6

3 i) $y = \pm\frac{3}{2}x$, ii) $y = \pm\frac{2}{3}$
5 $(-3, 0)$, $(3, 0)$

Exercise 12.7

1 1.5
2 2.00
3 2300, -160
4 1260, 0
5 0.05, 500
7 2000
8 0.04, 0.16
9 0.48, -2.25
10 0.049, -0.00125

Chapter 13

Exercise 13.1

1 i) scalar, ii) vector, iii) vector, iv) vector, v) scalar
2 2 m due North
3 i) ***CB***, ii) ***CB***, iii) ***AC***, iv) ***AD*** + ***CD*** + ***CF***, v) ***BE*** + ***FE***
5 i) true, ii) true, iii) false, iv) false, v) true

Exercise 13.2

1 i) 5 N, 8.66 N; ii) 4.95 ms^{-1}, 4.95 ms^{-1};
iii) 1.20 m, 3.29 m; iv) 5.64 ms^{-1}, 2.05 ms^{-1};
v) 12 N, 0 N
2 i) 4.33 N, 2.50 N; ii) 2.72 N, 1.27 N;
iii) 2.57 N, 3.06 N
3 i) 5 N, 36.9° above horizontal
ii) 9.43 N, 32.0° above horizontal

iii) $6.05\,\text{ms}^{-1}$, 34.2° below horizontal
iv) 5.83 N, 59.0° below horizontal
v) 7.27 m, 55.6° below horizontal
vi) $7.43\,\text{ms}^{-1}$, 49.6° above horizontal
vii) 8.49 N vertically upwards
viii) 4.88 m, 65.0° above horizontal

Exercise 13.3

1 i) 9.28 N, 16.7° above horizontal
ii) $6.89\,\text{ms}^{-1}$, 20.4° above negative horizontal
iii) 18.44 m, 62.1° below negative horizontal
iv) 8.12 N, 73.6° below horizontal

Exercise 13.5

1 $\boldsymbol{i}-\boldsymbol{j}-4\boldsymbol{k}$
2 $4\boldsymbol{i}+\boldsymbol{j}-6\boldsymbol{k}$
3 $-\boldsymbol{i}+7\boldsymbol{j}-\boldsymbol{k}$
4 $5(\boldsymbol{i}-2\boldsymbol{k})$
5 $-3\boldsymbol{i}+27\boldsymbol{j}-8\boldsymbol{k}$
6 $-5(\boldsymbol{i}-2\boldsymbol{k})$
7 $\boldsymbol{i}+7.5\boldsymbol{j}-4\boldsymbol{k}$
8 $20.5\boldsymbol{j}-10\boldsymbol{k}$
9 $3.6\boldsymbol{i}+4.4\boldsymbol{j}-6.9\boldsymbol{k}$
10 $2\boldsymbol{i}+40\boldsymbol{j}-43\boldsymbol{k}$

Exercise 13.6

1 4.12, 10.44, 7.21
2 9.54
3 $3\boldsymbol{i}+3\boldsymbol{j}+5\boldsymbol{k}$, 6.56
4 −8, −2
5 7.81, 10.82, 37.17
6 8.06 N, 60.3° to vertical
7 $2.69\,\text{ms}^{-1}$, 42° to vertical
8 $12.81\,\text{ms}^{-1}$, 051.3°, 160 m
9 191.3°, $31.50\,\text{ms}^{-1}$, 187.7°
10 $-100\boldsymbol{i}$, $-35.36(\boldsymbol{i}+\boldsymbol{j})$, $-120\boldsymbol{j}$, $-135.36\boldsymbol{i}-155.36\boldsymbol{j}$, 206 m, 9.64°

Chapter 14

Exercise 14.1

1 5
2 −43
3 −36
4 5
5 28
6 $\frac{1}{6}$
7 −28
8 60
9 206
10 0.405

Exercise 14.2

1 $x = 10, y = -9$
2 $x = -2, y = 13$
3 $x = 3, y = 3$
4 $x = 16, y = 10$
5 $x = 4, y = 10$
6 $x = -7, y = -2$
7 $x = -1, y = 2.5$
8 $x = 24, y = 4$
9 $x = 2, y = -1$
10 $x = 3, y = -0.5$

Exercise 14.3

1 2×2 **2** 2×3 **3** 3×2 **4** 2×1
5 3×3 **6** 2×3 **7** 2×3 **8** 1×1
9 1×3 **10** 2×2
11 Matrices from Questions **1** and **10**,
Matrices from Questions **2**, **6** and **7**

Exercise 14.4

1 $\begin{pmatrix} 9 & 3 \\ 5 & 4 \end{pmatrix}$ **2** $\begin{pmatrix} -5 & 9 \\ 5 & 14 \end{pmatrix}$ **3** $\begin{pmatrix} 13 & -3 \\ 7 & 4.5 \end{pmatrix}$

4 $\begin{pmatrix} -15 & 22 \\ 13 & 13 \end{pmatrix}$ **5** $\begin{pmatrix} 5 & -4 \\ -3 & 15 \end{pmatrix}$ **6** $\begin{pmatrix} 5 & -9 \\ -5 & -14 \end{pmatrix}$

7 $\begin{pmatrix} 9 & 3 \\ 5 & 4 \end{pmatrix}$ **8** $\begin{pmatrix} -32 & 44 \\ 10 & -3.5 \end{pmatrix}$ **9** $\begin{pmatrix} 11 & 9 \\ 10 & 13 \end{pmatrix}$

10 $\begin{pmatrix} -8 & 10 \\ 9 & -0.75 \end{pmatrix}$ **11** $\begin{pmatrix} 17 & -16 \\ -8 & -4 \end{pmatrix}$ **12** $\begin{pmatrix} 4 & 18 \\ -1 & -2.5 \end{pmatrix}$

13 $\begin{pmatrix} 3 & -9 \\ -0.5 & 4 \end{pmatrix}$ **14** $\begin{pmatrix} -55 & 80 \\ 34 & -11.5 \end{pmatrix}$ **15** $\begin{pmatrix} 23 & -34 \\ -9 & -12 \end{pmatrix}$

Exercise 14.5

AB*, *AC*, *BD*, *DA*, *DB*, *DC

Exercise 14.6

1 $\begin{pmatrix} 0 \\ 8 \end{pmatrix}$ **2** $\begin{pmatrix} -1 \\ 5 \end{pmatrix}$

3 $\begin{pmatrix} 15 \\ 3 \end{pmatrix}$ or $3\begin{pmatrix} 5 \\ 1 \end{pmatrix}$ **4** $\begin{pmatrix} 10.5 \\ 28 \end{pmatrix}$

5 $\begin{pmatrix} -7 & 12 \\ 21 & 8 \end{pmatrix}$ **6** $\begin{pmatrix} 5 & 9 \\ -6 & -2 \end{pmatrix}$

7 $\begin{pmatrix} 19 & 1 \\ -19 & -1 \end{pmatrix}$ **8** $\begin{pmatrix} -6 & 14 \\ 0 & 10 \end{pmatrix}$ or $2\begin{pmatrix} -3 & 7 \\ 0 & 5 \end{pmatrix}$

9 $\begin{pmatrix} 49 & 4 \\ 42 & -6 \end{pmatrix}$ **10** $\begin{pmatrix} 21.5 & 0.5 \\ 19 & 3 \end{pmatrix}$

Exercise 14.7

1 $\begin{pmatrix} 2 & 4 \\ -1 & 2 \end{pmatrix}$ $\Delta = 8$ **2** $\begin{pmatrix} 2 & 6 \\ 5 & 8 \end{pmatrix}$ $\Delta = -14$

3 $\begin{pmatrix} 3 & 6 \\ 10 & 6 \end{pmatrix}$ $\Delta = -42$ **4** $\begin{pmatrix} \frac{1}{2} & 1 \\ 3 & 2 \end{pmatrix}$ $\Delta = -2$

5 $\begin{pmatrix} 5 & 1 \\ 2 & 6 \end{pmatrix}$ $\Delta = 28$

6 $\begin{pmatrix} 11 & 4 \\ 5 & -2 \end{pmatrix}$ $\Delta = -42$

7 $\begin{pmatrix} 1 & 0 \\ 0 & 1 \end{pmatrix}$ $\Delta = 1$

8 $\begin{pmatrix} 3 & 4 \\ -2 & 1 \end{pmatrix}$ $\Delta = 11$

9 $\begin{pmatrix} 1 & -1 \\ 2 & -2 \end{pmatrix}$ $\Delta = 0$

10 $\begin{pmatrix} 1 & 2 \\ 4 & -\frac{1}{2} \end{pmatrix}$ $\Delta = -8.5$

Exercise 14.8

1 $\begin{pmatrix} 4 & -3 \\ -5 & 4 \end{pmatrix}$

2 $\frac{1}{2}\begin{pmatrix} 6 & -2 \\ -17 & 6 \end{pmatrix}$ or $\begin{pmatrix} 3 & -1 \\ -8.5 & 3 \end{pmatrix}$

3 $\frac{1}{4}\begin{pmatrix} 1 & 0 \\ 3 & 4 \end{pmatrix}$ or $\begin{pmatrix} 0.25 & 0 \\ 0.75 & 1 \end{pmatrix}$

4 $-\frac{1}{5}\begin{pmatrix} 1 & -2 \\ -4 & 3 \end{pmatrix}$ or $\begin{pmatrix} -0.2 & 0.4 \\ 0.8 & -0.6 \end{pmatrix}$

5 $\frac{1}{2}\begin{pmatrix} 3 & -2 \\ -5 & 4 \end{pmatrix}$ or $\begin{pmatrix} 1.5 & -1 \\ -2.5 & 2 \end{pmatrix}$

6 $\frac{1}{10}\begin{pmatrix} 3 & -4 \\ -5 & 10 \end{pmatrix}$ or $\begin{pmatrix} 0.3 & -0.4 \\ -0.5 & 1 \end{pmatrix}$

7 $\begin{pmatrix} 1 & 0 \\ 0 & 1 \end{pmatrix}$

8 $\begin{pmatrix} \frac{1}{2} & -3 \\ -1 & 8 \end{pmatrix}$

9 $-\frac{1}{40}\begin{pmatrix} -5 & -3 \\ -5 & 5 \end{pmatrix}$ or $\begin{pmatrix} 0.125 & 0.075 \\ 0.125 & -0.125 \end{pmatrix}$

10 $-\frac{1}{5}\begin{pmatrix} -\frac{1}{2} & -1 \\ -4 & 2 \end{pmatrix}$ or $\begin{pmatrix} 0.1 & 0.2 \\ 0.8 & 0.4 \end{pmatrix}$

Exercise 14.9

1 $x = 6, y = 5$
2 $x = 5, y = 1$
3 $x = 5, y = 2$
4 $x = \frac{1}{2}, y = 1$
5 $x = -\frac{1}{5}, y = 1$
6 $x = 3.8, y = 0.6$
7 $x = 0, y = 1$
8 no solution
9 $x = -\frac{10}{7}, y = \frac{29}{7}$
10 no solution

Exercise 14.10

1 $x = 3, y = 50$
2 $m = 0.25, c = 2$
3 $x = 13.89, y = 238.89$
4 $I_1 = 0.4$ A, $I_2 = 0.2$ A
5 $a = 400, b = -1200$

Chapter 15

Exercise 15.1

1 i) progression ii) series iii) neither
iv) progression v) neither vi) series
vii) series viii) neither
2 i) 7, 2 iv) 5, −1
3 18.5, 117.5 **4** 53 **5** 10.25
6 1.75, 1.25, 60.75 **7** 297 **8** −450, −300
9 2500, 625, 0 to 50 has only 25 odd numbers **10** 165, 465

Exercise 15.2

1 i) progression ii) neither iii) series
iv) progression v) progression vi) progression
vii) series viii) neither
2 i) 2, 2 iv) 1, −0.1 v) 36, $\frac{1}{3}$
vi) 1, −1
3 $0.\overline{074}$, 26.963 **4** 160 **5** 729
6 40.5, $1.\bar{3}$, 1092.1 **7** 7, 1093.5 **8** 92.23
9 i) no common ratio, no common difference
ii) no common ratio, no common difference
10 72, 30

Exercise 15.3

1 i) 2.5, ii) −0.625
2 There is a common ratio, $0.\bar{6}$
3 50, 5, 0.50, 0.05; $55.\bar{5}$
4 $0.1\bar{6}$
5 0.8

Exercise 15.4

1 2, 0.90, 0.40, 0.18; 3.61 m
2 351
3 50 000, 42 500, 36 125; just over $4\frac{1}{4}$ years
4 0.015 m, 42.85 m
5 i) 75.8 tonne, ii) 75.9 tonne

Chapter 16

Exercise 16.1

1 0.087 m^3, 0.950 m^2
2 0.25 m^3
3 0.63 m^2
4 0.173 m^3, 21, 19
5 0.216 m^3, 2.55 m^2, 0.20 m
6 0.36 m^3, 0.013 m^3, 0.35 m^3
7 0.014 m^3, 0.283 m^2, 70, 20 m^2, 5 *l*
8 4.62 m^2
9 5 m, 8.65 m, 5.77 m, 806 m^3
10 0.26 m^3, 2.36 m^2, 0.25 m^3, 0.01 m^3

Exercise 16.2

1 1.94 m^3, 69%, 25.4 m^2, 26.5 m^2
2 1.15 m^3, 5.62 m^2, 5.34 m^2
3 0.12 m^3
4 0.019 m^2, 0.045 m^3
5 0.05 m^3
6 0.94 m^2
7 0.25 m^3, 2.55 m^2
8 0.018 m^3, 8640 kg
9 0.34 m^3
10 0.075 m^3, 2 m^2

Exercise 16.3

1 21 $unit^2$
2 490 $unit^2$, $y = mx$
3 660 J
4 1 $unit^2$, 4 $unit^2$
5 3880 m^2, 1164 m^3
6 870 000 m^3
7 605 J
8 a) 6 $unit^2$, 14 $unit^2$,
b) 2.35 $unit^2$, 2.35 $unit^2$
9 30.1 m^2
10 8.76 s

Exercise 16.4

1 i) 113 unit, ii) 113 unit, 40.8 unit in each case
2 141.7 unit, 142.7 unit, 88.7 unit
3 circle, 28.3 $unit^2$, 9.4 $unit^2$, 18.9 $unit^2$
4 113 $unit^2$, 6 unit
5 25.8 $unit^2$

Exercise 16.5

1 0.03 m^3, 1.58 m^2
2 0.22 m^3, 4.74 m^2
3 0.04 m^3
4 1.45 m^3
5 0.94 m^3, 9.05 m^2
6 1.22 m^3, different centroid distance
7 0.84 m^3, 79%
8 0.15 m, 0.20 m, 0.17 m^3, 0.13 m^3
9 0.173 m, 0.058 m, 0.064 m^3, 1 m^3
10 0.88 m^3

Chapter 17

Exercise 17.1

1 discrete **2** continuous **3** continuous
4 discrete **5** discrete **6** continuous
7 continuous **8** discrete **9** continuous
10 continuous

Exercise 17.3

2 48% **3** 20% **4** 74% **5** 50%

Exercise 17.4

1 33 **2** 3 N **3** 4.3 Ω **4** 892 mm
5 0°C

Exercise 17.5

1 3 **2** 2 kg and 3 kg **3** no mode
4 8.51 **5** −0.2 J

Exercise 17.6

1 4 **2** 12.5 W **3** 53% **4** 0.3 μF
5 50.5

Exercise 17.7

1 3, 3, 3 all to nearest unit **2** 9, 10, 10 all to nearest hour
3 6, 6, 6 all to nearest minute **4** 25
5 1

Exercise 17.8

1 52.7%, 54.0%, 54.5% all to 1 decimal place
2 16.2, 14.5, 13.3 all hours to 1 decimal place
3 15.018, 15.018, 15.016 all Ω to 3 decimal places
4 124.6 mm
5 404.4 W

Exercise 17.9

1 21 **2** 2 N **3** 2.5 Ω **4** 823 mm
5 4°C

Exercise 17.10

1 i) 1.56, i.e. 2 units ii) 1.76, i.e. 2 hours iii) 1.91, i.e. 2 minutes iv) 0.89, i.e. 1 screw v) 1.22, i.e. 1 fault

2 i) 12% ii) 8.94 hours iii) 0.036 Ω iv) 0.76 mm v) 4.44 W

Exercise 17.11

1 $n = 7$, $\bar{x} = 32.7$, $\sigma = 20.8$

2 $n = 5$, $\bar{x} = 3.2$, $\sigma = 2.1$ N

3 $n = 5$, $\bar{x} = 4.3\,\Omega$, $\sigma = 2.5\,\Omega$

4 $n = 3$, $\bar{x} = 892$ mm, $\sigma = 823$ mm

5 $n = 5$, $\bar{x} = 0$°C, $\sigma = 3.6$°C

Exercise 17.12

1 $n = 150$, $\bar{x} = 3.\bar{3}$, $\sigma = 1.52$ units

2 $n = 1331$, $\bar{x} = 9.39$ hours, $\sigma = 1.71$ hours

3 $n = 100$, $\bar{x} = 5.66$ minutes, $\sigma = 1.88$ minutes

4 $n = 100$, $\bar{x} = 25.1$ screws, $\sigma = 1.52$ units

5 $n = 50$, $\bar{x} = 1.28$ faults, $\sigma = 1.18$ faults

6 $n = 50$, $\bar{x} = 52.7\%$, $\sigma = 12\%$

7 $n = 54$, $\bar{x} = 16.15$ hours, $\sigma = 8.94$ hours

8 $n = 60$, $\bar{x} = 15.018\,\Omega$, $\sigma = 0.036\,\Omega$

9 $n = 100$, $\bar{x} = 124.635$ mm, $\sigma = 0.757$ mm

10 $n = 100$, $\bar{x} = 404.44$ W, $\sigma = 4.419$ W

Exercise 17.13

1 i) 115, 110 or 111, 105 all to nearest mark

2 i) $\bar{x} = 155$ marks, $\sigma = 2.57$ marks
ii) 16.11 marks

3 i) 31, 32, 34 all to nearest mark
ii) 27, 32, 34 all to nearest mark

4 A: $\bar{x} = 2.0''$, $\sigma = 0.2''$, B: $\bar{x} = 1.7''$, $\sigma = 0.1''$, C: $\bar{x} = 2.0''\alpha$, $\sigma = 0.7''$

6 i) 4, 4, 4 ii) 1.8

7 i) 63.5 marks ii) *F* 13%, B(F/P) 13%, *P* 27%, *M* 35%, *D* 12%

8 i) 145.4, 141, 127 all hours ii) mean

9 i) $\bar{x} = 9.96$ mm, $\sigma = 0.076$ mm ii) Overhaul

Chapter 18

Exercise 18.1

1 2, 4.5

2 2.1, −2.1; 2.4, −1.6

3 0, i) 0.64, ii) −0.64, iii) 0

4 −10 000 Nm, −10 000 Nm, 0

5 -0.22°Cs^{-1}

Exercise 18.2

1 1 **2** 3 **3** $2x$ **4** $4t$

5 $2t+1$ **6** $4x-3$ **7** $3x^2$ **8** $-\frac{2}{x^2}$

9 $-\frac{8}{t^3}$ **10** $-\frac{2}{x^2}-3$

Exercise 18.3

1 0 **2** 3 **3** 3

4 $8x$ **5** $8t-3$ **6** $-3-8x$

7 $2x+1$ **8** $4x-3$ **9** $-6t^{-2}-4t^{-3}$

10 $-8t^{-3}$ **11** $-12x^{-3}-2x^{-2}$ **12** $3+2x^{-2}$

13 $-12x^{-7}-6x$ **14** $6x+2$ **15** $-3x^{-4}+2x^{-3}$

Exercise 18.4

1 $2x^{-\frac{1}{2}}$ **2** $1.75x^{-0.5}$ **3** $\frac{8}{3}x^{-\frac{2}{3}}$

4 $\frac{1}{24}x^{-\frac{7}{8}}$ **5** $2t^{-\frac{1}{2}}$ **6** $\frac{3}{2}x^{\frac{1}{2}}$

7 $\frac{2}{3}x^{-\frac{1}{3}}$ **8** $0.8(t^{-0.8}+t^{-0.6})$ **9** $3.5x^{-0.5}+x^{-1.5}$

10 $-\frac{1}{3}x^{-1.5}$ **11** $-\frac{1}{3}x^{-1.5}-\frac{1}{2}x^{-2.5}$ **12** $-\frac{1}{2}(x^{-\frac{4}{3}}+x^{-\frac{3}{2}})$

13 $\frac{1}{2}(t^{-\frac{1}{2}}+t^{-\frac{3}{2}})$ **14** $2x^{-\frac{1}{2}}+\frac{1}{4}x^{-\frac{3}{2}}$ **15** $-\frac{3}{2}x^{-\frac{5}{2}}-2x^{-3}$

Exercise 18.5

1 $-4\sin\theta-2\cos\theta$ **2** $2.4\cos\theta-3\sin\theta$ **3** $3.5e^t$

4 $\frac{5}{x}$ **5** $2.4\cos\theta+3\sin\theta$ **6** $2e^t-3\sin t$

7 $\frac{6}{x}$ **8** $3e^x+\frac{1}{x}$ **9** $\frac{1}{x}$

10 $\frac{3}{x}$

Exercise 18.6

1 15.4 **2** $2x^{-3.5}$, -0.18

3 2.866 **4** $-\frac{10}{\pi}x^{-2}-1$, -1.3

5 -2.53 **6** $15.6\,\text{ms}^{-1}$, $7.36\,\text{s}$

7 $2000\,\text{Nm}^{-1}$, $-2000\,\text{Nm}^{-1}$ **8** $0.0\bar{6}$ rad N^{-1}

9 $406.25°\text{s}^{-1}$ **10** $\frac{3t^2}{20}$, 31.6 hour

Chapter 19

Exercise 19.1

1 $7x+c$
2 $\frac{5}{2}x^2+c$
3 $7x-\frac{5}{2}x^2+c$
4 x^3+c
5 $t^3+\frac{5}{2}t^2-7t+c$
6 $7x-\frac{5}{2}x^2-x^3+c$
7 $\dfrac{x^5}{5}+x^2+c$
8 $\frac{4}{7}x^7-\dfrac{1}{2}x^4+c$
9 $-5t^{-1}-t^{-3}+c$
10 $-\frac{3}{2}t^{-2}+c$
11 $-5x^{-1}+x^{-3}+c$
12 $\frac{1}{2}t^4+3t^{-1}+c$
13 $-\frac{3}{8}x^{-4}-\frac{1}{2}x^6+c$
14 $3x+\frac{4}{3}x^3+c$
15 $-2x^{-1}+\frac{1}{2}x^{-2}+c$

Exercise 19.2

1 $4x^{\frac{3}{2}}+c$
2 $3.\bar{3}x^{1.5}+c$
3 $6t^{\frac{4}{3}}+c$
4 $\frac{4}{25}x^{\frac{5}{4}}+c$
5 $2.\bar{6}t^{\frac{3}{2}}+c$
6 $\frac{2}{5}x^{\frac{5}{2}}-6x+c$
7 $\frac{3}{5}x^{\frac{5}{3}}-4x+c$
8 $3.\bar{3}t^{1.2}+1.43t^{1.4}+c$
9 $6x^{1.5}+8x^{0.5}+c$
10 $3x^{0.5}+c$
11 $0.8x^{0.5}-0.4x^{-0.5}+c$
12 $\frac{15}{4}x^{\frac{2}{3}}+4x^{\frac{1}{2}}+c$
13 $2t^{\frac{3}{2}}-8t^{\frac{1}{2}}+c$
14 $\frac{16}{3}x^{\frac{3}{2}}+\frac{1}{2}x^{\frac{1}{2}}+c$
15 $-x^{-\frac{1}{2}}-2x^{-1}+c$

Exercise 19.3

1 $7\sin\theta+c$
2 $4e^x+c$
3 $\frac{1}{2}\ln x+c$
4 $-2\cos\theta+c$
5 $2e^x-3\cos x+c$
6 $5\sin x-\frac{1}{2}e^x+c$
7 $7\sin x+2\cos x+c$
8 $e^t-4\cos t+c$
9 $\ln x-x^{-2}+c$
10 $\frac{1}{2}x^2-3\sin x+c$

Exercise 19.4

1 15.40
2 1.46
3 19.17
4 5.50
5 -1.83
6 $94.\bar{6}$
7 -17.50
8 2
9 4.70
10 1.88

Exercise 19.5

1 4.27 unit^2
2 $3.\bar{3}$ unit^2
3 14.64 unit^2
4 0.693 unit^2
5 -1 unit^2
6 2.77 unit^2
7 16 unit^2
8 8 unit^2
9 a) $x = 1, 2$; $-0.1\bar{6}$ unit^2 b) $x = 1, 2$; $-0.1\bar{6}$ unit^2
10 i) $6.\bar{3}$ unit^2, ii) 8 unit^2, iii) $10.\bar{6}$ unit^2

Exercise 19.6

1 100 m
2 1.8 J
3 7.82×10^{-5} W
4 117.5 J
5 5.303×10^4, 2.75×10^4 J

Index